Chemical Communication in Crustaceans

Thomas Breithaupt • Martin Thiel
Editors

Chemical Communication in Crustaceans

With drawings by Jorge Andrés Varela Ramos

 Springer

Editors
Thomas Breithaupt
University of Hull
Department of Biological Sciences
Hull, HU6 7RX
UK
t.breithaupt@hull.ac.uk

Martin Thiel
Universidad Católica del Norte
Facultad Ciencas del Mar
Depto. de Biología Marina
Larrondo 1281
Coquimbo
Chile
thiel@ucn.cl

ISBN 978-0-387-77100-7   e-ISBN 978-0-387-77101-4
DOI 10.1007/978-0-387-77101-4
Springer New York Dordrecht Heidelberg London

Printed on acid-free paper

Springer is part of Springer Science+Business Media (www.springer.com)

To our families and friends.

Preface

Animal communication has fascinated biologists for centuries. This fascination has sustained many a scientific career as will be evident from the personal accounts by the contributors to this book. Chemical signaling is the most widespread form of communication in crustaceans. During the past two decades, there have been significant advances in our understanding of crustacean chemical ecology. Gathering this information in an edited volume was the next logical step.

This book covers a wide range of topics, including the identity, production, transmission, reception, and behavioral function of chemical signals in selected crustacean groups. The chapters are organized into five sections. The introductory section gives a brief overview of the main questions that are tackled in this volume and provides important definitions of signals, cues, and behavior. The next section on the transmission of chemical cues in the environment and on sensory biology is followed by a section on the behavioral contexts in which crustaceans use chemical communication, providing examples from the best studied taxa. Recent advances in the molecular identification of chemical signals are presented in the fourth section. The fifth and last section deals with the possible applications of pheromone research to aquaculture and pest management.

One of our goals as editors was to encourage contributors to identify similarities and differences in chemical communication by crustaceans and by other taxa and thereby address questions of general interest. We therefore invited experts on communication in insects, spiders, and fishes to contribute to this book. They readily and, at first perhaps, innocently accepted our invitation, unaware that we would ask them to integrate knowledge of crustaceans into their chapters. Thus, their tasks went beyond a synthesis of their own work and expertise and we feel that they all have done a marvelous job. We learned a lot from them and we now share some of their fascination for their favorite organisms and the excitement that comes from studying them. Within the same spirit, we hope that this book will attract readers who are interested in learning about crustaceans, but who study other taxa in their quest to understand the evolution and function of chemical communication.

There are several topics that we thought were ready for thorough review such as multimodal communication, deception, and pheromones in aquaculture and pest management, but are still beyond the mainstream of crustacean research. Several of our colleagues had some experience in these areas, and they were thus "naturals" to be invited for these contributions. Though reluctant at first, they accepted the challenge and their enthusiasm grew as they wrote.

In order to make this volume accessible to a broad audience that spans scientific and applied fields, we asked the authors to include a personal statement briefly describing why they entered their respective research fields. Such statements are not generally accepted in scientific writing. But we are most grateful that many of our authors adopted a more informal style and expressed their enthusiasm for their particular study species or research questions. We hope that our authors' enthusiasm is sufficiently infectious and that the scientific questions they raise in their contributions will stimulate future research. If only a few young scholars are infected by this excitement for crustacean chemical communication, this book has achieved its goal.

Hull, UK   Thomas Breithaupt
Coquimbo, Chile   Martin Thiel

Acknowledgments

We thank our teachers, collaborators, and students who have helped shape our understanding of crustacean communication during the past two decades. The contributors to this book deserve most of our gratitude – all this would not have been possible without their expertise and efforts. A very special thanks goes to the more than 80 reviewers for their time and suggestions, which were very helpful to us and the authors – we could not have done without their critical and constructive input. Iván A. Hinojosa provided expert help in the preparation of many of the figures in this book. Jorge A. Varela Ramos made many of the chemical interactions visible with his artistic drawings. TB would like to thank his wife Gabi and son Tobias for their endless patience during the many evenings and weekends when husband/daddy had to work on "the book" and was not available for family activities. MT thanks Taizhu for her continuous support and patience. His participation in this project would not have been possible without the unconditional support from Universidad Católica del Norte (UCN). Many of the chapters in this book were edited during a sabbatical stay at the Smithsonian Environmental Research Center (SERC) in Edgewater, Maryland. MT wishes to thank A.H. Hines for supporting his stay at SERC.

Contents

Contributors

Juan Aggio
Neuroscience Institute, Georgia State University, P.O. Box 5030
Atlanta, GA 30302, USA

Thomas C. Baker
Department of Entomology, 105 Chemical Ecology Laboratory,
Penn State University, University Park, PA 16802, USA

Assaf Barki
Aquaculture Research Unit, Volcani Center, Institute of Animal Science,
Agricultural Research Organization, P.O. Box 6, Bet Dagan 50250, Israel

Raymond T. Bauer
Department of Biology, University of Louisiana, Lafayette, LA 70504, USA

Thomas Breithaupt
Department of Biological Sciences, The University of Hull, Hull HU6 7RX, UK

Roy Caldwell
Department of Integrative Biology, University of California, Berkeley,
CA 94720-3140, USA

Ernest S. Chang
Bodega Marine Laboratory, University of California-Davis, P.O. Box 247
Bodega Bay, CA 94923, USA

John H. Christy
Smithsonian Tropical Research Institute, Apartado 0843-03092,
Balboa, Ancón, Panamá, Republic of Panama

Yu-Wen Chung-Davidson
Department of Fisheries and Wildlife, Michigan State University,
East Lansing, MI 48824, USA

Anthony S. Clare
School of Marine Science and Technology, Newcastle University,
Newcastle upon Tyne, NE1 7RU, UK

Charles D. Derby
Neuroscience Institute and Department of Biology, Georgia State University,
P. O. Box 5030, Atlanta, GA 30302, USA

Francesca Gherardi
Dipartimento di Biologia Evoluzionistica "Leo Pardi",
Università degli Studi di Firenze, Via Romana 17, 50125 Firenze, Italy

Eric Hallberg
Department of Biology, Lund University, Zoologihuset, HS 17 Sölvegatan 35,
SE-22362, Lund, Sweden

Bill S. Hansson
Department of Evolutionary Neuroethology, Max Planck Institute for
Chemical Ecology, Hans-Knöll-Street 8, Jena 07745, Germany

Joerg D. Hardege
Department of Biological Sciences, University of Hull, Hull, HU6 7RX, UK

Steffen Harzsch
Department of Cytology and Evolutionary Biology, University of Greifswald,
Zoological Institute and Museum, Johann-Sebastian-Bach-Street 11/12,
D-17487 Greifswald, Germany

Mark E. Hay
School of Biology, Georgia Institute of Technology, Atlanta, GA 30332, USA

Brian A. Hazlett
Department of Ecology and Evolutionary Biology, University of Michigan,
830 North University, Ann Arbor, MI 48109-1048, USA

Eileen A. Hebets
School of Biological Sciences, University of Nebraska, Lincoln,
NE 68588-0118, USA

Mar Huertas
Centre of Marine Sciences, University of Algarve, Campus de Gambelas,
8005-139 Faro, Portugal

Clive Jones
Department of Primary Industries and Fisheries, Northern Fisheries Centre,
P.O. Box 5396, Cairns Q 4870, Australia

Michiya Kamio
Department of Ocean Science, Tokyo University of Marine Science
and Technology 4-5-7 Konan, Minato-Ku, Tokyo 108-8477, Japan

Ilan Karplus
Aquaculture Research Unit, Volcani Center, Institute of Animal Science,
Agricultural Research Organization, P.O. Box 6, Bet Dagan 50250, Israel

Markus Knaden
Department of Evolutionary Neuroethology, Max Planck Institute
for Chemical Ecology, Hans-Knöll-Street 8, D-07745 Jena, Germany

Mimi A.R. Koehl
Department of Integrative Biology, University of California, Berkeley,
CA 94720-3140, USA

Rachel Lasley
School of Biology, Georgia Institute of Technology,
Atlanta, GA 30332-0230, USA

Weiming Li
Department of Fisheries and Wildlife, Michigan State University,
East Lansing, MI 48824, USA

Kristina Mead
Department of Biology, Denison University, Granville, OH 43023, USA

DeForest Mellon, Jr.
Department of Biology, University of Virginia, P.O. Box 400328,
Gilmer Hall Room 286, Charlottesville, VA 22903, USA

K. Håkan Olsén
School of Life Sciences, Södertörn University, SE-141 89, Huddinge, Sweden

Dan Rittschof
Marine Laboratory, Duke University, Beaufort, NC 28516-9721, USA

Aaron Rundus
School of Biological Sciences, University of Nebraska, Lincoln,
NE 68588-0118, USA

Manfred Schmidt
Neuroscience Institute and Department of Biology, Georgia State University,
P.O. Box 5030, Atlanta, GA 30302, USA

Malin Skog
Department of Biology, Lund University, Zoologihuset, HS 17,
Helgonavägen 3, SE-22362 Lund, Sweden

Terry Snell
School of Biology, Georgia Institute of Technology, Atlanta, GA 30332-0230, USA

Marcus Stensmyr
Department of Evolutionary Neuroethology, Max Planck Institute
for Chemical Ecology, Hans-Knöll-Street 8, D-07745 Jena, Germany

John A. Terschak
Department of Biological Sciences, University of Hull, Hull, HU6 7RX, UK

Martin Thiel
Universidad Católica del Norte, Facultad Ciencias del Mar,
Larrondo 1281, Coquimbo, Chile

Elena Tricarico
Dipartimento di Biologia Evoluzionistica "Leo Pardi",
Università degli Studi di Firenze, Via Romana 17,
50125 Firenze, Italy

Marc J. Weissburg
School of Biology, Georgia Institute of Technology, 310 Ferst Dr,
Atlanta, GA 30332, USA

Tristram D. Wyatt
Department of Zoology, The Tinbergen Building, South Parks Road,
Oxford, OX1 3PS, UK

Jeannette Yen
School of Biology, Georgia Institute of Technology,
Atlanta, GA 30332-0230, USA

Part I
Introductory Section

Chapter 1
Chemical Communication in Crustaceans: Research Challenges for the Twenty-First Century

Martin Thiel and Thomas Breithaupt

Abstract Chemical signals play an important role during various life stages of crustaceans. Settling of larvae, parent–offspring communication, mate finding, mate choice, aggressive contests, and dominance hierarchies are all mediated by chemical signals. Enormous advances have been made on understanding the function of chemical signals in crustaceans and we are on the doorstep of major advances in chemical characterization of pheromones. In many species urine is the carrier of chemical signals. Crustaceans control release and transfer direction of urine, but it is unknown whether crustacean senders can manipulate the composition of urineborne pheromones. Chemicals contained in the urine effectively convey information about conspecific properties such as sex, sexual receptivity, species identity, health status, motivation to fight, dominance, individual identity, and molt stage. In larger species (shrimp, crabs, lobsters, crayfish) signal delivery is often aided by self-generated fanning currents that flush chemicals towards receivers, which themselves might actively pull water towards their sensory structures. Antennal flicking also supports molecule exchange at the receptor level. Contact pheromones play a role in sex recognition in several crustacean taxa and in settlement of barnacles. Large crustacean species show little or no sexual dimorphism in receptor structures, but in smaller taxa, e.g. peracarids and copepods, males often have larger antennae than females. Whether differences in sexual roles have also resulted in sex-specific brain centers is not known at present. While pheromones play an important role in mate finding and species recognition, there are numerous examples from peracarids and copepods where males pursue or even form precopulatory pairs with females of closely related congeners. Differentiation of chemicals often appears to be insufficient to guarantee reproductive isolation. In many freshwater and coastal habitats, pollutants may also disrupt chemical communication in crustaceans, but the specific mechanisms of interference are not well understood. The chemical characterization of crustacean pheromones is viewed as a major step in improving our understanding of chemical communication.

M. Thiel (✉)
Universidad Católica del Norte, Facultad Ciencias del Mar,
Larrondo 1281, Coquimbo, Chile
e-mail: thiel@ucn.cl

T. Breithaupt and M. Thiel (eds.), *Chemical Communication in Crustaceans*,
DOI 10.1007/978-0-387-77101-4_1, © Springer Science+Business Media, LLC 2011

Knowing the chemical nature of pheromones in freshwater species will boost research on aquatic crustaceans. Interdisciplinary work between chemists (metabolomics), behavioral ecologists (bioassays), neurobiologists (chemoreception), and molecular biologists (genomics) promises to produce significant advances in our understanding of crustacean chemical communication during the coming decade.

He drew up plans, made lists, experimented with smells, traced diagrams, built structures out of wood, canvas, cardboard, and plastic. There were so many calculations to be made, so many tests to be run, so many daunting questions to be answered. What was the ideal sequence of smells? How long should a symphony last, and how many smells should it contain? What was the proper shape of the symphony hall? ... Should each symphony revolve around a single subject – food, for example, or female scents – or should various elements be mixed together? ...What difference did it make if he didn't fully understand? ... It might not have served any purpose, but the truth was that it was fun.

From *Timbuktu* by Paul Auster (1999)

1.1   Introduction

Crustaceans are found in all major environments in the oceans and on land. Given the diversity of habitats, they face numerous challenges in communicating with conspecifics. How does a female crab that is ready to reproduce find a male in the murky waters of a shallow estuary? She could roam in search of a male or she could stay put and wait for a male to find her. In both cases, her success in finding a mating partner would be enhanced by a chemical guidance system. If she searches for a male, it would be advantageous to sniff out the environment for chemical cues that would indicate the presence of a male. And if she waits for a male to find her, she could guide him towards her by releasing attractive chemicals. Regardless of the strategy, chemical stimuli enhance the probability of mate finding which is only one of many benefits offered by chemical communication.

Chemical signals play an important role during various life stages of crustaceans. Settling of larvae, parent–offspring communication, mate finding, mate choice, and aggressive contests are all mediated by chemical signals. Chemicals are ubiquitous messengers because they can effectively convey information about conspecific properties such as sex, sexual receptivity, species identity, health status, motivation to fight, dominance, individual identity, and molt stage. Not surprisingly, many crustaceans employ chemical communication to coordinate important life processes. At first glance, crustaceans do not seem to differ from many other animals such as insects or mammals in which chemical communication plays an important role. However, crustaceans have conquered a wider range of habitats than most other animals, inhabiting the deep abyss of the oceans, wave-battered shores, calm freshwater lakes, dark forests, and even dry deserts. Furthermore, the range of crustacean body sizes and shapes is unparalleled in many other animal taxa. And finally, the diversity of crustacean life styles is mind boggling even to well-seasoned crustacean researchers; tiny planktonic species share a common

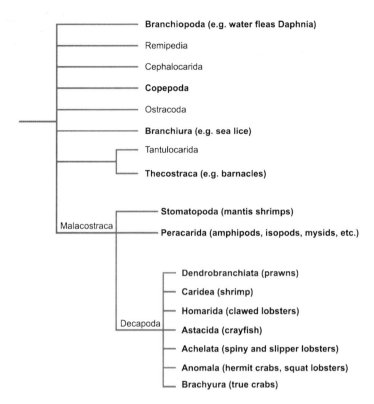

Fig. 1.1 Phylogeny of crustaceans, highlighting in bold the taxa that have been subject to research on chemical communication. Only those subtaxa of Malacostraca and Decapoda are shown that have been subject to chemical communication research. Phylogeny was modified after Tree of life (http://tolweb.org/Crustacea), and Dixon et al. (2003)

history with bulky crabs, colorful shrimp, and strange parasitic forms that can only be recognized as crustaceans during their larval stages.

Many of these species, regardless of habitat or morphology, communicate with their conspecifics via chemical substances. The crustacean species that have been subject to chemical communication research were drawn from six of the 12 classes of crustacea including Branchiopoda (water fleas), Copepoda, Branchiura (including fish lice), Thecostraca (including barnacles), and Malacostraca (the largest class including stomatopods, peracarids, and decapods) (Fig. 1.1). By far the greatest contribution to our understanding of chemical communication comes from research on decapod crustaceans including crabs, lobsters, and shrimps.

Many species from these groups employ chemical signals throughout or during parts of their lives. How do they do it and how have their phylogenetic histories and current environmental conditions shaped their communication systems? The contributions in this book offer answers to these questions and they also highlight fascinating challenges for the future.

1.2 Chemical Communication in Crustaceans – A Brief Literature Survey

1.2.1 Pheromone Signaling in Marine Invertebrates

In crustaceans, communication is mainly through the visual, chemical, and mechanical channels (see e.g., Mead and Caldwell, Chap. 11; Christy and Rittschof, Chap. 16; Clayton 2008). Whereas visual communication is mainly limited to species from terrestrial and clear-water environments, chemical communication can occur under most environmental conditions. Not surprisingly, studies on chemical communication dominate the literature. Of a total of 76 publications on crustacean communication (with the keywords communicat* and crustacea*) published between 1990 and 2010, 43 were on chemical communication, 24 on visual communication, and only 9 on mechanical/acoustic communication (Web of Science 2010).

Chemical communication may be prominent not only because it works under almost any environmental condition, but also because it may be subject to rapid evolutionary change (Symonds and Elgar 2008), possibly much more so than visual or mechanical communication, as was recently highlighted by Bargmann (2006): "The visual system and auditory system are stable because light and sound are immutable physical entities. By contrast, the olfactory system, like the immune system, tracks a moving world of cues generated by other organisms, and must constantly generate, test and discard receptor genes and coding strategies over evolutionary time." The high potential for specificity has been one of the main reasons that many species communicate via chemical signals. These are often employed to attract conspecifics or to convey particular messages.

The first unequivocal demonstration of pheromone use by a crustacean was presented by Ryan (1966) who showed that male Pacific crabs *Portunus sanguinolentus* display a typical courtship response when stimulated with female premolt water. Males did not display when the female's excretory pores were sealed. This paper was followed by several other studies confirming that crustaceans employ pheromones during mating interactions (e.g., Dahl et al. 1970, for amphipods; Atema and Engstrom 1971, for lobsters; Ameyaw-Akumfi and Hazlett 1975, for crayfish). Surprisingly, the first marine invertebrate for which the sex pheromone was chemically identified was the polychaete *Platynereis dumerilii* from the North Atlantic (Zeeck et al. 1988). Since then the chemical structure of pheromones has also been characterized for molluscs (Painter et al. 1998). Only during the past decade pheromones have been purified in several crustacean species (Kamio and Derby, Chap. 20; Hardege and Terschak, Chap. 19; Clare, Chap. 22; Rittschof and Cohen 2004).

Despite these advances, our knowledge about pheromone structure, production, and effects in marine invertebrates is scarce. A Boolean literature search from the past 20 years (1990–2009) showed that most pheromone studies with marine invertebrates have investigated crustaceans, polychaetes, and molluscs (Fig. 1.2). Especially during

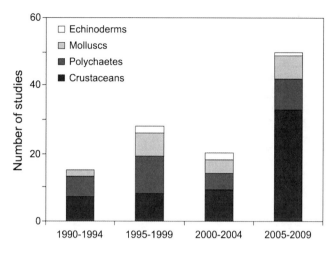

Fig. 1.2 Number of studies in pheromones in marine invertebrates during the period 1990 and 2009. Results based on Web of Science, Boolean search for pheromon* AND taxon*; freshwater and terrestrial taxa excluded

the pentad 2005–2009, there has been an increasing number of studies on crustacean pheromones, which most likely has been fostered by the beginning of the chemical characterization of pheromones in several species. Given recent advances in this field, it can be expected that this trend will continue in the future.

1.2.2   Crustaceans, Fish and Insects

Since most research on pheromones has been conducted in other taxa (e.g. insects, fish, and mammals), it is not surprising that crustacean researchers studying pheromones rely on this rich literature. Interestingly, not only do crustacean research-ers cite a comparatively large number of studies on other taxa, but their own studies are also cited by researchers studying a diverse range of other taxa (Fig. 1.3). Tradition-ally, crustacean researchers studying pheromones have been inspired by research on fish (living in water) and insects (arthropod relatives of the crustaceans). Whereas crustacean studies often integrate information from studies on other taxa, the corresponding proportion in fish and insect studies is <10% (Fig. 1.3). Also, recipro-cally, fish and insect studies are only rarely cited by pheromone studies on other taxa. Most likely, these differences between studies on crustaceans, fish, and insects are due to the fact that much more is known about pheromones in fish and insects than in crustaceans. Crustacean researchers might also cite studies on both aquatic (fish) and terrestrial (insects) taxa frequently because crustaceans have conquered both these environments. This integrative approach has always characterized studies on crusta-cean chemoreception (e.g., Weissburg 2000; Vickers 2000; Koehl 2001) and promises to do so in the future (see contributions in this volume).

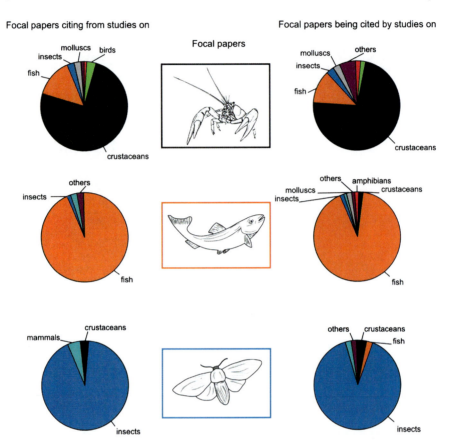

Fig. 1.3 Percentage of taxon-based studies cited by focal papers (*left column*) and citing the focal papers (*right column*). Results from Web of Science, based on the three most cited pheromone studies for each taxon (period 1990–2009)

1.3 Chemical Signals: Source and Identity

In crustaceans, chemical signals can be released to the surrounding liquid medium (soluble or volatile pheromones, "distance pheromones") or bound to the body surface ("contact pheromones"; see e.g. Bauer, Chap. 14, and Snell, Chap. 23). In decapod crustaceans, the pheromones are released through the excretory pores (nephropores) located in the head region (Atema and Steinbach 2007; see also Kamio and Derby, Chap. 20; Hardege and Terschak, Chap. 19; Breithaupt, Chap. 13). Urine is often, but not always, the carrier of the chemical signal (see e.g. Kamio et al. 2002). Urine is predestined as a source of information molecules as it contains body metabolites that mirror the internal processes involved in sexual maturation, aggression, and illness. Many of the hormones underlying behavioral and developmental processes are well known in crustaceans

(Chang, Chap. 21). Numerous studies on fish have shown that hormones, once released, assume a pheromonal role (Chung-Davidson et al., Chap. 24). Although this is likely the case in other animals as well, there are only few studies providing examples of hormonal pheromones in crustaceans (Chang, Chap. 21). The larger decapod crustaceans should be ideal model organisms to close the gap between endocrinology and chemical communication research.

Urineborne chemicals reveal crucial information about conspecifics that can provide the receiver with distinct advantages over competitors in feeding, reproduction, and dominance interactions. Early during the evolutionary history of chemical communication, individuals might have obtained information by spying on urine chemicals from conspecifics. If emitters of these chemicals had adaptive advantages in revealing their status to others, this may have led to the evolution of complex urine release pattern (see Fig. 2.4 in Wyatt, Chap. 2). An example would be the release of chemicals that permit individual recognition within dominance hierarchies, where senders and receivers benefit from recognizing conspecifics (Aggio and Derby, Chap. 12; Gherardi and Tricarico, Chap. 15).

It is unknown whether crustacean senders can manipulate the composition of urineborne pheromones. They do, however, have control over the timing of urine release (Breithaupt, Chap. 13) and are therefore able to adjust the signaling to their own benefit. This may include opportunities to manipulate the receiver by either falsely reporting or by withholding information (Christy and Rittschof, Chap. 16). Only in few examples has the chemical nature of distance pheromones been characterized. These studies employed behavioral assays that used a specific behavioral response in the receiver as an indicator for pheromonal activity (Kamio and Derby, Chap. 20; Hardege and Terschak, Chap. 19).

The slow progress in crustacean pheromone identification is due (1) to difficulties in designing appropriate bioassays for animals that are under conflicting motivational regimes such as fighting, mating, or escape (Breithaupt, Chap. 13; Hardege and Terschak, Chap. 19), (2) to the quick alteration and degradation of the chemical components by aquatic bacteria (Hay, Chap. 3; for a terrestrial example see Voigt et al. 2005), and (3) to analytical challenges particular to identification of marine pheromones such as the difficulty in extracting and separating small molecules from a salty medium (Hay, Chap. 3; Hardege and Terschak, Chap. 19).

The latter problem may also explain the bias towards freshwater species in fish pheromone studies. Hormonal pheromones (see Chung-Davidson et al., Chap. 24) were identified in goldfish, round goby, African catfish, and Atlantic salmons that all release the pheromones into a freshwater environment (Sorensen and Stacey 2004). Even in sea lampreys, the chemical nature of larval migratory pheromone attracting adults and of male sex pheromones attracting females was identified for components that are naturally emitted into the freshwater spawning environment (Chung-Davidson et al., Chap. 24). The difficulties inherent in identifying marine semiochemicals suggest that freshwater crustaceans such as amphipods and crayfish may be better model systems for chemical characterization of pheromone components (Fig. 1.4).

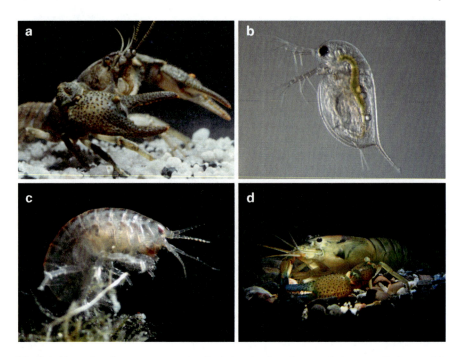

Fig. 1.4 Examples of crustacean species (in their natural environment) that are suited as models to address particular questions on crustacean chemical communication. (**a**) Crayfish *Austropota-mobius torrentium* (photograph courtesy of Dr. Michael van der Wall); (**b**) water flea *Daphnia pulex* (photograph courtesy of Linda C. Weiss); (**c**) amphipod *Hyalella costera*; (**d**) freshwater shrimp *Cryphiops caementarius* (photographs (**c**, **d**) courtesy of Iván A. Hinojosa)

Despite these medium-specific problems, recent progress in pheromone characterization in crustaceans including hair crabs (Asai et al. 2000), helmet crabs (Kamio et al. 2000, 2002), blue crabs (Kamio and Derby, Chap. 20), green crabs (Hardege and Terschak, Chap. 19), peppermint shrimp (Zhang et al. 2010a), and barnacles (Clare, Chap. 22) highlights the fact that some of the initial difficulties in chemical purification have now been overcome and that the door is open towards rapid progress in structural identification.

While almost all insect pheromones are fatty-acid-derived hydrocarbons (Baker, Chap. 27), crustacean pheromones are more diverse. They belong to various substance classes such as peptides (Rittschof and Cohen 2004), nucleotides (Hardege and Terschak, Chap. 19) or other small polar molecules (Kamio and Derby, Chap. 20), small nonpolar molecules (Ingvarsdóttir et al. 2002), and possibly to ceramides (Asai et al. 2000). The higher diversity of waterborne pheromones again reflects the physical differences between the two media, with solubility in water being much less restrictive for the evolution of signal molecules than volatility in air.

Contact pheromones were shown to play a role in sex recognition in copepods (Snell, Chap. 23) and in shrimp (Bauer, Chap. 14) as well as in inducing settlement in barnacles (Clare, Chap. 22). In copepods and barnacles, the molecules were identified as surface-bound glycoproteins (Snell, Chap. 23). Glycoproteins were also found on the surface of caridean shrimp from the genus *Lysmata*, but behavioral experiments on the role of these molecules as mate recognition pheromone in shrimps revealed contradictory results (Caskey et al. 2009; Zhang et al. 2010b), calling for additional studies. Chemical characterization of contact and distance pheromones remains one of the main challenges in crustacean chemical ecology and promises the greatest progress in this field.

1.4 Signal Transmission, Reception, and Processing

1.4.1 Signal Delivery

Pheromones released in an aquatic environment will be carried downstream by the ambient flow (see Weissburg, Chap. 4). In stagnant environments such as some lakes and ponds, odor dispersal will be slow. Signalers that are walking or swimming leave a scented trail behind that facilitates detection as it can be used by receivers to track and find the signaler (Yen and Lasley, Chap. 9; Weissburg, Chap. 4). Stationary senders generate their own water currents by ventilating or by fanning maxillipeds or pleopods to disperse the chemical signals (for lobsters see Atema and Steinbach 2007, and Aggio and Derby, Chap. 12; for crayfish see Breithaupt, Chap. 13; for blue crabs see Kamio and Derby, Chap. 20; for shrimp see Bauer, Chap. 14; for stomatopods see Mead and Caldwell, Chap. 11).

Actively flushing signals towards conspecifics appears to be a general strategy in many crustaceans as they are equipped with specialized fanning structures to generate water currents (see e.g., Breithaupt 2001; Cheer and Koehl 1987). Some insects (e.g., bees; Agosta 1992) and mammals (e.g., bats; Voigt and von Helversen 1999; and ring-tailed lemurs; Bradbury and Vehrencamp 1998) are also able to direct their chemical signals by using their wings (bees, bats) or tails (lemurs), but this strategy of dispersing odors is much less common in terrestrial animals than in aquatic organisms.

Terrestrial animals often display chemical signals by depositing gland excretions, urine, or feces to the substratum. There are numerous examples of terrestrial animals marking their territories using scent marks. Common examples are mammals such as badgers and mice where defecating or urinating appears to serve a territorial function (Roper et al. 1993; Hurst 2005), or female spiders giving away their reproductive status via chemicals in their web's silk (Roberts and Uetz 2005). Interestingly, in terrestrial isopods, burrows or communal dwellings also carry kin- or species-specific scents, while observations of aquatic amphipods could find no evidence for the existence of scent marks on dwellings (Borowsky 1989). The lack of scent marks in aquatic environments may be a consequence of the high solubility of even large molecules such as proteins in water causing any scent marks to be rapidly diluted by water movements. In addition, the ubiquitous bacteria in water may quickly attack and degrade any scent marks.

1.4.2   Reception and Processing of Pheromone Signals

Crustaceans perceive chemical signals with olfactory receptors – limited to the aesthetasc hairs that only contain chemoreceptor neurons and are located on the antennae – or with other chemoreceptors situated in setae that are distributed over the body surface ("distributed chemoreceptors" including contact chemoreceptors, Schmidt and Mellon, Chap. 7; Hallberg and Skog, Chap. 6). "Contact chemoreceptors" contain both chemoreceptor neurons and mechanoreceptive neurons (Schmidt and Mellon, Chap. 7).

The evolutionary transition from water to land has resulted in an expansion of the chemoreceptor genes, most likely in response to the multitude of airborne odorants (Bargmann 2006). Organisms that frequently change between aquatic and terrestrial environments (e.g., amphibians) appear to have chemosensory systems for perception of both water-soluble as well as volatile odorants (Freitag et al. 1995). Soluble and volatile chemicals can also be perceived by aquatic and terrestrial crustaceans, respectively (e.g., Hansson et al., Chap. 8). However, at least in terrestrial peracarids, taste reception of odorants appears to be mediated by liquids (Seelinger 1983; Holdich 1984), just as food-smelling of terrestrial mammals under water is mediated by air bubbles (Catania 2006).

So far, in decapod crustaceans, the receptor-bearing structures have not been shown to display strong sexual dimorphism as is found in many insects (Hallberg and Skog, Chap. 6). In insects, particularly in moths, males generally have much larger chemoreceptor-bearing antennae than females (Lee and Strausfeld 1990; Baker, Chap. 27). This dimorphism reflects the direction of sexual communication, with females generally being the pheromone emitter and males being the pheromone receiver as reported or inferred for many species (Fig. 1.5). There is a strong selective pressure on the males to detect minor amounts of female pheromones and track down the female that usually remains stationary while signaling (Phelan 1997). Concordant with the dimorphism in olfactory organ morphology, the dimorphism extends to sex-specific differences in the brain. In most insect genera where adults are terrestrial, a sexual dimorphism was found in olfactory brain centers. In contrast to females, males often possess a system of sex-specific brain centers that make up the "macroglomerular complex", which is involved in the processing of pheromone information (Strausfeld and Reisenman 2009). So far, no sexual dimorphism with respect to olfactory structures has been found in any decapod crustacean (Hallberg and Skog, Chap. 6). However, sexual dimorphism is evident in some peracarid crustaceans where males possess larger and more differentiated olfactory organs than females as well as exhibiting sex-specific olfactory centers (Johansson and Hallberg 1992; Hallberg and Skog, Chap. 6; Thiel, Chap. 10). It remains to be investigated whether the receptor dimorphism in peracarids is caused by sex-specific pheromones and whether it mediates sex-specific behaviors. In crustaceans with female sex pheromones and male-specific responses (Bauer, Chap. 14; Breithaupt, Chap. 13; Hardege and Terschak, Chap. 19; Kamio and Derby, Chap. 20; Yen and

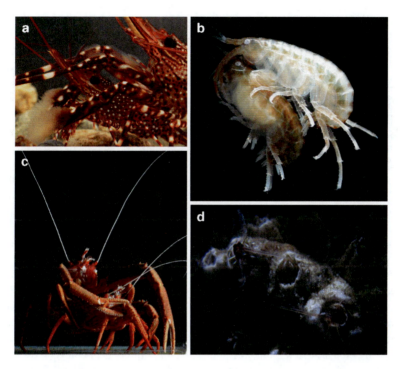

Fig. 1.5 Examples of mating interactions in several species of crustaceans where males are known or inferred to be receiver of female pheromones. (**a**) Rock shrimp *Rhynchocinetes typus*; (**b**) amphipod *Parhyalella penai*; (**c**) squat lobster *Cervimunida johni*; (**d**) barnacle *Balanus laevis*; (photographs courtesy of Iván A. Hinojosa)

Lasley, Chap. 9), males may have specific adaptations for neural processing of female chemical signals.

Sex recognition may also involve multiple sensory channels in some (many?) crustacean species (see Hebets and Rundus, Chap. 17), requiring more complex central processing of multimodal information. One of the future challenges to research on crustacean chemical communication is to enhance our understanding of the neuronal processing underlying pheromone perception (see Schmidt and Mellon, Chap. 7; Hansson et al., Chap. 8). Most importantly, the pheromone receptors need to be identified. This will then facilitate further investigation of the central neural pathways mediating chemical communication. Knowledge of pheromone receptor proteins will also open the door to sequencing of olfactory receptor genes.

1.4.3   Signal Enhancement

Crustaceans can actively enhance odor acquisition by creating water currents that draw the molecules towards them (in lobsters: Atema and Steinbach 2007; crayfish:

Breithaupt, Chap. 13, Denissenko et al. 2007; stomatopods: Mead and Caldwell, Chap. 11; copepods: Moore et al. 1999). Generation of microcurrents to obtain chemical information may be a strategy to enhance olfaction, as also described for some insects (silkworm moth: Loudon and Koehl 2000). While such activities that enhance odor transmission are limited to only few examples in insects and mammals, they are common strategies in crustaceans involving specific fanning appendages.

Active odor sampling by antennal flicking is an additional mechanism to enhance olfaction in crustaceans (see Koehl, Chap. 5). These active behavioral investments in fanning or flicking activities serving signal transmission and reception appear to be much more common in the aquatic environment than in the terrestrial realm, most likely due to the differences in density and viscosity. Given the importance of signal delivery and transport in chemical communication, this will be a promising research field for the future.

Aggregated individuals also produce stronger chemical signals than solitary individuals. This might explain density-dependent gregariousness in spiny lobsters (Ratchford and Eggleston 1998). This effect could also be at work in crustacean species that form large mating aggregations. Individuals might initially aggregate because concentrated chemical signals will then attract other conspecifics, including potential mates. For example, during the mating period male green crabs *Carcinus maenas* gather at particular mating grounds (van der Meeren 1994), to which females are then attracted. Male hotspots with larger numbers of males (and stronger chemical signals) might attract many receptive females in a given area, thus benefitting both females and males. Since various crustacean species are highly gregarious, it appears worthwhile to explore how density-dependent strength of chemical signals might affect the attraction of additional conspecifics to aggregations.

1.5 Speciation Processes and the Role of Chemical Communication

A variety of factors can modulate the pheromone blends produced by the signaler or the chemosensory system of the receiver, which may have important evolutionary implications (Symonds and Elgar 2008; Smajda and Butlin 2009). Divergence in chemical signals may be a driving force in speciation; variations in the composition of chemical blends lead to reproductive isolation as has been shown for a number of insect species (Smajda and Butlin 2009 and references therein). Ecological factors can play an important role in divergence of pheromone blends. For example, if pheromone communication is susceptible to eavesdropping and is selected to avoid exploitation by parasites or predators, this could provoke shifts in pheromone composition leading to divergence between populations frequently exposed to predators/parasites and those living in enemy-free environments (e.g., Symonds and Elgar 2008, and references therein). Compounds obtained with the diet also may influence the composition of pheromone blends (Bryant and Atema 1987; Symonds and Elgar 2008). Habitat or host preferences may affect the chemosensory system and result in divergent pheromone communication between populations from different habitats/hosts (e.g., Smajda and Butlin 2009, and references therein). It is likely that these processes are also at work in crustaceans, which are targets of diverse predators or parasites and utilize a variety of hosts as food or shelter (e.g., Poore et al. 2008).

Interestingly, several examples from crustaceans suggest that reproductive isolation via pheromones might be incomplete. For example, some congeneric amphipod species appear unable to distinguish between mates of closely related species (Kolding 1986). Also, in other peracarids species, males mistakenly pair up with closely related species or ecotypes (Mead and Gabouriaut 1977; Hargeby and Erlandsson 2006), which could be due to lack of species recognition via chemical stimuli. Possibly, the divergence of contact pheromones is lagging behind compared to other life-history traits that lead to reproductive isolation under natural conditions (e.g., size or habitat preferences). Sutherland et al. (2010) observed that males from a species complex of freshwater amphipods could only discriminate against females from genetically distinct clades, but not against those from more closely related clades. In an estuarine amphipod with populations from different habitats, males preferred females from their own habitat, but were unable to distinguish between female populations if these were raised on other food types (Stanhope et al. 1992). Interestingly, pheromone specificity is also low in planktonic copepods and males frequently pursue pheromone trails laid by females of other species; once reaching and grasping the female, most males recognized their mistake based on contact chemicals, but some males even placed a spermatophore on the female from a different species (Goetze 2008). In experiments with specific mate-attracting signals, Bublitz et al. (2008) found that males of several species of crabs were attracted in similar ways by uridine diphosphate (UDP). Similarly, cyprid larvae of the barnacle *Balanus amphitrite* settled preferentially on substrata treated with the settlement-inducing protein complex (SIPC) from conspecifics, but relatively high proportions of larvae also settled on surfaces treated with the SIPC of other barnacle species (Dreanno et al. 2007; Clare, Chap. 22).

These examples suggest that species recognition in many crustaceans is not based only on particular chemical stimuli. Species recognition may require specific pheromone blends as in many insects (e.g. Symonds and Elgar 2004), "combination-lock" cascades of chemical signals that must proceed in a specific order (Hay, Chap. 3), or multimodal communication where chemical signals are complemented by visual or mechanical signals (Hebets and Rundus, Chap. 17). Additionally, sex-specific differences in behavior may ensure reproductive isolation. For examples, males of two closely related species of spiders were unable to discriminate between chemical signals from females of their own and of a closely related species, but specific female responses to male courtship behaviors maintained reproductive isolation in these species (Roberts and Uetz 2004).

In summary, for chemical signals to play a relevant role in species recognition among crustaceans, they may have to be (1) complex blends, (2) arranged in particular sequences, or (3) complemented by other modes of communication. Given that chemical signals are not as easily recognizable by human observers as morphological (visual or mechanical) signals, their role in reproductive isolation may often be strongly underestimated. There is a high probability that closely related species that depend primarily on chemical signals to recognize conspecific mates are not distinguished by taxonomists (Bickford et al. 2006). This might be especially relevant for marine organisms (including crustaceans); marine animal taxonomists, unlike their terrestrial colleagues, have only limited chances to observe the living organisms (Bickford et al. 2006).

In addition to behavioral interactions mediated by chemicals, there is a wide range of chemical interactions at the sperm–egg level that can result in reproductive isolation. Divergence of proteins acting at the gamete level may be especially rapid (Swanson and Vacquier 2002). In *Drosophila*, the gametic interactions are mediated by accessory gland proteins (ACPs) released with the seminal fluids (Ravi Ram and Wolfner 2007). Similar interactions could be expected for many crustaceans, where spermatophores also contain appreciable quantities of seminal fluid. Especially in brachyuran crabs, seminal fluids that are injected together with sperm in the female's spermatheca may interact with female reproductive processes, causing sexual conflict (see also Christy and Rittschof, Chap. 16). Sexual conflict is viewed as an important selective force in the divergence of reproductive proteins (Clark et al. 2009). Given the theoretical (species divergence) and applied (aquaculture and fishery) implications, it appears especially rewarding to explore chemical communication at the gametic level in crustaceans.

1.6 Anthropogenic Impact on Crustacean Chemical Communication

The contributions in this book demonstrate that chemical communication fulfills various crucial functions in reproduction, aggregation, and resolution of conflicts among crustaceans from a wide range of habitats. These functions are marred if pheromone production, transmission, or reception is disturbed by chemical or physical disturbances such as changes in salinity, acidity, temperature, or chemical pollution. Olsén (Chap. 26) explores how anthropogenic pollutants such as heavy metals, steroids, and pesticides disrupt different stages of chemical communication in fish and crustaceans. These interferences can produce morphological, physiological, and behavioral effects (e.g., Lürling and Scheffer 2007) that may entail dramatic consequences on reproduction and individual survival. Research on pheromone disruption is still in its infancy, but is direly needed to understand ecological consequences of anthropogenic disturbances of aquatic ecosystems and implement regulatory measures to protect aquatic species.

Variations in sex-specific population densities, e.g., due to extraction by fisheries, may also affect the dynamics of chemical communication in crustaceans (e.g., van Son and Thiel 2007). At low population densities, members from the mate-attracting sex may have to invest more in chemical advertisement than at high densities. On the other hand, individuals from the mate-searching sex that are efficient and rapid in responding to chemical signals may be at an advantage at low population densities.

1.7   Applied Aspects

Once identified, pheromones may be artificially synthesized to be used in mass attraction of target organisms. For crustaceans, this might be useful in the fisheries and aquaculture where desired individuals could be lured into traps (e.g., Barki et al., Chap. 25). Invasive crustaceans could also be caught by mass trapping, but it needs to be kept in mind that trapping is only efficient in combination with other measures aiming at suppressing or eliminating the populations of undesired species (see Baker, Chap. 27). Furthermore, pheromone trapping, if employed, may cause artificial selection for individuals responding to or producing other pheromone blends than the majority of the target population (see, e.g., Symonds and Elgar 2008).

Many of the contributions in this book are concerned with pheromones, i.e., chemical stimuli that are employed by crustaceans to attract conspecifics. However, chemical communication also includes substances that are used to repel other organisms. These repellents could be especially useful in the aquaculture context, e.g., to repel parasites or fouling organisms. Many crustaceans are parasites of commercially important fish species (e.g., salmon) and both traps and repellents could be used in controlling infection levels (e.g., Mordue and Birkett 2009). Similarly, barnacles are abundant fouling organisms in suspended structures or seawater systems and developing techniques to suppress their recruitment is one of the main motivations behind the identification of settlement factors (Clare, Chap. 22).

1.8   Outlook

The past two decades have seen enormous advances in our understanding of crustacean chemical communication. Nevertheless, our knowledge is still in its infancy when compared to insects, fish, or mammals. Most research on crustacean behavior in response to chemical stimuli has been conducted in controlled laboratory environments, where often only one stimulus context is tested. For example, the responses of numerous crustacean species to chemicals emitted by potential mates, by conspecific aggressors, or by interspecific enemies have been tested in isolation. However, in the natural environment, crustaceans are often exposed to multiple stimuli at the same time (see, e.g., Fig. 18.2 in Hazlett, Chap. 18). How an organism reacts in such a situation depends not only on external stimuli, but also on intrinsic factors (Hazlett, Chap. 18). Before drawing conclusions about the function of pheromones, it would be valuable to perform field tests in which organisms are confronted with a complex set of stimuli from all sensory modes (e.g., Johnson and Li 2010). While this is logistically challenging, it is essential if we are to understand how crustaceans employ pheromones in the natural environment.

In the field of behavioral ecology, a number of topics appear to be interesting. We consider it most rewarding to formally examine the presence of multimodal communication and of deception in crustaceans. Hebets and Rundus (Chap. 17) presented a number of examples that suggest that multimodal communication is much more common in crustaceans than previously assumed. In particular, chemical communication appears to be often combined with other sensory channels. Christy and Rittschof (Chap. 16) present several cases of deception in the visual channel, and they speculate on instances where crustaceans might deceive during chemical communication. Based on their overview, they suggest that it might be much easier to fake visual signals than chemical signals, but this hypothesis remains untested.

The major challenge ahead of us is the chemical identification of crustacean pheromones. We consider that this is the most crucial step because it not only will help us to understand the role of these pheromones in speciation processes, but will also permit a whole suite of experimental approaches. Insect biologists have utilized synthetically produced pheromones in experimental studies for several decades now (see Baker, Chap. 27). Once the first crustacean pheromone is synthetically produced, we expect a boost of research activities on chemical communication in crustaceans.

There are several crustacean taxa that appear to be at the doorstep of having the chemical structure of their pheromones identified. Probably the most promising approach to achieve pheromone identification will be to combine chemical analyses with genomics. Based on the fact that freshwater environments are so much more amenable to the chemical analysis of dissolved substances (see above), we suggest to focus on freshwater crustaceans during the next phase of chemical communication research (Fig. 1.4). Several crustacean taxa offer themselves for this approach. Similar to cladocerans (*Daphnia* spp.), freshwater amphipods and isopods are relatively small and can be easily reproduced in the lab without major space requirements. Short generation times permit the production of different strains under different selective pressures. Freshwater crayfish, in contrast, have longer generation times and require more space, but they have other important advantages. For example, emission of pheromone-containing urine can be made visible, neurological methods can be easily applied, and finally they are economically important fisheries and aquaculture resources, i.e., the need for applied research is high.

This next phase of research on chemical communication probably will also see more interdisciplinary work. Future interactions should involve behavioral ecologists, chemists, and geneticists. For example, recently 58 chemoreceptor genes were identified for *Daphnia pulex*, but at present it is not well known which particular chemicals these detect; females had some receptor genes that were lacking in males, but since there were no male-specific genes their role in mate finding is unclear (Peñalva-Arana et al. 2009). Future genomics studies should incorporate behavioral assays. Several of the contributions in this book highlight the importance of behavioral assays in the study of chemical communication (Breithaupt, Chap. 13; Hardege and Terschak, Chap. 19; Kamio and Derby, Chap. 20). It requires careful observations to identify well-suited behavioral assays that can be employed in tests of chemical compounds. Similarly, the selective mechanisms that ultimately lead to the divergence of signaling or chemoreceptor genes can only be uncovered with the help of behavioral studies. First advances in this next phase of interdisciplinary research have been achieved (see, e.g., Kamio and Derby, Chap. 20); more progress is expected in the near future. The next two decades of research on crustacean chemical communication promise to be very exciting and we look forward to what is to come.

Acknowledgments We thank Iván A. Hinojosa for his help in preparing the final figures and Drs. Chuck Derby and Marc Weissburg for helpful comments on the manuscript.

References

Agosta WC (1992) Chemical communication – the language of pheromones. Scientific American Library, New York

Ameyaw-Akumfi C, Hazlett B (1975) Sex recognition in the crayfish *Procambarus clarkii*. Science 190:1225–1226

Asai N, Fusetani N, Matsunaga S, Sasaki J (2000) Sex pheromones of the hair crab *Erimacrus isenbeckii*. Part 1: isolation and structures of novel ceramides. Tetrahedron 56:9895–9899

Atema J, Engstrom DG (1971) Sex pheromone in the lobster, *Homarus americanus*. Nature 232:261–263

Atema J, Steinbach MA (2007) Chemical communication and social behavior of the lobster *Homarus americanus* and other decapod Crustacea. In: Duffy JE, Thiel M (eds) Evolutionary ecology of social and sexual systems – crustaceans as model organisms. Oxford University Press, New York, pp 115–144

Bargmann CI (2006) Comparative chemosensation from receptors to ecology. Nature 444:295–301

Bickford D, Lohman DJ, Sodhi NS, Ng PKL, Meier R, Winker K, Ingram KK, Das I (2006) Cryptic species as a window on diversity and conservation. Trends Ecol Evol 22:148–155

Borowsky B (1989) The effects of residential tubes on reproductive behaviors in *Microdeutopus gryllotalpa* (Costa) (Crustacea: Amphipoda). J Exp Mar Biol Ecol 128:117–125

Bradbury JW, Vehrencamp SL (1998) Principles of animal communication. Sinauer Associates, Sunderland

Breithaupt T (2001) Fan organs of crayfish enhance chemical information flow. Biol Bull 200:150–154

Bryant BP, Atema J (1987) Diet manipulation affects social behavior of catfish: importance of body odor. J Chem Ecol 13:1645–1661

Bublitz R, Sainte-Marie B, Newcomb-Hodgetts C, Fletcher N, Smith M, Hardege JD (2008) Interspecific activity of sex pheromone of the European shore crab (*Carcinus maenas*). Behaviour 145:1465–1478

Caskey JL, Watson GM, Bauer RT (2009) Studies on contact pheromones of the caridean shrimp *Palaemonetes pugio*: II. The role of glucosamine in mate recognition. Invertebr Reprod Dev 53:105–116

Catania KC (2006) Underwater "sniffing" by semi-aquatic animals. Nature 444:1024–1025

Cheer AYL, Koehl MAR (1987) Paddles and rakes: fluid flow through bristled appendages of small organisms. J Theor Biol 129:17–39

Clark NL, Gasper J, Sekino M, Springer SA, Aquadro CF, Swanson WJ (2009) Coevolution of interacting fertilization proteins. PLoS Genet 5(7):e1000570

Clayton D (2008) Singing and dancing in the ghost crab *Ocypode platytarsus* (Crustacea, Decapoda, Ocypodidae). J Nat Hist 42:141–155

Dahl E, Emanuelsson H, von Mecklenburg C (1970) Pheromone transport and reception in an amphipod. Science 170:739–740

Denissenko P, Lukaschuk S, Breithaupt T (2007) The flow generated by an active olfactory system of the red swamp crayfish. J Exp Biol 210:4083–4091

Dixon CJ, Ahyong ST, Schram FR (2003) A new hypothesis of decapod phylogeny. Crustaceana 76:935–975

Dreanno C, Kirby RR, Clare AS (2007) Involvement of the barnacle settlement-inducing protein complex (SIPC) in species recognition at settlement. J Exp Mar Biol Ecol 351: 276–282

Freitag J, Krieger J, Strotmann J, Breer H (1995) Two classes of olfactory receptors in *Xenopus laevis*. Neuron 15:1383–1392

Goetze E (2008) Heterospecific mating and partial prezygotic reproductive isolation in the planktonic marine copepods *Centropages typicus* and *Centropages hamatus*. Limnol Oceanogr 53:33–45

Hargeby A, Erlandsson J (2006) Is size-assortative mating important for rapid pigment differentiation in a freshwater isopod? J Evol Biol 19:1911–1919

Holdich DM (1984) The cuticular surface of woodlice: a search for receptors. Symp Zool Soc Lond 53:9–48

Hurst JL (2005) Scent marking and social communication. In: McGregor PK (ed) Animal communication networks. Cambridge University Press, Cambridge, pp 219–243

Ingvarsdóttir A, Birkett MA, Duce I, Mordue W, Pickett JA, Wadhams LJ, Mordue (Luntz) AJ (2002) Role of semiochemicals in mate location by parasitic sea louse, *Lepeophtheirus salmonis*. J Chem Ecol 28:2107–2117

Johansson KUI, Hallberg E (1992) Male-specific structures in the olfactory system of mysids (Mysidacea; Crustacea). Cell Tissue Res 268:359–368

Johnson NS, Li W (2010) Understanding behavioral responses of fish to pheromones in natural freshwater environments. J Comp Physiol A. 196:701–711

Kamio M, Matsunaga S, Fusetani N (2000) Studies on sex pheromones of the helmet crab, *Telmessus cheiragonus*. 1. An assay based on precopulatory mate-guarding. Zool Sci 6:731–733

Kamio M, Matsunaga S, Fusetani N (2002) Copulation pheromone in the crab *Telmessus cheiragonus* (Brachyura: Decapoda). Mar Ecol Prog Ser 234:183–190

Koehl MAR (2001) Fluid dynamics of animal appendages that capture molecules: Arthropod olfactory antennae. In: Fauci L, Gueron S (eds) Computational modeling in biological fluid dynamics. Springer, New York, pp 97–116

Kolding S (1986) Interspecific competition for mates and habitat selection in five species of *Gammarus* (Amphipoda: Crustacea). Mar Biol 91:491–495

Lee JK, Strausfeld NJ (1990) Structure, distribution and number of surface sensilla and their receptor cells on the olfactory appendage of the male moth *Manduca sexta*. J Neurocytol 19:519–538

Loudon C, Koehl MAR (2000) Sniffing by a silkworm moth: wing fanning enhances air penetration through and pheromone interception by antennae. J Exp Biol 203:2977–2990

Lürling M, Scheffer M (2007) Info-disruption: pollution and the transfer of chemical information between organisms. Trends Ecol Evol 22:374–379

Mead F, Gabouriaut D (1977) Chevauchée nuptiale et accouplement chez l'isopode terrestre *Helleria brevicornis* Ebner (Tylidae). Analyse des facteurs qui contrôlent ces deux phases du comportement sexuel. Behaviour 63:262–280

Moore P, Fields DM, Jen Y (1999) Physical constraints of chemoreception in foraging copepods. Limnol Oceanogr 44:166–177

Mordue (Luntz) AJ, Birkett MA (2009) A review of host finding behaviour in the parasitic sea louse, *Lepeophtheirus salmonis* (Caligidae: Copepoda). J Fish Dis 32:3–13

Painter SD, Clough B, Garden RW, Sweedler JV, Nagle GT (1998) Characterization of *Aplysia* attractin, the first water-borne peptide pheromone in invertebrates. Biol Bull 194:120–131

Peñalva-Arana DC, Lynch M, Robertson HM (2009) The chemoreceptor genes of the waterflea *Daphnia pulex*: many Grs but no Ors. BMC Evol Biol 9:79. doi:10.1186/1471-2148-9-79

Phelan PL (1997) Evolution of mate-signalling in moths: phylogenetic considerations and prediction from the asymmetric tracking hypothesis. In: Choe JC, Crespi BJ (eds) The evolution of mating systems in insects and Arachnids. Cambridge University Press, Cambridge, pp 240–256

Poore AGB, Hill NA, Sotka EE (2008) Phylogenetic and geographic variation in host breadth and composition by herbivorous amphipods in the family Ampithoidae. Evolution 62:21–38

Ratchford SG, Eggleston DB (1998) Size- and scale-dependent chemical attraction contribute to an ontogenetic shift in sociality. Anim Behav 56:1027–1034

Ravi Ram K, Wolfner MF (2007) Seminal influences: *Drosophila* Acps and the molecular interplay between males and females during reproduction. Integr Comp Biol 47:427–445

Rittschof D, Cohen JH (2004) Crustacean peptide and peptide-like pheromones and kairomones. Peptides 25:1503–1516

Roberts JA, Uetz GW (2004) Species-specificity of chemical signals: Silk source affects discrimination in a wolf spider (Araneae: Lycosidae). Insect Behav 17:477–491

Roberts JA, Uetz GW (2005) Discrimination of female reproductive state from chemical cues in silk by males of the wolf spider, *Schizocosa ocreata* (Araneae, Lycosidae). Anim Behav 70:217–223

Roper TJ, Conradt J, Butler JE, Ostler CJ, Schmid TK (1993) Territorial marking with faeces in badgers (Meles meles): a comparison of boundary and hinterland latrine use. Behaviour 127:289–307

Ryan FP (1966) Pheromone: evidence in a decapod crustacean. Science 151:340–341

Seelinger G (1983) Response characteristics and specificity of chemoreceptors in *Hemilepistus reaumuri* (Crustacea, Isopoda). J Comp Physiol A 152:219–229

Smajda C, Butlin RK (2009) On the scent of speciation: the chemosensory system and its role in premating isolation. Heredity 102:77–97

Sorensen PW, Stacey NE (2004) Brief review of fish pheromones and discussion of their possible uses in the control of non-indigenous teleost fishes. N Z J Mar Freshwater Res 38:399–417

Stanhope MJ, Connelly MM, Hartwick B (1992) Evolution of a crustacean chemical communication channel: behavioral and ecological genetic evidence for a habitat-modified, race-specific pheromone. J Chem Ecol 18:1871–1887

Strausfeld N, Reisenman CE (2009) Dimorphic olfactory lobes in the Arthropoda. Ann N Y Acad Sci 1170:487–496

Sutherland DL, Hogg ID, Waas JR (2010) Phylogeography and species discrimination in the *Paracalliope fluviatilis* species complex (Crustacea: Amphipoda): can morphologically similar heterospecifics identify compatible mates? Biol J Linn Soc 99:196–205

Swanson WJ, Vacquier VD (2002) The rapid evolution of reproductive proteins. Nat Rev Genet 3:137–144

Symonds MRE, Elgar MA (2004) The mode of pheromone evolution: evidence from bark beetles. Proc Biol Sci 271:839–846

Symonds MRE, Elgar MA (2008) The evolution of pheromone diversity. Trends Ecol Evol 23:220–228

Van der Meeren GI (1994) Sex- and size-dependent mating tactics in a natural population of shore crabs *Carcinus maenas*. J Anim Ecol 63:307–314

Van Son TC, Thiel M (2007) Anthropogenic stressors and their effects on the behavior of aquatic crustaceans. In: Duffy JE, Thiel M (eds) Evolutionary ecology of social and sexual systems: Crustaceans as model organisms. Oxford University Press, New York, pp 413–441

Vickers NJ (2000) Mechanisms of animal navigation in odor plumes. Biol Bull 198:203–212

Voigt CC, von Helversen O (1999) Storage and display of odour by male *Saccopteryx bilineata* (Chiroptera, Emballonuridae). Behav Ecol Sociobiol 47:29–40

Voigt CC, Caspers B, Speck S (2005) Bats, bacteria, and bat smell: sex-specific diversity of microbes in a sexually selected scent organ. J Mammal 86:745–749

Weissburg MJ (2000) The fluid dynamical context of chemosensory behavior. Biol Bull 198:188–202

Zeeck E, Hardege JD, Bartels-Hardege HD, Wesselmann G (1988) Sex pheromone in a marine polychaete: determination of the chemical structure. J Exp Zool 246:285–292

Zhang D, Lin J, Harley M, Hardege JD (2010a) Characterization of a sex pheromone in a simultaneous hermaphroditic shrimp, *Lysmata wurdemanni*. Mar Biol 157:1–6

Zhang D, Zhu J, Hardege JD, Lin J (2010b) Surface glycoproteins are not the contact pheromones in the *Lysmata* shrimp. Mar Biol 157:171–176

Chapter 2
Pheromones and Behavior

Tristram D. Wyatt

Abstract Chemical communication is widely used by crustaceans, for example, in sexual interactions, larval release, and planktonic settlement. However, we know the identity of very few of the molecules involved. In this chapter, I introduce pheromones and contrast them with signature mixtures. Pheromones are molecules that are evolved signals, in defined ratios in the case of multiple component pheromones, which are emitted by an individual and received by a second individual of the same species, in which they cause a specific reaction, for example, a stereotyped behavior or a developmental process. Signature mixtures are variable chemical mixtures (a subset of the molecules in an animal's chemical profile) learned by other conspecifics and used to recognize an organism as an individual (e.g., lobsters, mammals) or as a member of a particular social group such as a family, clan, or colony (e.g., mammals, desert woodlouse *Hemilepistus reaumuri*, ants, bees). A key difference between pheromones and signature mixtures is that in all taxa so far investigated it seems that signature mixtures need to be learnt (unlike most pheromones). These signature mixtures may be best thought of as cues. Pheromones evolve from molecules which give a selective advantage to the receiver and signaler. The evolution of pheromones is facilitated by the combinatorial basis of the olfactory system found in crustaceans and other animals. In crustaceans, some pheromones are also detected by the distributed chemosensory system. Crustaceans have great potential as model organisms for chemical communication research, in particular now that the *Daphnia pulex* genome has been sequenced.

2.1 Introducing Chemical Ecology

Like other animals, crustaceans live in a chemosensory world full of chemical information and signals from conspecifics, prey, and predators (Fig. 2.1). To a human

T.D. Wyatt (✉)
Department of Zoology, The Tinbergen Building, South Parks Road,
Oxford, OX1 3PS, UK
e-mail: tristram.wyatt@zoo.ox.ac.uk

T. Breithaupt and M. Thiel (eds.), *Chemical Communication in Crustaceans*,
DOI 10.1007/978-0-387-77101-4_2, © Springer Science+Business Media, LLC 2011

Fig. 2.1 Just as peacock males display their fan tails, the male lobster (*left*) displays with pheromones in the directed jet he sends toward the female. Drawing by Jorge A. Varela Ramos (Picture inspired by the mouse pheromone "peacock tail" in Penn and Potts 1998)

it may be strange to think of a world of smells underwater (as we experience smells in air) but chemical communication is widely used by aquatic animals including crustaceans. However, the types of molecules used by crustaceans for communication underwater are likely to be different from those used in terrestrial environments by most insects and mammals. For aquatic animals, solubility of signal molecules is the equivalent of volatility for airborne messages.

The division of chemical senses in vertebrates into taste and smell does not work well for crustaceans and other invertebrates. Like vertebrates, crustaceans have olfactory receptor neurons (ORNs) (these are also known as olfactory sensory neurons) with one end exposed to the chemical world outside the animal and the other leading to the brain. These ORNs in the aesthetasc sensilla of crustaceans approximate to vertebrate "olfaction" as the ORNs project to glomeruli in the crustacean olfactory lobe, analogous to the organization of the brain in vertebrates (Caprio and Derby 2008; Derby and Sorensen 2008; Schmidt and Mellon, Chap. 7). Long distance sex pheromones and chemical cues for social interactions tend to be processed by the olfactory/aesthetasc pathways, for example, spiny lobster responses to conspecific urine signals (Horner et al. 2008). However, crustaceans also have "distributed chemoreception" (Schmidt and Mellon, Chap. 7) which goes beyond vertebrate taste. Distributed chemoreceptors are typically packaged with mechanosensors into sensilla over other parts of the body and have a nonglomerular organization in the brain/ganglia. While distributed chemoreception does include the equivalent of taste, with contact chemoreception for food for example, it also includes other chemical senses such as control of antennular grooming and the coordination of mating and copulation (Schmidt and Mellon, Chap. 7). Schmidt and Mellon point out that integrated inputs from sensilla in both the olfactory and distributed chemoreceptor systems may participate together in the control of complex behaviors, including associative odor learning.

As you will see in the rest of this book there is already an impressive understanding of crustacean chemical communication – but really we are just at the beginning of the exploration. In the coming years, as demonstrated in the chapters of this book, chemical communication by crustaceans will be seen to be just as spectacular as the currently well-known examples of moth males flying upwind to find the female moth releasing pheromone.

What is surprising is that only a small number of pheromones have been chemically identified for crustaceans compared with the thousands for insects. Part of the difference may be the importance of moths as agricultural pests, which has prompted governments to provide funding, but I wonder if the disparity also suggests that crustaceans are harder to study. Is it the type of molecules used by crustaceans or the current difficulties of working with the chemical communication of aquatic animals? Hardege and Terschak (Chap. 19) discuss the importance of discriminating bioassays for identifying pheromones. It may be that further work needs to be done on bioassays that yield discriminating data for crustaceans.

In this chapter I would like to introduce concepts of chemical ecology, to discuss pheromones and signature mixtures, and contrast the research on terrestrial pheromones with research on aquatic pheromones.

2.2   Definition of Pheromones

Pheromones are chemical signals used to communicate within a species. Pheromones are "molecules that are evolved signals, in defined ratios in the case of multiple component pheromones, which are emitted by an individual and received by a second individual of the same species, in which they cause a specific reaction, for example, a stereotyped behavior or a developmental process" (Wyatt 2010, modified after Karlson and Lüscher 1959). The word is derived from the Greek *pherein*, to transfer and *hormon*, to excite. Karlson and Lüscher intended that pheromones should include chemical signals in animals of all kinds and they included Crustacea (the prawn *Palaemon serratus*), as well as fish, mammals, and insects, among their examples.

Karlson and Lüscher anticipated that species would not necessarily have exclusive use of a particular molecule as their pheromone. We now know many examples of shared use of molecules (Kelly 1996; Wyatt 2003, 2010). One of the most spectacular is the way that the female Asian elephant shares its small molecule sex pheromone, (Z)-7-dodecen-1-yl acetate, with some 140 species of moth (Rasmussen et al. 1996). Because the moths use defined combinations of molecules, of which (Z)-7-dodecen-1-yl acetate is just one, there is not likely to be confusion between elephants and moths. Because the defined combination is different for each moth species there is no confusion between the moth species: in cases where the combination is the same, the species "call" with their pheromones at different times or have nonoverlapping habitats.

Mice and some bark beetle species also share some terpene compounds as pheromones. There are a number of reasons why these overlaps occur. The first is that all animals share a common heritage and so have basically the same metabolic pathways and molecular capabilities. Second, the number of potential small volatile molecules, which are also chemically stable and nontoxic, is limited. Incidentally, this may be less of a problem for underwater communication as even very large molecules can be soluble. We might expect some aquatic species to have unique large pheromone molecules used singly, for example species-specific peptides with different amino acid sequences, rather than blends of common small molecules as in moth sex pheromones.

2.2.1   Signals vs. Cues

Pheromones are signals as defined by Maynard Smith and Harper (2003, p3) in their book *Animal Signals*: "any act or structure which alters the behavior of other organisms, which evolved because of that effect, and which is effective because the receiver's response has also evolved." The definition is further discussed by Scott-Phillips (2008) in a thoughtful paper which makes explicit that the response itself has to have evolved to be affected by the act or structure concerned (not just evolved to respond to something else). For something to be a signal, both the emission and response need to have evolved.

This distinguishes signals from cues. Cues are things such as the CO_2 released by an animal as it breathes, and used as a cue by a blood-sucking insect to find its host. The mosquito's response is certainly evolved (and indeed it has highly specialized receptors to detect CO_2), but the release of CO_2 by the host does not evolve to have the effect of attracting mosquitoes so CO_2 release does not count as a signal. The odors used by mammals and crustaceans when recognizing kin or familiar animals are also probably best seen as cues rather than signals, in my opinion (see below).

How do we know that a chemical signal has evolved for an effect? In moth females we can see the specialized pheromone glands evolved for secreting and releasing the female sex pheromones – for the effect of signaling – and in the male, the highly evolved feathery antennae and specialized olfactory sensory cells on those antennae for detecting the pheromones, combined with the specialized sex-specific brain structures, the macroglomerular complex (MGC), in the male brain for processing the data (Hansson 2002; Wyatt 2010). Often the evidence of evolved structures and processes for signal reception is more subtle. For example, male and female *Drosophila* antennae are morphologically very similar. The evolved effects of pheromones in *Drosophila* are at the receptor level and circuits of the brain of the male and female (Vosshall 2008). In lobsters, pheromones may be added to the urine before release rather than there being a special gland visible. Pheromone components can be acquired as well as synthesized by the emitter. For example, to attract females, male euglossine bees in the American tropics must collect perfume oils such as mono- and sesqui-terpenes from orchid flowers (see Wyatt 2010).

2.2.2 Pheromone Types by Function

Pheromones are often described by function, by the effect they have. For example, sex pheromones describe those involved in mate-finding or attraction. Others include aggregation, alarm, and trail pheromones. Some responses are context-specific – for example in some ant species alarm pheromones cause ants to disperse if released far from the nest but to attack if released close to the nest. In a way, these descriptors allow us simply to describe the range of behaviors mediated by pheromones and then generalize from these to describe patterns in use, say, of sex pheromones across taxa.

Our definitions reflect our attempts to make sense of the world and identify patterns. Useful definitions enable us to generalize and predict characteristics. However, no definition is going to be a perfect match to the range of observed scenarios. There will always be fuzzy areas and difficult cases. Part of the problem may be trying to include all phenomena in the same definition. Another problem is that we may be describing pheromones at different stages of evolution – it is likely that many pheromones start as chemical cues and only later evolve into signals, an evolutionary process known as ritualization (Maynard Smith and Harper 2003, p.15).

2.2.3 Releaser vs. Primer Effects

Pheromones may elicit a behavioral response (releaser effects), longer-lasting developmental (primer effects) mediated via hormones, or both (see Fig. 2.2). So for example, prior exposure for some days to the female pheromone(s) in the premolt urine of the shore crab *Carcinus maenas* primes his sexual behavioral responses such as cradling when he is later exposed to the female premoult urine (Ekerholm and Hallberg 2005).

2.2.4 Semiochemicals vs. Infochemicals

As the variety of chemical interactions between organisms became better understood during the 1960s and 1970s, many authors attempted to classify the interactions and the chemical agents involved. If pheromones describe intraspecific signals, what terms should be used for interspecific chemical interactions? In the recent aquatic chemosensory literature, authors seem to be almost equally divided between using "semiochemicals" and "infochemicals."

Semiochemicals (from the Greek *semeion* mark or signal) are the chemicals, acquired or produced by the sender, involved in the chemical interactions between organisms (Nordlund and Lewis 1976) (Table 2.1). In their original definition, Law and Regnier (1971) used the phrase "chemical *signals* for transmitting information between individuals" [my italics], which is fine for pheromones but causes a

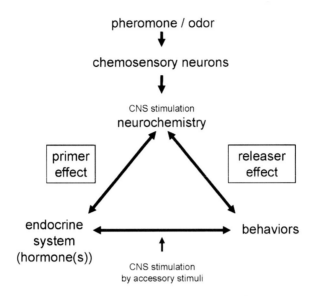

Fig. 2.2 Pheromones can be a stimulus leading to a prompt behavioral response by nerve impulses from the brain (CNS, central nervous system) (*releaser effects*) or can act indirectly by stimulation of hormone secretion resulting in physiological changes, "priming" the animal for a different behavioral repertoire (*primer effects*). A given pheromone can of course have primer or releaser effects at the same time or in different contexts. The distinction between primer and releaser effects has become blurred as we understand more about the links, interactions, and feedback loops in the sequence from odor to behavioral and endocrine effects. Hormonal effects can be rapid, and memories, sometimes facilitated by local neurochemistry changes, can be long-lasting. The diagram brings together ideas from figures in Wilson and Bossert (1963) and Sachs (1999). Text and diagram adapted from Wyatt (2003, Fig 1.10, p. 18)

problem for many interspecific interactions as we now reserve the terms "communication" and "signals" for mutually evolved signals. Nordlund and Lewis (1976) recognized the problem and reworked their definition of semiochemicals to remove "signals."

Pheromones are thus semiochemicals used within a species. Interspecific semiochemicals are *allelochemicals*. They are further subdivided depending on the costs and benefits to the sender and the receiver (Nordlund and Lewis 1976). If the sender benefits at the expense of the receiver, for example a bolas spider releasing moth-identical pheromones to lure a male moth, the spider's chemical message is an *allomone*. If a prey species such as the gammarid *Gammarus roeselii* avoids its predators' odors, then the gammarid is using those odors as a *kairomone* (Hesselschwerdt et al. 2009). If both partners benefit, as in the case of mutualisms between sea anemones and anemone clown-fish, the chemicals involved are termed *synomones*. The multiplicity of terms is only useful as shorthand and they are clearly overlapping, not mutually exclusive (a molecule used as a pheromone within a species can be used as a kairomone by its predator, for example).

In proposing the word kairomone, Brown et al. (1970) had in mind the word's Greek root *kairos*, especially in its senses of "opportunistic" and

Table 2.1 Definitions of chemical mediators important in chemical interactions among organisms, based on Nordlund and Lewis (1976) and Wyatt (2010)

A. Hormone: A chemical agent, produced by tissue or endocrine glands, which controls various physiological processes within an organism (Nordlund and Lewis 1976)

B. Semiochemical: A chemical involved in the chemical interaction between organisms (Nordlund and Lewis 1976)

   1. Pheromone: Molecules that are evolved signals, in defined ratios in the case of multiple component pheromones, which are emitted by an individual and received by a second individual of the same species, in which they cause a specific reaction, for example, a stereotyped behavior or a developmental process (Wyatt 2010, modified after Karlson and Lüscher 1959)

   2. Signature Mixture: A variable chemical mixture (a subset of the molecules in an animal's chemical profile) learned by other conspecifics and used to recognize an animal as an individual (e.g., lobsters, mammals) or as a member of a particular social group such as a family, clan, or colony (e.g., *Hemilepistus reaumuri,* ants, bees) (Wyatt 2010, derived from Wyatt 2005 "signature odor" and Johnston's "mosaic signal" 2003)

   3. Allelochemical: Chemical significant to organisms of a species different from their source, for reasons other than food as such (Nordlund and Lewis 1976)

      a. Allomone: A substance, produced or acquired by an organism, which, when it contacts an individual of another species in the natural context, evokes in the receiver a behavioral or physiological reaction adaptively favorable to the emitter but not to the receiver (Nordlund and Lewis 1976)

      b. Kairomone: A substance, produced, acquired by, or released as a result of the activities of an organism, which, when it contacts an individual of another species in the natural context, evokes in the receiver a behavioral or physiological reaction adaptively favorable to the receiver but not to the emitter (Nordlund and Lewis 1976)

      c. Synomone: A substance produced or acquired by an organism, which, when it contacts an individual of another species in the natural context, evokes in the receiver a behavioral or physiological reaction adaptively favorable to both emitter and receiver (Nordlund and Lewis 1976)

      d. Apneumone: A substance emitted by a nonliving material that evokes a behavioral or physiological reaction adaptively favorable to a receiving organism, but detrimental to an organism, of another species, which may be found in or on the nonliving material (Nordlund and Lewis 1976)

"Infochemical" as an alternative to "semiochemical" was proposed by Dicke and Sabelis (1988) though its main change was to replace "produced or acquired by" with "pertinent to biology of" in each case for allelochemics. Apneumone is rarely used. Paragraphs from Nordlund and Lewis (1976) with permission

"exploitative" – which is just how a predator uses a kairomone. Kairomones have not evolved as signals and this is reflected in my usage of kairomone usually in the context of "using [molecule x] as a kairomone" rather than defining a particular molecule itself as a kairomone.

"Infochemical" was proposed by Dicke and Sabelis (1988) as a term to replace "semiochemical" at the head of the classification. They argued that the cost-benefit of the interaction should be the only criterion, not the origin of the chemicals concerned. This was prompted perhaps by their work on tritrophic interactions, which have volatiles produced by the herbivores themselves and also induced in their host plants. However, indirectly produced or induced volatiles were already included by Nordlund and Lewis (1976) in their definition for kairomone which includes substances "released as a result of the activities of an organism" (Table 2.1).

On balance I feel the best solution is to stick with "semiochemicals," but with a relaxation of the definition of pheromone to include, where needed, compounds produced by symbionts or other associated organisms (Wyatt 2010).

2.2.5   Pheromones, Signature Mixtures, and Recognition

One of the most contentious problems in chemical communication has been what to name the variable odors used by mammals to distinguish individuals of their own species and used by social insects to distinguish fellow colony members from "foreigners" (and probably used by lobsters and hermit crabs to recognize individuals in fights, see Atema and Steinbach 2007; Gherardi and Tricarico, Chap. 15). These chemical cues do not fit the pheromone definition of a "defined combination of molecules causing a specific behavior or response" (Table 2.1; Wyatt 2010). Instead, the variation in chemical mixtures between individuals or colonies *is* the message. I propose that we use the term "signature mixtures" to describe these (Wyatt 2010) (Table 2.1). These signature mixtures are learnt (in contrast to pheromones, which tend not to need learning for response) (Table 2.2).

For kin recognition, animals learn any cues (signature mixtures) that give a statistical probability allowing recognition (Sherman et al. 1997). Sometimes these signature mixtures are produced by the animal itself but in some species they may be acquired from the shared local environment instead, for example in wood frog tadpoles *Rana sylvatica* (Sherman et al. 1997). Desert woodlice (*Hemilepistus reaumuri*) show family recognition by cuticular chemical signatures, which appear to be a mixture of compounds from all family members; signature mixtures have to be reacquired after each molting event (Linsenmair 1987, 2007). The recognition by guards in *Hemilepistus* is clearly highly evolved (as it is in honey bees). However, the cuticular molecules in *Hemilepistus*, perhaps with a waterproofing

Table 2.2 Contrasting pheromones with signature mixtures

	Molecule(s)	Receiver system/perception
Pheromones *Anonymous*	Species specific e.g., Sex pheromones e.g., "Category" pheromones (such as immature, adult)	"Hardwired," learning not usually required
Signature Mixtures *Variable*	Individual, family, colony odors	Learning required

Hölldobler and Carlin (1987) introduced the idea of anonymous signals (pheromones) contrasted with variable signature mixtures (though their terminology was different). The anonymous pheromone signals are uniform throughout a category (e.g., species, male, female, and perhaps molt state and dominance status). In contrast, signature mixtures vary and can be used to recognize the organism as an individual or member of a particular social group such as a family or colony. A more extensive comparison is given in Wyatt (2010)

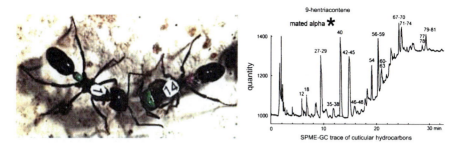

Fig. 2.3 Most animals have both anonymous pheromone signals and signature mixtures (Wyatt 2010). This is demonstrated in the cuticular hydrocarbons of the "queenless" ant *Dinoponera quadriceps* (Monnin et al. 1998). A gas chromatographic analysis of the cuticular hydrocarbons on the alpha female, the only fertilized and egg-laying individual in the colony, shows the anonymous hydrocarbon pheromone 9-hentriacontene (indicated by the *asterisk*) characteristic of alpha females in all colonies, together with the diverse range of other hydrocarbons which make up the colony odor. Her fellow colony members have the colony odor but lack the 9-hentriacontene. Ants photograph, courtesy Thibaud Monnin. Gas chromatogram, Monnin et al. (1998), with permission

function, might have been coopted for a signature function. It is possible that complexity of signatures might later be selected, under natural selection, for ease of recognition (Tibbetts and Dale 2007). Do the eusocial sponge-living shrimps use odor in colony recognition in a similar way to *Hemilepistus,* ants and bees?

A given animal will produce both pheromones and signature mixtures (Fig. 2.3). In addition to its sex pheromone(s), each lobster has its own highly individual odor mixture and this complex bouquet can be used by other lobsters for recognition (see Atema and Steinbach 2007).

It is likely that signature mixtures, as they involve learning, are processed differently from pheromones though this is still not fully understood (Wyatt 2010). The neurobiology of the olfactory imprinting of signature mixtures by adult mammals is well studied in mice (for the Bruce effect, the odors of her mate are learnt in her accessory olfactory lobe) and sheep (the mother learns the odors of her lamb in her main olfactory lobe) (Brennan and Kendrick 2006). Individual recognition in lobsters is mediated by the esthetic/olfactory pathways (Johnson and Atema 2005) though currently not enough is known about olfactory processing in lobsters to know if pheromones are processed differently.

2.3 Evolution of Pheromones

Pheromones are used right across the animal kingdom because any molecule that gives a selective advantage can potentially evolve into a pheromone. This is in part a consequence of the combinatorial mechanisms of olfaction. Invertebrate and vertebrate olfaction, despite all the superficial differences between crustacean aesthetascs and mammalian noses, works in roughly the same way. Crustaceans, with an impermeable exoskeleton, have evolved olfactory "windows" in the exoskeleton.

Odor molecules interact with olfactory receptor proteins on the surface of each ORN. In arthropods there are hundreds of different olfactory receptor proteins depending on species, each with a broad sensitivity/range of molecules which will interact with (Hallem and Carlson 2006). Each receptor's sensitivities overlap with those of other receptor proteins, so that a huge "olfactory world" can be covered. Each ORN presents just one of the different olfactory receptor proteins. Chemosensory systems are discussed in chapters by Hallberg and Skog (Chap. 9), and Schmidt and Mellon (Chap. 7). Derby and Sorensen (2008) remind us that detectors of pheromones and olfactory social cues are not exclusively located in the aesthetasc/olfactory pathway, giving as an example male crayfish which have sensors on their claws that detect female odors.

All the ORNs carrying the receptor of a particular type feed to the same glomerulus (neuropil) on each side of the olfactory lobe (Caprio and Derby 2008). The number of glomeruli (and by implication the number of different ORs) in Crustacea ranges from about 150 to 1,300 (Caprio and Derby 2008). The brain builds up an olfactory picture of the world from these, integrating the responses across the different glomeruli. So for example, hypothetically a molecule might stimulate ORN types 1, 3, and 25. A different molecule might stimulate ORN types 2 and 91. This combinatorial processing allows organisms to discriminate and distinguish innumerable molecules, including ones never encountered before.

Thus the olfactory system is primed to respond to any odor chemical(s) in the environment that give a selective advantage, for example detecting females about to produce eggs. Selection can start on the receiver to become more sensitive, with more selective olfactory receptor proteins, more expressed on each ORN and more of the increasingly specific ORNs in the animal's chemosensory organ(s). In this way we can evolve the sensitive pheromone detection system of the male crab. There will be corresponding selection on the emitter to release more of the molecules, now becoming an evolved signal or pheromone. A hypothetical scenario, from "spying" to an evolved signal, is shown in Fig. 2.4.

Pheromones can often be related to their likely original source or function, though they may have been modified somewhat to give specificity. For example, many fish sex pheromones are or have evolved from steroid and prostaglandin hormones (Stacey and Sorensen 2006). Alarm pheromones in ants often appear to have evolved from defensive compounds used by that species – and presumably evolved the pheromone function as these compounds would be released during combat. The Settlement-Inducing Protein Complex in barnacles appears to have evolved from α2-macroglobulin-like proteins which occur in the barnacle cuticle (perhaps through duplication of an ancestral barnacle A2M gene) (Dreanno et al. 2006; Clare, Chap. 22).

In aquatic systems, solubility takes the place of volatility. Aquatic pheromones can be large molecules so long as they are soluble in water (though contact pheromones, such as those used in mate recognition in shrimp and copepods are presumably almost insoluble – see e.g., Bauer, Chap. 14 and Snell, Chap. 23). Many aquatic organisms use polypeptides as pheromones, for example the crab "pumping pheromone" (Rittschof and Cohen 2004). Pheromone specificity can be gained by

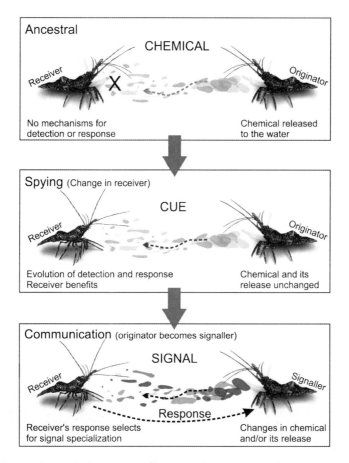

Fig. 2.4 Proposed stages in the evolution of a communication function for molecules released by an individual. The X in the upper panel indicates that the receiving individual has no adaptations to receive the cue. In the "spying phase," only the receiver benefits. The transition to bilateral benefit to both sender and receiver could occur later if there is a selective advantage to the sender. In the "spying phase" there need not be changes in the molecules released by the sender. An original figure by Ivan Hinojosa, inspired by and with text from Stacey and Sorensen (2006, Fig. 9.1, p. 363) (with permission from the authors)

unique amino acid sequences. Peptides have short half-lives due to rapid consumption by microbes and this can be a useful property for some signals (Rittschof and Cohen 2004) though it may be one reason why territorial marking by scent marks does not seem to be found in aquatic organisms (Thiel and Duffy 2007).

As the pheromones of fish and crustaceans become better understood, species specificity through multicomponent blends may be revealed (Stacey and Sorensen 2006). Multicomponent synergy, with two or more pheromone components needed together, has been demonstrated to stimulate a gonadotropin surge in male goldfish.

2.3.1 Pheromones and Speciation

For moths we know enough about the pheromones across species and families to analyze the way pheromones change in speciation. Closely related sympatric moth species have multicomponent pheromone blends that differ by added, lost, or changed components (Wyatt 2010). In some species we can see the changes at the gene level (e.g., Xue et al. 2007). Sex pheromones are used widely by crustaceans but copepod cuticular pheromones aside (Snell, Chap. 23) they have so far resisted identification (though we are close, see Hardege and Terschak, Chap. 19; Kamio and Derby, Chap. 20). We therefore do not yet know enough about the pheromones of any group of Crustacea to look at the variation between species. Over time I am confident we will be able to do this.

2.4 Peculiarities of Chemical Communication as Compared with Visual and Acoustic Communication

The key difference between chemical communication and visual and acoustic communication is the need for molecules to travel from signaler to receiver. This requires either diffusion, likely to be important only for small organisms at the scale of millimeters (see Yen and Lasley, Chap. 9) or flow of currents (see Koehl, Chap. 5). Chemical signals are thus rarely almost instantaneous in the way that visual and acoustic signals can be. On the other hand, chemical signals can be long-lasting. The chemical cues from barnacle cuticular proteins, for example, resist a variety of chemical attacks including acids and caustic alkalis (Knight-Jones 1953; Clare, Chap. 22).

Specificity is achieved in different ways in chemical communication compared with visual and acoustic routes, which are a continuous range of spectra, differing by wavelength and temporal structure. Chemical communication offers ways of differing in many dimensions, and in particular the opportunity to combine different molecules to give specificity. A further difference may be the honesty of chemical signals. Deception in signaling, and contrasts between visual, acoustic, and chemical signals, are discussed by Christy and Rittschof (Chap. 16).

2.5 Novel Techniques and Questions in Chemical Communication Research

For research on pheromones, the bioassay is crucial and is the first step to establish that there is a chemically mediated effect to be studied. The bioassay test for activity can be behavioral or physiological depending on the pheromone being studied. The gold standard or pheromone equivalent of "Koch's postulates" requires the isolation, identification, synthesis, and bioassay confirmation of activity (Wyatt 2009). Demonstration of activity at the biologically relevant concentration is also important.

Chemical communication is more difficult to investigate than sound and visual signals because molecules are involved rather than being phenomena amenable to spectral analysis. First there is the challenge of identifying the molecules involved (which may be at low concentrations, near the limits of instrumentation) and second there is the problem of playback. Whereas video or an MP3 player can reproduce visual or sound signals, and many signals can be created by software and then replayed, for studies of chemical communication the exact molecules have to be offered. Part of the difficulty is that the olfactory system is sensitive to every aspect of the precise identity of molecules, including their chirality.

Simple bioassay-guided fractionation approaches separate a biological sample into fractions based on a number of properties, including solubility in solvents of different polarity, molecular mass, and molecular charge (Derby and Sorensen 2008; Hardege and Terschak, Chap. 19; Kamio and Derby, Chap. 20). However, many pheromones are multicomponent, requiring a number of molecules to be present in the correct ratio and concentration for activity, so simple fractionation may not work as the components may end up in different fractions, inactive by themselves (we are very lucky that the first pheromone identified, bombykol, had a main component that was active by itself, otherwise Butenandt's bioassays would not have revealed the molecule). This problem is made more difficult because the active molecules may not be the ones present in the largest quantities. One approach to tackle the problem of bioassaying multicomponent pheromones is to use a subtractive-combination bioassay in which a complete putative mixture is tested with successively more fractions or molecules missing, to see which ones can be removed without extinguishing the signal (Byers 1992; Kamio and Derby, Chap. 20).

Progress in chemical techniques of extraction, analysis, and synthesis are likely to enable many more pheromones to be identified. The history of pheromone research, like many other areas of science, shows the importance of new technologies and methods including gas chromatography (so helpful for work on insect pheromones), mass spectroscopy, and, more recently, the use of molecular biological techniques. The new technique of metabolomics has promise to identify biomarkers, molecules enriched in a particular context and thus candidate pheromone components (Derby and Sorensen 2008; Kamio and Derby, Chap. 20). The problem of the highest peaks, not necessarily being the pheromones, still remains though.

Recording from the animals' own sensors as a pheromone-identifying tool shows promise though the antennogram is problematic in aquatic crustaceans because of the technical difficulty of isolating the recording electrodes from the stimulus water. There has, however, been extensive single cell recording of chemosensory organs of crustaceans (Caprio and Derby 2008).

2.6 The Promise of Crustaceans as Model Systems
for Chemical Communication Research

Crustaceans could provide a new range of model organisms contributing to an understanding of chemical communication and chemoreception in general. However, currently a key challenge to using crustaceans as model systems in chemical communication is the lack of knowledge about the chemical identity of the pheromones involved. As more crustacean pheromones are identified, a rich range of research questions will come within reach. Nonetheless there is much that can be done in the meantime even if the chemical identities of the molecules have not been confirmed. Behavioral work in particular can go ahead using active biological samples or extracts as the chemical stimuli.

With the *Daphnia pulex* water flea genome now available (wFleaBase), genetic dissection of chemoreception in Crustacea becomes possible (e.g., Peñalva-Arana, et al. 2009). As wFleaBase's introduction points out: "*Daphnia* shares with *C. elegans* an interesting and rich gene structure and wealth of gene duplications. It has closest gene homology to the insects, *Apis* and *Drosophila*. However, it has nearly as many unique or strong gene homologies to Mouse as to insects" (wFleaBase 2009).

Many crustaceans can be cultured in controlled laboratory environments (Barki, Jones, and Karplus, Chap. 25). While many large decapods have planktonic larval stages, the peracarids (which include isopods and amphipods) are direct developers, with the eggs brooded by the female, facilitating lab culture (Thiel, Chap. 10). One possible area of research could use such direct developing species to generate particular strains, for example that have grown in particular selective environments.

One topic which is already showing the important contribution of work on crustaceans is the orientation behavior of animals to chemical stimuli. How do animals follow an odor plume to its source? Orientation behavior by crustaceans is explored in Weissberg (Chap. 4). One advantage of decapods is that they are large enough to carry tracking devices. There seem to be intriguing differences between the mechanisms used by different crustacean groups, even within the decapods.

2.7 Conclusions and Outlook

Chemical communication is the commonest method of communication across the animal kingdom. Crustaceans provide some of the most interesting examples of this, from sex to settlement, but with few exceptions, we do not know the molecules that are being used. Crustaceans have the potential to be among the key model organisms for work on chemoreception in the future, in particular if their pheromones can be identified and synthesized.

One area of debate, for chemical communication in all animals, is the definition of pheromone and whether the signature mixtures used for recognition should count as pheromone signals or cues (Wyatt 2010). While this is partly of semantic interest, the question is also important because of the generalizations that come from definitions.

A key priority for the future is the continued collaboration between crustacean researchers and chemists – for both pheromone identification and synthesis. There are good signs of this happening. The next decades may be among the most exciting yet.

Acknowledgments   I thank the late Martin Birch for his inspiring lead on insect pheromones, Oxford University Continuing Education (in particular Dr G.P. Thomas) and Zoology Departments for supporting me over many years, and the editors and chapter referees for their suggestions.

References

Atema J, Steinbach MA (2007) Chemical communication and social behavior of the lobster *Homarus americanus* and other decapod Crustacea. In: Duffy JE, Thiel M (eds) Evolutionary ecology of social and sexual systems: crustaceans as model organisms. Oxford University Press, New York, pp 115–144

Brennan PA, Kendrick KM (2006) Mammalian social odours: attraction and individual recognition. Philos Trans R Soc B 361:2061–2078

Brown WL, Eisner T, Whittaker RH (1970) Allomones and kairomones: transpecific chemical messengers. Bioscience 20:21–22

Byers JA (1992) Optimal fractionation and bioassay plans for isolation of synergistic chemicals: the subtractive-combination method. J Chem Ecol 18:1603–1621

Caprio J, Derby CD (2008) Aquatic animal models in the study of chemoreception. In: Firestein S, Beauchamp GK (eds) The senses: a comprehensive reference, volume 4 olfaction & taste. Academic Press, San Diego, pp 97–134

Derby CD, Sorensen PW (2008) Neural processing, perception, and behavioral responses to natural chemical stimuli by fish and crustaceans. J Chem Ecol 34:898–914

Dicke M, Sabelis MW (1988) Infochemical terminology: based on cost-benefit analysis rather than origin of compounds? Funct Ecol 2:131–139

Dreanno C, Matsumura K, Dohmae N, Takio K, Hirota H, Kirby RR, Clare AS (2006) An alpha2-macroglobulin-like protein is the cue to gregarious settlement of the barnacle *Balanus amphitrite*. Proc Natl Acad Sci USA 103:14396–14401

Ekerholm M, Hallberg E (2005) Primer and short-range releaser pheromone properties of premolt female urine from the shore crab *Carcinus maenas*. J Chem Ecol 31:1845–1864

Hallem EA, Carlson JR (2006) Coding of odors by a receptor repertoire. Cell 125:143–160

Hansson BS (2002) A bug's smell – research into insect olfaction. Trends Neurosci 25:270–274

Hesselschwerdt J, Tscharner S, Necker J, Wantzen K (2009) A local gammarid uses kairomones to avoid predation by the invasive crustaceans *Dikerogammarus villosus* and *Orconectes limosus*. Biol Invasions 11:2133–2140

Hölldobler B, Carlin NF (1987) Anonymity and specificity in the chemical communication signals of social insects. J Comp Physiol A 161:567–581

Horner AJ, Weissburg MJ, Derby CD (2008) The olfactory pathway mediates sheltering behavior of Caribbean spiny lobsters, *Panulirus argus*, to conspecific urine signals. J Comp Physiol A 194:243–253

Johnson ME, Atema J (2005) The olfactory pathway for individual recognition in the American lobster *Homarus americanus*. J Exp Biol 208:2865–2872

Johnston RE (2003) Chemical communication in rodents: from pheromones to individual recognition. J Mammal 84:1141–1162

Karlson P, Lüscher M (1959) 'Pheromones': a new term for a class of biologically active substances. Nature 183:55–56

Kelly DR (1996) When is a butterfly like an elephant? Chem Biol 3:595–602

Knight-Jones EW (1953) Laboratory experiments on gregariousness during settling in *Balanus balanoides* and other barnacles. J Exp Biol 30:584–598

Law RH, Regnier FE (1971) Pheromones. Annu Rev Biochem 40:533–548

Linsenmair KE (1987) Kin recognition in subsocial arthropods, in particular in the desert isopod *Hemilepistus reaumuri*. In: Fletcher DJC, Michener CD (eds) Kin recognition in animals. Wiley, Chichester, pp 121–208

Linsenmair KE (2007) Sociobiology of terrestrial isopods. In: Duffy JE, Thiel M (eds) Evolutionary ecology of social and sexual systems: crustaceans as model organisms. Oxford University Press, New York, pp 339–364

Maynard Smith J, Harper D (2003) Animal signals. Oxford University Press, Oxford

Monnin T, Malosse C, Peeters C (1998) Solid-phase microextraction and cuticular hydrocarbon differences related to reproductive activity in queenless ant *Dinoponera quadriceps*. J Chem Ecol 24:473–490

Nordlund DA, Lewis WJ (1976) Terminology of chemical releasing stimuli in intraspecific and interspecific interactions. J Chem Ecol 2:211–220

Peñalva-Arana DC, Lynch M, Robertson HM (2009) The chemoreceptor genes of the waterflea *Daphnia pulex*: many Grs but no Ors. BMC Evol Biol 9:79–90

Penn D, Potts WK (1998) Chemical signals and parasite-mediated sexual selection. Trends Ecol Evol 13:391–396

Rasmussen LEL, Lee TD, Roelofs WL, Zhang AJ, Daves GD (1996) Insect pheromone in elephants. Nature 379:684

Rittschof D, Cohen JH (2004) Crustacean peptide and peptide-like pheromones and kairomones. Peptides 25:1503–1516

Sachs BD (1999) Airborne aphrodisiac odor from estrous rats: implication for pheromonal classification. In: Johnston RE, Müller-Schwarze D, Sorensen PW (eds) Advances in chemical signals in vertebrates. Plenum, New York, pp 333–342

Scott-Phillips TC (2008) Defining biological communication. J Evol Biol 21:387–395

Sherman PW, Reeve HK, Pfennig DW (1997) Recognition systems. In: Krebs JR, Davies NB (eds) Behavioural ecology. Blackwell Science, Oxford, pp 69–96

Stacey NE, Sorensen PW (2006) Reproductive pheromones. Fish Physiol 24:359–412

Thiel M, Duffy JE (2007) The behavioral ecology of crustaceans: a primer in taxonomy, morphology, and biology. In: Duffy JE, Thiel M (eds) Evolutionary ecology of social and sexual systems: crustaceans as model organisms. Oxford University Press, New York, pp 3–28

Tibbetts EA, Dale J (2007) Individual recognition: it is good to be different. Trends Ecol Evol 22:529–537

Vosshall LB (2008) Scent of a fly. Neuron 59:685–689

wFleaBase. 2009. *Daphnia pulex* genome http://wfleabase.org Accessed 20 April 2009

Wilson EO, Bossert WH (1963) Chemical communication among animals. Recent Prog Horm Res 19:673–716

Wyatt TD (2003) Pheromones and animal behaviour: communication by smell and taste. Cambridge University Press, Cambridge

Wyatt TD (2005) Pheromones: convergence and contrasts in insects and vertebrates. In: Mason RT, LeMaster MP, Müller-Schwarze D (eds) Chemical signals in vertebrates 10. Springer, New York, pp 7–20

Wyatt TD (2009) Fifty years of pheromones. Nature 457:262–263

Wyatt TD (2010) Pheromones and signature mixtures: defining species-wide signals and variable cues for individuality in both invertebrates and vertebrates. J Comp Physiol A 196:685–700

Xue BY, Rooney AP, Kajikawa M, Okada N, Roelofs WL (2007) Novel sex pheromone desaturases in the genomes of corn borers generated through gene duplication and retroposon fusion. Proc Natl Acad Sci USA 104:4467–4472

Part II
General Overview of Signal Characteristics and Reception

Chapter 3
Crustaceans as Powerful Models in Aquatic Chemical Ecology

Mark E. Hay

Abstract Crustaceans use chemical cues to find and assess mates, signal dominance, recognize known conspecifics, find favored foods and appropriate habitats, and assess threats such as the presence of predators. The behavior demonstrating these cues is dramatic (e.g., males of some species will guard and try to copulate with sponges, rocks, or even other males treated with water holding receptive, conspecific females), but the specific chemicals eliciting these behaviors have rarely been unambiguously described. Here I discuss the known compounds that affect crustacean behaviors and note that identifying active chemical signals and cues may be a challenge because these chemicals may: (1) occur at low concentrations in a media holding many other metabolites, (2) be complex blends instead of particular pure compounds – and it may be the blend itself (the specific ratio of different signaling molecules) that carries information, (3) be "combination-lock" cascades of chemical signals and cues that have to occur in the correct order to generate a critical behavior, and (4) have been selected as critical molecules eliciting behavior because they rapidly degrade, thus preventing maladaptive behavior based on old data. These potential traits will make separation and structural determination difficult. Despite many signals and cues being delivered via water, there are several examples of cuing compounds being lipid-soluble; it is important for biologists to realize that even lipids dissolve to some degree in water and can thus function as waterborne cues. Because of the critical role that crustaceans play in marine and freshwater food webs, the effects of chemical mediation of crustacean behavior can reach far beyond the direct effects on crustaceans to impact community organization and even ecosystem-level processes.

3.1 Introduction

Humans sense the world primarily via vision and auditory cues; we go to art museums and concerts, but rarely, if ever, to a museum or a performance of

M.E. Hay (✉)
School of Biology, Georgia Institute of Technology, Atlanta, GA 30332, USA
e-mail: mark.hay@biology.gatech.edu

T. Breithaupt and M. Thiel (eds.), *Chemical Communication in Crustaceans*,
DOI 10.1007/978-0-387-77101-4_3, © Springer Science+Business Media, LLC 2011

"smells." In contrast, many organisms lack eyes and ears; they sense the world primarily via chemical signals and cues. When they encounter another organism, they decide to eat it, run from it, fight it, or mate with it based primarily on olfactory cues. Even humans do this, but usually unknowingly. Odors from women alter the menstrual cycles of other women (Stern and McClintock 1998); women find the body odor of men with symmetrical faces (a proxy for good genes) to be more pleasant (Thornhill and Gangestad 1999); and men tip exotic dancers much more when they are in estrus than when they are not – dancers using contraceptive pills show no estrous earnings peak (Miller et al. 2007). Thus, the behaviors of even "visually-biased" organisms like humans and crustaceans can be strongly affected by chemical signals and cues (for definitions of signals and cues see Chap. 2).

Thus, to be good ecologists, behaviorists, or evolutionary biologists who more fully understand how organisms perceive and react to their environment, we need to understand chemical signals and stimuli and their ecological roles in different intra- and interspecific interactions. If the Doctor Dolittle of Hugh Lofting's children's books could truly "talk to the animals," then it is more likely that he was a chemical ecologist than a linguist.

In aquatic systems, chemical cues and signals determine feeding, habitat, and mating choices. They stabilize dominance hierarchies. They determine whether animals forage actively to acquire lunch or stay hidden and endure hunger to avoid becoming lunch. Chemical stimuli regulate the behavior of not only higher animals, but also the behavior of plants and microbes. As an example, when the bloom-forming phytoplankton *Phaeocystis globosa* chemically senses its neighbors being attacked by ciliates that feed on single cells, it shape-shifts and enhances growth in colonial forms that are too big for ciliates to consume (Long et al. 2007). In contrast, when its neighbors are attacked by copepods that feed on colonies, it suppresses colony formation and grows more frequently as single cells that are too small to interest copepods (Fig. 3.1). These shifts may alter energy and nutrient flow, and possibly carbon sequestration, across oceanic regions. Thus, chemical

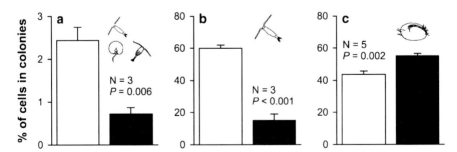

Fig. 3.1 Percentage of total *Phaeocystis globosa* cells within colonies following exposure to chemical cues from grazer-free (*open bars*) or grazer-containing *P. globosa* cultures (*dark bars*). Grazers were: (**a**) natural mixtures of mesozooplankton, (**b**) the copepod *Acartia tonsa*, or (**c**) the ciliate *Euplotes* sp. From Long et al. (2007) Copyright (2007) National Academy of Sciences, USA

signals and cues affect not only individual behavior and population-level processes, but also community organization and ecosystem function (Hay 1996; Hay and Kubanek 2002; Pohnert et al. 2007).

In this chapter, I focus on some of the more prominent ways that chemical signals and cues affect crustacean behavior and ecology. Where it is known, I discuss the types of compounds involved, but in most instances of chemically mediated interactions involving crustaceans the critical metabolites are unidentified. In a few instances, I use noncrustacean examples (rotifers, seaweeds, sea-slugs) where these organisms illustrate concepts or interactions that likely occur for crustaceans, but have yet to be demonstrated.

It is well established that crustaceans such as crabs, lobsters, crayfish, and amphipods use chemical stimuli to find and assess mates, signal dominance, recognize individual conspecifics, find favored foods and appropriate habitats, and assess threats such as the presence of predators. However, the specific chemicals mediating these behaviors have rarely been rigorously and unambiguously described. In the past, many chemicals-mediating behavior were often discussed as being signals that were purposely sent, even in cases where it was difficult or impossible to suggest a pressure selecting for this "sending." Here I refer to chemical stimuli as signals only if they have been selected to be sent by the producer and elicit a response in the receiver that is adaptive to the signaler. If the response of the receiver is not adaptive for the producer, the chemicals are referred to as cues (further discussion of cues and signals see Chap. 2). Below, I address what is known about the nature of the various chemical signals and cues, but a few generalities can be mentioned here.

First, there is a common misconception that chemical stimuli transmitted through the water need to be water-soluble or polar as opposed to lipid-soluble or nonpolar. While more polar compounds will better dissolve in water, even lipid-soluble metabolites have some degree of solubility in water and some nonpolar lipids mediate behaviors via transmission through the water. Second, it has been common to search for "the" chemical producing a specific behavior. It may be common for the signal to be a mixture of numerous compounds and for the ratio of the mixture to convey information. Mixtures of chemicals could be similar to an alphabet where the letters carry more information when mixed to form words and the words mixed to form sentences. If chemical mixtures are similarly critical for meaning, then it will be impossible to isolate "the" cue. Third, for critically important processes such as mating, recruiting to specific hosts, or some aspects of feeding, there may be cascades of chemical signals that assure appropriate behavior and prevent mistakes that would severely lower fitness. As examples, for consumers that feed only on specific hosts, there may be a distance cue that stimulates search for the host, a contact cue that stimulates biting, an additional cue that stimulates chewing, and a final cue that stimulates swallowing. Such cascades of chemical cues are known to control feeding by some species of specialist insects. Similar types of cascades could assure appropriate mating, habitat selection, etc. I term such critical series of cues or signals as "chemical combination locks" where a combination of chemicals must occur in the correct order or the full behavioral cascade will not occur. These pose a challenge for thoroughly understanding the ecology and evolution of chemically mediated behavior.

Below, I provide a few examples of chemical cues and signals involved in basic life processes affecting crustacean individuals and populations, as well as the effects of these chemically mediated interactions on community structure. More detailed, in-depth accounts of these issues for specific crustacean groups are available in the other chapters of this volume.

3.2 Chemical Mediation of Basic Life Processes

3.2.1 Food

Generalist consumers should respond to primary metabolites (sugars, proteins, amino acids, products of respiration, or metabolic wastes, etc.) because these indicate the presence of food resources in general – the particulars (which type of food) may be acquired via interpretation of unique ratios of these primary metabolites indicating which prey the consumer may be sensing. Crustaceans are often attracted to amino acids and there is a rich literature based on laboratory experiments (Rittschof and Cohen 2004), but the notion that amino acids or similar small, polar metabolites (e.g., peptides) are the attractants actually used by foraging crustaceans in nature has rarely been rigorously tested under field conditions. Blends vary considerably across the species producing them and may vary as a function of the metabolic state (well-fed or starved; healthy or sick) or activity level of the producer. If it is the particular ratio and blend of signals that the crustacean is tracking to its food source, then determining the particular compounds stimulating search behaviors becomes difficult to track down via bioassay-guided fractionation because this approach separates these mixtures and may thus uncouple critical aspects of the chemical blends stimulating food search. In the lab or in various field observations, peptides have repeatedly been shown to attract crustaceans in ways that suggest that they are commonly used to find foods or other resources (shells for hermit crabs, settlement sites for barnacle larvae) (Rittschof and Cohen 2004). Consistent with critical cues being blends rather than specific molecules within those blends, crustacean response to particular peptides are rarely as robust as to natural blends (Rittschof and Cohen 2004).

In contrast to generalist consumers, specialists that recruit to and feed only from certain hosts may be stimulated by unique metabolites produced only by that host. This is true for the crab *Caphyra rotundifrons* which feeds only on the chemically defended tropical seaweed *Chlorodesmis fastigiata*. The cytotoxic diterpenoid chlorodesmin (Fig. 3.2a), the major secondary metabolite of *Chlorodesmis*, deters fishes from consuming this alga, but stimulates feeding by the specialist crab when it is applied to a red alga (Fig. 3.3) that the crab will not normally consume (Hay et al. 1989). There are numerous similar cases where specialist crabs, amphipods, or

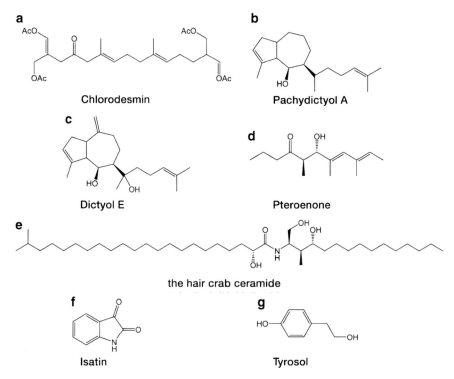

Fig. 3.2 Metabolites known to elicit crustacean behavior. Letters of compounds (**a–g**) will be referred to in the text when these metabolites are mentioned

isopods are stimulated to feed, live in, or decorate with prey by the specific prey metabolites that deter other consumers (Hay 1996; Stachowicz and Hay 1999b). There is also a general trend of crustacean tolerance to chemical defenses being negatively correlated with mobility; in numerous cases, less mobile crustaceans that live in association with specific hosts (such as branching corals) may consume most competing organisms from the host's surface – even when these competitors are chemically defended (Stachowicz and Hay 1999a).

In most investigations to date, generalist crustaceans have been studied in flumes or other lab settings to determine how they navigate to macerated body fluids of prey or specific water-soluble compounds like various peptides (see Chap. 4); the effects of these metabolites on crustacean feeding as opposed to attraction are less well investigated. In contrast, most studies on specialist crustaceans have focused on metabolites (often lipid-soluble terpenoids) stimulating feeding once prey have been contacted; compounds responsible for attraction from a distance have rarely been investigated, but a few examples exist. The salmon louse, *Lepeophtheirus salmonis*, is a specialist parasite on salmon; this copepod is stimulated to initiate search behavior by the lipophilic metabolites isophorone and 6-methyl-5-hepten-2-one given off by salmon,

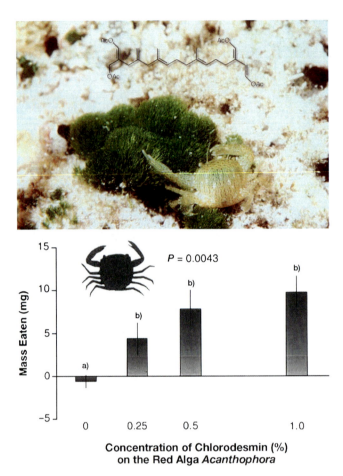

Fig. 3.3 The crab *Caphyra rotundifrons* on its host alga, *Chlorodesmis fastigiata*, which produces the cytotoxic diterpenoid chlorodesmin. When a nonhost (*Acantophora* sp.) is treated with chlorodesmin, the crab consumes the nonhost. Data redrawn from Hay et al. (1989). With kind permission from Springer Science + Business Media

but is repelled by metabolites from a nonhost fish (Bailey et al. 2006). A specialist sea slug that feeds only on a specific, chemically defended green alga shows a similar pattern; the metabolite that attracts it from a distance is lipid-soluble rather than water-soluble in seawater (D. Rasher, S. Engle, M.E. Hay work in progress). Both of these examples demonstrated that distance cues need not be polar or water-soluble in the traditional sense. More efforts that simultaneously study distance chemical attractants and contact stimulants or deterrents could be useful, not only in studies of feeding, but also in other contexts such as mate selection, host selection, or settlement site selection.

3.2.2   Clothing

Several crustaceans wrap themselves in, or decorate their carapace with, specific species of seaweeds, anemones, or sponges. This behavior is chemically mediated, but vision also can play a role. The Caribbean amphipod *Pseudamphithoides incurvaria* constructs a domicile from the chemically defended seaweed *Dictyota bartayresii*. The diterpene alcohol, pachydictyol-A (Fig. 3.2b), produced by this seaweed deters feeding by fishes, but stimulates domicile building by the amphipod; when in this domicile, the amphipod is immune to fish predation, but the amphipod is rapidly consumed when removed from the domicile or when forced to build its domicile from a palatable seaweed lacking potent chemical defenses (Hay et al. 1990). This safety from predation occurs even when the amphipod in its domicile is fed to fishes specifically trained to take foods from investigators; thus, the domicile may not only hide the amphipod from visually searching predators, but also causes the amphipod to be rejected unharmed even when it is initially drawn into the fish's mouth. Similarly, juveniles of the Atlantic decorator crab *Libinia dubia* selectively decorate their carapaces with the chemically defended seaweed *Dictyota menstrualis* (Fig. 3.4). This seaweed produces multiple related diterpene alcohols, but the single compound dictyol E (Fig. 3.2c), which most strongly deters local fishes, is the compound that the crab uses to select favored decoration materials (Stachowicz and Hay 1999b). When tethered in the field, crabs decorated with *Dictyota* are rarely consumed while those decorated with seaweeds lacking chemical defenses are rapidly consumed. In both of these examples, the specific compound that deters fish feeding is the compound stimulating crustacean use of that alga. This suggests that the behavior has been selected due to its value as a means of deterring predators rather than as some side effect of feeding or nutritional needs. In fact, in the case of the herbivorous decorator crab, the compound that stimulates decorating behavior deters the crab from feeding on that seaweed. In a compelling example from Antarctica (McClintock and Janssen 1990; Yoshida et al. 1995), the pelagic amphipod *Hyperiella dilatata* grasps the chemically defended pteropod *Clione antartica*, holds it on its dorsal surface, and by doing so becomes protected from fish predation. Local fishes reject the pteropod, reject the amphipod when it holds the pteropod, but rapidly consume the amphipod if the pteropod is removed. The polypropionate-derived metabolite, pteroeone (Fig. 3.2d), produced by this pteropod is the metabolite that deters fish feeding. In each of these examples, a crustacean acquires a chemical defense by behaviorally sequestering the body or tissue of a chemically defended organism and "clothing" itself in the living symbiont. Evidence to date suggests that most of these chemicals are recognized by the crustaceans on contact rather than dispersed in the water where they could be detected from a distance; this also appears to be the case for predators, which will take "clothed" amphipods into their mouths, but then rapidly reject them unharmed. Limited tests show that some crabs can identify and move toward their host alga by vision alone, but the contact chemical cues are then critical for stimulating feeding (Hay et al. 1989).

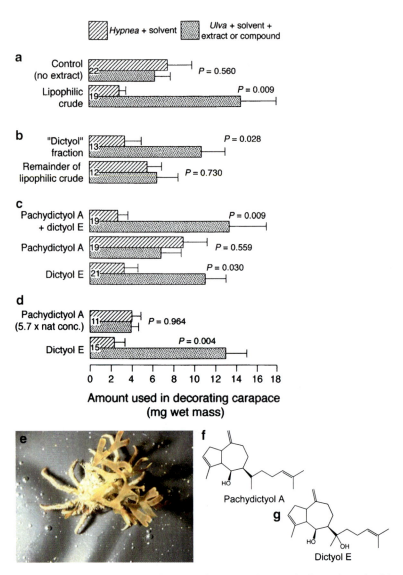

Fig. 3.4 The decorator crab *Libinia dubia* camouflages with the chemically noxious alga *Dictyota menstrualis*, which produces the diterpene alcohols dictyol E and pachydictyol. Extract and pure metabolites of *D. menstrualis* were tested by coating them onto the surface of other, less-preferred, algae (e.g., *Ulva* sp.) offered to the crab. The effect of solvent was tested by coating it onto the alga *Hypnea* sp. offered to the crab. (**a–c**) A bioassay-guided fractionation demonstrating that the lipid-soluble extract from *Dictyota* (**a**), the "dictyols" portion of that extract (**b**), and dictyol E alone (**c**) cue this decoration behavior. (**d**) Tests the two dictyols at the same concentration, other assays are at natural concentrations. Number of crabs per assay is given at the base of each pair of bars. (**e**) *Libinia dubia* with *Dictyota menstrualis* on its carapace and (**f, g**) chemical structure of the active compounds. Data from Stachowicz and Hay (1999b) with permission from Ecology

3.2.3   Shelter and Settlement Site Cues

Just as the above crustaceans seek protection by covering themselves with chemically noxious species, Caribbean spiny lobsters seek protection via clustering together in reef caves and crevices during the day, the combined defense of their many spiny antennae presumably being a better defense than the two antennae of lone individuals. Recruitment to occupied shelters (and thus gregarious occupancy) is chemically mediated with greater recruitment to shelters where conspecific urine is being released (Horner et al. 2004; see Chap. 12). Although urine is a demonstrated attractant, the specific molecules mediating this response are not known.

For more than 50 years, it has been clear that barnacles settled gregariously and in response to chemical cues indicating adult presence, but the chemical nature of the cue remained unidentified despite considerable effort. However, Dreanno et al. (2006) recently reported that the cue for settlement of *Balanus amphitrite* is a previously undescribed glycoprotein that shares a 30% sequence homology with the thioester-containing family of proteins that includes the alpha(2)-macroglobulins. Although the compound is produced by adults and stimulates larval settlement, it is also expressed in larvae and juveniles. As larvae explore the substratum in anticipation of attaching and metamorphosing, they leave small "tracks" of this compound that can then act as a stimulant to other larvae. For sessile internal fertilizers like barnacles, such gregarious cues are critical because adults do not release gametes into the water column to achieve fertilization, but rather must be close enough for direct insemination. Chapter 22 provides a more extensive treatment of this issue.

3.2.4   Sex

Chemical signals often guide sex from the levels of individuals finding and evaluating mates, to facilitating or deterring copulation once they are in contact, to sperm–egg chemical signaling that promotes or prevents fertilization after copulation (Kamio et al. 2002; Raffell et al. 2004; see also Chaps. 19, 20, and 23). In crustaceans, one potential mate commonly detects and tracks the other from a distance using chemical signals (pheromones). Behavioral responses to water that previously held potential mates clearly demonstrate that crustaceans respond to pheromones, but the specific chemicals stimulating the various mating responses have rarely been identified. Problems of signal identification include minute quantities dispersed in a salty medium (making extracting, desalinization, and purification difficult) and the fact that the chemical should be built to oxidize or be rapidly degraded by bacteria so that mates do not respond to "old information" that is no longer useful. Initial success in identifying such compounds in insects involved dissecting out pheromone glands (sometimes by the tens of thousands), investigating the pheromones in these concentrated chemical resources, and then testing mate responses to the identified chemicals in the field or in wind tunnels (see Chaps. 2 and 27). Similar approaches may be needed with copepods and other crustaceans, but dissecting as yet unidentified pheromone glands from many thousands of 1 mm long copepods is not a project that will entice many investigators. Such studies will be more feasible with large crustaceans like crabs, lobsters, or crayfishes.

For numerous species of crustaceans, mating is restricted to a brief reproductive season and furthermore a short temporal window after mature females molt and while the carapace is soft. These issues limit the opportunities for chemical investigations, but suggest why pheromones might be strongly selected among crustaceans. In many instances, males detect impending molting of females via chemical signals and hold, or guard, the female for a period before, during, and after the molt. This behavior gives these males preferential access to receptive mates and may protect the postmolt females until their carapace hardens. However, given that crabs are commonly cannibalistic, managing these cannibalistic urges during mate carrying is critical. To achieve this, the female emits a pheromone that suppresses feeding in males (Hayden et al. 2007). Chemical communication is so powerful in determining mating interactions that male crabs will move toward, carry and stroke, and often try to copulate with other males, or even sponges, rocks, or golf balls if they have been treated with female urine that carries the pheromone signal (see Chap. 19). That sex pheromones affect mate finding, recognition, and various stages in mating behavior and mate defense has been recognized for diverse groups of crustaceans for many decades, but few of the specific metabolites stimulating these behaviors have been identified. Recently, Asai et al. (2000) studied the brachyuran crab *Erimacrus isenbeckii* and identified the metabolites stimulating mate guarding and copulatory behavior of male crabs as ceramides consisting of linear sphingosines and branched fatty acids (Fig. 3.2e). Males exposed to sponges soaked with these compounds carried the sponges and attempted to copulate with them. Similar behaviors were also observed in males of other species that were offered sponges with female pheromones (Fig. 3.5).

Hardege et al. (2002) demonstrated that precopula females of the crab *Carcinus maenas* released a pheromone that stimulated mating behavior of the males. When males were exposed to other males or even to stones that had been exposed to the female substance, the assay male grasped the "pheromone-treated" male or stone to test the hardness of the cuticle and invariably proceeded to manipulate the pseudofemale beneath its abdomen as in precopula. Urine from precopula females, as well as "culture water" that had held these females, elicited a positive response; similar samples collected from intermolt females did not elicit such behavior. Preliminary characterization of the sex pheromone(s) indicated that it was smaller than 1,000 Da molecular weight and sparingly soluble in organic solvents (Hardege et al. 2002). Further bioassay-driven purification identified the female chemical as the nucleotide UDP, released in the urine of freshly molted females (see Chap. 19).

Hassler and Brockmann (2001) found that female horseshoe crabs (belonging to the Chelicerata) release pheromones that attract males during breeding. If a sponge is soaked in the water from beneath a breeding female, this sponge attracts males, but the chemical eliciting this behavior is uninvestigated. Several studies with

Fig. 3.5 When sponges are treated with ceramides and pheromones from receptive females, males of *Erimacrus isenbeckii*, *Telmessus cheiragonus*, and other species guard and attempt to copulate with the sponges. Drawing by Jorge A. Varela Ramos

crustaceans have documented the presence of pheromones stimulating mate searching from a distance. Fewer studies have investigated chemical stimulation of copulatory behavior once mates are in close contact. Kamio et al. (2002) found that the crab *Telmessus cheiragonus* produced a distance pheromone in the urine that stimulated tracking and courtship behavior, but this pheromone did not stimulate copulation. There was an additional pheromone released from postmolt females that evoked copulation in male crabs. The copulation pheromone was waterborne with a molecular weight of less than 1 kDa. Such multistage chemical signaling may be common. Mating mistakes can be minimized if there are distance stimuli, followed by separate contact signals that assure adequate evaluation of appropriate mates.

Sato and Goshima's (2007) investigation of the stone crab *Hapalogaster dentata* found that females selected larger males based on chemical signals, but could also distinguish between similar-sized males based on whether or not they had sufficient vs. depleted sperm reserves. This suggests that chemical stimuli not only indicate size and dominance, but even sperm availability.

Paralleling the findings for crabs, Stebbing et al. (2003) demonstrated that sexually mature female crayfish, *Pacifastacus leniusculus*, released a pheromone that stimulated mating behavior in males. If water conditioned by mature females was passed through an air-stone into containers holding males, these males commonly seized, mounted, and in a few cases deposited spermatophores onto the air-stones. Such behaviors did not occur if the air-stones carried water conditioned by immature females or control water that had not held crayfish (see also Chap. 13).

For alpheid shrimp, males guard premolt females for a considerable period because mating is restricted to a short time after the female's molt. Y-maze experiments demonstrated a distance chemical attraction of males for premolt females in *Alpheus angulatus*. Males were attracted to water that had held premolt females, but repulsed by water that had held intermolt males and females. In mate choice experiments allowing contact, significantly more males paired with premolt females than with postmolt females (Mathews 2003).

Snell and Morris (1993) found that male copepods track females from a distance suggesting a diffusible pheromone, but the chemical appears to lack species specificity in the species investigated (see also Chaps. 9 and 23). Rotifer males do not show distance chemo-attraction to their females, but both copepods and rotifers do exhibit contact chemoreception of a species-specific signal identifying appropriate mates. In the rotifers, the contact compound is a surface-associated glycoprotein; if small beads are treated with this pheromone, male rotifers attempt to mate with the beads. The signal for copepods is unknown, but a chemical signal is clearly present (see Chap. 9). When males of the copepod *Centropages typicus* cross a pheromone trail left by a female, they increase their swimming speed by 3–6 times, zigzag rapidly along the exact path taken by the female (Fig. 3.6), rapidly overtake her, and attempt to mate (Bagoien and Kiorboe 2005). They successfully follow even very contorted trails that are up to 31 s old. Models suggest that pheromones are small molecules (possibly amino acids), and that investment in pheromone production represents only a small fraction of the females' ingestion and metabolic rate (Bagoien and Kiorboe 2005).

3.2.5   Managing Juveniles

Once fertilization occurs, crustacean embryos are commonly carried and tended by the female. In some shrimp and lobster species, the embryos are chemically protected from microbial pathogens by symbiotic microbes covering the embryo surfaces and producing chemicals that prevent attack by harmful microbes. In the shrimp *Palaemon macrodactylus* and the lobster *Homarus americanus,* symbiotic bacteria on the embryo surfaces produce 2,3-indolinedione (isatin) and

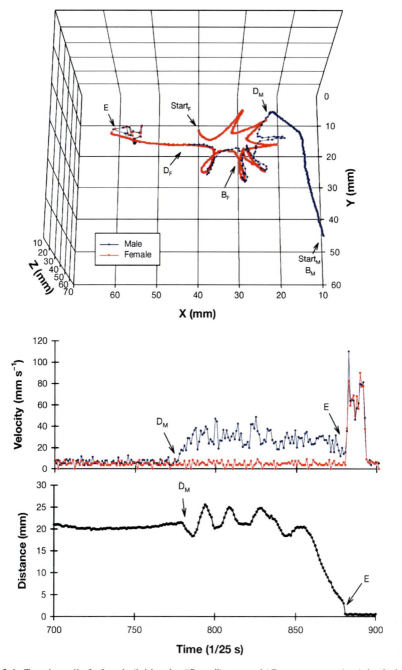

Fig. 3.6 *Top*: the trail of a female (initiated at "Start$_F$") copepod (*Centropages typicus*) that is then tracked by a male (Start$_M$) that crossed her path 22 s later. Positions of female and male before signal detection are marked as B$_F$ and B$_M$, positions when male detection occurs are indicated by D$_F$ and D$_M$. E indicates male–female encounter. *Middle*: swimming speeds showing the increased speed of the male after detecting (D$_M$) the female trail. *Bottom*: the distance between the male and female over the monitored period. Reproduced from Bagoien and Kiorboe (2005) with permission from Inter-research

4-hydroxyphenethyl alcohol (tyrosol), respectively (Fig. 3.2f, g); these compounds defend the embryos from pathogenic fungi (Gil-Turnes et al. 1989; Gil-Turnes and Fenical 1992).

When embryos mature, female crabs and lobsters release the larvae via abdominal extensions and pleopod pumping that helps to break open the egg membranes and ensures that larvae hatch in synchrony. This behavior is stimulated by chemical signals from the developing embryos. Tankersley et al. (2002) demonstrated that water from containers holding mature embryos induced this behavior in females holding eggs, that compounds from homogenized eggs produced a similar effect, and that water that had held late-stage embryos was more potent than water from early-stage embryos.

Ziegler and Forward (2007) suggested that crabs and lobsters might both release larvae based on small peptides emanating from the developing embryos. They found that pleopod pumping in a lobster was stimulated by di- and tripeptides with a neutral amino acid at the amino terminus and a basic amino acid at the carboxy terminus and also by the basic–basic dipeptide Lys-Arg. All carboxy-terminal arginine peptides tested produced a significant pumping response, with the exception of Trp-Ile-Arg. Their findings support the hypothesis that larval release in crustaceans may be controlled by small peptides acting as pheromones. It should be noted, however, that these assays were based on the hypothesis that these types of compounds stimulated larval release, and testing was done with metabolites from the chemical stock room, rather than with metabolites isolated from the embryos. Because bioassay-guided separations of sometimes unstable natural pheromones from small amounts of embryos or similar small samples is difficult, it has been common for investigators to make reasonable assumptions about what types of compounds may be active, acquire these from other sources, and then test them to see if they produce the behavior of interest. When these compounds do seem to cause the targeted behavior, it need not be the case that the "stock-room" compounds producing that behavior are the same as the natural compounds cuing that behavior. Pohnert et al. (2007) provide a good discussion of the costs and benefits of the bioassay-guided vs. stock-room approach under different constraints and list several examples of chemical "discoveries" using the stock-room approach that turned out to be in error.

3.2.6 Dominance

Communication of status can lessen costs of confrontation over resources for both dominant and subdominant individuals. If one must fight to establish relative status, then once this is decided, it is to the advantage of both dominant and subdominant to remember and correctly identify individuals so that the physical confrontation need not be repeated for each encounter. Chemical signals play important roles in reducing such costs. Additionally in aquatic systems where visibility may be limited and where vegetation or other physical barriers impede vision, chemical stimuli may be especially important in that they can flow around obstacles and deliver information over both short and longer distances.

Lobsters, stomatopods, hermit crabs, etc. can chemically recognize individuals that they have previously encountered and behave appropriately given their relative dominance. Compounds involved in this are unknown, but in some cases are present in the urine. The recognition is likely to be based on blends of common primary metabolites (waste products) that may degrade quickly or be rapidly metabolized. This rapid degradation will not prevent them serving as good dominance signals in close, near-contact, situations, but this trait and the likely importance of mixtures of metabolites will make identification of the specific active blends difficult.

Moore (2007) and Atema and Steinbach (2007) provide overviews of chemical communication regarding dominance among crustaceans (see also Chap. 13), so this will be covered only briefly here. Crayfish communicate socially via signals in the urine. Urine is released preferentially during fights and other social encounters, with eventual winners releasing more than eventual losers. There are multiple glands that can empty into the urine stream and gill currents can carry urineborne signals toward opponents. Individuals can shoot gill water several body lengths in specific directions, making this an effective communication strategy.

Chemicals may not only indicate the status of senders, but also influence the status of receivers (Moore and Bergman 2005). When crayfish are exposed for 5 days to the signals of dominant crayfish, they tend to act like subordinates and lose future fights. When exposed for 5 days to signals from subordinate crayfish, they tend to take on the traits of dominant crayfish and win future fights.

When crustaceans are prevented from producing (via blocking urine release) or receiving chemical stimuli (via blocking of receptors) during fights, the fights last longer, are more intense, and the predictability of the eventual victor is diminished. It is thus clear that chemical communications via urine play an important role in determining the outcome of antagonistic interactions. It appears that winning and losing change the neurochemistry of individual crayfish and that these neuro-chemical differences are sent to conspecifics during social encounters and that the receivers use these signals to judge appropriate behavior given the status of the sender.

Bushmann and Atema (1996) discuss similar patterns in which lobsters indicate status via chemical signals in released urine. Male urine plays a role in the determination of dominance and in female choice of dominant males, while female urine reduces the incidence of male aggressive behavior and induces male mating behavior. Clusters of nephropore rosette glands are organized into lateral and medial gland complexes lying alongside each ureter. Some of these glands have ducts terminating at the bladder, while others terminate outside the animal adjacent to the site of urine release. These two duct systems could allow glands to release their products into the environment with or without concomitant urine release. As such, these glands may be productive targets for chemists seeking to isolate and identify compounds cuing critical behaviors.

3.3   The Smell of Death

The smell of death comes in multiple forms, each of which can directly and
indirectly affect crustacean behavior, ecology, and the role of crustaceans in community organization and ecosystem function. The smell of conspecific death is a
powerful determinant of individual behavior that can cascade up to have larger
scale consequences. Additionally, microbes can use chemical signals to gain preferential access to nutritionally rich, food-fall resources and make these less available
to crustacean scavengers. Each of these situations is discussed briefly below.

3.3.1   Crustaceans as the Producers of the Smell of Death

When copepods consume the phytoplankter *Phaeocystis globosa*, these prey
chemically sense both that their conspecifics are being attacked and that the attack
is by copepods instead of smaller consumers (ciliates) and respond with morphological shifts that make them less susceptible to copepod attack (Long et al. 2007).
The compounds signaling this are unknown, but are lipid-soluble and appear to
degrade within a few days. In a similar way, waterborne cues from the copepod
Acartia tonsa induce paralytic shellfish toxin (PST) production in the bloom-
forming dinoflagellate *Alexandrium minutum* (Selander et al. 2006). Induced
A. minutum contain up to 2.5 times more toxins than controls and are more resistant
to copepod grazing. In both of the above examples, chemicals from grazing
copepods produce defensive responses in common, bloom-forming prey, and
these responses lower susceptibility of the prey to future attack. These chemically
stimulated responses could play important roles in facilitating bloom formation.
Both of these phytoplankton form harmful algal blooms that can kill thousands to
millions of fishes and other nontarget organisms, potentially altering the structure
and function of local food webs.

Similar prey responses to nearby crustacean predators also occur in benthic
systems. When Toth (2004) exposed 12 species of Swedish seaweeds to effluents
from neighbors being attacked by the herbivorous isopod *Idotea granulosa*, 4 of the
12 species induced greater distastefulness. Other investigations have sometimes
shown similar patterns when neighbors are attacked by amphipods. Bivalve prey
can also chemically sense the proximity of nearby consumers and respond appropriately (Smee and Weissburg 2006a, b). Blue crabs and whelks both prey on clams,
but blue crabs move rapidly and can track clams only in lower velocity, less
turbulent flows while whelks move slowly, integrate chemical cues over longer
periods of time, and can therefore track clams in higher velocity, more turbulent
flows. When in low velocity flows, clams quit pumping, or "clam-up," to reduce
chemical cues to their presence in response to the upstream presence of either
predator (see Chap. 4), but in high velocity flows, they respond only to the whelk,
which is able to track to prey under these conditions of flow. Additionally, when the

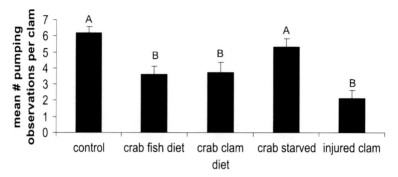

Fig. 3.7 Clam response to an upstream odor source. They suppress feeding in response to damaged conspecifics or to crabs that have consumed either fish or clams, but not to starved crabs. From Smee and Weissburg (2006b) with kind permission from Springer Science + Business Media

cues of the consumers are removed, the clams restart their feeding more rapidly following removal of crab cues than following removal of whelk cues. This appears adaptive because of the slower speed with which whelks leave the area. The clams reduce feeding if exposed to damaged conspecifics or to crabs that have recently fed on either clams or fish; they do not express this behavior if exposed to starved crabs (Fig. 3.7). This suggests that cues from the crabs may be normal products of digestion instead of specific cues indicating that crabs fed on conspecifics.

3.3.2 Crustaceans as the Receivers of the Smell of Death

Crustaceans can also sense body fluids from damaged conspecifics and avoid areas where other species similar to themselves have been attacked (see Chap. 18). When crab traps are baited with fish (control), fish plus an undamaged blue crab, fish plus a newly damaged blue crab, or fish plus a newly damaged stone crab, blue crabs avoid traps with the damaged blue crab (Fig. 3.8) relative to the other bait types (Ferner et al. 2005). The chemical signal is unknown, but is species-specific and degrades within 18–22 h.

Similarly, crayfish reduce feeding and movement when exposed to hemolymph of damaged conspecifics (Acquistapace et al. 2005). Hemolymph loses its bioactivity when tested 24 h after its extraction, but activity is maintained if it is treated with the antioxidant L-ascorbic acid. This suggests that crayfish alarm molecules are degraded by oxidation. Microbial activity alone did not rapidly degrade the alarm substances because hemolymph activity still declined after 24 h even if extracted and preserved in sterile conditions. However, when hemolymph molecules of less than 5 kDa were fractionated from hemolymph, they retained strong bioactivity even after 24 h at 20°C, possibly because the 5 kDa fractioning eliminates enzymes. It seems probable that alarm substances may be

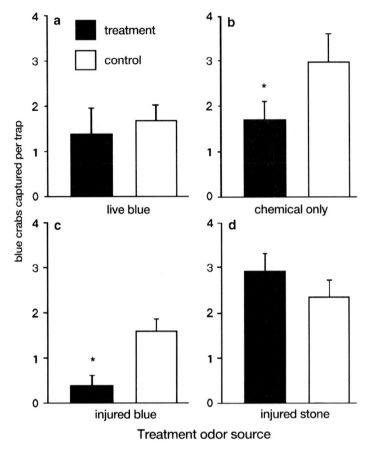

Fig. 3.8 Blue crabs avoid traps holding bait plus a damaged blue crab or water that has held a damaged blue crab, but not traps holding a live blue crab or a damaged stone crab. They thus chemically differentiate between damaged crabs and damaged conspecific crabs. Figure reproduced from Ferner et al. (2005) with permission from Inter-research. (a–d) all contained fish, but in addition (**a**) contained a live blue crab, (**b**) contained the chemical signal of a damaged blue crab, (**c**) contained a damaged blue crab, and (**d**) contained a damaged stone crab

degraded through enzymatic reactions. Like with many other signaling molecules, it is possible that these are peptides.

Many crustaceans are scavengers, making them and microbes potential competitors for carcasses. Following up on Daniel Janzen's suggestion that microbes chemically spoil meat as a way of competing with larger animals, Burkepile et al. (2006) tested the hypothesis that marine microbes on carcasses produce chemicals that deter scavengers from attraction to and feeding on these resources. They found that fish or oyster flesh allowed 1–2 days of microbial colonization attracted significantly fewer total consumers to crab traps than did fresh carcasses, that most crab species preferred the fresh flesh, but that the lesser blue crab *Callinectes similis* did not differentiate between rotted and fresh flesh (Fig. 3.9). When offered foods in the lab, the stone crab

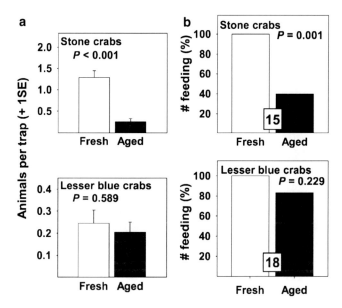

Fig. 3.9 (**a**) Crab traps baited with fresh fish caught nearly 260% more total animals than those baited with microbe-laden fish. All crabs but lesser blue crabs avoided the microbe-colonized bait. (**b**) In feeding assays, stone crabs and hermit crabs rejected microbe-colonized food which lesser blue crabs did not. Data redrawn from Burkepile et al. (2006) with permission from Ecology

Menippe mercenaria was less willing to eat microbe-laden meat and also was deterred by lipid-soluble extracts from microbe-laden meat. The extract fraction that strongly deterred stone crab feeding was composed primarily of nonesterified fatty acids, but the particular compounds suppressing feeding could not be identified. Manipulations with antibiotics indicated that the deterrent compounds were likely of microbial origin. The microbe produced smell of death deterred most crustaceans, but did not deter *C. similis*. These investigations involve how the "smell of death" affects crustacean attraction to a food resource. In contrast to these findings, some hermit crabs are attracted to the smell of dead gastropods (Tricarico and Gherardi 2006; see Chap. 15) because these represent newly available shells, which are an important resource for hermit crab populations.

3.3.3 Advantages of Crustaceans

Despite their visual acuity, crustaceans are built as chemical receivers (see other chapters of this volume); they have thousands of chemical receptors scattered over their bodies, they can spread these in flowing waters to rapidly track to mates or food, and their high metabolic rates, limited mating periods, and short life spans have imposed the need to rapidly respond to chemical cues and signals. These traits make them good biosensors to study the ecological role of chemical sensing.

Additionally, their rapid maturation, small size, and ease of maintenance in the lab make them excellent assay animals where one can study not only behavioral response to various chemical signals, but even measure the costs and benefits of this in terms of survivorship, growth, and fecundity (Cruz-Rivera and Hay 2003) – such studies would be prohibitive with larger consumers.

3.4 Summary

Crustaceans commonly demonstrate chemically mediated behaviors involved in finding foods, finding mates, assessing social status of conspecifics, and sensing and avoiding dangers. However, the specific chemicals eliciting these behaviors have rarely been determined – in part because they may occur at low concentrations, may be complex blends instead of particular pure compounds, and may degrade rapidly (to prevent miscommunication due to old data), all making their separation and structural determination difficult. Despite many cues and signals being delivered via water, there are several examples of cues, or known cuing compounds, being lipid-soluble. It is important for biologists to realize that even lipids dissolve to some degree in water and can thus function as waterborne cues. Because of the critical role that some crustaceans play in marine and freshwater food webs, the effects of chemically mediated crustacean behavior can reach far beyond the direct effects of crustaceans to impact community organization and even ecosystem-level processes.

Acknowledgments My research has been supported primarily by the U.S. National Science Foundation and the Fogarty International Center of the National Institutes of Health.

References

Acquistapace P, Calamai L, Hazlett BA, Gherardi F (2005) Source of alarm substances in crayfish and their preliminary chemical characterization. Can J Zool 83:1624–1630

Asai N, Fusetani N, Matsunaga S, Sasaki J (2000) Sex pheromones of the hair crab EP *Erimacrus isenbeckii*. Part 1: isolation and structures of novel ceramides. Tetrahedron 56:9895–9899

Atema J, Steinbach MA (2007) Chemical communication in the social behavior of the lobster *Homarus americanus* and other decapod crustacea. In: Duffy E, Thiel M (eds) Ecology and evolution of social behavior: crustaceans as model systems, pp. 115–144. Oxford University Press, New York, p 520

Bagoien E, Kiorboe T (2005) Blind dating – mate finding in planktonic copepods. I. Tracking the pheromone trail of Centropages typicus. Mar Ecol Progr Ser 300:105–115

Bailey RJE, Birkett MA, Ingvarsdottir A, Mordue (Luntz) AJ, Mordue W, O'Shea B, Pickett JA, Wadhams LJ (2006) The role of semiochemicals in host location and non-host avoidance by salmon louse (*Lepeophtheirus salmonis*) copepodids. Can J Fish Aquat Sci 63:448–456

Burkepile DE, Parker JD, Woodson CD, Mills HJ, Kubanek J, Sobecky PA, Hay ME (2006) Chemically-mediated competition between microbes and animals: microbes as consumers in food webs. Ecology 87:2821–2831

Bushmann PJ, Atema J (1996) Nephropore rosette glands of the lobster *Homarus americanus*: possible sources of urine pheromones. J Crust Biol 16:221–231

Cruz-Rivera E, Hay ME (2003) Prey nutritional quality interacts with chemical defenses to affect consumer feeding and fitness. Ecol Monogr 73:483–506

Dreanno C, Kirby RR, Clare AS (2006) Locating the barnacle settlement pheromone: spatial and ontogenetic expression of the settlement-inducing protein complex of *Balanus amphitrite*. Proc R Soc Lond B Biol Sci 273:2721–2728

Ferner MC, Smee DL, Chang YP (2005) Cannibalistic crabs respond to the scent of injured conspecifics: danger or dinner? Mar Ecol Prog Ser 300:193–200

Gil-Turnes MS, Fenical W (1992) Embryos of *Homarus americanus* are protected by epibiotic bacteria. Biol Bull 182:105–108

Gil-Turnes MS, Hay ME, Fenical W (1989) Symbiotic marine bacteria chemically defend crustacean embryos from a pathogenic fungus. Science 246:116–118

Hardege JD, Jennings A, Hayden D, Muller CT, Pascoe D, Bentley MG, Clare AS (2002) Novel behavioural assay and partial purification of a female-derived sex pheromone in *Carcinus maenas*. Mar Ecol Prog Ser 244:179–189

Hassler C, Brockmann HJ (2001) Evidence for use of chemical cues by male horseshoe crabs when locating nesting females (*Limulus polyphemus*). J Chem Ecol 27:2319–2335

Hay ME (1996) Marine chemical ecology: what is known and what is next? J Exp Mar Biol Ecol 200:103–134

Hay M, Kubanek J (2002) Community and ecosystem level consequences of chemical signaling in the plankton. J Chem Ecol 28:1981–1996

Hay ME, Pawlik JR, Duffy JE, Fenical W (1989) Seaweed-herbivore-predator interactions: host-plant specialization reduces predation on small herbivores. Oecologia 81:418–427

Hay ME, Duffy JE, Fenical W (1990) Host-plant specialization decreases predation on a marine amphipod: an herbivore in plant's clothing. Ecology 71:733–742

Hayden D, Jennings A, Muller C, Pascoe D, Bublitz R, Webb H, Breithaupt T, Watkins L, Hardege J (2007) Sex-specific mediation of foraging in the shore crab, *Carcinus maenas*. Horm Behav 52:162–168

Horner AJ, Weissburg MJ, Derby CD (2004) Dual antennular chemosensory pathways can mediate orientation by Caribbean spiny lobsters in naturalistic flow conditions. J Exp Biol 207:3785–3796

Kamio M, Matsunaga S, Fusetani N (2002) Copulation pheromone in the crab *Telmessus cheiragonus* (Brachyura: Decapoda). Mar Ecol Prog Ser 234:183–190

Long JD, Smalley GW, Barsby T, Anderson JT, Hay ME (2007) Chemical cues induce consumer-specific defenses in a bloom-forming marine phytoplankton. Proc Nat Acad Sci USA 104:10512–10517

Mathews LM (2003) Tests of the mate-guarding hypothesis for social monogamy: male snapping shrimp prefer to associate with high-value females. Behav Ecol 14:63–67

McClintock JB, Janssen J (1990) Pteropod abduction as a chemical defense in a pelagic Antarctic amphipod. Nature 346:462–464

Miller G, Tybur JM, Jordan BD (2007) Ovulatory cycle effects on tip earnings by lap dancers: economic evidence for human estrus? Evol Hum Behav 28:375–381

Moore PA (2007) Agonistic behavior in freshwater crayfish: the influence of intrinsic and extrinsic factors on aggressive behavior and dominance. In: Duffy JE, Thiel M, Duffy JE, Thiel M (eds) Evolutionary ecology of social and sexual systems: crustacea as models organisms. Oxford University Press, New York, pp 90–114

Moore PA, Bergman DA (2005) The smell of success and failure: the role of intrinsic and extrinsic chemical signals on the social behavior of crayfish. Integr Comp Biol 45:650–657

Pohnert G, Steinke M, Tollrian R (2007) Chemical cues, defence metabolites and the shaping of pelagic interspecific interactions. Trends Ecol Evol 22:198–204

Raffell JA, Krug PJ, Zimmer RK (2004) The ecological and evolutionary consequences of sperm chemoattraction. Proc Nat Acad Sci USA 101:4501–4506

Rittschof D, Cohen JH (2004) Crustacean peptide and peptide-like pheromones and kairomones. Peptides 25:1503–1516

Sato T, Goshima S (2007) Female choice in response to risk of sperm limitation by the stone crab, *Hapalogaster dentate*. Anim Behav 73:331–338

Selander E, Thor P, Toth G, Pavia H (2006) Copepods induce paralytic shellfish toxin production in marine dinoflagellates. Proc R Soc B Biol Sci 273:1673–1680

Smee DL, Weissburg MJ (2006a) Clamming up: environmental forces diminish the perceptive ability of bivalve prey. Ecology 87:1587–1598

Smee DL, Weissburg MJ (2006b) Hard clams (*Mercenaria mercenaria*) evaluate predation risk using chemical signals from predators and injured conspecifics. J Chem Ecol 32:605–619

Snell TW, Morris RD (1993) Sexual communication in copepods and rotifers. Hydrobiologia 255:109–116

Stachowicz JJ, Hay ME (1999a) Reduced mobility is associated with compensatory feeding and increased diet breadth of marine crabs. Mar Ecol Prog Ser 188:169–178

Stachowicz JJ, Hay ME (1999b) Reducing predation through chemically-mediated camouflage: indirect effects of plant defenses on herbivores. Ecology 80:495–509

Stebbing PD, Bentley MG, Watson GJ (2003) Mating behaviour and evidence for a female released courtship pheromone in the signal crayfish *Pacifastacus leniusculus*. J Chem Ecol 29:465–475

Stern K, McClintock MK (1998) Regulation of ovulation by human pheromones. Nature 392:177–179

Tankersley RA, Bullock TM, Forward RB, Rittschof D (2002) Larval release behaviors in the blue crab *Callinectes sapidus*: role of chemical cues. J Exp Mar Biol Ecol 273:1–14

Thornhill R, Gangestad SW (1999) The scent of symmetry: a human sex pheromone that signals fitness? Evol Hum Behav 20:175–201

Toth GB (2004) Screening for induced herbivore resistance in Swedish intertidal seaweeds. Mar Biol 151:1597–1604

Tricarico E, Gherardi F (2006) Shell acquisition by hermit crabs: which tactic is more efficient? Behav Ecol Sociobiol 60:492–500

Yoshida WY, Bryan PJ, Baker BJ, McClintock JB (1995) Pteroenone: a defensive metabolite of the abducted Antarctic pteropod *Clione antartica*. J Org Chem 60:780–782

Ziegler TA, Forward RB (2007) Larval release behaviors in the caribbean spiny lobster, *Panulirus argus*: role of peptide pheromones. J Chem Ecol 33:1795–1805

Chapter 4
Waterborne Chemical Communication: Stimulus Dispersal Dynamics and Orientation Strategies in Crustaceans

Marc J. Weissburg

Abstract Many animals obtain information about conspecifics or heterospecifics that is transmitted in waterborne chemical plumes or trails. The ability to use chemicals for distance communication minimally requires that the receiver of a chemical signal be able to detect and identify the chemical constituents. Since animals often use plumes or trails to find conspecifics, they frequently must also be able to extract directional and distance information from odor plumes and trails. As odor plumes are strongly affected by flow dynamics, whether animals can even detect chemical signals, or use them to navigate towards a source, are strongly contingent on the fluid physical environment. Here I review basic information on how to quantify the relevant fluid physical aspects of the environment, and discuss major findings concerning the relationship between odor-guided navigation, odor plume structure, and the fluid dynamic environment. A major result is that greater flow or more properly, fluid mixing, diminishes the ability of crustaceans to extract the information required for efficient navigation. Despite this straightforward conclusion, the consequences of turbulent mixing for chemical communication are not as easy to predict for several reasons. First, animal size and mobility are an important factor in how animals respond to changes in odor plume structure. Thus, the consequences of increased turbulence depend on the interaction between animal size and scale, and scales of spatial and temporal variation in odor signal structure; not all animals will be affected equally by turbulent mixing. Second, increased flow or mixing will cause plumes to be dispersed more widely, potentially increasing the active space of the signal if concentrations remain high enough to be detected. Finally, chemical signaling often may involve reciprocal information transfer, where a given animal acts as both a sender and a receiver. Differences in the perceptive abilities of animals may make it difficult to predict how fluid mixing affects this process. Given that much of what we know about the mechanisms and ecological consequences of waterborne chemical communication has been derived from examining predator–prey interactions, there is considerable potential for

M.J. Weissburg (✉)
School of Biology, Georgia Institute of Technology, 310 Ferst Dr., Atlanta, GA 30332, USA
e-mail: marc.weissburg@biology.gatech.edu

T. Breithaupt and M. Thiel (eds.), *Chemical Communication in Crustaceans*,
DOI 10.1007/978-0-387-77101-4_4, © Springer Science+Business Media, LLC 2011

similar investigations in systems where social communication is mediated by the transmission of chemicals via flow.

4.1 Introduction

It is commonly said that a picture is worth a thousand words, a truism I learnt to appreciate as a finishing PhD student charting my postgraduate career. I was swayed, not by a piece of experimental data, but rather, a simple figure showing that my previous research perspective might be on the wane. My thesis involved tests of optimal foraging theory (OFT) in crustaceans, and Pyke (1984), in his critical review, noted a dramatic decline in optimal foraging studies subsequent to 1980. This was a jarring observation to a soon-to-be-minted PhD – would my new and sparkling studies be met with a gigantic yawn of disinterest? Besides evoking panic, Pyke's review made me confront an important weakness of OFT; it does not easily incorporate the mechanisms used by animals to perceive their environment. I was left with the strong sense that understanding how animals acquire and respond to information was a key ingredient in relating animal behavior to animal ecology. This effort required a synthesis of behavior, sensory physiology, and physics (a sad fact for someone who could have paid more attention to this subject as an undergraduate!). Fortunately, such a synthesis was occurring in the sensory biology of aquatic animals (e.g., Atema et al. 1988), an effort that I embraced enthusiastically through a series of collaborations with experts (and wonderful mentors) in these areas. I am particularly pleased that we are now at a point where our understanding of waterborne chemical signaling is making important contributions to our understanding of ecology.

Dissolved chemicals are ubiquitous in aquatic habitats and provide a major source of information for crustaceans immersed in water. This chemical information is used to regulate nearly every critical activity, and often involves cues or signals (i.e., substances released unintentionally vs. intentionally by a sender; Dusenbery 1992) that reach the receiver via transmission through water. Crustaceans use waterborne chemical trails or plumes to find food (Weissburg 2000), recognize and locate potential mates (e.g., see Yen and Lasley, Chap. 9, Bauer, Chap. 14, and Hardege and Terschak, Chap. 19), or find suitable habitats (Ratchford and Eggleston 2000; Horner et al. 2006). Organisms use chemicals to determine dominant vs. subordinate individuals during aggressive encounters (see Breithaupt, Chap. 13) or for sibling and individual recognition (see Gherardi and Tricarico, Chap. 15).

Navigation to chemical cues occurs when chemicals are released by the transmitting agent, transported through the fluid medium, and ultimately received by an agent that responds by trying to locate the source. The first two steps are dominated by flow physics, and a variety of techniques and approaches have investigated chemical plume structures and their corresponding information content. Behavioral investigations have evaluated how animals use this information to find a source. Field experiments have attempted to examine these processes in natural environments. Taken together, this research program has made clear that: (1) the physical processes set fundamental limits on animal performance, (2) successful use of chemically-mediated orientation is strongly contingent on the prevailing fluid regime, and (3) sensory strategies and capabilities reflect the local sensory environment produced by these physical processes. The primary goal of this chapter is to discuss these findings, and how they allow us to identify the ecological implications of waterborne communication.

Despite the wide range of vital tasks that crustaceans solve via chemically-mediated guidance, our consideration of this problem has been largely through the lens of predator–prey interactions. Orientation to food sources often has motivated these studies, guided the selection of experimental conditions, and been the context in which the ecological implications of chemosensory orientation have been examined. However, many of the insights from these studies have important implications for chemical communication via waterborne signals. The physical processes studied in prey location also are applicable to chemical communication, so the receiver of a chemical signal emitted by a conspecific will face similar challenges as the individual searching for prey.

4.2   The Structure of Chemical Plumes and Trails

4.2.1   Characterizing the Flow Environment

Understanding chemically-mediated communication requires knowledge of fluid properties and flow because the fluid environment determines the rate of delivery and the spatial and temporal patterns of dissolved chemicals presented to the receiver (Weissburg 2000). A detailed summary of this literature is beyond the scope of this chapter, and has been covered in a number of recent reviews (Weissburg 2000; Koehl 2006; Webster and Weissburg 2009). Rather, my goal here is to articulate a number of findings that relate basic properties of chemical signal structure to very general aspects of the fluid environment.

The transmission of chemical signals is controlled by flow physics. The mode of transport is indexed by a combination of the Reynolds number:

$$Re = \frac{\rho U L}{\mu} = \frac{U L}{v}, \tag{4.1}$$

and Péclet number:

$$Pé = \frac{U L}{\Gamma}. \tag{4.2}$$

These numbers provide a rough guide to the flow regime, and hence the general character of the resulting chemical signal. *Re* is the ratio of inertial to viscous forces in the flow, whereas Pé is a ratio of flow-driven transport to molecular diffusion (Table 4.1). Molecular diffusion is the dominant mode of transport in the absence of fluid motion and controls the final delivery of chemicals to chemical sensors. I exclude molecular diffusion from further discussion since flow transports chemical signals to most aquatic crustaceans.

Fluid motion for $Re = 1$–100 indicates that the flow is laminar and transmission is characterized by laminar-advective transport. A low flow velocity or small length scale also means that molecular diffusion can be an important transport process relative to advective transport (i.e., Pé is low). Flow streamlines in low Re environments do not cross and the odor plume or trail consequently is relatively coherent and characterized by sharp gradients in chemical concentration, particularly in the axial (cross-flow) direction. This regime is important for smaller animals, particularly those in the plankton. Cruising copepods typically have a $Re < 10$, for instance, and the wake retains high concentration of "scent" in an isolated region due to the slow diffusive spread and dilution in the laminar regime (Yen et al. 1998).

Large values of Re denote a regime where turbulence is the predominant process distributing odorants. The precise value that demarcates this transition is dependent on specific aspects of the flow geometry, but $Re > 10^4$–10^6 generally indicate turbulent conditions. Many crustaceans, including the iconic crabs, lobsters, and crayfish live in such a regime, a state characterized by unpredictable velocity fluctuations. This leads to random fluctuations of the chemical concentration, producing a high degree of spatial and temporal variance that becomes the defining aspect of turbulent plumes.

The dynamics of turbulent plumes relevant to most crustaceans are complicated by the fact that many are produced in boundary layer flows. A crustacean moving across the substratum does so in a velocity gradient characterized by no motion of fluid in contact with the substratum and a nominal or "free-stream" velocity at some distance away. The region in between is characterized by a roughly logarithmic velocity profile that comprises approximately 30% of the water depth (Schlichting 1987).

Table 4.1 Physical constants

Symbol	Meaning
D	Sediment grain size diameter-a specific length scale used in calculation of roughness Reynolds number (Re*)
L	Characteristic length or length scale associated with fluid physical calculations
ρ	Fluid density
Pé	Péclet number-a ratio of the magnitude of bulk flow driven and diffusive transport
Γ	Diffusion constant of a fluid
Re	Reynolds number-the ratio of inertial to viscous forces in a fluid, and a rough guide to the fluid physical regime
U	Fluid or relative velocity
u*	Boundary layer shear velocity-an indicator of the shear stress of an object on the substratum in flow, and the velocity scale used in calculation of roughness Reynolds number (Re*)
μ	Dynamic viscosity
ν	Kinematic viscosity (equals μ/ρ)

Boundary layer flows often are characterized by the shear velocity, u*, or the roughness Reynolds number: $Re* = u*D/v$, where D represents the grain size of the bed. Shear velocity is generally calculated from the law-of-the-wall equation (Schlichting 1987; Weissburg and Zimmer-Faust 1993) that relates velocity to distance above the bed.

These two quantities reflect the degree to which turbulent velocity fluctuations reach the bed, and the magnitude of correlated velocity fluctuations (i.e., turbulent structures) impinging on the bed. Outer boundary layer regions begin to feel the effects of turbulence when $3.5 < Re* < 6$ (transitional flows) and turbulence extends to the bottom when $75 < Re* < 100$ (fully rough flows) (Schlichting 1987; Jackson et al. 2007). The presence of a length scale, D, indicates that rougher substrata (e.g., larger diameter) increases turbulent mixing independent of velocity changes, which means that sediment type and surface features (animal mounds, sand ripples) can have considerable impact on the chemical cue structure that confronts the receiver, even under conditions of constant velocity.

It is important to note that the relationship between measures of boundary layer structure and actual flow dynamics are most accurate in well controlled laboratory conditions (so-called fully developed, equilibrium boundary layers). Even then, Re provides only an approximate guide to the flow dynamics, because the specific geometry of the flow will influence the amount of turbulence at any given Re, as mentioned above. Consequently, lab measurements provide excellent assessments of the relationship between animal performance and flow and chemical plume structure, but limited insight into flow dynamics in the field even when conducted under nominally similar conditions (e.g., flow velocity, substratum size). In fact, oscillating tidal flows, waves, changing current direction, and varied topography makes $Re*$ and u* extremely crude predictors of flow effects on chemical signal structure in the field. In light of these considerations, investigators sometimes characterize the flow by computing a measure of the intensity of velocity fluctuations, the root mean square (RMS, standard deviation):

$$\sqrt{(RMS_u^2 + RMS_v^2 + RMS_w^2)},$$

or the turbulence intensity, the RMS scaled to the average:

$$TI = \frac{\sqrt{(RMS_u^2 + RMS_v^2 + RMS_w^2)}}{\sqrt{(u^2 + v^2 + w^2)}},$$

where u, v, and w are the velocity components in the x, y, and z dimensions respectively, and RMS_u, RMS_v, RMS_w are the RMS (SD) of each velocity component.

A variety of experiments indicates clearly that animal orientation performance varies in different regimes as a result of changes in odor plume structure. This appears to reflect both the homogenization of the plume and the dilution of chemical concentration as the plume expands.

4.2.2   Plume Structure

Odor plumes are complex and time-varying structures that present challenges for animals attempting to extract information on odor source location. Although empirical work generally supports relatively simple approximations for along- and across-stream chemical average concentrations within turbulent plumes, these statistical or ensemble properties reflect averages observed only over large time intervals, not the instantaneous "snapshot" obtained by an animal moving through this plume (Fig. 4.1). Average concentrations across the plume width are Gaussian, whereas average concentration decreases rapidly downstream with a power law rate approximately in proportion to $(distance)^{1.5}$ close to the source and a somewhat slower rate downstream (Rahman and Webster 2005). These relationships are roughly valid for airborne plumes as well, suggesting similarity in structure despite the differences in fluid properties of air vs. water. There are two cases where the ensemble or statistical

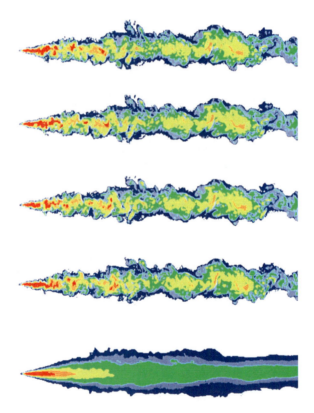

Fig. 4.1 Instantaneous vs. time average views of odor signal structure. Photograph is a series of laser-induced fluorescence images of dye concentration expressed as color (*red* is highest concentration, *blue* is lowest). The top 4 images are each taken on a rapid time scale (4 ns per frame, 30 ms per successive frame), and show the fine scale instantaneous structure. Note how coherent structures move downstream. The bottom panel is an average of 1,000 frames, and gives a result similar to employing a diffusion equation. This plume is approximately 1×0.15 m. Reproduced from Weissburg (2000) with permission from The Biological Bulletin

plume description may accurately reflect the information available to a searcher. The first is where Re is low (e.g., the flow environment is in the laminar-advective or diffusive regime). The second is when the animal sensors themselves act as time-averagers and integrate local concentration over long time intervals (see below).

Direct visualization using laser-induced dye fluorescence of plumes incorporating dye tracers reveals much about plume dynamics, and highlights the presence of regions of extremely high odor concentration (particularly near the source) inter-spersed with clean water (Fig. 4.1). Odor concentrations are extremely variable in space and time as "filaments" move past a fixed location. For this reason, distribu-tions of instantaneous concentrations (or other time varying properties) are more relevant descriptors of information content than simple averages. Note also that the highly filamentous nature of the plume means that there are large differences between average concentration and filament concentrations; the latter may exceed the former by 10–100 times in some plume regions (see also Baker, Chap. 27). The filamentous character of many plumes also severely complicates the calculation of active space (i.e., the region where chemical concentration is above threshold for a behavioral response), as many animals seem to access the information from individual filaments as opposed to average concentrations (see below). Thus, averages are only approxi-mate guides to the physical location and intensity of the signal, and measures of active space for most animals will be more biologically realistic if based on filament properties. In fact, turbulent diffusion equations underpredict the active space for moth pheromone signaling (Elkinton et al. 1984). Animals may draw information from average instantaneous filament concentration (i.e., conditional averages that include only nonzero measurements) or probability of encountering a filament above some threshold (e.g., the detection limit of the animal). As expected based on intuition and common experience (e.g., stirring cream into your morning coffee), greater turbulent mixing enhances the rate of dilution and spread of the plume. In other words, turbulent mixing accelerates the rate at which local instantaneous concentrations converge on the average.

Plume properties change as a function of both downstream distance from the source and distance above the bed. Downstream changes are particularly well-known, having been recognized as an important phenomenon since the early and seminal papers on insect odor-mediated guidance (Murlis et al. 1992). Although plume width is an abstraction given the large spatial variance, it can be defined statistically (Webster et al. 2001; Jackson et al. 2007). Plumes widen as they evolve downstream from the source approximately in proportion to $(distance)^{0.75}$ near the source, and $(distance)^{0.50}$ far from it (Rahman and Webster 2005). Plume width increases with turbulence. By example, investigators often use plumes generated from a small diameter source and alter turbulent mixing by changing the bed substratum. Plumes created with large gravel (Median diameter $= 21.5$ mm, $u^* = 5.82$ mm s^{-1}, $Re^* = 350$) had nearly double the width 1 m downstream from the source when compared to plumes created over sand (Median diameter $= 2.5$ mm, $u^* = 3.44$ mm s^{-1}, $Re^* = 12$) (Rahman and Webster 2005; Jackson et al. 2007).

A principle factor governing plume structure is the interaction between scales of individual odor filaments and scales of turbulent energy. It is essential to appreciate that turbulent flows contain eddies of a variety of sizes. Individual odor filaments are small, often because they are emitted from a small orifice or diffuse passively from the transmitter. A turbulent eddy larger than the filament length cannot break the filament apart – it can only move it. Thus, plumes close to the source may meander (e.g., the centerline may shift from side to side) or become distorted until the filaments diffuse to a size that is larger than the smallest turbulent eddies in the flow (the Kolmogorov scale). Further downstream, turbulent eddies can rapidly homogenize or repackage the larger odor filaments, which widens the entire plume and decreases concentration within the plume. Greater turbulent energy of the flow decreases the Kolmogorov limit and results in more rapid plume homogenization (Schlichting 1987).

Changes in the three-dimensional structure of odor plumes are less well-known. Here, the vertical variation in velocity results in a vertical variation in shear intensity that has dramatic effects on chemical plume structure (Rahman and Webster 2005). In particular, if odor sources are released near the bed, peak shear intensity close to the substratum results in near-bed signals that can be relatively homogenous, spatially coherent, but somewhat dilute relative to signals above the bed. Benthic animals have appendages located at different elevations above the bed, or are able to scan through the vertical domain, and may therefore take advantage of height-dependent changes in plume chemical structure to locate a source (e.g., Jackson et al. 2007).

One important caveat is that much of our understanding of odor plume dynamics comes from a limited set of conditions with respect to flow and odor signal geometry. The most common experiments use small odor sources that do not disrupt the flow, and release rates that result in little momentum relative to the ambient flow (e.g., isokinetic release) – this both simplifies the flow problem and closely represents challenges faced by many foraging animals. However, this may not embody all communication processes in aquatic realms, where some signaling may be in the form of jets that contain appreciable momentum (Breithaupt and Eger 2002), and that result in relatively more rapid homogenization of odor signal structure (Webster and Weissburg 2001). This may limit the information available to animals utilizing chemical plumes for orientation, particularly when animals rely on instantaneous structure for navigation. Even if odorants are released passively, animals produce considerable turbulence in their wake. The odorant plume in the wake of a tracking blue crab is greatly homogenized and dispersed (Weissburg et al. 2003) relative to the plume before the blue crab has moved through it. The presence, and orientation of objects releasing odor also can have a considerable impact on the downstream plume (Keller et al. 2001).

4.3   Orientation Strategies

The recent availability of detailed data on chemical plume signal structure, and the relationship of this structure to the fluid environment, has lead to tractable and testable hypotheses on presumptive orientation strategies. These hypotheses center on the availability of spatial vs. temporal information, the role of accessory information in addition to odor, and the influence of organism size and scale. Investigators have recognized that these hypotheses must be consistent with observed changes in animal performance in different flow environments, given the intimate connection between flow and chemical signal structure.

4.3.1   Temporal Sampling

The current consensus is that navigation involves sampling the temporal domain relatively coarsely, since odor concentrations tend to fluctuate greatly at any given sampling location. Although average properties theoretically could provide information on distance and direction to the source, the time required to estimate average concentration at a particular location greatly exceeds the total time required for source location. For instance, blue crabs can take less than 20 s to navigate to a food source 1.5 m away, whereas reliable estimates of average concentration at a *single location* require approximately 100 s (Webster and Weissburg 2001). Properties of individual odor filaments also could provide information about the relative distance of animals to the odor source. Filament peak height and onset slope increase with decreasing distance to the source. However, receptor contacts with individual odor filaments are very brief, lasting generally less than a few 100 ms (Webster and Weissburg 2001), so that adequately resolving onset slopes or peak concentrations is probably beyond the temporal sampling ability of crustacean chemosensors, which operate at around 5 Hz (Gomez and Atema 1994). Additionally, the most informative filaments (those with very high peaks and sharp onsets) have a low probability of occurrence. Most decapod crustaceans do not appear to sample long enough to obtain information from these rare events (Webster and Weissburg 2001).

    The most likely information gained from temporal sampling is an indication of odor presence that is used to move upstream in response to flow direction (Fig. 4.2). This is consistent with the inability of some crustaceans to follow odor signals in the absence of flow (Weissburg and Zimmer-Faust 1993), and observations indicating that animals in experimentally pulsed plumes walk slower and pause longer during tracking than animals in continuously released plumes (Keller and Weissburg 2004). This hypothesis is also consistent with decreased performance in more turbulent flows (Weissburg and Zimmer-Faust 1994), which cause greater dilution of filament concentration. The polarization of movement in response to flow (odor-gated rheotaxis) is similar to the strategy proposed for insect navigation to pheromone plumes, where reiterative contact with odor filaments induces upstream surges in response to wind (odor-gated amenotaxis; Vickers 2000), and also has been documented in fish (Baker et al. 2002). Interestingly, the addition of oscillatory flow (as in a wavy marine environment) does not necessarily increase turbulent mixing. Oscillatory flow did not decrease navigational performance in marine stomatopods (Mead et al. 2003). Some investigations in crayfish have suggested that turbulence facilitates navigation (Moore and Grills 1999), but the lack of signal structure data makes interpretation of these results difficult.

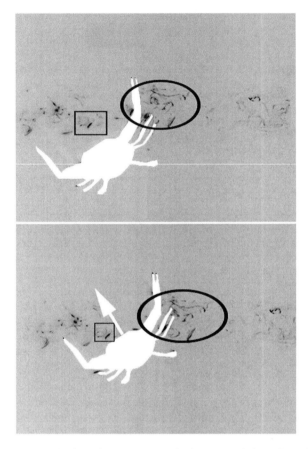

Fig. 4.2 Real-time visualization of a blue crab navigating to a turbulent plume. Flow is *left* to *right*. Panels show two successive frames illuminated by a thin laser sheet approximately 3 cm above the bed, separated by 200 ms. Contact with small odor filaments at the antennules (*boxed region*) guides upstream progress and turning is mediated by the spatial asymmetry, e.g., more filaments arriving at the right legs (*circled region*) compared to left legs. This animal surged upstream and to the *right* (*arrow*) subsequent to this encounter. Modified after Dickman et al. (2009)

Drawing straightforward connections between hydrodynamic measurements and chemical plume structure is not always possible, especially if the environment is not sufficiently well-controlled or produced (e.g., nonequilibrium boundary layers). Additionally, using a limited range of flow conditions makes it challenging to discern overall relationships; some turbulent mixing may be beneficial by transporting odorants into regions where otherwise they would not be accessible by animal sensors, especially those close to or far from the bed. An important caveat is to not let the perfect be the enemy of the good; even qualitative visualizations of odor plume structure (e.g., Breithaupt and Eger 2002; Keller and Weissburg 2004) are enormously helpful in interpreting behavioral patterns and discerning the potential effects of mixing.

4.3.2 Spatial Sampling

Spatial sampling potentially provides navigating animals with information about their location relative to the plume, and allows them to remain in contact with it as they progress towards the source. The presumptive mechanism involves simultaneous comparisons of chemical signal intensity between chemosensory receptors distributed across the body surface. These spatial comparisons are enabled by the general crustacean body plan, which includes multiple chemosensory bearing and paired appendages (e.g., claws, legs, antennae and antennules), as well as individual appendages (antennules) that can sample over a broad expanse (e.g., in spiny lobsters). Spatial sampling permits rapid information gathering since odor plumes in flow exhibit much more consistent spatial vs. temporal patterns. Differences in filament arrival between spatially separated sensors rapidly (e.g., within seconds) provide information on which side of the body is closer to the main axis of the plume (Webster et al. 2001; Fig. 4.2). These comparisons are facilitated when animals reside in regions of maximal contrast, that is, when they span the plume "edge."

Lobsters and crabs both appear to use spatial sampling to detect asymmetry in chemical signal intensity and keep within or near the odor plume. Blue crabs position themselves such that interior sensors (those closer to the plume axis) are much more likely than exterior sensors to encounter individual odor filaments, and often straddle the plume "edge" as they move upstream (Zimmer-Faust et al. 1995; Jackson et al. 2007). Lobsters receiving bilaterally asymmetric patterns of chemical signal intensity due to antennule lesions turn towards the side of greatest stimulation (Reeder and Ache 1980; Devine and Atema 1982). Note again, however, that plume edges are statistical abstractions (see above and Fig. 4.1). Thus, the failure of animals to always move along the time averaged plume edge (e.g., Moore and Grills 1999) is not clear cut evidence for a lack of spatial sampling.

The necessity to detect a high degree of contrast in order to maintain plume contact is consistent with diminished performance of animals navigating in more turbulent conditions (e.g., Weissburg and Zimmer-Faust 1993; Jackson et al. 2007). Turbulent mixing homogenizes the plume, and widens it, thereby decreasing both instantaneous filament and average concentrations. This reduces signal contrast, and regions of maximal contrast (the "edge") occur at greater distances from the plume centerline. Animals make increasingly dramatic lateral excursions, and are farther from the plume centerline in more turbulent conditions, likely associated with the greater lateral spread of the plume (see also Baker, Chap. 27). Lateral movements and distance to centerline also are exaggerated in animals farther downstream, which is attributed to plumes widening as they evolve. Note that natural plumes (i.e., those not confined by the walls of a flow tank) may become so broad and diffuse that signal contrast is not readily available, either requiring a new navigational strategy or severely compromising spatial sensing mechanisms. A reasonable (but untested) hypothesis is that indirect guidance mechanisms may be employed far downstream when other strategies are unprofitable. The simplest indirect guidance strategy, analogous to chemosensory navigation in bacteria (Dusenbery 1992), is to maintain a consistent direction when stimulus strength is increasing (or above a threshold) and otherwise randomly change direction.

4.3.3   Role of Size and Scale

The size and scale of animal sensory arrays relative to plume width sets limits on the ability of navigating animals to determine concentration differences in space, whereas the sampling frequency of animals constrains the ability to determine the local average concentration (Weissburg 2000). Most detailed behavioral studies involve rapidly moving crustaceans that are large relative to the plume. The hypotheses on proposed navigational strategies derived from these cases probably represent only a subset of available mechanisms, and *a priori* considerations suggest a variety of strategies that may be predicted based on the relationships between temporal and spatial scales of animals and plumes (Weissburg 2000). There has been remarkably little work on these scaling issues, despite the important role of size and scale in explaining both physical phenomena and animal responses to flow and odor properties.

Empirical and theoretical analysis of turbulent plumes indicates that local concentrations converge; plumes exhibit smooth gradients in chemical signal intensity when measurements are averaged over sufficiently long intervals, and concentrations are as predicted by diffusion models. However, the mismatch between long time scales of convergence in plumes vs. short sampling periods in rapidly moving animals (e.g., blue crabs, lobsters) means that odor signals contain little information, particularly with respect to the distance between source and animal. Using odor to evoke an upstream response to flow becomes the only viable strategy to move towards the source (see also Baker, Chap. 27). In contrast, an animal (or sensor) would be able to resolve averages by sampling over a longer period. This ability permits using successive sampling to determine whether average concentrations are increasing, and hence, to establish progress to the source. Large gastropods such as whelks move at speeds on the order of mm s^{-1} (Ferner and Weissburg 2005), 10–100 times slower than crustaceans used in chemosensory navigation studies discussed above. These animals appear to use a time-averaging strategy during navigation and are less sensitive to turbulence than crustaceans (Ferner and Weissburg 2005). We might also expect slow moving crustaceans to utilize this mechanism. For instance, stone crabs (genus *Menippe*) move rather slowly (at least, compared to blue crabs; *pers. obs.*) and might utilize a time-averaging mechanism. Note, though that the issue is the relative scaling of convergence time to sampling period; animals foraging in largely homogeneous plumes may be more likely to develop time averaging strategies.

The use of stimulus asymmetry as a source of information also is contingent on animal size. Creatures with small sensor spans relative to plume width may be unable to position the sensors in such a way that asymmetry is detectable, and larger sensor spans relative to plume width increase the ability to detect spatial variance in odor signal structure (Weissburg et al. 2002; Jackson et al. 2007). Insects are quite a bit smaller than the airborne pheromone plumes they travel through (Murlis and Jones 1981). Perhaps for this reason, flying insects do not seem to steer relative to the odor plume as do the larger crustaceans. Rather, they steer relative to the wind direction in response to pheromone presence; contact with odor filaments produces a surge in movement upwind. This behavior, combined with an endogenous program of alternate left–right turns (again determined with respect to wind direction), elicits the characteristic across-plume tracking and subsequent upstream progress to the source (Vickers 2000). One wonders whether crustaceans that also are small relative to the plume width (e.g., amphipods or isopods) might employ similar strategies, but we have yet to examine closely the behavior of animals in this situation.

4.4 Ecological Implications of Chemical Information Transfer

The necessity of successful mating clearly suggests a large selective advantage for effective means of locating and identifying mates. Chemical cues, which can propagate for long distances, are especially well-suited as mechanisms for mate finding, particularly when other modalities (e.g., vision, mechanosensation) may be limited. Unfortunately, we have very little reliable information on effective signaling distances for conspecific chemical communication. Most previous studies have been geared to demonstrate chemical communication, and often do not incorporate naturally relevant stimulus concentrations or flows. Thus, we are currently unable to say much about whether mating distances may be limited in the field, or how flow velocity, substratum size, emergent vegetation etc. may change these distances in particular habitats. However, the fact that substratum size (gravel or sand) affects plume structure suggests that habitat characteristics also influence chemical communication.

Chemical trails vastly increase the effective signaling distance in planktonic crustaceans. Many species of copepods can follow pheromonal trails for 10–100 body lengths (e.g., Doall et al. 1998), which is considerably larger than the signaling distance for fluid mechanical cues (Yen et al. 1998). In the tube dwelling amphipod *Corophium volutator*, male attraction to female pheromones occurs over a distance of at least 34 cm (Krång and Baden 2004), although the use of a Y-maze, and undefined stimulus concentration relative to natural conditions suggests a cautious interpretation of this signaling distance. *Panulirus argus* can easily locate shelters 1.5–2 m distant by following the scent of conspecific urine in hydrodynamically well-controlled flume trials (Horner et al. 2006), even when the urine was diluted 100-fold. Ratchford and Eggleston (1998, 2000) reported similar results, albeit in a Y-maze with unknown flow characteristics. Thus, this signaling distance probably is a conservative estimate under benign hydrodynamic conditions.

Despite the absence of studies on signaling distance in conspecific communication, some general trends may be discerned by examining orientation of predators to potential prey. The roles of fluid forces and chemical signal structure in mediating attraction are generalizable, even if the specific behaviors evoked by these chemical signals are not. However, there are remarkably few field studies examining chemosensory navigation even in the relatively well studied area of predator–prey relationships.

Finelli et al. (2000) examined blue crab responses to natural and synthetic odor sources mimicking release from carrion in tidal salt marsh channels. Hydrodynamic conditions (u*, $Re*$) were within ranges of those used in flume studies. Blue crabs seemed to respond from a distance of 2–3 m, although the methods were not precise enough to determine accurately a perceptive distance. There was little evidence of a decreased olfactory performance (e.g., less direct search paths, lower response rate) at higher flow speeds or turbulence levels (10–15 cm s^{-1}, $Re* > 100$), in contrast to results from laboratory studies in similar conditions (Jackson et al. 2007). Finelli et al. (2000) speculated that their use of a concentrated cue might have resulted in a strong signal even in well mixed conditions, and that using lower concentrations or release rates more characteristic of laboratory studies would have revealed the effect of flow. This conclusion points to the obvious need to define both stimulus levels emitted by the sender and behavioral dose-response functions of the receiver in order to have any insights into responses in the field. Foraging crabs also performed most poorly in flows of less than 1 cm s^{-1}, both in terms of success rates and path efficiency. This observation is consistent with the necessity of flow as a cue during olfactory-guided search. Consider also that slow flows prevent odor signals from propagating downstream, and that the horizontal (cross-stream expanse) of plumes increases with the degree of mixing so that slow flows produce narrower plumes that expand more slowly as they propagate downstream (Fig. 4.3). Thus, the probability that a downstream searcher intersects the plume may actually be diminished in less turbulent conditions, and illustrates a somewhat subtle trade-off in the effect of flow/turbulent mixing; a potentially positive increase in downstream and cross-stream propagation that is opposed by a negative effect due to odor dilution. Here again, more concentrated signals will require an increased level of turbulence to produce a pernicious effect on chemosensory navigation.

Assaying chemosensory orientation in flow in the field can be done, as above, using focal observations of individual behavior (conditions permitting) or by examining the end point of the behavioral process (e.g., the incidence of successful mating or predation events). We have developed methods to examine blue crab predation and its relationship to flow structure in the field (Smee et al. 2008, 2010) in the area surrounding Wassaw Sound, GA, USA. These studies consist of monitoring predation intensity in experimental plots and measuring flow structure and

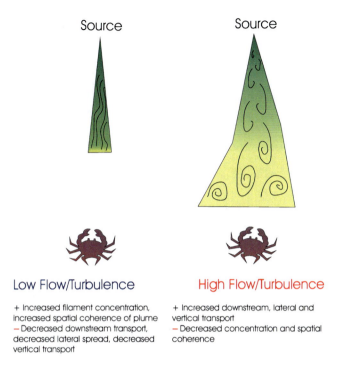

Fig. 4.3 Positive and negative effects of flow/turbulence on olfactory orientation

other biological and/or physical variables. Experiments utilized naturally occurring variation in flow, and took place in areas ranging from semiexposed sounds to moderate sized tidal channels (10–15 ft wide, 6–8 ft deep) distant from the open coast. Blue crabs seem primarily responsible for predation as judged by the prevalence of broken/crushed shells or missing clams in experimental plots, and long-term monitoring of this predator suggest no systematic differences in crab density. Additionally, it appears that bivalve prey detect blue crab predators using chemosensation, and cease their pumping when they detect predator scent (Smee and Weissburg 2006). This behavior decreases the likelihood that crab predators can orient to potential bivalve prey and helps increase bivalve survivorship in the field. Thus, each participant is both a sender and a receiver of chemical messages, much like many social communication systems.

These studies not only show patterns somewhat consistent with laboratory results, but also illustrate an additional level of complexity in signaling processes when communication is bidirectional. Predation peaks at an intermediate level of flow/turbulence (Fig. 4.4), which probably reflects some of the physical processes outlined above. In addition, the dynamics of this communication system also appear contingent on the balance between abilities of predator and prey to detect each other (Smee and Weissburg 2006). Slow and less turbulent flows potentially not only allow predators to navigate effectively, but also allow prey to detect predator scent

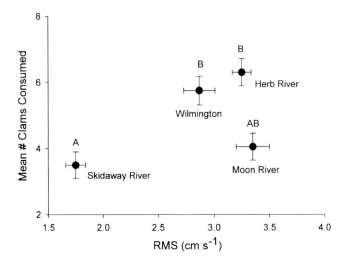

Fig. 4.4 Relationship of flow and turbulence to predatory success of naturally foraging blue crabs. The number of clams consumed in plots of bivalves placed in the field was maximal at intermediate turbulence intensities as measured over several tidal cycles with an accoustic Doppler velocimeter. Numbers indicate mean flow velocity (± 1 SD, which is equivalent to RMS turbulence), and letters indicate treatments that are significantly different based upon a one-way ANOVA with Tukey-Kramer post hoc tests. Reproduced from Smee et al. (2010) with permission from Ecology

and respond by becoming cryptic. Although more turbulent flows suppress predator detection they also may compromise prey ability to detect the predator more greatly, so predation intensity rises. Turbulent mixing ultimately becomes so great that predator navigation is substantially diminished, and predation intensity falls off. These considerations are important for defining environments that constitute a refuge from predators via disruption of olfactory orientation (i.e., a turbulence refuge). Studies focused on predator abilities lead to the (implicit) hypothesis that the refuge potential was a monotonic function of increasing turbulence; that is, increased turbulence leads generally to decreased predation pressure (Weissburg and Zimmer-Faust 1993). This simple view clearly does not depict the dynamics of the crab-bivalve system; complicated interactions ensue because each participant in this interaction is both a sender and receiver of chemical signals (Fig. 4.5) and therefore, engage in reciprocal information transfer.

4.5   Conclusions and Prospectus

Taken together, research presented in this chapter has made clear that: (1) the physical environment can promote or retard distance chemical signaling by affecting the structure of turbulent plumes; (2) organisms have developed strategies to obtain information, but these strategies do not work equally well under all conditions;

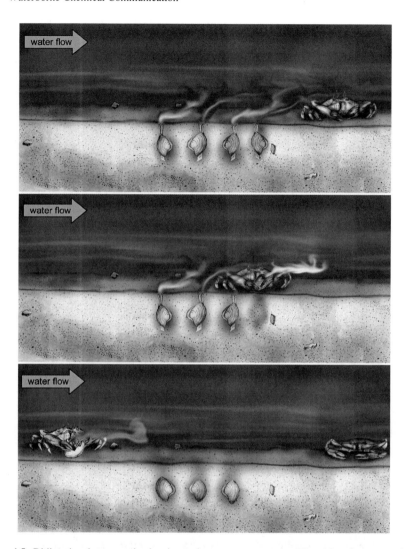

Fig. 4.5 Bidirectional communication in predator–prey systems mediated by chemosensation. Foraging crabs locate bivalves by homing in on the metabolites contained in the plume created by the excurrent flow (*top*). Chemical cues are released by foraging crabs, as well as by injured bivalves (*middle*). These cues suppress the response of other bivalves, and which makes them undetectable to downstream predators (*bottom*). Drawing courtesy to Jorge Andrés Varela Ramos

(3) interactions between the environment and animal sensory abilities cause local differences in the efficiency of chemical communication. Although our current understanding of these processes is mostly in the context of predator–prey relationships (except for behavioral studies on mate location in plankton), the universality of basic principles governing signal transmission and reception means that insights gleaned from investigating orientation to prey odor plumes can serve to

drive studies examining conspecific chemical communication. This, in fact, is exactly the reverse of the history of inquiries examining insect chemosensory-orientation, where studies on mate tracking have made substantial contributions to our understanding of olfactory orientation (even in crustaceans).

Our present understanding of predator–prey relationships suggests a number of intriguing issues, particularly with respect to the ecological implications of chemosensory orientation.

1. *What habitats are most conducive to long-distance social signaling?* Although laboratory studies have suggested that turbulence diminishes olfactory orientation, this effect is not seen in the field, which instead has shown more dome-like relationships between orientation and flow or mixing. This observation may reflect the trade-off between positive and negative influences of turbulence on orientation (Fig. 4.3), and highlights the fact that both signal levels and local flow geometry make it difficult to extrapolate directly from laboratory to field environments. Spatial and temporal variation also may be extremely important. Current speeds and turbulence may fluctuate by an order of magnitude over the tidal cycle and these differences within a site may be as great as those occurring between sites (e.g., Fig. 4.4). Animals needing to orient to conspecifics may be constrained to do so only at particular times or particular places. It is important to keep in mind that a suitable habitat is defined only in relationship to the orientation strategy employed, so that potential differences in animal size and scale (see above) may translate to different habitat specificities among various animal groups.

2. *What are the trade-offs between constraints imposed by fluid-physical vs. other factors?* Successful social interactions (e.g., mating) often involve other events aside from locating a conspecific. For instance, crustaceans often guard mates for relatively long periods during which time they may be vulnerable to predators. Habitats that provide structural protection also may create undesirable flow dynamics. For instance, emergent vegetation (sea or marsh grasses) may either substantially impede flow or increase turbulent mixing depending on its expanse and density (Lightbody and Nepf 2006). Although these habitats may provide protection, this may come at a cost of reduced communication range and effectiveness. The notion of trade-offs here is similar to that when examining compromises between intake rate and predation risk – a foraging animal may forgo getting dinner in order to avoid being something else's lunch.

3. *What is the role of sender behavior?* Boundary layers represent a barrier to chemical transmission, as do slow flows. Just as prey species may use behavioral or other mechanism to disguise their chemical signatures, we might expect animals engaged in social signaling to employ dispersal strategies that widely advertise their presence or status. Moths release pheromones in pulses (Conner et al. 1980), and modeling this release pattern suggests it increases the range of attraction (Dusenbery 1989). Recent studies indicate that the distinctive pleopod fanning behavior of the male blue crab mating display serves to increase the transmission of scent away from the animal (Kamio et al. 2008). This, or similar

mechanisms may be particularly widespread when animals are in habitats that inhibit chemical transmission (e.g., see above). For instance, the need to minimize predation risk during mating may require animals to utilize structurally complex habitats such as grass beds, which may limit chemical signaling. Many other crustaceans release their odor in jets or with currents created by appendage movements (e.g., Breithaupt and Eger 2002). There may be considerable behavioral flexibility to direct currents in a manner that maximizes transmission, or to produce pulses that modulate the active space. Animals may simply elect to signal only during particular periods (e.g., certain phases of the tidal cycle) in order to maximize their signaling capacity. These behaviors would restrict potential mating periods and may therefore impact population dynamics.

4. *Are there trade-offs or conflicts in bi-directional communication systems?* The observation that properties of predator–prey systems may be strongly structured by the differential impact of turbulence on both participants (Fig. 4.4) raises the specter of similar effects in some social communication systems. For instance, both male and female blue crabs release pheromones (Gleeson and Adams 1984; Kamio et al. 2008). Sex-specific differences in animal size, chemical release rate, or olfactory sensitivity all may produce a situation where flow environments favoring the transmission of signals from one sex may be less favorable (or unfavorable) from the perspective of the other. Just as how we have been led to revise our interpretation of what constitutes a "turbulence refuge" by examining both predators and prey (see above), our understanding of what constitutes an effective sensory environment may depend on analysis of both partners in the mating duet.

As these examples show, there is considerable interplay between fluid dynamics, information gathering abilities, and the dynamics of populations in the field. Examining the environmental impacts on chemosensory orientation may therefore provide an important link between the sensory capabilities of animals and their ecology.

References

Atema J, Fay RR, Popper AN, Tavolga WN (1988). Sensory Biology of Aquatic Animals, pp. 945. New York/Berlin/Heidelberg/London/Paris/Tokyo: Springer-Verlag

Baker CF, Montgomery JC, Dennis TE (2002) The sensory basis of olfactory search behavior in the banded kokopu (*Galaxias fasciatus*). J Comp Physiol A 188:533–560

Breithaupt T, Eger P (2002) Urine makes the difference: chemical communication in fighting crayfish made visible. J Exp Biol 205:1221–1231

Conner WE, Eisner T, Vander Meer RK, Guerrero A, Meinwald J (1980) Sex attractant of an arctiid moth (*Utetheisa ornatrix*): a pulsed chemical signal. Behav Ecol Sociobiol 7:55–63

Devine DV, Atema J (1982) Function of chemoreceptor organs in spatial orientation of the lobster, *Homarus americanus*: differences and overlap. Biol Bull 163:144–153

Dickman BD, Webster DR, Page JL, Weissburg MJ (2009) Three-dimensional odorant concentration measurements around actively tracking blue crabs. Limnol Oceanogr Meth 7:96–108

Doall MH, Colin SP, Strickler JR, Yen J (1998) Locating a mate in 3D: the case of *Temora longicornis*. Phil Trans Roy Soc B 353:681–689

Dusenbery D (1989) Calculated effect of pulsated pheromone release on range of attraction. J Chem Ecol 15:971–977

Dusenbery DB (1992) Sensory ecology: how organisms acquire and respond to information. W.H. Freeman and Co., New York

Elkinton JS, Carde RT, Mason CJ (1984) Evaluation of time-average dispersion models for estimating pheromone concentration in a deciduous forest. J Chem Ecol 10:1081–1108

Ferner MC, Weissburg MJ (2005) Slow-moving predatory gastropods track prey odors in fast and turbulent flow. J Exp Biol 208:809–819

Finelli CM, Pentcheff ND, Zimmer RK, Wethey DS (2000) Physical constraints on ecological processes: a field test of odor-mediated foraging. Ecology 81:784–797

Gleeson RA, Adams MA (1984) Characterization of a sex pheromone in the blue crab *Callinectes sapidus*: crustecdysone studies. J Chem Ecol 10:913–921

Gomez G, Atema J (1994) Frequency filter properties of lobster chemoreceptor cells determined with high-resolution stimulus measurement. J Comp Physiol A 174:803–811

Horner AJ, Nickles SP, Weissburg MJ, Derby CD (2006) Source and specificity of chemical cues mediating shelter preference of Caribbean spiny lobsters (*Panulirus argus*). Biol Bull 211:128–139

Jackson JL, Webster DR, Rahman S, Weissburg MJ (2007) Bed roughness effects on boundary-layer turbulence and consequences for odor tracking behavior of blue crabs (*Callinectes sapidus*). Limnol Oceangr 52:1883–1897

Kamio M, Reidenbach MA, Derby CD (2008) To paddle or not: context dependent courtship display by male blue crabs, *Callinectes sapidus*. J Exp Biol 211:1243–1248

Keller TA, Weissburg MJ (2004) Effects of odor flux and pulse rate on chemosensory tracking in turbulent odor plumes by the blue crab, *Callinectes sapidus*. Biol Bull 207:44–55

Keller TA, Tomba AM, Moore PA (2001) Orientation in complex chemical landscapes: spatial arrangement of odor sources influences crayfish food finding efficiency in artificial streams. Limnol Oceangr 46:238–247

Koehl MAR (2006) The fluid mechanics of arthropod sniffing in turbulent odor plumes. J Chem Ecol 31:93–105

Krå ng SA, Baden SP (2004) The ability of the amphipod *Corophium volutator* (Pallas) to follow chemical signals from con-specifics. J Exp Mar Biol Ecol 310:195–206

Lightbody A, Nepf HM (2006) Prediction of velocity profiles and longitudinal dispersion in emergent salt marsh vegetation. Limnol Oceangr 51:218–228

Mead KS, Wiley MB, Koehl MAR, Koseff JR (2003) Fine-scale patterns of odor encounter by the antenule of the mantis shrimp tracking turbulent plumes in wave-affected and unidirectional flow. J Exp Biol 206:181–193

Moore PA, Grills JL (1999) Chemical orientation to food by the crayfish, *Orconectes rusticus*, influence by hydrodynamics. Anim Behav 58:953–963

Murlis J, Jones CD (1981) Fine-scale structure of odour plumes in relation to insect orientation to distant pheromone and other attractant sources. Physiol Entomol 6:71–86

Murlis J, Elkinton JS, Cardé RT (1992) Odor plumes and how insects use them. Ann Rev Entomol 37:505–532

Pyke GL (1984) Optimal foraging theory: a critical review. Ann Rev Ecol Syst 15:523–575

Rahman S, Webster DR (2005) The effect of bed roughness on scalar fluctuations in turbulent boundary layers. Exp Fluids 38:372–384

Ratchford SG, Eggleston DB (1998) Size- and scale-dependent chemical attraction contribute to an ontogenetic shift in sociality. Anim Behav 56:1027–1034

Ratchford SG, Eggleston DB (2000) Temporal shift in the presence of a chemical cue contributes to a diel shift in sociality. Anim Behav 59:793–799

Reeder PB, Ache BW (1980) Chemotaxis in the Florida spiny lobster, *Panulirus argus*. Anim Behav 28:831–839

Schlichting H (1987) Boundary layer theory. McGraw-Hill, New York

Smee DL, Weissburg MJ (2006) Clamming up: environmental forces diminish the perceptive ability of bivalve prey. Ecology 87:1587–1598

Smee DL, Ferner M, Weissburg MJ (2008) Environmental conditions alter prey reactions to risk and the scales of nonlethal predator effects in natural systems. Oecologia 156:399–409

Smee DL, Ferner M, Weissburg MJ (2010) Hydrodynamic sensory stressors produce nonlinear predation patterns. Ecology 91:1391–1400

Vickers NJ (2000) Mechanisms of animal navigation in odor plumes. Biol Bull 198:203–212

Webster DR, Weissburg MJ (2001) Chemosensory guidance cues in a turbulent odor plume. Limnol Oceangr 46:1048–1053

Webster DR, Weissburg MJ (2009) The hydrodynamics of chemical cues among aquatic organisms. Ann Rev Fluid Mech 41:73–90

Webster DR, Rahman S, Dasi LP (2001) On the usefulness of bilateral comparison to tracking turbulent chemical odor plumes. Limnol Oceangr 46:1048–1053

Weissburg MJ (2000) The fluid dynamical context of chemosensory behavior. Biol Bull 198:188–202

Weissburg MJ, Zimmer-Faust RK (1993) Life and death in moving fluids: hydrodynamic effects on chemosensory-mediated predation. Ecology 74:1428–1443

Weissburg MJ, Zimmer-Faust RK (1994) Odor plumes and how blue crabs use them to find prey. J Exp Biol 197:349–375

Weissburg MJ, Dusenbery DB, Ishida H, Janata J, Keller T, Roberts PJW, Webster DR (2002) A multidisciplinary study of spatial and temporal scales containing information in turbulent chemical plume tracking. J Environ Fluid Mech I2:65–94

Weissburg MJ, James CP, Webster DR (2003) Fluid mechanics produces conflicting constraints during olfactory navigation in blue crabs, *Callinectes sapidus*. J Exp Biol 206:171–180

Yen J, Weissburg MJ, Doall MH (1998) The fluid physics of signal perception by mate tracking copepods. Phil Trans Roy Soc B 353:787–804

Zimmer-Faust RK, Finelli CM, Pentcheff ND, Wethey DS (1995) Odor plumes and animal navigation in turbulent water flow. A field study. Biol Bull 188:111–116

Chapter 5
Hydrodynamics of Sniffing by Crustaceans

Mimi A.R. Koehl

Abstract Chemical signals are dispersed in aquatic environments by turbulent water currents. The first step in smelling these signals is the capture of odor molecules from the water around an organism. Olfactory antennules of crustaceans are used to study the physical process of odor capture because they are external organs protruding into the water where researchers can measure how they interact with their fluid environment. The antennules of lobsters, crabs, and stomatopods, which bear chemosensory hairs ("aesthetascs"), flick through the water. For any array of small hairs, there is a critical velocity range above which the array is "leaky" and fluid can flow between the hairs, and below which fluid barely moves through the spaces between the hairs. When antennules flick they move faster than the critical velocity and water flows into the spaces between aesthetascs. In contrast, during the return stroke the antennule moves more slowly than the critical velocity and the water sampled during the flick is trapped between the aesthetascs until the next flick. Odorant molecules in the water trapped between the aesthetascs during the return stroke and interflick pause diffuse to the surfaces of the aesthetascs, before the next flick traps a new parcel of water. Therefore, each antennule flick is a "sniff," taking a discrete sample of the odor plume in space and time.

5.1 Introduction

Many animals communicate via chemical signals (odors) released into the surrounding fluid (water or air). The first step in smelling chemical signals is the capture of odor molecules from the fluid around an organism. Therefore, to understand how organisms capture odors, scientists need to investigate how olfactory organs interact with the water or air around them. The olfactory antennules of

M.A.R. Koehl (✉)
Department of Integrative Biology, University of California at Berkeley,
Berkeley, CA 94720-3140, USA
e-mail: cnidaria@berkeley.edu

T. Breithaupt and M. Thiel (eds.), *Chemical Communication in Crustaceans*,
DOI 10.1007/978-0-387-77101-4_5, © Springer Science+Business Media, LLC 2011

crustaceans provide useful systems for studying the physical process of odor capture because they are external organs protruding into the water where researchers can see how they interact with their fluid environment.

The olfactory organs of malacostracan crustaceans (e.g., lobsters, shrimp, crabs, stomatopods) are the lateral branches (called "lateral filaments") of the antennules, which bear chemosensory hairs (called "aesthetascs") (Fig. 5.1a, c, e) (reviewed by

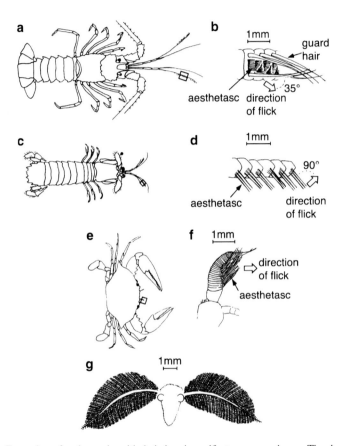

Fig. 5.1 Examples of arthropods with hair-bearing olfactory appendages. The *boxes* around antennules in (**a**, **c**, and **e**) indicate the region of the antennule diagrammed in (**b**, **d**, and **f**), respectively. (**a**) Spiny lobster, *Panulirus argus*. (**b**) Magnified view of a section of the lateral filament of a *P. argus* antennule. The lateral filament flicks downward, with the aesthetascs at an angle of ~35° to the direction of motion (Gleeson et al. 1993). (**c**) Stomatopod ("mantis shrimp"), *Squilla empusa*. (**d**) Magnified view of a section of the aesthetasc-bearing filament of a stomatopod antennule, *Gonodactylus mutatus*. The antennule flicks laterally, with the aesthetascs perpendicular to the direction of movement (Mead et al. 1999). (**e**) Blue crab, *Callinectes sapidus*. (**f**) Magnified view of the tip of the antennule of a *C. sapidus*. The antennule can flick in many directions, but the aesthetascs point in the direction of motion during a flick (Martinez, Lee, and Koehl, unpublished data). (**g**) Head of a male silkworm moth, *Bombyx mori,* showing the olfactory antennae. When the male fans his wings, air moves from front to back across the antennae (Loudon and Koehl 2000) (figure reprinted from Koehl 2001a, with kind permission of Springer Science+Business Media)

Ache 1982; Koehl 2006) (see Hallberg and Skog, Chap. 6). Although nonaesthetasc chemosensory hairs on antennules or legs of some species can also be involved in olfaction (reviewed in Koehl 2006), I will focus here on the aesthetasc-bearing lateral filaments of the antennules to explain the physics of odor capture. Odor molecules in the water around an animal must reach the surfaces of those aesthetascs to be sensed, so understanding the fluid mechanics of arrays of hairs is critical to deciphering how these crustaceans catch chemical signals.

The diversity of antennule morphology and deployment is intriguing. For example, the arrangement of aesthetascs on the lateral filaments differs between species, ranging from the complex arrays on lobster antennules (Fig. 5.1b) to the simple rows on stomatopod antennules (Fig. 5.1d) and the dense brushes on crab antennules (Fig. 5.1f). Do these differences in morphology affect odor capture? Furthermore, many malacostracans flick the lateral filaments of their antennules through the surrounding water (Fig. 5.2). How does flicking affect water motion around the aesthetascs, and thus odor capture?

My interest in the hydrodynamics of molecule capture by flicking crustacean antennules grew from our studies of the hydrodynamics of hair-bearing suspension-feeding appendages and the physical mechanisms they use to capture food particles from the surrounding water (reviewed in Koehl 1995). Those studies, which revealed how difficult it is to get water to flow between tiny hairs, sparked my curiosity about how the chemosensory hairs on insect antennae (Fig. 5.1g) (Loudon and Koehl 2000) and crustacean antennules (Fig. 5.1b, d, f) (Koehl 2001a) can capture molecules from the surrounding fluid. The basic physical rules we discovered about how arrays of hairs interact with fluids (Cheer and Koehl 1987a; Koehl 1992) predicted that the size and spacing of aesthetascs as well as the velocity of antennule flicking should make a big difference to the flow of odor-bearing water into arrays of these chemosensory hairs (Koehl 1996).

Another research path also led me to crustacean antennules. Years of field research in coastal marine habitats made me realize the importance of understanding the physical environment of an organism on the spatial and temporal scales

Fig. 5.2 Diagram of the spiny lobster, *Panulirus argus*, flicking the aesthetasc-bearing lateral filaments of its olfactory antennules. Drawing by Jorge A. Varela Ramos

experienced by that organism (which are not necessarily the scales at which we humans experience the environment). For example, the hydrodynamic forces that can rip a sea anemone off a wave-swept shore depend on the instantaneous water velocities and accelerations it encounters just a few centimeters above the sub-stratum as it sits among its neighbors, not on the much faster freestream water flow across the habitat (Koehl 1977). The waterborne chemical signals that benthic crustaceans encounter in their natural habitats are dispersed from odor sources by messy turbulent water currents. Early models of how animals search for the source of a chemical signal assumed that such turbulent odor plumes are diffuse clouds that become wider and more dilute with distance from the source (reviewed in Koehl 2006). My field experience studying flow microhabitats, however, led me to ask what patterns of odor concentration are actually inter-cepted by crustaceans navigating in marine habitats. To answer this question I would have to figure out what the *instantaneous* odor concentrations are *in the small slices of water sampled by the olfactory antennules* as they flick in natural environments.

5.2   Physical Mechanisms of Odor Capture

What are the physical mechanisms that olfactory antennules use to capture chemical signals from the surrounding water? Odor molecules diffuse in a fluid via Brownian motion. The time required for molecules to travel through a fluid by Brownian motion increases as the square of the distance (Vogel 1994), therefore molecular diffusion is only important in moving odors over very short distances (e.g., from the water surrounding an aesthetasc to the receptors). Turbulent water currents in the environment transport chemical signals from a source to a crustacean's olfactory antennule, while small-scale water motion near the aesthetascs carries signal-laden water close enough to the surfaces of these chemosensory hairs that odor molecules can diffuse to the olfactory receptors (Koehl 1996, 2001a, b, 2006). Thus, under-standing how water samples are moved into the spaces between aesthetascs is an important part of deciphering the process of capturing chemical signals from the environment.

5.2.1   Antennule Flicking

A number of researchers have suggested that when malacostracan crustaceans flick the lateral filaments of their antennules, they increase the penetration of ambient water into the spaces between aesthetascs, and thus bring odor-carrying water closer to the receptor cells in those chemosensory hairs (Snow 1973; Schmitt and Ache 1979; Atema 1985; Gleeson et al. 1993; Koehl 1995, 1996).

Early evidence for this idea was provided by Schmitt and Ache (1979), who found that the response to changes in odor concentration by olfactory receptor neurons in lobster antennules was enhanced if the antennule flicked. The idea that this enhanced response was due to improved water flow into the aesthetasc array was supported by Moore et al. (1991), who found that when they squirted water onto lobster antennules (to mimic flicking), the penetration into the aesthetasc array of tracer molecules carried in the water was increased. How does water flow through an aesthetasc array during a flick, and how does it depend on antennule morphology and motion?

5.2.2 Fluid Flow Through Arrays of Hairs

Fluid flow around a hair in an array depends on the relative importance of inertial and viscous forces, as represented by the Reynolds number (Re $= ul\rho/\mu$), where u is velocity, l is hair diameter, ρ is fluid density, and μ is the dynamic viscosity of the fluid (viscosity is the resistance of the fluid to being sheared; a fluid is sheared when neighboring layers of fluid move at different velocities). We humans are big (high l), rapidly moving (large u) organisms who experience high Re turbulent flow dominated by inertia. In contrast, very small (low l) structures such as aesthetascs operate at low Re, where fluid motion is smooth and laminar because viscous forces damp out disturbances to the flow.

Fluid in contact with the surface of a moving body does not slip relative to the body. Therefore, a velocity gradient develops in the fluid next to the body. The lower the Re (i.e., the slower or smaller the body, or the higher the viscosity of the fluid), the thicker this boundary layer of sheared fluid is relative to the size of the body. If the boundary layers around cylinders (i.e., hairs) in an array are thick relative to the gaps between neighboring cylinders, then fluid tends to move around rather than through the array. Using a mathematical model to calculate the velocities of fluid flow around and between cylinders in arrays (Cheer and Koehl 1987a, Cheer and Koehl 1987b; Koehl 1992, 1995, 1996; Koehl 2001a, b), we discovered that hair arrays undergo a transition between nonleaky behavior (where little fluid flows between adjacent hairs) and leaky, sieve-like behavior (where fluid flows between hairs) as Re is increased. Our model predicts that, for very closely-spaced hairs (like aesthetascs on many crustacean antennules), this transition occurs at Re's of about 1, where the leakiness of the hair array is very sensitive to velocity. We found that the sensillae on the antennae of male silkworm moths (*Bombyx mori*) (Fig. 5.1g) operate in this transitional Re range. Although *B. mori* rarely fly, males exposed to female sex pheromone fan their wings. Wing fanning that raises air speed past a walking male moth by 15-fold can increase the velocity through the antennae by a factor of 560 and pheromone interception rates by an order of magnitude (Loudon and Koehl 2000). Similarly, to understand odor capture by aesthetascs in arrays on antennules, their Re's when the antennules flick must be determined.

5.3   Hydrodynamics of Flicking Antennules

We made high-speed videos of flicking antennules of lobsters (Goldman and Koehl 2001), shrimp (Mead 1998), stomatopods (Mead et al. 1999), and crabs (Koehl 2001a). By digitizing the position of an antennule lateral filament in each video frame, we determined its velocity during flicks and return strokes. For all these animals the flick down stroke or outstroke was much faster than the return stroke, and the Re's of the aesthetascs were in the range where the leakiness transition should occur. Our models predicted that water should flow between the aesthetascs during the rapid flick, but not during the slower return stroke.

5.3.1   *Dynamically-Scaled Physical Models of Antennules Reveal When Fluid Flows into Arrays of Aesthetascs*

We tested these predictions using dynamically-scaled physical models of lobster (Reidenbach et al. 2008), crab (Waldrop, Reidenbach, and Koehl, unpublished data), and stomatopod (Mead and Koehl 2000) antennule lateral filaments. The relative magnitudes of flow velocities measured at different positions in the fluid around a dynamically-scaled model are the same as the relative magnitudes of flow velocities measured at comparable positions around a real antennule (e.g., Koehl 2003). Therefore, we can use dynamically-scaled models to work out the detailed flow velocity maps around and through arrays of aesthetascs (water velocities that would be very difficult to measure around such tiny chemosensory hairs on real flicking antennules). Dynamically-scaled models are geometrically-similar to real antennules and operate at the same Re's. We used large (higher l) models, but operated them at the same Re's as flicking antennules by moving them more slowly (lower u) through mineral oil, a fluid that is more viscous (higher μ) than water.

We used a technique called "particle image velocimetry" (PIV) to determine the flow velocity maps around and through aesthetasc arrays. We marked the oil with neutrally-buoyant particles, and visualized one plane of fluid at a time by illuminating it with a thin sheet of laser light. By moving a video camera at the same speed as a towed antennule model, we could record fluid motion relative to the aesthetascs. Analyzing these videos, we calculated the water velocity vector fields relative to real antennules during their rapid flick and slower return stroke. Such PIV studies of dynamically-scaled physical models revealed that water does flow through the aesthetasc array during the rapid flick down stroke, but not during the slower return stroke for spiny lobsters (Koehl 2001a, b; Reidenbach et al. 2008), stomatopods (Mead and Koehl 2000; Mead and Caldwell, Chap. 11), and crabs (Waldrop, Reidenbach and Koehl, unpublished data).

5.3.2 Water Flow Through Aesthetasc Arrays of Lobsters and Crabs

The water velocity profile around the lateral filament of the olfactory antennule of the spiny lobster, *Panulirus argus*, is very different during the flick downstroke than during the return stroke (Fig. 5.3). Although the velocity of the flick is about four times the speed of the return stroke, the water velocity between the aesthetascs during the flick (Fig. 5.3, middle panel), is roughly twenty five times faster than during the return (Fig. 5.3, right panel) because the flick Re is above the Re of the leakiness transition and the return Re is below it (Reidenbach et al. 2008). We found that the flick down stroke lasts just long enough to allow complete replacement of the water in the spaces between the aesthetascs. In contrast, the water between the aesthetascs during the return stroke is essentially trapped there. By working with physical models, we could manipulate the morphology and orientation of an

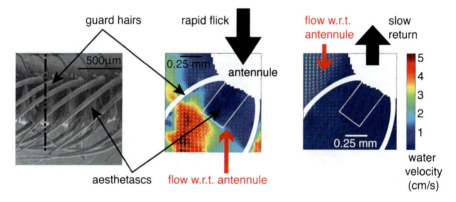

Fig. 5.3 Lateral filament of the antennule of the spiny lobster, *Panulirus argus*. The photograph on the left is a scanning electron micrograph (SEM) of a portion of the lateral filament, showing the chemosensory hairs (aesthetascs) and guard hairs attached to the stalk of the filament (segments of which are visible at the top of the photograph). The *dashed line* shows where a cross-section is taken through the lateral filament, and diagrams of that cross-section are shown in the *middle* and *left panels* of this figure. These diagrams are maps of water velocities *relative to* the lateral filament when it rapidly flicks downward (*middle panel*) and slowly returns upward (*right panel*). The stalk of the lateral filament is labeled "antennule" and the direction of its motion is indicated by the *large black arrow*. The *large red arrow* indicates the direction of water motion relative to the antennule lateral filament. The *white box* outlines the region occupied by the array of aesthetascs, and the position of the guard hairs is labeled. The scale to the right of the diagrams indicates water velocity: *red areas with yellow velocity vectors* show the fastest flow *relative to* the antennule, yellow/green areas with shorter velocity vectors indicate less rapid flow, and *blue areas without velocity vectors* are regions of the slowest water movement. The downstroke is about 4 times faster than the return stroke, but the velocity of the water between the aesthetascs is approximately 25 times faster during the downstroke than during the return stroke (Reidenbach et al. 2008) (SEM by J. Goldman; water velocity maps calculated from PIV measurements around dynamically-scaled physical models of the antennules; modified after Reidenbach et al. 2008)

antennule to explore how such features affect water flow near aesthetascs. These experiments revealed that the complex zigzag arrangement of aesthetascs on the antennules of spiny lobsters (Fig. 5.3, left panel) and their orientation relative to the flicking direction produce uniform flow velocities along the length of the aesthetascs when the antennule flicks.

While a lobster antennule is long and bears a complex array of aesthetascs and nonchemosensory guard hairs, a crab antennule is short and bears a dense cluster of aesthetascs, like a toothbrush (Fig. 5.4, left panel). The aesthetascs on the antennule of a blue crab, *Callinectes sapidus*, are flexible. Ferner and Gaylord (2008) found that if cylinders in a row are flexible and experience fluid flow at right angles to their length at very low Re's (10^{-5}–10^{-3}), then increasing their speed reduces their already low leakiness as the hairs are bent over and moved closer together. What happens to the leakiness of the dense brush of flexible aesthetascs operating at Re's near 1 on a flicking crab antennule? During the rapid flick down stroke when the aesthetascs are on the upstream side of the antennule, they splay apart such that the gaps between neighboring aesthetascs become wider, while during the slower return stroke when the aesthetascs are on the downstream side of the antennule, they are pushed together and the gaps between the hairs become narrower (Koehl 2001a). At Re's near 1 the leakiness of a hair array is very sensitive not only to flow velocity, but also to gap width (Cheer and Koehl 1987a). PIV experiments with dynamically-scaled physical models of *C. sapidus* antennules (Waldrop, Reidenbach and Koehl, unpublished data) showed that water flows through the gaps between aesthetascs during the flick, but not during the return stroke (Fig. 5.4). Because we conducted our experiments with physical models, we could vary the hair spacing, antennule orientation, and antennule speed independently to measure the effects of each. We found that both hair splaying and rapid motion during a flick contribute to the increase in leakiness of the crab aesthetasc array, while both hair clumping and slower motion during the return stroke contribute to the decrease in leakiness of the aesthetasc tuft. As we saw for the lobster, the duration and speed of a crab flick are large enough that much of the water in the aesthetasc array is flushed out and replaced by a new sample of water during the flick (Fig. 5.4, bottom row).

5.3.3   Sniffing

Because these diverse crustaceans flick their antennules in the Re range at which the leakiness of their hair arrays is sensitive to speed, they are able to take fluid samples into their aesthetasc arrays during the rapid down stroke of a flick when the aesthetasc array is leaky. They then retain that captured water within the hair array during the slower return stroke and subsequent stationary pause of the antennule when the aesthetasc array is not leaky. During the next rapid flick down stroke, that water sample is flushed away and replaced by a new one. Therefore, antennule flicking permits these animals to take discrete samples in space and time of their odor environment. In other words, a flick is a sniff (reviewed in Koehl 2006).

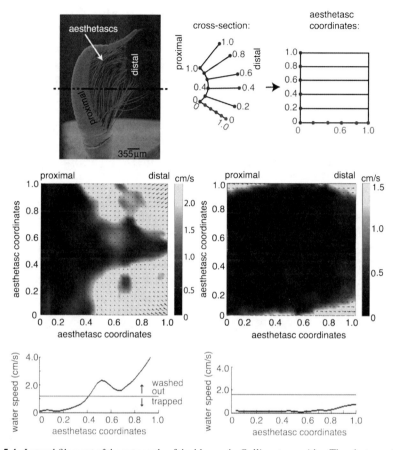

Fig. 5.4 Lateral filament of the antennule of the blue crab, *Callinectes sapidus*. The photograph on the *top left* is a SEM of the lateral filament, showing the chemosensory hairs (aesthetascs) attached to the stalk of the filament. The proximal and distal ends of the aesthetascs are labeled. The *dashed line* shows where a cross-section is taken through the lateral filament, and a diagram of that cross-section is shown just to the *left* of the SEM. Since the aesthetascs splay apart during the rapid flick downstroke and collapse together during the slower return stroke, we have transformed the actual coordinates of the aesthetascs into a rectilinear grid (the "aesthetasc coordinates" shown on the *left* of the *top row*) so that the comparison of the flow between them during the down and return strokes is easier to see. The diagrams in the *middle row* are maps of water velocities *relative to* the lateral filament when it rapidly flicks (*left diagram*) and slowly returns (*right diagram*). These velocity maps are plotted on "aesthetasc coordinates". In the flick downstroke diagram, the antennule is shown moving from left to right, so the water flow *relative to* the antennule is right to left. In the return stroke diagram, the antennule is moving right to left, so the flow relative to the antennule is left to right. The scale to the right of these flow maps indicates water velocity: areas of *white* and *pale gray* show the fastest flow *relative to* the antennule, while the *darkest areas* are regions of the slowest water movement. The graphs at the bottom of the figure show the water velocity relative to the aesthetascs at different positions along the length of the aesthetascs for the flick down stroke (*left*) and return stroke (*right*). The *line* across each graph indicates the water speed necessary for the water in the middle of the aesthetasc array to be washed out of the hair array during the stroke. During the down stroke, most of the water between the aesthetascs is flushed out of the array, whereas during the return stroke, the water is trapped between the aesthetascs (SEM by M. Martinez; water velocity maps calculated from PIV measurements around dynamically-scaled physical models of the antennules by L. Waldrop, M. Reidenbach, and M. Koehl)

5.4   How Odor Plumes are Sampled by Flicking Antennules

When the lateral filament of an antennule flicks, it samples a small slice of the water
in a crustacean's environment. What are the patterns of odor concentrations in the
water samples captured by flicking antennules as the animals move through habitats
exposed to ambient water motion?

5.4.1   Odor Concentrations in Turbulent Odor Plumes

Water currents in aquatic habitats are turbulent. As a chemical signal from an odor
source is carried across the environment by a turbulent current, the signal-bearing
water is also stirred into the surrounding odor-free water by swirling eddies.
Although early models of how animals search for the source of a chemical signal
assumed that such odor plumes are diffuse clouds that become wider and more
dilute with distance from the source (reviewed in Koehl 2006), scientists are now
able to map the fine-scale spatial distribution of odor concentrations in turbulent
moving water using a technique called "planar laser-induced fluorescence" (PLIF).
If a chemical cue is labeled with a fluorescent dye and allowed to ooze from a
source in a flume (a long tank in which water flows), investigators can see how that
dye is dispersed by turbulent water currents or waves by illuminating a slice of the
odor plume with a sheet of laser light. The laser light makes the dye glow, and the
brightness of the dye is proportional to the odor concentration.

Videos of PLIF experiments have revealed that turbulent odor plumes are quite
complex and beautiful, and that they are full of holes (i.e., strips of odor-free fluid)
(Fig. 5.5). When a turbulent water current flows past an odor source, the water next
to the source that contains a high concentration of odor is sheared into filaments.
These odor filaments are stretched and rolled up with layers of odor-free water by
swirling eddies of various sizes, producing a spatially complex and temporally
varying distribution of signal concentrations in an odor plume that becomes wider
as it is carried away from the source and meanders across the habitat (reviewed by
Weissburg 2000; Moore and Crimaldi 2004; Koehl 2006; Weissburg, Chap. 4).
Many shallow coastal marine habitats are subjected to the back-and-forth water flow
of waves as well as water currents. Flume PLIF experiments in which waves were
superimposed on a water current showed that odor filaments tend to be wider and to
be carried to greater heights above the substratum than in the unidirectional current
without waves, and that animals navigating near an odor source in wavy flow
encounter odor filaments more often than in unidirectional flow (Mead et al. 2003).

In both waves and unidirectional currents, the spatial distribution of odor filaments and odor-free water in a turbulent plume changes with distance from the
source of the chemical signal (details reviewed in Koehl 2006; see also Weissburg,
Chap. 4). For example, in a plume near the odor source the concentration gradients
at the edges of odor filaments are steeper, the concentrations are generally higher,

Fig. 5.5 Frames of videos of a robotic spiny lobster, *Panulirus argus*, flicking a real antennule lateral filament at a position 1 m downstream from an "odor" (i.e., fluorescent dye) source in a flume in which a turbulent water current of 0.10 m/s was flowing. The *top image* shows how a sheet of laser light reflected off a mirror on the floor of the flume illuminates the water both above and below a flicking antennule. The light swirls in the water are filaments of dye, and their brightness is proportional to concentration. The *middle* and *lower images* show close-up views of a lateral filament during the rapid flick downstroke. In the *middle image*, filaments of dye can be seen flowing into the aesthetasc array. In contrast, in the lower image, the antennule encounters an odor-free "hole" during the down stroke (frames of videos taken by M. Koehl and J. Koseff)

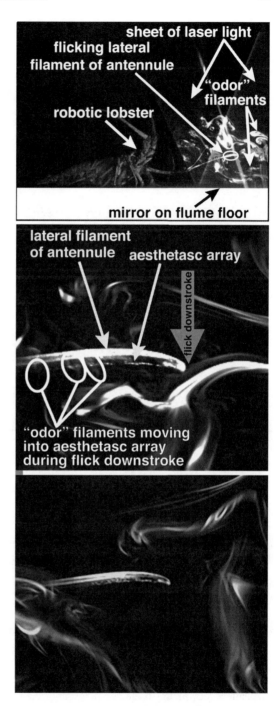

the odor filaments and the gaps between them tend to be narrower, and the variation in concentration between filaments is greater than they are in that plume at a greater distance downstream from the odor source. Furthermore, the frequency of encounters with odor filaments at the edge of an odor plume is lower than along its midline, although the odor concentrations can be similar. Therefore, the fine-scale patterns of odor concentration in the water contain information about position relative to the source of that odor. Can a flicking antennule capture these fine-scale aspects of odor plume structure?

5.4.2 Patterns of Odor Concentrations Captured by Flicking Antennules

A flicking antennule samples only the thin slice of water through which it sweeps. If that slice of water can be illuminated by a sheet of laser light, PLIF can be used to measure the pattern of odor concentrations in the water in the aesthetasc array of the lateral filament of a crustacean antennule. A challenge to this approach is getting a crustacean to flick its antennule in the plane of laser light. We overcame this challenge by using a robotic lobster to flick real antennule lateral filaments in a sheet of laser light shining through a turbulent dye plume in a flume (Koehl et al. 2001) (Fig. 5.5, top). We used fresh antennules from spiny lobsters, *P. argus*, and the robot flicked them using the kinematics we had measured for antennules of that species (Goldman and Koehl 2001). We made high-speed videos of the robot-flicked antennules, and in each video frame we measured the brightness of dye within the aesthetasc array to determine how those dye (i.e., odor) concentrations changed over time.

Our PLIF experiments using the robotic lobster revealed a number of surprises. For example, since an odor plume is full of aroma-free holes, the flicking antennule of a lobster standing in the middle of a plume sometimes encounters filaments of chemical signal (Fig. 5.5, middle), and sometimes it does not (Fig. 5.5, bottom). Furthermore, when a flicking antennule does run into odor filaments, water and the fine filaments of dye (i.e., odor) it carries flow through the spaces between aesthetascs during the rapid down stroke without being stirred up (Fig. 5.5, middle). Then the spatial pattern of odor concentration peaks and valleys that happen to be in the aesthetasc array at the very end of the down stroke are trapped there during the slow return stroke and the stationary pause before the next flick. The odor filaments in the plume around the antennule flow past the lateral filament in the ambient current, but the stripes of chemical signal and of odor-free water within the aesthetasc array stay in place until the next flick, when the old water sample is flushed away and a new sample is trapped between the aesthetascs (Koehl et al. 2001; Koehl 2006). PLIF measurements of the odor samples captured by flicking stomatopod antennules yielded similar results (Mead et al. 2003; Caldwell and Mead, this volume). These experiments indicate that each time the lateral filament of an antennule flicks, it captures a snapshot of the fine-scale odor concentration patterns in a small slice of the odor plume.

5.5 Flux of Odorant Molecules to Aesthetasc Surfaces

Chemical signals in the water trapped in an aesthetasc array disperse across the small distances between these chemosensory hairs via molecular diffusion. Calculation of the diffusion of odorant molecules carried in the water between aesthetascs on spiny lobster antennules during the slow return stroke (Fig. 5.3, right panel) and interflick pause indicate that the duration of the return stroke and the pause before the next flick is long enough for odor molecules in that water sample to diffuse to aesthetasc surfaces (Reidenbach et al. 2008). Similarly, a mathematical model of the advection (i.e., transport by moving water) and diffusion (i.e., Brownian motion) of odorant molecules in aroma filaments encountered by flicking stomatopod antennules showed that the flux (number arriving per area per time) to aesthetasc surfaces of molecules in odor filaments that have been carried into the spaces between the aesthetascs is high (Stacey et al. 2002). In contrast, the model showed that the flux of signal molecules from odor filaments carried past antennules in the ambient current during the slow return stroke (when water does not flow between the aesthetascs) is very low. These calculations, in combination with the PLIF measurements described above, indicate that a sample of the odor plume is captured within the aesthetasc array during the flick down stroke, and the odorant molecules in that trapped sample diffuse to the chemosensory aesthetascs before that sample is shed and the next sample taken by the subsequent flick.

5.6 Effects of Ambient Flow, Locomotion, and Size on Odor Sampling

Since the leakiness of an array of chemosensory hairs depends on Re, odor sampling by antennules can be affected by the fluid velocity (u) relative to the antennule and by the diameter of the aesthetascs (l). Therefore, ambient water flow and animal locomotion (both affecting u), as well as body size (affecting l) can influence odor capture by antennules.

The ways in which crustaceans deploy their antennules in ambient currents can affect water flow through the aesthetasc arrays. In our flume experiments with lobsters (Koehl et al. 2001), an ambient water current of 10 cm/s did not force water and odor filaments into the aesthetasc arrays on antennules held parallel to the flow direction, whereas water and odor samples did move into the arrays during flick down strokes of 6 cm/s (during a down stroke where the water flow relative to the antennule is perpendicular to the long axis of the antennule). This suggests that if crustaceans hold their antennules perpendicular to ambient currents with the aesthetascs facing upstream, water should penetrate into the hair array if the ambient flow is fast enough. Are there ambient current velocities above which crustaceans cease antennule flicking because the water motion in the environment drives fluid through their aesthetasc arrays? If so, can the animals track an odor plume to its source as well as they do when they can sniff (i.e., take a discrete odor sample in space and time with each antennule flick)?

Water also moves relative to the antennules of crustaceans when they run (e.g., crabs, 11 cm/s, Martinez et al. 1998) or swim (e.g., mysids, 10–18 cm/s, Cowles and Childress 1988; amphipods 4–14 cm/s, Sainte-Marie 1986; isopods, 8–30 cm/s, Alexander and Chen 1990) through the water. What are the orientations of the antennules when crustaceans locomote, and how does the water movement relative to them affect the leakiness of their arrays of aesthetascs?

Since the leakiness of a hair array depends on size (1), an intriguing aspect of odor capture by crustacean antennules is the ontogeny of sniffing. Malacostracan crustaceans grow from microscopic larvae into large adults. How do antennule morphology and kinematics change during the ontogeny of a crustacean as it changes size, and how does that affect how they take odor samples from the surrounding water? Comparison of different sizes of stomatopods (Mead et al. 1999) and lobsters (Goldman and Koehl 2001) revealed that small animals have larger aesthetascs relative to body size than do big animals, and move rapidly enough during the flick downstroke that they can sniff. Future research should extend these studies to smaller sizes to explore how the morphology and kinematics of the antennules of microscopic larvae and tiny juveniles affect how they sample their odor environment. Is there a lower limit to antennule size for sniffing to be possible?

Our predictions about how Re affects flow through arrays of chemosensory hairs, and thus odor capture, suggest other comparative studies. For example, how do the fluid dynamics of odor sampling by the olfactory organs of the larvae and juveniles mentioned above compare with those of other small crustaceans, such as the planktonic copepods that follow scent trails to find mates (e.g., Weissburg et al. 1998), or the deep sea amphipods that use odors to locate carrion (e.g., Premke et al. 2003)? How do their swimming behaviors or feeding currents affect flow across their olfactory organs, and do they flick? Antennule "sweeps" have been reported from lysianassid amphipods and it was suggested that sweeps facilitate water exchange around the aesthetasc-bearing callynophores (Kaufmann 1994).

5.7 Crustaceans as Model Systems to Study Odor Capture

There are a number of advantages of using crustaceans as systems to study the physical process of odor capture. Antennules are olfactory organs that protrude into the environment, so their interactions with the surrounding odor-bearing fluid are much easier to study than are the fluid mechanics of chemosensory surfaces hidden within internal nasal passages, such as those of vertebrates. Furthermore, the diversity of antennule morphologies shown by different species of crustaceans enables us to explore the functional consequences of different "designs" of external olfactory organs. Many other types of invertebrate animals have external chemosensory organs that bear hair-like sensillae (reviewed in Koehl 2001a, 2006), so the principles about odor capture learned by studying crustacean antennules can be useful for understanding the function of these other "noses" as well. Another advantage of the species of crustaceans that we have been using as study organisms (e.g., lobsters, crabs, stomatopods) is that they live in accessible shallow marine habitats where we can measure the hydrodynamic conditions that they experience in their natural habitats and that disperse the chemical signals they encounter. Such information enables us to design biologically relevant laboratory studies of antennule function.

Since crustaceans are used to study the neurobiology of olfaction and the behavioral uses of chemical signaling, they provide a system with the potential of enabling us to relate the biophysics of odor capture to how animals process that information, and to how they react to the spatio-temporal patterns of odor information they capture.

5.8 Future Directions

Several important questions about crustacean odor capture that remain unanswered provide promising directions for future research. One such question is whether crustaceans "use" the fine-scale spatio-temporal information their antennules are physically able to capture when they flick in turbulent odor plumes. This question can be addressed at the neurobiological level (Do fine-scale patterns of odor concentrations captured by the antennules affect patterns of neuron firing in the olfactory lobe?) and at the behavioral level (Do fine-scale patterns of odor concentrations captured by the antennules affect plume-searching behavior?). For example, to determine whether different spatio-temporal odor-concentration patterns alter neuron firing, the standard olfactometers (e.g., Y-tube flow set-ups; see also Fig. 10.2 in Thiel, Chap. 10) used to deliver odors to neurobiological preparations could be replaced by odor-delivery systems that mimic the realistic spatial and temporal patterns of odor delivery that antennules experience when flicking at different positions in an odor plume. Similarly, video records of the movements of crustaceans searching in odor plumes visualized by PLIF (Fig. 5.5) enable us to correlate the fine-scale patterns of concentrations of signal captured by right and left antennules on each flick with the subsequent behaviors of the animals (Mead et al. 2003; Mead and Caldwell, Chap. 11). Such studies of *what antennules actually sample* as animals navigate in odor plumes should also enable us to work out search algorithms that were not possible to recognize when whole plumes were visualized and flicking was ignored.

While the olfactory antennules of large malacostracan crustaceans have been used as research systems to elucidate mechanisms by which chemical signals in the environment are sampled, less is known about the hydrodynamics of the chemosensory hairs on other parts of crustacean bodies. The approaches used and the principles elucidated by studying flow through arrays of aesthetascs on antennules can also shed light on how the morphology of other types of chemosensory hairs, as well as their arrangement in arrays and their positions on the body affect their odor-capturing performance. Another important avenue for future neurobiological and behavioral research is to explore how information from chemosensors on the body and legs is coupled with information sampled by the antennules to inform an animal's behavior.

Comparisons of the morphology and kinematics of aquatic olfactory organs with those that operate in air (such as the antennules of terrestrial hermit crabs or the antennae of insects) would provide an interesting test of our ideas about mechanisms of sniffing. The density (ρ) of water is about 1,000 times higher than that of air, but the viscosity (μ) is only about 56 times greater (Vogel 1994), so a structure of a given size must move about 18 times faster in air than in water to achieve the same Re. Therefore, I would expect the leakiness transition for an array of chemosensory hairs of a given size to occur at higher speeds in air. A more striking difference between water and air, however, is that the diffusivity (D, a molecule's propensity to diffuse in a particular fluid) of molecules in air is about 10,000 times greater than in water (Vogel 1994). Because odors can move a given distance via molecular diffusion through air much more rapidly than through water, the filaments in odor plumes are wider (reviewed in Koehl 2006), and the time required for odor molecules to diffuse from sampled air to the surfaces of chemosensory sensillae is much shorter in air than in water. Analysis of the aerodynamics of odor capture by the antennules of intertidal or terrestrial crustaceans (e.g., brachyuran or hermit crabs) could be done using approaches similar to those described in this chapter to determine how antennule morphology and kinematics determine if and how they sniff.

5.9 Summary and Conclusions

Chemical signals are dispersed in aquatic environments by turbulent water currents. The lateral filaments of the antennules of malacostracan crustaceans, which bear arrays of chemosensory hairs (aesthetascs), are an important research system for studying the physics of how olfactory organs bearing hair-like olfactory sensilla capture such chemical signals from the environment. On the scale of an antennule an odor plume is not a diffuse cloud, but rather is a series of fine filaments of aroma-bearing water swirling in scent-free water; the spatio-temporal patterns of these filaments depend on distance from the odor source. When a lobster, stomatopod, or crab flicks the lateral filament of its antennule, water and any odor filaments it is carrying flow through the spaces between the aesthetascs during the rapid down stroke, but not during the slower return stroke or during the stationary pause before the next flick. For any array of small hairs, there is a critical velocity range above which the array is "leaky" and fluid can flow between the hairs, and below which fluid barely moves through the spaces between the hairs. The striking difference in flow through aesthetasc arrays during the down stroke vs. the upstroke occurs because the down stroke velocities are above and the return stroke velocities are below the speeds at which this transition in leakiness occurs. Odorant molecules in the water trapped between the aesthetascs during the return stroke and pause diffuse to the surfaces of the aesthetascs, before the next flick traps a new parcel of water. Therefore, each antennule flick is a "sniff," taking a discrete sample of the odor plume in space and time. Our work has shown that the size and arrangement of chemosensory hairs on olfactory organs like antennules or antennae, as well as their velocity relative to the surrounding fluid, affect the temporal patterns of odor delivery to the chemosensory hairs. Thus, the physics of odor capture provides the first step in filtering olfactory information from the environment.

Acknowledgments My research reported here was supported by grants from the James S. McDonnell Foundation and from the Office of Naval Research (USA), a John D. and Catherine T. MacArthur Foundation Fellowship, and the Virginia G. and Robert E. Gill Chair (University of California, Berkeley). I thank the coauthors on my papers cited in this chapter for the discussions and collaborations that led to the ideas presented here.

References

Ache BW (1982) Chemoreception and thermoreception. In: Atwood HL, Sandeman DC (eds) The biology of the crustacea. Academic Press, New York, pp 369–393

Alexander DE, Chen T (1990) Comparison of swimming speed and hydrodynamic drag in two species of *Idotea* (Isopoda). J Crust Biol 10:406–412

Atema J (1985) Chemoreception in the sea: Adaptations of chemoreceptors and behavior to aquatic stimulus conditions. Soc Exp Biol Symp 39:387–423

Cheer AYL, Koehl MAR (1987a) Fluid flow through filtering appendages of insects. I.M.A. J Math Appl Med Biol 4:185–199

Cheer AYL, Koehl MAR (1987b) Paddles and rakes: fluid flow through bristled appendages of small organisms. J Theor Biol 129:17–39

Cowles DL, Childress JJ (1988) Swimming speed and oxygen consumption in the bathypelagic mysid *Gnathophausia ingens*. Biol Bull 175:111–121

Ferner MC, Gaylord B (2008) Flexibility foils filter function: structural limitations on suspension feeding. J Exp Biol 211:3563–3572

Gleeson RA, Carr WES, Trapido-Rosenthal HG (1993) Morphological characteristics facilitating stimulus access and removal in the olfactory organ of the spiny lobster, *Panulirus argus*: insight from the design. Chem Senses 18:67–75

Goldman JA, Koehl MAR (2001) Fluid dynamic design of lobster olfactory organs: High-speed kinematic analysis of antennule flicking by *Panulirus argus*. Chem Senses 26:385–398

Kaufmann RS (1994) Structure and function of chemoreceptors in scavenging lysianassoid amphipods. J Crust Biol 14:54–71

Koehl MAR (1977) Effects of sea anemones on the flow forces they encounter. J Exp Biol 69:87–105

Koehl MAR (1992) Hairy little legs: feeding, smelling, and swimming at low Reynolds number. Fluid dynamics in biology. Contemp Math 141:33–64

Koehl MAR (1995) Fluid flow through hair-bearing appendages: feeding, smelling, and swimming at low and intermediate Reynolds number. In: Ellington CP, Pedley TJ (eds) Biological fluid dynamics. Soc Exp Biol Symp 49, pp 157–182

Koehl MAR (1996) Small-Scale fluid dynamics of olfactory antennae. Mar Fresh Behav Physiol 27:127–141

Koehl MAR (2001a) Fluid dynamics of animal appendages that capture molecules: arthropod olfactory antennae. In: Fauci L, Gueron S (eds) Computational modeling in biological fluid dynamics. IMA Series # 124, pp 97–116

Koehl MAR (2001b) Transitions in function at low Reynolds number: hair-bearing animal appendages. Math Methods Appl Sci 24:1523–1532

Koehl MAR (2003) Physical modelling in biomechanics. Phil Trans Roy Soc B 358:1589–1596

Koehl MAR (2006) The fluid mechanics of arthropod sniffing in turbulent odor plumes. Chem Senses 31:93–105

Koehl MAR, Koseff JR, Crimaldi JP, McCay MG, Cooper T, Wiley MB, Moore PA (2001) Lobster sniffing: antennule design and hydrodynamic filtering of information in an odor plume. Science 294:1948–1951

Loudon C, Koehl MAR (2000) Sniffing by a silkworm moth: wing fanning enhances air penetration through and pheromone interception by antennae. J Exp Biol 203:2977–2990

Martinez MM, Full JR, Koehl MAR (1998) Underwater punting by an intertidal crab: A novel gait revealed by the kinematics of pedestrian locomotion in air vs. water. J Exp Biol 201:2609–2623

Mead KS (1998) The biomechanics of odorant access to aesthetascs in the Grass Shrimp, *Palaemonetes vulgaris*. Biol Bull 195:184–185

Mead KS, Koehl MAR (2000) Stomatopod antennule design: The asymmetry, sampling efficiency, and ontogeny of olfactory flicking. J Exp Biol 203:3795–3808

Mead KS, Koehl MAR, O'Donnell MJ (1999) Stomatopod sniffing: the scaling of chemosensory sensillae and flicking behavior with body size. J Exp Mar Biol Ecol 241:235–261

Mead KS, Wiley MB, Koehl MAR, Koseff JR (2003) Fine-scale patterns of odor encounter by the antennules of mantis shrimp tracking turbulent plumes in wave-affected and unidirectional flow. J Exp Biol 206:181–193

Moore PA, Atema J, Gerhardt GA (1991) Fluid dynamics and microscale chemical movement in the chemosensory appendages of the lobster, *Homarus americanus*. Chem Senses 16:663–674

Moore P, Crimaldi J (2004) Odor landscapes and animal behavior: tracking odor plumes in different physical worlds. J Mar Syst 49:55–64

Premke K, Muyakshin S, Klages M, Wegner J (2003) Evidence for long-range chemoreceptive tracking of food odour in deep-sea scavengers by scanning sonar data. J Exp Mar Biol Ecol 285:283–294

Reidenbach MA, George NT, Koehl MAR (2008) Antennule morphology and flicking kinematics facilitate odor sampling by the spiny lobster, *Panulirus argus*. J Exp Biol 211:2849–2858

Sainte-Marie B (1986) Feeding and swimming of lysianassid amphipods in a shallow cold-water bay. Mar Biol 91:219–229

Schmitt BC, Ache BW (1979) Olfaction: responses of a decapod crustacean are enhanced by flicking. Science 205:204–206

Snow PJ (1973) The antennular activities of the hermit crab, *Pagurus alaskensis* (Benedict). J Exp Biol 58:745–765

Stacey M, Mead KS, Koehl MAR (2002) Molecule capture by olfactory antennules: mantis shrimp. J Math Biol 44:1–30

Vogel S (1994) Life in moving fluids. Princeton University Press, Princeton

Weissburg MJ (2000) The fluid dynamical context of chemosensory behavior. Biol Bull 198:188–202

Weissburg MJ, Doall MH, Yen J (1998) Following the invisible trail: mechanisms of chemosensory mate tracking by the copepod *Temora*. Phil Trans Roy Soc B 353:701–712

Chapter 6
Chemosensory Sensilla in Crustaceans

Eric Hallberg and Malin Skog

Abstract Crustaceans, like most other animals, have two types of chemosensory organs. In crustaceans these organs consist of sensilla that differ structurally as well as functionally. Unimodal olfactory sensilla are usually considered as more long-range and bimodal chemo- and mechanosensory sensilla as short-range or contact chemosensory sensilla. All chemosensory sensilla are characterized by the presence of ciliated bipolar sensory cells. The bimodal sensilla are unevenly distributed over the entire body, with dense arrays on the mouthparts and walking legs. These sensilla contain both chemosensory and mechanosensory cells. The chemosensory cells contain one transformed cilium (dendritic outer segment) each. Bimodal sensilla have a relatively densely arranged cuticle and also feature an apical pore. The most common unimodal olfactory sensilla in crustaceans are the aesthetascs, which vary in structure, number, and distribution. Another type of unimodal olfactory organ is the male-specific sensilla, found basally on the first antenna in some isopods, mysids, and amphipods. Excluding these groups, sexual dimorphism of the olfactory organs is not as pronounced in crustaceans as in some insects. The chemosensory cells of both aesthetascs and male-specific sensilla have two transformed cilia each, thus differing from gustatory chemosensory cells which have only one. Further, the cuticle of aesthetascs and male-specific sensilla is "spongy," possibly functioning as a molecular sieve. Crustaceans molt in order to grow, and there are two known patterns of molting of the chemosensory sensilla: the first retains the chemosensory ability through maintained contact between the old and the new sensillum (the "mysid-type" of molt) and the second where there is no such contact and thus possibly no chemosensory ability (the "isopod-type"). Previous studies have mostly described the general morphology of the chemosensory sensilla, which is well-known in crustaceans. Future studies should focus on the function of these sensilla in an ecological and behavioral context.

E. Hallberg (✉)
Department of Biology, Lund University Zoologihuset, HS 17 Sölvegatan 35,
SE-22362, Lund, Sweden
e-mail: Eric.hallberg@cob.lu.se

T. Breithaupt and M. Thiel (eds.), *Chemical Communication in Crustaceans*,
DOI 10.1007/978-0-387-77101-4_6, © Springer Science+Business Media, LLC 2011

6.1   Introduction

Crustaceans are fascinating creatures, found in habitats ranging from hydrothermal vents at the ocean seafloors via polar waters, freshwater lakes, and hypersaline ponds to arid deserts. In order to ensure proper functionality, the structure of chemosensory sensilla needs to adapt to the specific environmental conditions. For example, the chemosensory organs of land-living coconut crabs look different from the chemoreceptors of oceanic krill. The investigation of the morphology of chemosensory organs contributes to a better understanding of the physiology, behavior, and ecology of crustaceans.

The study of chemosensory sensilla morphology may provide answers to questions about how the chemosensory organs function and why they function as they do. The main questions in this research field are: What do crustacean chemosensory sensilla look like (exterior and interior)? Why are they similar or dissimilar? How can the morphology be correlated to function? What are the functional modalities of these sensory organs?

The idea that chemosensory organs could be found on the first antenna (more specifically on the lateral antennular branch) in decapods was introduced by Leydig in 1860 (cited in Balss 1944) and the histology of the lateral branch was later investigated in different decapods by Doflein (1910) and Marcus (1911). The introduction of the electron microscope and related techniques resulted in an increased knowledge of the fine structure of the different sensory organs (Laverack 1968, 1988). Complemented by physiological and behavioral approaches, the morphological study of chemoreceptors contributes greatly to an understanding of the mechanisms and adaptations of the sensory biology of crustaceans.

Arthropod sensory systems has always been Eric Hallberg's (EH) greatest research interest, ranging from mysid compound eyes via pheromone glands in pest insects (Lepidoptera) to the fine structure of chemosensory sensilla and olfactory lobes in a number of crustacean taxa (Johansson and Hallberg 1992; Johansson et al. 1996; Hallberg et al. 1997). Hallberg et al. (1992) defined "the aesthetasc concept" – common structural characteristics of olfactory sensilla that can be used throughout the Crustacea. EH's interest in morphological and behavioral aspects of pheromone communication in crustaceans (Ekerholm and Hallberg 2005) was shared by Malin Skog (MS), who joined the project as a doctoral student in 2003, and studied communication during aggressive and reproductive behaviors in the European lobster (*Homarus gammarus*) (Skog, 2009a, 2009b; Skog et al. 2009) as well as the morphology, potential sex differences, and molting of the aesthetascs in *H. gammarus* and the European shore crab *Carcinus maenas* (MS and EH, unpublished data).

6.2   The Chemical Senses in Crustacea

All arthropods have an integument characterized by a tough cuticle, impeding diffusion of chemical substances from the environment. The chemosensory organs have evolved special morphological solutions, sensilla, to provide direct access to

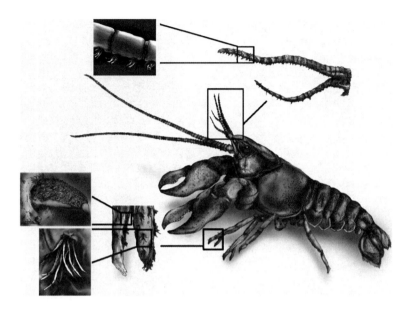

Fig. 6.1 Artist's drawing of a crayfish showing some locations for important chemosensory sensilla. *Above the crayfish:* the antennule is where the unimodal (olfactory) chemosensory aesthetascs are situated. *To the left of the crayfish:* on the tips of the walking legs, bimodal chemo- and mechanosensory sensilla may be present e.g., as hedgehog hairs (*above left*) and smooth, squamose setae (*below left*). Bimodal sensilla show a considerable structural variety and can also be found e.g., on the mouthparts of the animal (not shown). Drawing by artist Jorge A. Varela Ramos

the stimulating molecules that are in some cases similar in insects, chelicerates, and crustaceans. Sensilla are small, often hair-like cuticular organs containing sensory cells and enveloping cells. Chemosensory sensilla are found on many parts of the crustacean body. Important sites for chemoreception in all crustaceans are the antennae, antennules and mouthparts, and in benthic crustaceans also the tips of the walking legs (Fig. 6.1). There are two main types of chemosensory sensilla in crustaceans: (1) bimodal chemo- and mechanosensory sensilla found on the mouthparts, walking legs, and spread over the body surface (Fig. 6.2), and (2) unimodal olfactory sensilla called aesthetascs generally found on the antennules (Fig. 6.3). Bimodal sensilla are thought to mediate chemical stimuli from chemical sources that are in direct contact with the animal (similar to vertebrate taste/gustatory receptors) while unimodal chemosensory sensilla appear to mediate reception of molecules that have traveled a distance through the water. The latter are therefore comparable to olfactory receptors in vertebrates. The distinction between olfaction and gustation is debated both for aquatic animals and invertebrates in general (see Schmidt and Mellon, Chap. 7, for a critical assessment of this distinction).

The fine structure of crustacean chemosensory sensilla has been comprehensively described during the last 50 years (Laverack 1968, 1988; Altner and Prillinger 1980;

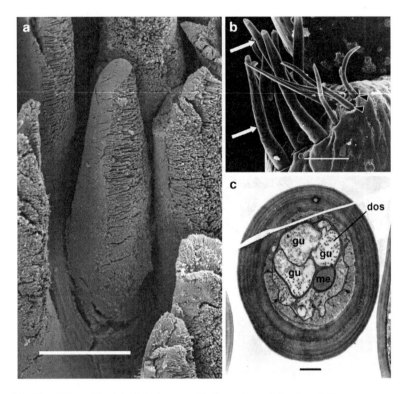

Fig. 6.2 Bimodal sensilla. (**a**) Hedgehog sensilla from the walking leg of *Homarus gammarus*. Scale bar: 100 μm. (**b**) Antennules of *Pacifastacus leniusculus* with both unimodal chemosensory aesthetascs (*arrows*) and slender bimodal chemo- and mechanosensory sensilla (*arrowheads*). Scale bar: 50 μm. (**c**) Transverse section through the dendritic outer segments (dos) of a bimodal sensillum from the antennule of *Lophogaster typicus*. There are three gustatory cells (gu) and one mechanosensory cell (me). The latter has a dense array of microtubules. Scale bar: 500 nm

Ache 1982; Derby 1982; Hallberg and Chaigneau 2004). Both bimodal and unimodal sensilla contain the following structural elements: (1) a cuticular hair-like structure, (2) a number of sensory cells (either chemosensory cells only or combined with mechanosensory cells), and (3) enveloping cells which surround the sensory cells (Fig. 6.4). These three structures will now be described further, first in general and then specifically for the two different types of sensilla, clarifying the differences between them.

The cuticular hair can be of varying length and thickness, and usually has a smooth surface, but the cuticular structure depends on the type of chemosensory sensillum; some have cuticular superficial structures. For example, the hedgehog hairs of *H. americanus* are fringed (Fig. 6.2) (Derby 1982), whereas the hooded sensilla of *Panulirus argus* have leaf-like setules (Cate and Derby 2002).

In terrestrial habitats, the risk of desiccation through diffusion of water to the environment makes robust sensory structures necessary, where pore systems provide access for stimulatory molecules into the insect sensilla (Altner and Prillinger 1980).

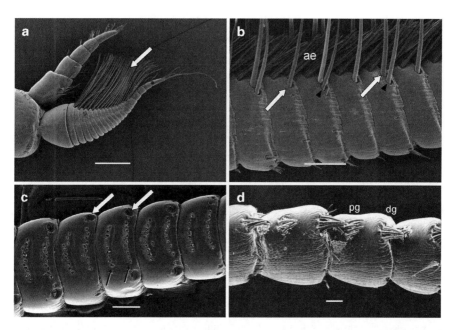

Fig. 6.3 Decapod antennules and sensilla. (**a**) Antennule of the crab *Carcinus maenas*. The aesthetascs (*arrow*) form a dense array on the lateral filament. Scale bar: 500 µm. (**b**) Part of antennule of the European lobster *Homarus gammarus*. Unimodal aesthetascs (ae) are surrounded by bimodal guard hairs (*arrows*) and companion hairs (*arrowheads*). Scale bar: 500 µm. (**c**) The aesthetascs on the antennule of *H. gammarus* are arranged in two rows (*black arrows*) on each annulus. The sockets of the guard and companion hairs surround the aesthetascs (*white arrows*). Scale bar: 200 µm. (**d**) Part of antennule of the crayfish *Pacifastacus leniusculus*. There is one distal group (dg) and one proximal group (pg) of aesthetascs on each annulus. Scale bar: 100 µm

In contrast, the aquatic crustaceans have aesthetascs with a "spongy" cuticle, allowing the passage of small molecules (see below).

Terrestrial crustaceans preserve the general design of the aquatic chemosensory organs, but the chemoreceptive sensilla have a thicker cuticle to reduce desiccation (Hansson et al., Chap. 8). The ultrastructure of the sensilla is more similar to that of insects than that of aquatic crustaceans. Many terrestrial and semiterrestrial decapods (e.g., fiddler crabs) have switched to auditory and/or visual cues for social communication (Christy and Rittschof, Chap. 16), and exhibit structural adaptations of their aesthetascs such as shorter hairs and thicker cuticle of the noninnervated aesthetasc surface (Ache 1982). On the other hand, terrestrial isopods *Oniscus* sp. use airborne pheromones for chemical communication, and both the anomuran *Coenobita* sp. and the giant robber crab *Birgus latro* locate food sources on land entirely by smell (Ache 1982; Stensmyr et al. 2005; Hansson et al., Chap. 8). In desert isopods, both olfactory and gustatory organs are used in the perception of social communication chemicals (often perceived via physical contact using the tips of the antennae) (Seelinger 1983; Lefebvre et al. 2000).

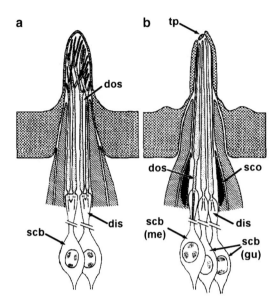

Fig. 6.4 Schematic comparison of one unimodal chemosensory aesthetasc hair and one bimodal chemo- and mechanosensory sensillum (after Schmidt and Gnatzy 1984, with kind permission from Springer Science + Business Media). (**a**) In the aesthetasc, the chemosensory cell bodies (scb) have one dendritic inner segment (dis) each. In the distal part of the dis, two dendritic outer segments (dos) arise, which are present within the entire cuticular hair, and they may or may not be branched. (**b**) In the bimodal gustatory/tactile sensilla only one dendritic outer segment (dos) is connected to each dendritic inner segment (dis). The dos are unbranched and terminate below the terminal pore (tp) of the hair. Two types of sensory cells are found: mechanosensory (me) with a prominent rootlet system and a long ciliar region, and gustatory (gu) with a shorter ciliar region and rootlet system. An electron dense structure, scolopale (sco) is found in the innermost enveloping cells and is typical for mechanosensitive units

    The chemosensory cells responsible for the chemoreception in both unimodal and bimodal sensilla are almost identical in structure. The chemosensory cells are bipolar, with a cell body containing the nucleus, and a proximal axon connecting the sensory cells with the central nervous system (Fig. 6.4). A dendrite emanates from the distal part of the cell body. The dendrite consists of two parts: one dendritic inner segment and one (in bimodal sensilla) or two (in unimodal sensilla) dendritic outer segments projecting into the cuticular hair. The inner and outer segments of the dendrite are connected by a ciliar region, which is characterized by basal bodies in the distal part of the dendritic inner segment, and a ciliar arrangement of microtubules in the proximal part of the outer segment (Fig. 6.4).

    A rootlet system is usually present in the terminal part of the dendritic outer segment. The rootlets appear as electron dense structures which have a striated pattern, when seen in longitudinal sections. These rootlets may proceed a considerable distance proximally in the dendritic inner segments. The dendritic outer segment, which is either branched or unbranched, can thus be interpreted as a transformed cilium arising from the dendritic inner segment.

The enveloping cells, which may be numerous, surround the sensory cells in a concentric manner. They form a fluid-filled cavity around the basal parts of the dendritic outer segments and also proceed distally into the cuticular hair. The ionic composition of the sensillum lymph can be regulated by translocation of certain ions from the hemolymph via extracellular spaces between the cells of the sensillum (Gleeson et al. 1993). During ontogeny, enveloping cells form the cuticular hair and they are also responsible for forming new cuticular hairs below the old ones at the molt. After molting, the enveloping cells usually withdraw from the hair. Molting of sensilla is further discussed later in this chapter.

6.3   The Sensilla Types

6.3.1   Functional Anatomy of the Bimodal Chemo- and Mechanosensory Sensilla

The bimodal sensilla are distributed over the entire crustacean body with the highest concentrations found on both pairs of antennae, mouthparts, and legs (Derby 1982). The cuticular hairs of the bimodal sensilla are either short or long and slender. The short, more robust sensilla (e.g., hedgehog hairs – Derby 1982, funnel-canal organs – Schmidt and Gnatzy 1984, and hooded sensilla – Cate and Derby 2001) are found on other parts of the animals such as the legs and carapace (Fig. 6.2). The longer hairs (e.g., guard hairs, companion hairs – Laverack and Ardill 1965, seta 1, 2, 3 – Garm et al. 2003), are typically found on the mouthparts, antennules, and antennae (Fig. 6.3b).

In the longer hairs, the inner part has a homogenous cuticle, and the outer portion is built up by cuticle arranged in a spiral pattern of varying electron density and there is an apical pore in the sensillum which most probably allows access to the stimuli. The pattern of the cuticle likely reveals zones of different pliability, making the hair both flexible and strong which is an advantage for a hair with a contact chemosensory function (Garm et al. 2003).

There are two different types of enveloping cells in the bimodal sensilla; the innermost is characterized by an electron dense substance (scolopale) present between longitudinally arranged microtubules in the cytoplasm (Fig. 6.4b), whereas the other outer enveloping cells lack scolopales. Scolopale and nonscolopale enveloping cells are arranged in a concentric pattern.

There are also two types of bipolar sensory cells represented: chemosensory and mechanosensory, which differ both in modality and morphology (fine structure) (Fig. 6.2c). One type, believed to be the mechanosensory unit, contains a slender dendritic outer segment with densely arranged microtubules throughout and a well-developed rootlet system, combined with scolopale enveloping cells. The other type, interpreted as the chemosensory unit, has a less densely arranged microtubular array and a less developed rootlet system and lacks scolopale structures in the adjoining cytoplasm of the innermost enveloping cell (Altner et al. 1983; Schmidt and Gnatzy 1984). This division is based on the fact that the scolopale type of enveloping cells and sensory cells with densely arranged microtubules are also found in internal mechanosensory organs in both crustaceans and insects (Hallberg and Hansson 1999).

The bimodal sensilla are mainly involved in close-range location and evaluation of food, but may also be active when contact pheromones are used in social contexts (Bauer, Chap. 14). Bimodal sensilla are also found in insects, with a morphology slightly diverging from crustaceans (Zacharuk 1985).

6.3.2   Functional Anatomy of the Unimodal Olfactory Sensilla

The olfactory sensilla in crustaceans are called aesthetascs, and are usually found on the lateral filament of the first antennal pair (antennula) (Figs. 6.1 and 6.3). The antennula of most crustaceans is branched into two filaments (lateral and medial), but in stomatopods and certain decapods the lateral filament gives rise to an extra branch laterally, resulting in three antennular filaments (Mead and Caldwell, Chap. 11). The aesthetascs in euphausiids and remipedians are located instead on the peduncle of the antennula, not its filament (Hallberg et al. 1992; Felgenhauer et al. 1992). The functional explanation for this localization of the aesthetascs is presently unknown. The lateral filaments of the antennules are usually directed outwards obtaining a large sampling space for olfactory stimuli. The sampled volume can be further enhanced by self-generated currents, flicking, and locomotion of the entire animal (Atema 1985, see also Koehl, Chap. 5, and Breithaupt, Chap. 13).

The aesthetasc hairs (Fig. 6.5) have been defined in the following way: (Hallberg et al. 1992): (1) they have a thin-walled tube-like cuticular hair, (2) the number of sensory cells varies between a few and several hundred, (3) the sensory cells have a dendrite from which usually two transformed cilia (dendritic outer segments) arise, (4) the dendritic outer segments may, or may not, split into many branches, (5) the sensory cells are surrounded by enveloping cells in the basal region. The fine structure of aesthetascs has been most thoroughly described in decapods (Laverack and Ardill 1965; Ghiradella et al. 1968; Snow 1973), but there are also descriptions from other crustaceans (reviewed in: Hallberg and Chaigneau 2004). The aesthetasc cuticle, especially the distal part, appears loosely arranged when viewed under the electron microscope, and is evidently functioning as a molecular sieve, allowing the passage only of molecules <8.5 kDa in molecular weight (Derby et al. 1997).

The definition above allows for a most varied ultrastructure of aesthetascs, which is needed to account for the considerable morphological variation between the most distantly related crustacean taxa. In euphausiids, there are annular swellings along the aesthetasc (Hallberg et al. 1992). In isopods and amphipods, the hairs have a bipartite appearance, with a larger diameter of the outer part than of the inner part (Heimann 1984; Hallberg and Chaigneau 2004) and in most decapods,

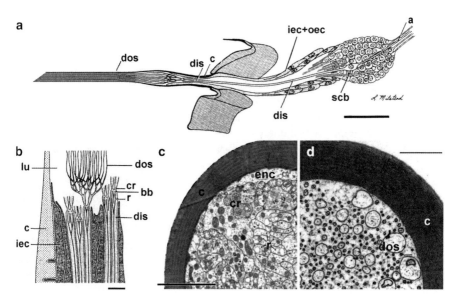

Fig. 6.5 The structure of the aesthetasc. (**a, b**) Schematic longitudinal section through an aesthetasc of *Panulirus argus* (after Grünert and Ache 1988, with kind permission from Springer Science + Business Media). The bipolar sensory cell bodies (scb) are located well below the antennular cuticle (c), sending axons (a) to the brain. Dendritic inner segments (dis) are surrounded by inner and outer enveloping cells (iec, oec) and proceed into the basal part of the cuticular hair, where they give rise to the dendritic outer segments (dos). The distal part of the dis contains two rootlets (r) from which the basal bodies (bb) arise (two per dis), each connecting in turn to the ciliar region (cr) of one dos. The dos in turn may branch extensively above the cr and are surrounded by a liquid-filled lumen (lu) inside the cuticular hair. (**c, d**) Transverse sections through a single aesthetasc hair from the crab *Carcinus maenas*. (**c**) Section through the transition zone between dis/dos. A ciliar region (cr) and rootlets (r) can be seen. The cuticle (c) of the aesthetasc hair encloses the enveloping cells (enc) and sensory cells. Scale bar: 2 μm. (**d**) More distal section showing branched dendritic outer segments (dos) without enveloping cells inside the cuticle (c). Scale bar: 1 μm

the aesthetascs are usually long and slender. One notable exception to this last statement is the aesthetascs of freshwater crayfish, which are rather few and short, probably an adaptation to the low osmotic pressure of freshwater (Tierney et al. 1986). In brachyurans and anomurans, the aesthetascs are long with transverse furrows along the hair (Fontaine et al. 1982).

Aesthetasc length is somewhat correlated to the total body size. Generally, the large decapods, like crabs and lobsters, have the longest aesthetasc hairs (600–1,400 μm) but freshwater crayfish have considerably shorter aesthetascs (60–150 μm). In the peracarids, which are usually smaller than the decapods, aesthetascs are normally shorter (120–400 μm), and the shortest cuticular hairs (10–70 μm) are found in the tiny nonmalacostracans, such as Cladocera and Notostraca (Hallberg et al. 1992).

The number of aesthetasc hairs is also variable. In some groups there are just a few aesthetascs present, as in the ostracod *Notodromas monachus* (Andersson 1975) that has only one aesthetasc on each second antenna. This sensillar type is, however, a special type of compound sensilla, "grouped setae" that is typical for ostracodes (Kaji and Tsukagoshi 2008). The lowest number of olfactory sensilla (~15 aesthetascs at the tip of the antennula) in decapods is found in the burrowing decapods *Upogebia* spp. and *Calocaris* spp. In *Callianassa australiensis* the number of aesthetascs is also low, 22 (Beltz et al. 2003). Whether these burrowing crustaceans rely less on olfaction than others is not known.

Decapods have the highest known number of aesthetascs (being more moderate in some species than others), and the number may increase with age (and size) of the individual (Sandeman and Sandeman 1996). In *C. maenas* the number of aesthetascs is estimated to be 150–200 per antennula, in *H. gammarus* to about 800–1,000 (Skog, unpublished data), and *P. argus* has 1,200–2,400, depending on the size of the individual (Spencer and Linberg 1986; Beltz et al. 2003).

Many decapods have very dense, brush-like arrays of olfactory sensilla and this more or less dense packing of aesthetascs inhibits water flow between them. Thus, the aesthetascs are embedded in a viscous boundary layer of water. Exchange of the boundary layer water is achieved by flicking, an intermittent powerful beat of the entire antennula that allows rapid odor access to the animal's aesthetasc tufts by splaying out the aesthetasc sensilla and substituting the water lodged there with new water (Schmitt and Ache 1979; Koehl, Chap. 5).

The dendritic outer segments of crustacean chemosensory cells may be branched or unbranched; in *P. argus* the branching is extensive, resulting in 20–30 branches from each sensory cell (Grünert and Ache 1988). The branching of the dendritic outer segments will result in a larger chemosensitive surface area. Unbranched dendritic outer segments are present in e.g., the crayfish *Orconectes propinquus* (Tierney et al. 1986). In most crustaceans, each dendritic inner segment gives rise to two outer segments, but the isopod *Asellus aquaticus* has either two or only one outer segment (Heimann 1984).

In some limnic and terrestrial crustaceans, the ciliar region of the dendritic segments is situated inside the antennula instead of being located in the cuticular hair as in marine species (Ghiradella et al. 1968; Hallberg et al. 1997). This is believed to be an ecological adaptation of the sensilla to prevent desiccation in the terrestrial species and osmotic shock in the limnic ones.

6.3.3   Sexual Dimorphism and Male-Specific Sensilla

A morphologically distinct type of unimodal olfactory chemosensors is the "male-specific sensilla" (Johansson and Hallberg 1992) (Fig. 6.6) found on the base of the first antenna in (male) mysids and some other peracarid crustaceans. Whereas aesthetasc hairs are ubiquitous, male-specific sensilla are found only in a few groups and, as the name implies, only in the males. In mysids, female antennae are very similar to those of the males except for the absence of male-specific

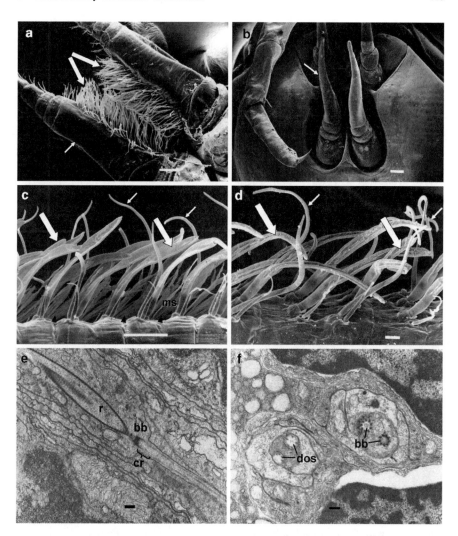

Fig. 6.6 Male-specific sensilla (ms). (**a**) Antennules (*small arrow*) of an amphipod *Hyperia galba* male with dense medial arrays of ms (*large arrows*). Scale bar: 100 µm. (**b**) Antennules (*small arrow*) of an amphipod *H. galba* female without ms. Scale bar: 100 µm. (**c**) Antennule of a mysid *Lophogaster typicus* male with *ms* between aesthetascs (*large arrows*) and bimodal sensilla (*small arrows*). Scale bar: 50 µm. (**d**) Antennula of a mysid *L. typicus* female without ms. Aesthetascs (*large arrows*) and bimodal sensilla (*small arrows*) are present. Scale bar: 10 µm. (**e**) Longitudinal section through the transition zone between dendritic inner/outer segments of ms in the mysid *Praunus flexuosus*. The basal body (bb) is connected to a rootlet (r), which proceeds proximally, and a ciliar region (cr) representing the most proximal part of the unbranched dendritic outer segment. Scale bar: 200 nm. (**f**) Transverse section through the transition zone of two neighboring ms of *P. flexuosus*. Two basal bodies (bb) are seen in one of them, and the proximal parts of the dendritic outer segments (dos) in the other. Scale bar: 200 nm

sensilla (6.6c, d). In contrast, the sexual dimorphism in the amphipod *Hyperia galba* is greater, encompassing the entire first antenna. Male *H. galba* have long antennules (10–12 mm) with a dense array of male-specific sensilla (Fig. 6.6a), whereas female antennules (Fig 6.6b) are very short (0.5–2 mm) and have no sensilla at all. Based on the ultrastructural characteristics, the male-specific sensilla are considered to be chemosensory units.

Sexual dimorphism of the sensory organs is common in many insects. For example, male moths may have extremely well-developed (about 10 mm long and ~3 mm wide) feather-shaped olfactory antennae, while the female antennae are unfeathered and half as long as those of the males (<5 mm long). This type of sexual dimorphism in the chemosensory organs is not as widespread among the crustaceans, but appears to be common in some groups. For example, in a number of amphipods, mysids, cumaceans, isopods, and shrimps a sensory organ called callynophore can be found on the outer antennular flagella. The callynophore is a field of densely arranged aesthetascs. In some amphipod species it is present in both sexes and is probably used for food detection through chemoreception, while in other species it is only present in sexually mature males and is presumably used for chemosensory detection of females (Lowry 1986). Many isopods display sexual dimorphism in the number of aesthetascs found on the antennula (Wilson and Hessler 1974) and in the isopod *Asellus aquaticus,* males have longer antennae than females relative to body size (Bertin and Cézilly 2003).

As mentioned above, some crustacean groups also exhibit sexual dimorphism in the types of chemosensory sensilla present on the antennulae. In these species the males (but not females) have an array of slender sensory hairs, called male-specific sensilla, on the peduncle of the antennule (Fig. 6.6a, b). These male-specific sensilla appear to be characteristic for some taxa within the order Peracarida, e.g., mysids, certain amphipods, and possibly cumaceans (Johansson and Hallberg 1992).

Male *H. galba* have about 200 male-specific sensilla on the medial side of both antennulae, each hair being about 150 μm long. These sensilla are innervated by only one to three sensory cells. Female *H. galba* live inside jellyfish medusa together with their offspring after these have left the female marsupium. The females appear to be inept swimmers and are probably more or less stationary in their medusa. The males are better swimmers and probably use their well-developed chemosensors to find females (Dittrich 1988; Hallberg et al. 1997).

Most mysids have a bush of long, slender male-specific sensilla on the specialized protrusion called lobus masculinus on the antennular peduncle. The number of male-specific sensilla in mysids is often high, about 2,000 in *Praunus flexuosus*. Each sensillum in mysids is innervated by a single sensory cell which sends two unbranched dendritic outer segments into the hair (Fig. 6.6e, f). The hair is flattened in cross section and its walls appear to consist of loosely arranged cuticle (Johansson and Hallberg 1992). The lophogastriid mysid, *Lophogaster typicus,* also has male-specific sensilla, but these are instead distributed among the aesthetascs on the antennulae (Fig. 6.6c) (Johansson et al. 1996).

6.4   Molting and Development of Sensilla

Molting is a prerequisite for growth in arthropods. It is an energy-demanding process, and often makes the animal very vulnerable. Further, sensory input may be severely impaired during molting, including from the chemosensory organs.

In insects and arachnids, there are two main molting processes of the sensilla; one which retains a dendritic connection between the old and the new sensillum, thus retaining the chemosensory function, and one where there is no such connection and the chemosensory function is presumably temporarily lost during the molt. The type without loss of chemosensory function is the more predominant one (Guse 1983).

Similarly, two different types of molting of the chemosensory organ have been described in crustaceans: the mysid type and the isopod type (Guse 1983). In both types, a new sensillum is formed under the old one with enveloping cells forming the shaft of the new cuticular hair in an invaginated sleeve around the sensory cells (Fig. 6.7). In the mysid type of molt, the dendritic outer segments of the sensory cells proceed through the apical pore of the developing hair to the old cuticular hair. This connection provides a route for stimulus propagation to the central nervous system throughout the molt (Figs. 6.7a and 6.8a), maintaining the chemosensory capacity. This type of molt can be seen in, for example, the mysid *Neomysis integer* and in the brachyuran crab *Carcinus maenas* (Fig. 6.8b–d). In contrast, the developing cuticular hair is lacking an apical pore in the isopod type of molt, as exemplified by the isopod *Idotea* sp. (Fig. 6.7b). In this type of molt, the old sensillum becomes disconnected from the sensory cells below and sensory input is most likely nonexisting during this type of molt process. As soon as the new aesthetasc is fully developed, the perception of chemical stimuli should be regained. How long the possible loss of chemosensory input lasts is not known.

Most crustaceans are very vulnerable at the molt and will have an advantage if they can detect predators chemically. Further, reproduction in many crustaceans is tightly linked to the female molt, and it may be important for the female to be able to find and/or evaluate conspecific males by chemical cues during the courtship. This is only possible for crustaceans using the mysid-type of sensilla molting unless chemosensory information can be obtained from other sources than the chemoreceptive sensilla. In crustaceans with the isopod type of molt, reproductive communication and predator detection through chemical cues may be impossible during the molt. It is not known at present whether this affects their behavior during the molt.

During ontogeny, there is a massive turnover of aesthetascs. In spiny lobsters, new aesthetascs are formed proximally on the antennule and the old ones degenerate distally (Derby et al. 2003). Thus, there is a possibility that the tuning of the chemosensory system changes during the life-span of the individual. In crustaceans with terminal molting, there is no possibility of regeneration of the chemosensory structures. As in large decapods, which may molt very seldom, we expect that the sensory structures become increasingly fouled by epibionts and their function may

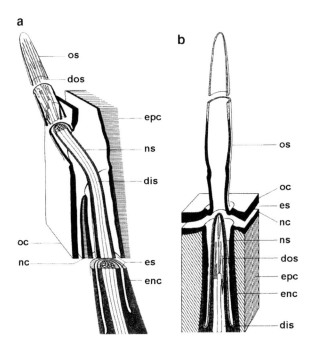

Fig. 6.7 Molting of the aesthetascs (after Guse 1983, with kind permission from Springer Science + Business Media). (**a**) Mysid type of molt. The "old" sensillum (os) contains sensory cells with dendritic outer segments (dos) which are continuous with the dendritic inner segments (dis) present in the developing sensillum within the antennule. The new shaft (ns) of the developing sensillum is present in the exuvial space (es). The new cuticle (nc) of the basal parts of the new shaft is formed by the enveloping cells (enc) in an invagination of the exuvial space. Epidermal cells (epc) are found beneath the new cuticle. (**b**) Isopod type of molt. The new aesthetasc shaft (ns) is formed by the enveloping cells (enc) inside the antennular surface around an invagination of the exuvial space (es). The dendritic outer (dos) and inner (dis) segments of the sensory cells are confined to the space inside the shaft of the new sensillum. The old sensillar hair (os) is connected to the old cuticle (oc) and contains no sensory elements. The developing sensillum is surrounded by epidermal cells (epc)

be impaired. However, crustaceans generally groom their antennules, a behavior that reduces fouling (Bauer 1981).

6.5   Challenges for Future Studies of Crustacean Chemosensory Organs

Crustaceans are found in many different habitats with a range of physical challenges (e.g., salinity, temperature, air/water) for the chemosensory organs, which also show a considerable morphological variation among crustacean taxa. The general ultra-structure and fine structure of crustacean chemosensory sensilla have been

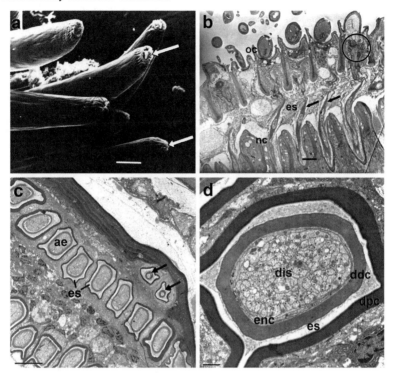

Fig. 6.8 Molting of aesthetascs. (**a**) Apical pores (*arrows*) at the tips of the aesthetascs of *Pacifastacus leniusculus*, probably functioning as molting pores. (**b**) Oblique section through the antennule of *Praunus flexuosus* during preecdysis. The old cuticle (oc) and its aesthetascs are separated from the new cuticle (nc) by the exuvial space (es). The tips of new aesthetascs (*arrows*) are anchored to the bases of the old aesthetascs by branch-like structures (*encircled area*). (**c**) Section through the antennule of *Carcinus maenas* during preecdysis. The new aesthetascs (ae) are formed inside the invaginated cuticle and surrounded by exuvial space (es). Two companion hairs (*arrows*) are seen distal to the aesthetasc row. (**d**) One developing aesthetasc. The dendritic inner segments (dis) are seen inside the hair cuticle which is secreted by the enveloping cells (enc). The cuticle surrounding the dendritic inner segments will give rise to the distal part of the hair after ecdysis (developing distal cuticle, ddc). The cuticle forming the outer delimitation of the exuvial space (es) will end up as the proximal part of the hair (developing proximal cuticle, dpc)

thoroughly studied in many taxa, but the relation between the morphology of sensory structures and the animal's ecology, environment, and behavior are often unknown.

Foraging (e.g., food finding, evaluation and acquisition), agonistic and reproductive behaviors are contexts where chemosensory cues and chemosensory sensilla are essential for most crustaceans. Of those three topics, the role of chemoreception and chemical signals in reproductive behaviors is least understood and needs more research effort in the future. In particular, the role of pheromone receptors in mate-finding and evaluation, courtship, copulation, and brood care is underexplored in crustaceans compared to the vast knowledge we have about sex pheromone receptors in insects.

Amphipods and mysids with male-specific sensilla evidently show sexual dimorphism in the olfactory sensilla. These sensilla are ideal for studies of the correlation between sensilla morphology and functional reproductive behaviors like mate location. Mating has barely been studied at all in mysids (Clutter 1969) but is apparently a very rapid process that takes place immediately after the female molt. Sex attraction by males to molting females is species-specific and stimulus acquisition and mating-related behaviors by the male can indicate the role of the male-specific sensilla in mysid reproduction.

Even in those crustacean species where visible morphological sex differences in the chemosensory sensilla are small or imperceptible, variations of aesthetasc structure, size and number, sensory cell morphology, and spatial arrangement can still occur. Any such sexual dimorphism could help enlighten the sex-specific role of these chemosensory structures in, for example, reproductive behaviors.

Morphological and behavioral investigations should be supplemented by e.g., studies of perception of relevant cues by the sense organs and electrophysiological studies of the male-specific sensilla. This is a true challenge due to the tiny size of the animals where this structure is present and the electrophysiological techniques available today may need some refinement. Some amphipod or isopod species are more robust than most other species possessing male-specific sensilla, and may be better suited for this type of experimentation.

6.6 Summary

Crustacean sensilla are sensory organs, usually consisting of hair-like cuticular structures. The sensilla contain basally located sensory cells that extend sensory dendrites into the hair and axons that transmit signals to the central nervous system.

The chemosensory organs of crustaceans are divided into bimodal chemo- and mechanosensory sensilla (found e.g., on mouthparts and legs) and unimodal olfactory sensilla (usually found on the first antenna). Although the general morphology of these structures has often been thoroughly described, less is known about the function of the sensilla in an ecological and behavioral context.

Acknowledgements We would like to thank NFR, Formas and the Royal Physiographic Society in Lund for supporting this research. Rolf Elofsson and Ali Steinbrecht have been inspiring and most helpful during the years. The long and fruitful cooperation with Rita Wallén, a most competent and enthusiastic technician in the field of electron microscopy, is deeply acknowledged.

References

Ache BW (1982) Chemoreception and thermoreception. In: Atwood HL, Sandeman DC (eds) The biology of Crustacea, Vol. 3. Academic Press, New York, pp 369–398

Altner H, Prillinger L (1980) Ultrastructure of invertebrate chemo-, thermo-, and hygroreceptors and its functional significance. Int Rev Cytol 67:69–139

Altner I, Hatt H, Altner H (1983) Structural properties of bimodal chemo- and mechanosensitive setae on the pereiopod chelae of the crayfish, *Austropotamobius torrentium*. Cell Tiss Res 228:357–374

Andersson A (1975) The ultrastructure of the presumed chemo receptor aesthetasc of a cypridid ostracod. Zool Script 4:151–158

Atema J (1985) Chemoreception in the sea: adaptations of chemoreceptors and behaviour to aquatic stimulus conditions. In: Laverack MS (ed) Society of experimental biology symposium 39: physiological adaptations of marine animals. Cambridge University Press, Cambridge, pp 387–423

Balss H (1944) Decapoda. In: Bronns HG (ed) Klassen und Ordnungen des Tierreichs (Band 5, Abteilung 1, Buch 7). Leipzig: Akademische Verlagsgesellschaft Geest & Portig Kg, pp 321–480. (Reference to Leydig (1870) pp 363–364)

Bauer RT (1981) Grooming behavior and morphology in the decapod Crustacea. J Crust Biol 1:153–173

Beltz BS, Kordas K, Lee MM, Long JB, Benton JL, Sandeman DC (2003) Ecological, evolutionary, and functional correlates of sensilla number and glomerular density in the olgfactory system of decapod crustaceans. J Comp Neurol 455:260–269

Bertin A, Cézilly F (2003) Sexual selection, antennae length and the mating advantage of large males in *Asellus aquaticus*. J Evol Biol 16:491–500

Cate HS, Derby CD (2001) Morphology and distrbution of setae on the antennules of the Caribbean spiny lobster *Panulirus argus* reveal new types of bimodal chemo-mechanosensilla. Cell Tissue Res 304:439–454

Cate HS, Derby CD (2002) Ultrastructure and physiology of the hooded sensillum, a bimodal chemo-mechanosensillum of lobsters. J Comp Neurol 442:293–307

Clutter RI (1969) The microdistribution and social behaviour of some pelagic mysid shrimps. J Exp Mar Biol Ecol 3:125–155

Derby CD (1982) Structure and function of cuticular sensilla of the lobster *Homarus americanus*. J Crust Biol 2:1–21

Derby CD, Cate HS, Gentilcore LR (1997) Perireception in olfaction: Molecular mass sieving by aesthetasc sensillar cuticle determines odorant access to receptor sites in the Caribbean spiny lobster *Panulirus argus*. J Exp Biol 200:2073–2081

Derby CD, Cate HS, Steullet P, Harrison PJH (2003) Comparison of turnover in the olfactory organ of early juvenile stage and adult Caribbean spiny lobsters. Arthr Struct Dev 31:297–311

Dittrich B (1988) Studies on the life cycle and reproduction of the parasitic amphipod *Hyperia galba* in the North Sea. Helgoländ. Meeresuntersuch 42:79–98

Doflein F (1910) Lebensgewohnheiten und Anpassungen bei decapoden Krebsen. Festschrift Richard Hertwig zum 60. Geburtstag. Band 3:215–292

Ekerholm M, Hallberg E (2005) Primer and short-range releaser pheromone properties of premoult female urine from the shore crab, *Carcinus maenas*. J Chem Ecol 31:1845–1864

Felgenhauer BE, Abele LG, Felder DL (1992) Remipedia. In: Harrison FW, Humes AG (eds) Microscopic anatomy of invertebrates 9. Wiley-Liss, New York, pp 225–247

Fontaine MT, Passelecq E, Bauchau AG (1982) Structures chemoréceptrices des antennules du crabe *Carcinus maenas* (L.) (Decapoda, Brachyura). Crustaceana 43:271–283

Garm A, Hallberg E, Høeg JT (2003) Role of maxilla 2 and its setae during feeding in the shrimp *Palaemon adspersus* (Crustacea: Decapoda). Biol Bull 204:126–137

Ghiradella H, Case J, Cronshaw J (1968) Structure of aesthetascs in selected marine and terrestrial decapods: chemoreceptor morphology and environment. Am Zool 8:603–621

Gleeson RA, Aldric HC, White JF, Trapido-Rosenthal HG, Carr WES (1993) Ionic and elemental analyses of olfactory sensillar lymph in the spiny lobster, *Panulirus argus*. Comp Biochem Physiol 105A:29–34

Grünert U, Ache BW (1988) Ultrastructure of the aesthetasc (olfactory) sensilla of the spiny lobster, Panulirus argus. Cell Tiss Res 251:95–103

Guse GW (1983) Ultrastructure, development, and moulting of the aesthetascs of *Neomysis integer* and *Idotea baltica* (Crustacea, Malacostraca). Zoomorphol 103:121–133

Hallberg E, Chaigneau J (2004) The non-visual sense organs. In: Forest J, von Vaupel Klein JC (eds) Treatise on zoology – anatomy, taxonomy, biology. The crustacea, revised and updated from the Traité de Zoologie. Brill, Leiden, pp 301–386

Hallberg E, Hansson BS (1999) Arthropod sensilla: morphology and phylogenetic considerations. Micr Res Tech 47:428–439

Hallberg E, Johansson KUI, Elofsson R (1992) The aesthetasc concept: structural variations of putative olfactory receptor cell complexes in Crustacea. Microsc Res Tech 22:325–335

Hallberg E, Johansson KUI, Wallén R (1997) Olfactory sensilla in crustaceans: Morphology, sexual dimorphism, and distribution patterns. Int J Insect Morphol Embryol 26:173–180

Heimann P (1984) Fine structure and moulting of aesthetasc sense organs on the antennules of the isopod, *Asellus aquaticus* (Crustacea). Cell Tiss Res 235:117–128

Johansson KUI, Hallberg E (1992) Male-specific structures in the olfactory system of mysids (Mysidacea: Crustacea). Cell Tiss Res 268:359–368

Johansson KUI, Gefors L, Wallen R, Hallberg E (1996) Structure and distribution patterns of aesthetascs and male-specific sensilla in *Lophogaster typicus* (Mysidacea). J Crust Biol 16:45–53

Kaji T, Tsukagoshi A (2008) Origin of the novel chemoreceptor aesthetasc "Y" in Ostracoda: morphogenetic thresholds and evolutionary innovation. Evol Dev 10:228–240

Laverack MS (1968) On the receptors of marine invertebrates. Oceanogr Mar Biol Annu Rev 6:249–324

Laverack MS (1988) The diversity of chemoreceptors. In: Atema J, Fay RR, Popper AN, Tavolga WN (eds) Sensory biology of aquatic animal. Springer, Heidelberg, pp 287–312

Laverack MS, Ardill DJ (1965) The innervation of the aesthetasc hairs of *Panulirus argus*. Quart J Micr Sci 106:45–60

Lefebvre F, Limousin M, Caubet Y (2000) Sexual dimorphism in the antennae of terrestrial isopods: a result of male contests or scramble competition? Can J Zool 78:1987–1993

Lowry JK (1986) The callynophore, a eucaridean/peracaridean sensory organ prevalent among the Amphipoda (Crustacea). Zool Script 15:333–349

Marcus K (1911) Über Geruchsorgane bei decapoden Krebsen aus der Gruppe der Galatheiden. Zeitschr Wiss Zool 97:511–545

Sandeman RE, Sandeman DC (1996) Pre-and postembryonic development, growth and turnover of olfactory receptor neurons in crayfish antennules. J Exp Biol 199:2409–2418

Schmidt M, Gnatzy W (1984) Are the funnel-canal organs the 'campaniform sensilla' of the shore crab, *Carcinus maenas* (Decapoda, Crustacea)? II. Ultrastructure. Cell Tiss Res 237:81–93

Schmitt BC, Ache BA (1979) Olfaction: responses of a decapod are enhanced by flicking. Science 205:204–206

Seelinger G (1983) Response characteristics and specificity of chemoreceptors in *Hemilepistus reaumuri* (Crustacea, Isopoda). J Comp Physiol 152:219–229

Skog M (2009a) Intersexual differences in European lobster (*Homarus gammarus*): recognition mechanisms and agonistic behaviours. Behaviour 146:1071–1091

Skog M (2009b) Male but not female olfaction is crucial for intermolt mating in European lobsters (*Homarus gammarus* L.). Chem Senses 34:159–169

Skog M, Chandrapavan A, Hallberg E, Breithaupt T (2009) Maintenance of dominance is mediated by urinary chemicals in male European lobsters, *Homarus gammarus*. Mar Freshw Behav Physiol 42:119–133

Snow PJ (1973) The antennular activities of the hermit crab, *Pagurus alaskensis*. J Exp Biol 58:745–765

Spencer M, Linberg KA (1986) Ultrastructure of aesthetasc innervation and external morphology of the lateral antennule setae of the spiny lobster *Panulirus interruptus* (Randall). Cell Tiss Res 245:69–80

Stensmyr MC, Erland S, Hallberg E, Wallén R, Greenaway P, Hansson BS (2005) Insect-like olfactory adaptations in the terrestrial giant robber crab. Curr Biol 15:116–121

Tierney AJ, Thompson CS, Dunham DW (1986) Fine structure of aesthetasc chemoreceptors in the crayfish *Orconectes propinquus*. Can J Zool 64:392–399

Wilson GD, Hessler RR (1974) Some unusual Paraselloidea (Isopoda, Asellota) from the deep benthos of the Atlantic. Crustaceana 27:47–67

Zacharuk RY (1985) Antennae and sensilla. In: Kerkut GA, Gilbert LI (eds) Comprehensive insect physiology, biochemsitry and pharmacology, vol 6. Pergamon Press, Oxford, pp 1–69

Chapter 7
Neuronal Processing of Chemical Information in Crustaceans

Manfred Schmidt and DeForest Mellon, Jr.

Abstract Most crustaceans live in aquatic environments and chemoreception is their dominant sensory modality. Crustacean chemoreception is mediated by small cuticular sense organs (sensilla) occurring on all body parts, with the antennules (first antennae), second antennae, legs, and mouthparts representing the major chemosensory organs. Chemoreceptive sensilla of crustaceans are divided into *bimodal sensilla* which comprise a few mechano- and some chemoreceptor neurons and occur on all appendages and *aesthetascs* which are innervated by 40–500 olfactory receptor neurons and exclusively occur on the antennular outer flagellum. Olfactory receptor neurons differ from chemoreceptor neurons of bimodal sensilla in having spontaneous activity, inhibitory responses, and autonomous bursting, but both types of receptor neurons mainly respond to small water-soluble molecules such as amino acids. The dichotomy in sensilla structure is reflected in the organization of the associated CNS pathways. Olfactory receptor neurons selectively innervate a synaptic region in the midbrain, the olfactory lobe, which is organized into dense substructures called glomeruli. As is typical of the first synaptic relay in the central olfactory pathway across metazoans, olfactory information processing in glomeruli is based on multiple types of inhibitory local interneurons and on projection neurons ascending to higher brain areas. Receptor neurons from bimodal sensilla target synaptic areas that are distributed throughout the brain and ventral nerve cord and contain arborizations of motoneurons innervating muscles of the segmental appendages that provide the chemo- and mechanosensory input. Based on the matching dichotomy of sensilla construction and of sensory pathway organization, we propose that crustacean chemoreception is differentiated into two fundamentally different modes: "olfaction" – chemoreception mediated by the aesthetasc–olfactory lobe pathway, and "distributed chemoreception" – chemoreception mediated by bimodal sensilla on all appendages and the associated synaptic areas serving as local motor centers. In decapod crustaceans, pheromone

M. Schmidt (✉)
Neuroscience Institute and Department of Biology, Georgia State University,
P.O. Box 5030, Atlanta, GA 30302, USA
e-mail: mschmidt@gsu.edu

T. Breithaupt and M. Thiel (eds.), *Chemical Communication in Crustaceans*,
DOI 10.1007/978-0-387-77101-4_7, © Springer Science+Business Media, LLC 2011

detection and processing of pheromone information are not mediated by dedicated sensilla and CNS pathways, respectively, but seem to be integral components of olfaction and distributed chemoreception. Aesthetascs mediate responses to distant pheromones, whereas bimodal sensilla located on the appendages touching the conspecific partner are likely responsible for the detection of contact pheromones.

List of Abbreviations

AL	Accessory lobe
CNS	Central nervous system
CRN	Chemoreceptor neuron
dCRN	"Distributed" chemoreceptor neuron of bimodal sensillum
DC	Deutocerebral commissure
DCN	Deutocerebral commissural neuropil
LAN	Lateral antennular neuropil
LN	Local interneuron
MAN	Median antennular neuropil
MRN	Mechanoreceptor neuron
OGT	Olfactory globular tract
OGTN	Olfactory globular tract neuropil
OL	Olfactory lobe
ORN	Olfactory receptor neuron (CRN of aesthetasc)
PN	Projection neuron

7.1 Introduction

For most crustaceans, chemoreception is the dominant sensory modality. My (MS) interest in crustacean chemoreception started when I realized, during my diploma thesis, that most sensilla of crustaceans contain chemoreceptor neurons and occur in many different types on all appendages. Since then, trying to understand how behavior is controlled by these sensilla has been a cornerstone of my research. I (DM) have been fascinated by crayfish and their behaviors since, as a young boy, I would capture them in a creek near my home and keep them in an aquarium in my bedroom. We have both since then gone long ways trying to unravel the neuronal basis of chemoreception.

Numerous behavioral studies have shown that chemoreceptors in crustaceans occur on all appendages and many parts of the body. Among these, the antennules (first antennae), second antennae, legs, and mouthparts are the major chemosensory organs (Brock 1926; Spiegel 1927; Hindley 1975) (Fig. 7.1). Based on the perceived duplicity of chemical senses in vertebrates, several attempts were made to

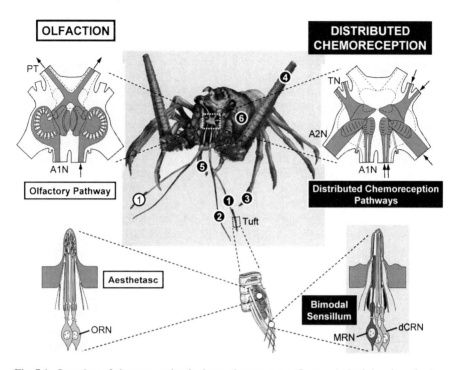

Fig. 7.1 Overview of chemoreception in decapod crustaceans. *Center:* Artists' drawing of spiny lobster showing the location of chemoreceptive sensilla (*white dot:* aesthetascs; *black dots:* bimodal sensilla) on different body parts (*1:* lateral flagellum of antennule; *2:* medial flagellum of antennule; *3:* walking legs; *4:* second antenna; *5:* third maxillipeds; *6:* gill chamber). Tuft region of lateral flagellum shown in higher magnification (modified from Grünert and Ache, 1988). Lobster drawing by Jorge A. Varela Ramos. *Left:* Organization of "Olfaction." *Bottom:* Ultrastucture of aesthetasc (ORN: olfactory receptor neuron) (modified from Schmidt and Gnatzy, 1984); *Top:* Olfactory pathway in the brain with arrows indicating direction of information flow (*A1N* antennular nerve; *PT* protocerebral tract to lateral protocerebrum in eyestalk ganglia). *Right:* Organization of "Distributed Chemoreception." *Bottom:* Ultrastucture of bimodal sensillum (*MRN* mechanoreceptor neuron; *dCRN* "distributed" chemoreceptor neuron) (modified from Schmidt and Gnatzy, 1984); *Top:* Distributed chemoreception pathways in the brain with arrows indicating sensory inputs (*A2N* antennal nerve; *TN* tegumentary nerve)

categorize crustacean chemoreception into "olfaction" and "taste." For crustacean chemoreception this distinction has been based on differences in sensitivity (Laverack 1968), on differences in the elicited behaviors and the appendages mediating them (Atema 1980), and finally on how chemoreception is interwoven with mechanoreception (Laverack 1988). However, none of these attempts fully matches the available data and none has found general acceptance. We propose that chemoreception in crustaceans in fact comprises multiple chemical senses and that the distinction of only two – "olfaction" and "taste" – failed because it is incomplete.

7.2   Olfaction and Distributed Chemoreception

Crustaceans detect chemicals with small cuticular sense organs called sensilla. Crustacean chemoreceptive sensilla – in spite of their diversity in outer structure – fall in two main classes: (a) *Bimodal sensilla* are innervated by 1–3 mechanoreceptor neurons (MRNs) and 1–22 chemoreceptor neurons (CRNs) whose unbranched dendrites run in a narrow canal to a terminal pore at the tip of a thick-walled hair shaft (seta); (b) *Aesthetascs* are innervated by 40–500 CRNs whose branched dendrites fill the wide lumen of a thin-walled tube-like seta (Schmidt and Gnatzy 1984; Hallberg and Skog, Chap. 6) (Fig. 7.1). Bimodal sensilla of numerous structural types occur on the body and all appendages and in their construction correspond to insect contact chemoreceptors. Aesthetascs are morphologically homogeneous, exclusively occur on the outer flagellum of the antennules, and represent olfactory sensilla corresponding to those of insects in ultrastructure. Hence, we will use the term olfactory receptor neurons – ORNs – for the CRNs they contain.

The dichotomy in the structure of crustacean sensilla is reflected in the organization of the associated sensory pathways within the CNS. The crustacean brain consists of three paired ganglia, the protocerebrum receiving input from the compound eyes, the deutocerebrum receiving input from the antennules, and the tritocerebrum receiving input from the second antennae and the head (Sandeman et al. 1992). The aesthetasc ORNs selectively innervate a neuropil (= region where neurons interact via synaptic contacts) of the deutocerebrum, the olfactory lobe (OL). The OL is organized into glomeruli (= small clumps of particularly dense neuropil) which are characteristic of the first synaptic relay of the central olfactory pathway across higher metazoans (Sandeman et al. 1992; Schachtner et al. 2005) (Figs. 7.1 and 7.2b). The glomerular organization of the OL contrasts with a striated (perpendicular to the long axis) organization of the neuropils receiving input from bimodal sensilla, the second antenna neuropils, the lateral antennular neuropils (LAN), and the leg neuromeres (Tautz and Müller-Tautz 1983; Schmidt and Ache 1996a) (Figs. 7.1 and 7.2a). The perpendicular striation appears to result from a topographical representation of sensory input and an interdigitating arborization pattern of motoneurons driving the muscles of the segmental appendage providing the sensory input. Thus, these neuropils act as local motor centers of the segmental appendages and mediate direct sensory-motor integration.

The matching dichotomy of sensilla construction and neuroanatomical organization of sensory neuropils suggests that in crustaceans chemical information is received and processed in two fundamentally different modes. One mode is "Olfaction" which we define as chemoreception mediated by the aesthetasc – OL pathway; the second mode is "Distributed Chemoreception," which we define as chemoreception mediated by bimodal sensilla on all appendages and the associated striated neuropils that serve as local motor centers. Distributed chemoreception not only comprises "taste," which we define as contact chemoreception in the context of

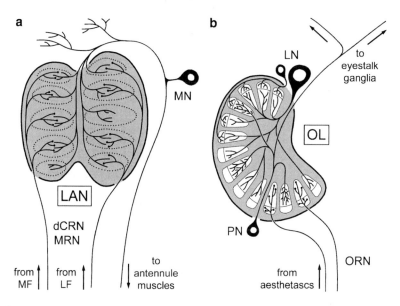

Fig. 7.2 Schematic representations of chemosensory pathways in the brain. (**a**) Lateral antennular neuropil (LAN) pathway. dCRNs and MRNs from sensilla on the lateral (LF) and medial (MF) antennular flagellum project somatotopically into both lobes of the LAN and couple directly to antennular motoneurons (MN). (**b**) Olfactory pathway. ORNs project nontopographically into the olfactory lobe (OL) which is organized into glomeruli. Information is processed by various types of multiglomerular local interneurons (LN) and output is provided by multiglomerular projections neurons (PN) ascending to the lateral protocerebrum in the eyestalk ganglia

food detection and control of ingestion, but also other "chemical senses" with specialized function, like the control of antennular grooming (Schmidt and Derby 2005) and the coordination of mating and copulation (Belanger and Moore 2006; Bauer, Chap. 14; Snell, Chap. 23).

Behavioral studies have shown that particular chemically-elicited behaviors are mediated exclusively by certain sensilla types. Aesthetascs mediate courtship behavior in response to distant sex pheromones, agonistic interactions, aggregation behavior, and avoidance behavior in response to conspecific alarm cues (Shabani et al. 2008, and references therein). Asymmetric setae – bimodal sensilla neighboring the aesthetascs (Fig. 7.3a–d) – mediate antennular grooming in response to chemical stimulation (Schmidt and Derby 2005). On the other hand, aesthetascs and bimodal sensilla on the antennules both contribute to complex chemically-elicited behaviors such as food search (Steullet et al. 2001; Horner et al. 2004) and associative odor learning (Steullet et al. 2002). This suggests that different sensilla types not only serve specific behavioral functions and thus likely differ in their physiological properties, but also that different sensilla types participate in the control of complex behaviors indicating that they have overlapping functional properties.

7.3 Physiological Methods Used for Analyzing Crustacean Chemoreception

The electrophysiological analysis of crustacean chemoreceptors is mainly based on extracellular recordings from bundles of receptor neuron axons which usually do not allow attributing axon activity to a specific sensillum type. This is particularly problematic for the analysis of ORNs. Their axons are extremely thin (diameter 0.1–0.3 μm; Spencer and Linberg 1986; Mellon et al. 1989) and intermingle with thicker axons from neighboring bimodal sensilla (Fig. 7.3e–g) making it more likely that the latter are the source of discriminable action potentials in the antennular nerve. Only the ORNs of the aesthetascs of *P. argus* have been studied by patch clamp recordings from their somata (pioneered by Anderson and Ache 1985) leaving no ambiguity about the origin of the responses. The processing of

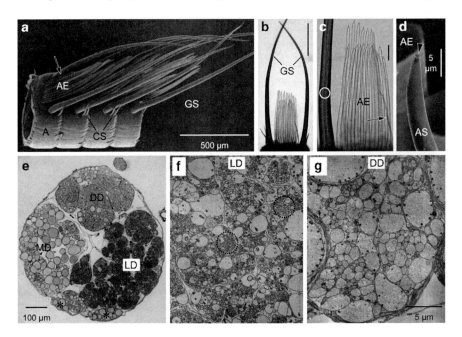

Fig. 7.3 Morphology of crustacean chemoreceptive sensilla and sensory nerves. (**a** – **d**) Tuft region of the lateral flagellum of *P. argus* (scale bars: *B* = 500 μm, *C* = 100 μm). Each annulus (A) bears two rows of aesthetascs (AE) which are associated with three types of bimodal sensilla: long guard setae (GS) with a thin central canal (*circle* in **c**), small companion setae (CS), and one asymmetric seta (AS) (*arrow* in **a** and **c**) with a terminal pore (*arrowhead* in **d**). (**e–g**) Organization of the antennular nerve of *P. argus* (**f**, **g**: TEM micrographs at the scale shown in **g**). (**e**) The antennular nerve is organized into a lateral division (LD) containing axons of sensilla on the lateral flagellum, a dorsal division (DD) containing axons of sensilla on the medial flagellum, a medial division (MD) containing axons of sensilla on the basal segments, and two bundles of motoneuron axons (*asterisks*). (**f**) Axon bundle in the lateral division contains numerous very thin ORN axons (*dotted circles*) mixed with axons of larger diameter. (**g**) Axon bundle in the dorsal division contains axons of various sizes. (**a**, **d**) from Schmidt and Derby 2005; (**e**) from Schmidt et al. 1992

chemosensory input in the CNS of crustaceans has been investigated using intracellular recordings and labeling of the recorded neurons by intracellular markers (Ache and Sandeman 1980; Arbas et al. 1988; Wachowiak and Ache 1994; Mellon and Alones 1995; Schmidt and Ache 1996a, b). Most of the available knowledge is from the antennular-deutocerebral pathway of three species of decapod crustaceans, the spiny lobster *P. argus*, and the crayfish *P. clarkii*, and *Cherax destructor*.

7.4 Physiology of Pheromone Reception in Crustaceans

Most of what is known about crustacean chemoreception is in the context of food detection and ingestion; only very little is known in other behavioral contexts, such as predator avoidance, orientation in the habitat, interactions with symbiotic partners, or intraspecific communication by pheromones. From recent behavioral and neuroanatomical studies, two important concepts about pheromone reception in crustaceans have emerged:

(A) In crustaceans, as insects, pheromones act on two distinct spatial scales, distance and contact. This dichotomy is well established for sex pheromones, some of which mediate distant attraction while others control copulatory behavior upon body contact (e.g., Kamiguchi 1972; Kamio et al. 2002; Bauer, Chap. 14). Responses to distant pheromones (mating, aggregation, agonistic interactions, individual recognition) or alarm cues released by conspecifics are mediated by the aesthetascs (Shabani et al. 2008, and references therein). Behavioral observations suggest, sensilla mediating responses to contact pheromones are located on the appendages touching the conspecific partner, usually the second antennae, and/ or the legs (Bauer, Chap. 14) and hence must be bimodal sensilla.

(B) In decapod crustaceans, neither the sensilla mediating responses to pheromones nor the pathways in the CNS that are implicated in processing pheromone information show specializations indicating that pheromones are detected or processed differently than "ordinary" chemical stimuli. This is obvious in sex pheromone detection where such specializations would be apparent as sex-specific differences in sensilla endowment and/or the neuroanatomy of sensory pathways in the brain as found in many insects. In decapods, sex-specific differences in the number of aesthetascs and of bimodal sensilla on the second antennae only occur in few species of shrimp with males having more sensilla than females (Kamiguchi 1972; Bauer, Chap. 14). However, sex-specific sensilla types or sex-specific differences in the organization of neuropils receiving chemosensory input have not been reported for any decapod.

7.5 Physiology of Chemoreceptor Neurons

Chemosensitivity has been electrophysiologically confirmed for all behaviorally identified major chemosensory organs of crustaceans, namely, the antennules (both flagellae), second antennae, legs, and mouthparts (Caprio and Derby 2008) (Fig. 7.1).

7.5.1   Location of Bimodal Sensilla and Aesthetascs

Bimodal sensilla have been identified on both flagella of the antennules (Cate and Derby 2001, 2002), on the *second antennae* (Voigt and Atema 1992), on the third maxillipeds and other mouthparts (Derby 1982; Garm and Hoeg 2006), and on the distal leg segments in particular the dactyl including – in clawed crustaceans – the chelae (Derby 1982; Altner et al. 1983; Schmidt and Gnatzy 1984). In contrast, aesthetascs only occur on the lateral flagella of the antennules. In *P. argus*, *Panulirus interruptus*, and *H. americanus*, in which most electrophysiological studies were performed, 1,000–1,500 aesthetascs form a regular distal tuft on the lateral flagellum (Figs. 7.1 and 7.3a–c). Each aesthetasc is innervated by approximately 300 ORNs whose somata form a dense cluster below the aesthetasc seta (Spencer and Linberg 1986; Grünert and Ache 1988; Cate and Derby 2001). The aesthetascs are associated with several types of bimodal sensilla, among them are: guard setae, companion setae, and asymmetric setae (Fig. 7.3a–d) (Cate and Derby 2001, 2002).

7.5.2   Response Properties of dCRNs and ORNs

To determine how different CRNs respond to chemical stimuli, the following response properties are determined: (1) general firing properties including spontaneous activity, direction of response (excitation or inhibition), and rhythmic activity, (2) coding of stimulus quality measured as specificity and spectral tuning in response to different single chemical compounds, and (3) coding of stimulus intensity measured as sensitivity in response to different concentrations of one chemical compound.

(A) Bimodal sensilla: dCRNs of bimodal sensilla have very little or no spontaneous activity and respond with an increase in the frequency of action potentials, i.e., an excitation, to chemical stimulation (Figs. 7.4a and 7.6a, c). Typically, dCRNs have a phasic-tonic temporal response profile characterized by a strong and transient excitation at the onset of a stimulus (phasic component) and a lower excitation maintained throughout the stimulus duration (tonic component). Some dCRNs, however, respond purely tonic or purely phasic (e.g., Derby and Atema 1982a, b; Hatt and Bauer 1982; Schmidt and Gnatzy 1989; Garm et al. 2005). In addition to dCRNs, bimodal sensilla contain 1 – 3 MRNs which typically respond directionally to touch or water movements (Derby 1982; Altner et al. 1983; Cate and Derby 2002).

To analyze the spectral tuning of dCRNs, food extracts or mixtures of chemicals mimicking such extracts have been used as "search" stimuli and this limits our knowledge about stimulatory chemicals for crustacean dCRNs to mainly food-related substances. The chemicals most widely tested and found to be stimulatory are L-amino acids followed by ammonium chloride, betaine, nucleotides, amines, sugars, pyridines, organic acids, and small peptides. Only rarely, other chemicals, including the putative pheromone ecdysone (Spencer and Case 1984), were tested

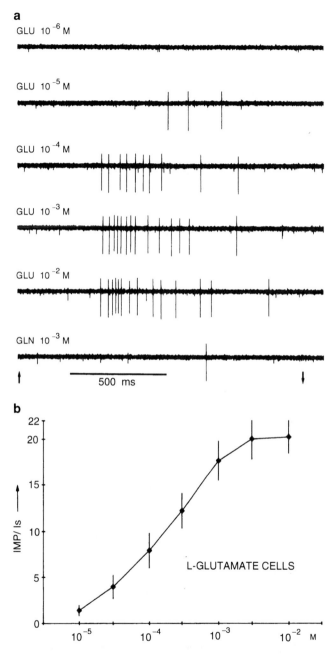

Fig. 7.4 Physiological properties of dCRNs on the legs (in funnel-canal organs of the shore crab, *C. maenas*). (**a**) Responses of an L-glutamate-best cell to stimulation with ascending concentrations of L-glutamate. (**b**) Mean dose–response curve of 13 L-glutamate-best cells. The working range (3×10^{-5} to 3×10^{-3} M) indicates that under natural conditions these cells only respond when contacting food. From Schmidt and Gnatzy 1989

and found to activate dCRNs. In breadth of tuning, dCRNs range from very narrowly tuned responding strongly to only one chemical to more broadly tuned responding strongly to several chemicals (e.g., Tierney et al. 1988; Schmidt and Gnatzy 1989; Corotto, Voigt and Atema 1992; Voigt and Atema 1992; Voigt et al. 1997; Garm et al. 2005) (Fig. 7.5). Between different chemosensory appendages, dCRNs do not differ systematically in breadth of tuning.

Determination of the sensitivity of dCRNs to single chemicals revealed that they mostly have threshold concentrations (minimal concentration eliciting a significant response above control) between 10^{-9} and 10^{-6} M for the most stimulatory chemicals with a working range (concentration range between threshold and saturation) of 2–3 orders of magnitude (Derby and Atema 1982a, b; Hatt and Bauer 1982; Tierney et al. 1988; Voigt and Atema 1992) (Figs. 7.4b and 7.6b, d). However, some dCRNs located on the legs as well as on both flagella of the antennules have either significantly higher ($5 \times 10^{-6} - 5 \times 10^{-4}$ M) or lower ($<10^{-12}$ M) threshold concentrations (Thompson and Ache 1980; Derby and Atema 1982a; Carr et al. 1986; Schmidt and Gnatzy 1989). dCRNs located on mouthparts generally have a high threshold concentration of $10^{-6} - 10^{-5}$ M and a working range of about 3 orders of magnitude (Corotto et al. 1992; Garm et al. 2005).

(B) Aesthetascs: The general response properties of ORNs have three unique features: (1) most ORNs show spontaneous tonic activity ranging from 0.1 to ca. 8 Hz, (2) ORNs are excited by some and inhibited by other chemical stimuli indicating that two transduction pathways are present in each ORN, and (3) some ORNs (ca. 30%) show spontaneous, rhythmic bursts of action potentials (Fig. 7.7) (Anderson and Ache 1985; Michel et al. 1991; Michel et al. 1993; Bobkov and Ache 2007).

The specificity of single ORNs has not been analyzed systematically. In two studies, almost all tested amino acids caused excitatory responses in some ORNs but L-proline, L-arginine, and L-cysteine also elicited inhibitory responses in many ORNs (Michel et al. 1991, 1993) (Fig. 7.7b). Since the same amino acid excited some but inhibited other ORNs, the response type (excitation or inhibition) was not a property of the stimulant but depended on the ORN.

The sensitivity of ORNs has not been analyzed systematically. One study found that the threshold concentration of some amino acid-sensitive ORNs ranged from 10^{-6} to 10^{-4} M with a working range of 1–3 orders of magnitude (Michel et al. 1993) suggesting that ORNs may be less sensitive than CRNs of neighboring bimodal sensilla.

7.6 Projections of Antennular Chemoreceptors in the Brain

Axons from chemo- and mechanosensory sensilla on the antennular flagella course within branches of the antennular nerve to three major target areas in the deutocerebrum: The olfactory lobes (OL), the lateral antennular neuropil (LAN), and the median antennular neuropil (MAN). Backfilling techniques in spiny lobsters have established that ORNs connect with synaptic targets exclusively within the OL, while dCRN and MRN axons target the LAN and MAN (Schmidt et al. 1992;

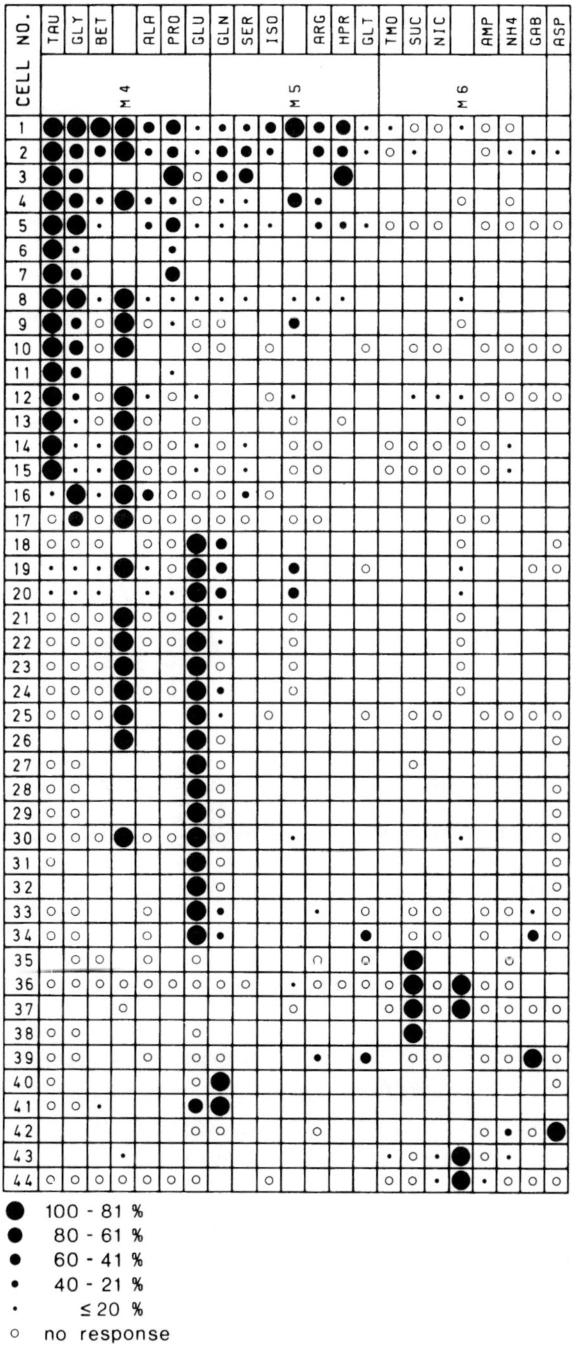

Fig. 7.5 Physiological properties of dCRNs on the legs (in funnel-canal organs of the shore crab, *C. maenas*). Response spectra of 44 cells to stimulation with 19 single chemicals. Some dCRNs are broadly tuned but most are narrowly tuned with L-glutamate-best and taurine-best cells being most frequent. From Schmidt and Gnatzy 1989

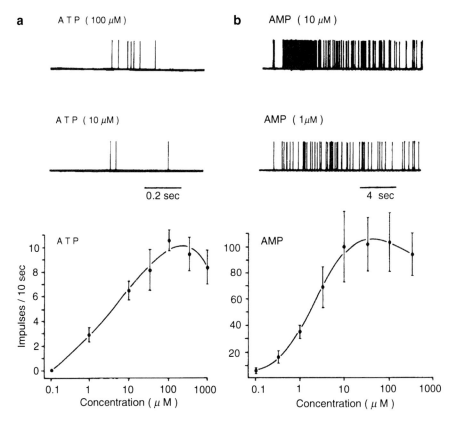

Fig. 7.6 Physiological properties of dCRNs on the antennules. Responses of an ATP-best (**a**) and an AMP-best cell (**c**) on the lateral flagellum of *P. argus* and mean dose–response curves of 11 ATP-best cells (**b**) and 7 AMP-best cells (**d**). The working range of these cells (10^{-7}–10^{-4} M) indicates that under natural conditions they may not respond to distant stimuli. From Carr et al. 1986

Schmidt and Ache 1992) (Fig. 7.2a, b). In crayfish the innervation of neither the MAN nor LAN has been specifically examined, but ORN axons have been successfully traced to the OL (Sandeman and Denburg 1976; Mellon et al. 1989; Mellon and Munger 1990; Mellon and Alones 1993; Sandeman and Sandeman 1994). In spiny lobsters and crayfish, ORN axons penetrate individual OL columnar glomeruli from the periphery, arborize extensively in the peripheral cap region, and a few branches run axially to the base of the columns, making terminal and *en passant* synaptic connections along the way (Figs. 7.2b and 7.10). Electron microscopical studies in the crayfish OL suggest that ORN axons form synapses with higher-order neurons primarily in the cap region, whereas synaptic connections between higher-order neurons predominate in the subcap and base (Mellon and Alones unpublished observation).

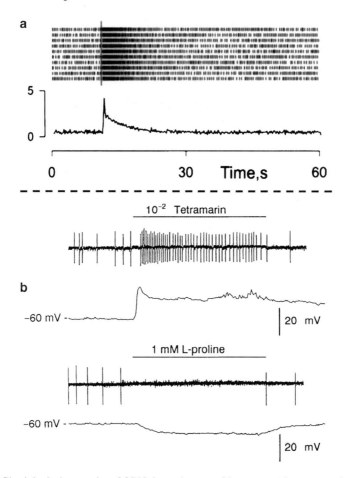

Fig. 7.7 Physiological properties of ORNs in aesthetascs of *P. argus*. (**a**) Responses of an ORN to repetitive stimulation with a food extract and average response (number of spikes/200 ms). Note the ORN-typical high spontaneous activity. (**b**) Coexistence of excitatory and inhibitory responses in a single ORN (top traces – action potentials, bottom traces – membrane potential). The spontaneously active ORN is excited by stimulation with a food extract (Tetramarin) but inhibited by stimulation with L-proline. (**a**) from Bobkov and Ache 2007; (**b**) from Michel et al. 1991

Current interpretations of ORN projections across taxa suggest that each ORN expresses just one type of receptor protein, and all ORNs expressing identical receptor proteins on their dendrites terminate within the same glomerulus. In mammals, each class of ORNs (about 1,000) which expresses the same receptor protein on its dendrites, terminates in a specific glomerulus, or pair of glomeruli, in the olfactory bulb (Munger et al. 2009); this principle, however, may not apply as strictly to insects or to other arthropods.

7.7   Central Olfactory Pathways in Crayfish and Spiny Lobster

7.7.1   *Organization of Deutocerebral and Projection Pathways*

In both crayfish and spiny lobsters, in addition to the paired OLs, the deutocerebrum comprises the paired LANs and the unpaired MAN as well as paired accessory lobes (ALs), olfactory globular tract neuropils (OGTNs), and deutocerebral commissural neuropils (DCNs). While paired ALs are present in crayfish, spiny lobsters, and clawed lobsters, they are highly reduced in size in brachyurans and anomurans, and they are totally absent in the Stomatopoda, the Dendrobranchiata, the Caridea, and the Stenopodidea (Sullivan and Beltz 2004). The AL is organized into small glomeruli and does not receive primary sensory inputs of any modality. It does, however, receive secondary or tertiary multimodal inputs from several sensory systems, including the olfactory system, the visual system, and somatosensory systems (Sandeman et al. 1995; Wachowiak et al. 1996; Mellon 2000). These findings are consistent with a role for higher order multimodal processing in the AL. A deutocerebral commissure (DC) connects the two ALs; its axons terminate bilaterally in AL glomeruli. In *P. argus* local interneurons interconnect the OL with the AL (Wachowiak and Ache 1994; Wachowiak et al. 1996).

Projection neurons (PNs) are the output pathways from the OL and AL and convey olfactory and multimodal information to the lateral protocerebrum. The somata of PNs receive their inputs from either the OL or AL and their axons ascend the olfactory globular tract (OGT) coursing within the eyestalks to target neurons in neuropils of the lateral protocerebrum, the medulla terminalis or the hemiellipsoid body, respectively (Mellon et al. 1992; Sullivan and Beltz 2001, 2005a) (Fig. 7.2b). The evolutionary history of the OL/AL/lateral protocerebrum axis exhibits extreme plasticity, confounding our current, imperfect understanding of the functional significance of its major and minor subdivisions. The hemiellipsoid body of *P. clarkii* is bilobed, and its principal local interneurons, the parasol cells, are anatomically and functionally separate in the two lobes. It has been suggested that common targeting of parasol cells in one lobe by bilateral AL inputs and of those in the other lobe by exclusively ipsilateral AL inputs could be the basis for comparing odorant signal strength at the two antennules (McKinzie et al. 2003). As the parasol cells of the *Procambarus* hemiellipsoid body receive their input primarily if not, exclusively, from the ALs, it should not be surprising that they are driven not only by exposure of the lateral antennular flagella to odors (Mellon and Alones 1997; Mellon and Wheeler 1999), but also by visual and tactile inputs (Mellon 2000). It can be speculated, therefore, that the hemiellipsoid bodies may be important in interpreting odors on the basis of environmental contexts, and in this respect they would be similar to the mushroom bodies in the insect brain that have a similar position within the central olfactory pathway.

7.7.2   *Electrophysiology of Deutocerebral LNs and PNs*

Electrophysiological studies of deutocerebral structures have been initiated in both crayfish and spiny lobster. Several types of LNs invest dendritic arbors within the OL and LAN of both crustaceans (Arbas et al. 1988; Mellon and Alones 1995; Schmidt and Ache 1996a, b). All of those observed so far are multiglomerular and respond to broad-spectrum complex odorants via their heavily arborized dendrites within the OL glomeruli (Fig. 7.8). In both species some of these LNs respond also to hydrodynamic stimulation of the antennules presumably via additional dendritic arbors that invest the LAN (Mellon and Alones 1995; Schmidt and Ache 1996a, b; Mellon 2005; Mellon and Humphrey 2007). Integration of odorant and

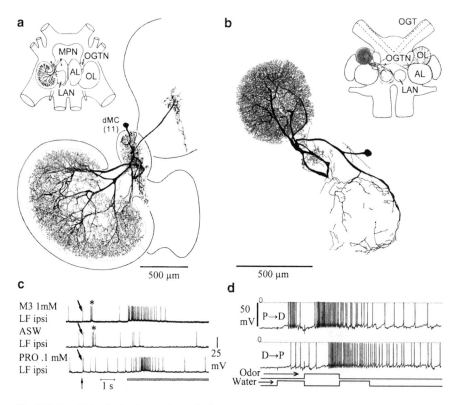

Fig. 7.8 Dendritic arborizations and electrical responses to odors of similar classes of multiglomerular local interneurons in the deutocerebrum of *P. argus* (**a, c**) and *P. clarkii* (**b, d**). Note that in both crustaceans the dendrites invest both the OL and LAN, and both respond not only to odorants but also to the onset of hydrodynamic flow past the antennular flagellum. (**a, c**) from Schmidt and Ache 1996b; (**b**) from Mellon and Alones 1995; (**d**) from Mellon and Humphrey 2007

hydrodynamic inputs is clearly a common property of the deutocerebrum. LNs are anatomically divided into those that enter the core of the OL and radiate their dendritic branches outwards through the glomeruli toward the cap and others that send their dendrites around the periphery of the OL and penetrate the glomerular individual columns through the cap region (Mellon and Alones 1995; Schmidt and Ache 1996b) (Fig. 7.2b).

Electrophysiological studies of OL PNs in both spiny lobster and crayfish suggest that the individual neurons receive their inputs from several different glomeruli (up to 80% of the available glomeruli in the spiny lobster) and thus the majority if not all of these PNs must be driven by a variety of odorant inputs (Wachowiak and Ache 1994; Schmidt and Ache 1996b; Sandeman and Mellon 2002). In a few successful penetrations with glass microelectrodes within the crayfish OL, electrical responses of PNs were obtained to stimulation of the ipsilateral antennule with single amino acids and simple mixtures (Mellon unpublished observations). The responses to odorant stimulation consisted of an initial brief excitation followed by a much more prolonged inhibition (Fig. 7.9). In *P. argus*, stimulation of the antennular nerve by single electrical shocks generated similarly patterned responses in individual PNs (Wachowiak and Ache 1994, 1998). Assuming that ORN input directly excites PNs and that the subsequent inhibition is secondary, a model for possible OL network properties can be proposed (Fig. 7.10). In this scheme individual ORNs send their axon terminals to a specific glomerulus dependent upon the olfactory determinant sensed by their dendritic receptor proteins. Several different PNs, on the other hand, may supply dendrites to any specific glomerulus; while some of these may receive strong excitatory ORN inputs in any one glomerulus, other PN dendrites within the same glomerulus may be excited less strongly but may themselves be more activated by ORNs in a different or several other glomeruli. Thus, no matter what the nature of the odorant (assumed to be a complex mix of olfactory determinants), all PNs will receive global inhibition following the selective excitation of only a few. Not included in the model is the evidence for widespread presynaptic inhibition of ORN terminals within the OL (Wachowiak and Ache 1998), possibly mediated by LNs whose dendrites course around the rim (glomerular cap region) of the OL, and presumably having a similar functional significance to the global postsynaptic inhibition as the one proposed in Fig. 7.10. This scheme predicts that individual ORNs and their target glomeruli represent individual olfactory determinants, while individual PNs are excited by mixtures of odor determinants, dependent upon the combination of glomeruli from which they receive strong inputs. The secondary inhibition provides an assurance that only the strongest ORN-PN input pathways are expressed as PN output spikes. The network would permit transient excitation of large numbers of different combinations of PNs, while suppressing other possible combinations, thereby forming a computational basis for identification of novel or previously learned combinations of olfactory determinants ("odors") sensed by the ORN array.

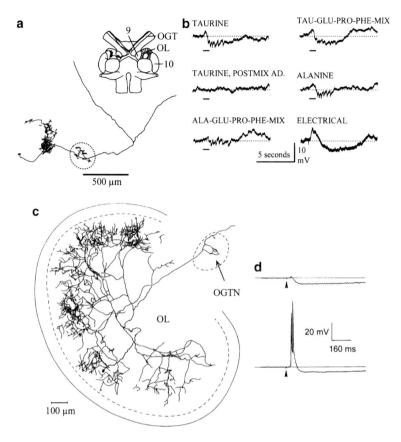

Fig. 7.9 Dendritic distribution and electrical responses of OL projection neurons (PNs) in *P. clarkii* (**a, b**) and *P. argus* (**c, d**). In both crustaceans the PNs are multiglomerular and respond with characteristic electrical profiles to broad-spectrum chemical stimulation (**b**) or to electrical stimulation (**d**). An initial excitatory phase is truncated by secondary inhibition which in *P. clarkii* does not persist. (**a, b**) from Sandeman and Mellon 2002; (**c, d**) from Wachowiak and Ache 1994

7.8 Persistent Neurogenesis in the Deutocerebrum

Beginning in 1997, using the mitotic incorporation of the thymidine analog 5-bromo-2'-deoxyuridine (BrdU), reports began to emerge of persistent neurogenesis in the brains of adult crustaceans (Schmidt 2007, and references therein; Sullivan et al. 2007, and references therein). In crayfish, spiny lobsters, and clawed lobsters persistent neurogenesis is a feature of PNs in cell cluster 10 and LNs in cell cluster 9 of the deutocerebrum. In the crabs *Cancer pagurus* and *Carcinus maenas*, (both brachyurans having highly reduced ALs), and in the shrimp *Sicyonia brevirostris* (having no ALs), neurogenesis was observed in adult animals in cell cluster

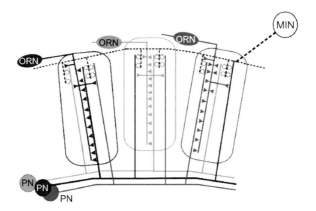

Fig. 7.10 Conceptual model to explain response profiles of crustacean OL PNs. Different classes of ORNs are tuned to specific olfactory determinants and each class terminates within a different glomerulus. OL PNs are less specific and present dendritic arbors across large numbers of glomeruli; however, it is hypothesized that each glomerulus is supplied by a preferred class of PN (same shade in the diagram) as well as a few other classes, so that the preferred PN receives most of the excitatory synaptic input from the ORNs in that glomerulus. In each glomerulus, the ORN input makes excitatory connections not only with PNs but also with multiglomerular inhibitory neurons (MINs) that cause feedback inhibition to the intraglomerular PNs and feedforward inhibitory input to all other PNs. Output of the PNs is thus dependent upon matching their preferences with that of the array of glomeruli that they innervate. This scheme omits the evidence for presynaptic inhibition of individual ORN terminals within the OL, the functional significance of which could be similar to that of the MINs

10. In the crayfish *Cherax destructor,* convincing evidence has been obtained from PNs double labeled for BrdU and dextran that the nascent cells are incorporated into neuronal circuits of the OL and the AL (Sullivan and Beltz 2005b). In *Cancer pagurus, Carcinus maenas,* and the hermit crab *Pagurus bernhardus,* an additional site of neurogenesis was found in a cell cluster close to the hemiellipsoid body in the lateral protocerebrum (Schmidt 2007). In the brachyuran crab *Libinia emarginata,* adult neurogenesis has been found in cell clusters 10 and 9 and also near the hemiellipsoid body (Sullivan et al. 2007). This last finding is of particular interest because *L. emarginata* has a terminal adult molt, after which no new ORNs are incorporated into the antennule; nonetheless, neurogenesis in the brain continues throughout the animal's life, arguing against a role for nascent PNs as accommodation to the continuous addition of new ORNs. It would appear, therefore, that neuronal proliferation in cluster 10 is a general feature found in all decapods, whereas proliferation in cluster 9 and in a soma cluster close to the hemiellipsoid body only occurs in some species.

Why is persistent neurogenesis seen primarily in the central olfactory pathway of adult crustaceans, and what factors are involved in its control? Most, but not all crustaceans continue to grow throughout their lives which, in some cases, span decades. During this time, new sensilla of all types, including aesthetascs, are added to the antennules and other appendages and, as a result, new sensory neurons must be incorporated within the framework of the OLs and other CNS regions. It has been argued (Schmidt 2007) that unlike the addition of somatotopic inputs to the structured brain neuropils of the deuto- and tritocerebrum, where simple branching of intrinsic neurons can accommodate new inputs in an orderly fashion, in the olfactory system novel combinations of ORN inputs that may be critical for the animals' survival can better be accommodated by additional classes of PNs that selectively provide an output pathway from those novel combinations of glomeruli that cannot be predicted. New PNs thus would form the substrate by which novel odors could be "learned" throughout the animals' lifetime. Although the number of classes of different ORNs may remain constant and the specific kinds of olfactory determinants the ORN array can detect is fixed throughout an animal's lifetime, new combinations of these determinants, i.e., complex environmental odors, will be interpreted through brain rewiring.

Several factors have been identified that impact persistent neurogenesis in the brains of decapod crustaceans. Environmental richness has been shown to influence not only proliferation in clusters 9 and 10 of the crayfish *C. destructor,* but also neuronal survival (Sandeman and Sandeman 2000). Young animals were confined either communally or individually, exposed to the same water conditions and maintained on the same feeding regimen. BrdU staining of nascent brain neurons indicated that brains of animals that had lived under "enriched conditions" had significantly more BrdU+ neurons than the brains of animals kept for the same amount of time in the "impoverished condition." The data suggest, therefore, that social interactions may play a critical role in regulating neurogenesis, at least in juvenile animals. Other influences on persistent neurogenesis in crustaceans have also been examined, including a possible role of sensory afferents (Schmidt 2007), circadian time (Goergen et al. 2002), and dietary omega-3 fatty acids (Beltz et al. 2007). The role of adult neurogenesis in the crustacean olfactory pathway is a relatively young field of study, but its implications are sufficiently profound that it will occupy an increasingly prominent place in the olfactory literature for the foreseeable future.

7.9 Summary and Outlook: What Can We Learn by Studying Crustacean Chemoreception?

In spite of more than 100 years of research, the physiology of crustacean chemoreception is far from being understood in detail. Several factors have delayed progress:

1. The aesthetascs are olfactory sensilla and usually all chemoreception mediated by the lateral antennular flagella (or the entire antennules) was attributed to them and hence considered as "olfaction." This interpretation is flawed because other chemosensory sensilla are associated with the aesthetascs and likely the activity of dCRNs in these bimodal sensilla was analyzed in most axon recordings from the lateral flagella.

2. Numerous attempts have been made to divide crustacean chemoreception in "olfaction" and "taste" along the lines of the perceived duplicity of chemical senses in mammals. These attempts have not been successful and in the meantime, the conceptual basis for grouping crustacean chemoreception into two chemical senses has become obsolete because the mammalian nose turned out to be a compound sense organ housing several chemosensory systems with distinct physiology and function (Munger et al. 2009).

3. The analysis of chemoreception in crustaceans has been focused almost exclusively on food-related chemicals. However, chemoreception is an important sensory channel in many other behavioral contexts such as social interactions and the assessment of danger from predation. Very likely different chemicals are important in these other contexts, but only recently progress was made in the identification of such chemicals and the sensilla detecting them.

4. The available data sets about the physiological properties of dCRNs and ORNs and about the functional organization of chemosensory pathways in the CNS are incongruent. While much more is known about the physiology of dCRNs of bimodal sensilla than of aesthetasc ORNs, the olfactory pathway in the brain is far better analyzed than any pathway receiving chemosensory input from bimodal sensilla.

Nevertheless, the following conclusion of broader significance emerges from research on crustacean chemoreception: Chemoreception in crustaceans is organized into two "modes," for which we propose the terms "Olfaction" and "Distributed Chemoreception" (Fig. 7.1). This distinction is based on the profound difference in the construction of the two classes of crustacean chemosensory sensilla (aesthetascs vs. bimodal sensilla) and the equally profound difference in the neuroanatomical organization of the neuropils receiving their respective input (OL vs. sensory-motor neuropils). Since the aesthetasc-OL pathway shows obvious parallels to the olfactory pathway of vertebrates and insects (structure of receptor neurons, first synaptic relay with glomerular organization and adult neurogenesis), it seems justified to continue using the term "Olfaction" for chemoreception mediated by this pathway. The second mode, "Distributed Chemoreception," however, cannot simply be equated with "taste" or with insect "contact chemoreception" which it parallels in terms of sensilla construction and functional neuroanatomy. "Distributed Chemoreception" comprises not only food localization and control of ingestion by bimodal sensilla on legs and mouthparts – i.e., taste – but also other bimodal sensilla on diverse appendages mediating different behaviors such as food search, grooming, or mating. The low threshold sensitivity of many dCRNs indicates that chemicals from distant sources can be detected (e.g., Thompson and Ache 1980; Derby and Atema 1982a) and the MRNs of bimodal sensilla are not exclusively tactile but in some cases are more sensitive and respond to water vibrations or flow (e.g., Derby 1982; Schmidt and Ache 1996a). Thus dCRNs and MRNs apparently not always require contact with a stimulus source but can be activated by distant stimulus sources as well.

To understand why in the marine environment two distinct modes of chemoreception evolved in crustaceans, we have to delineate what the unique properties of them are. As has been detailed above, "Olfaction" and Distributed Chemoreception" differ in: (A) the number of "chemosensory channels" (aesthetascs house several hundred ORNs, whereas bimodal sensilla typically contain fewer than 10 dCRNs), (B) the packaging of dCRNs with MRNs and the "topographic logic" information processing in the target neuropils (ORNs are not packaged with MRNs, their projections are nontopographic, and information processing in the OL is likely based on an odotopic map rather than a somatotopic map; Schmidt, 2007), (C) the integration with motor control (the target neuropils of dCRNs but not the OL contain motoneurons innervating the appendage providing the chemosensory input), and (D) adult neurogenesis in the target neuropils (in the OL but not in the target neuropils of bimodal sensilla, new interneurons are continuously generated in adults, although the sensory input from both aesthetascs and bimodal sensilla increases with each molt). Together these points suggest that the essence of "Olfaction" is to provide a detailed representation of the complex chemical environment integrating chemical signals from a variety of potentially interesting sources (food, conspecifics, predators, symbiotic partners, landscape) without reference to the location of stimuli and not to generate directed behavioral output by exerting motor control. In contrast, the essence of "Distributed Chemoreception" is to form representations of only few key chemicals (food-related chemicals, pheromones) within a somatotopic context provided by mechanoreception. The integration of chemo- and mechanosensory information permits pinpointing the location of chemical stimuli and can be utilized to control movements of the stimulated appendage via direct sensory-motor coupling and to orchestrate this with movements of other appendages.

As the functional organization of chemoreception in fish and crustaceans clearly demonstrates, the differentiation of chemoreception into distinct chemical senses must have its evolutionary origin in the aquatic environment. In light of this scenario, many apparently distinguishing features of taste and olfaction in terrestrial vertebrates and insects appear to be adaptations to the terrestrial environment, raising the question about the ultimate causes for the differentiation of chemoreception into different "modes." This fundamental question can only be answered by further studying the functional organization and behavioral function of chemoreception in aquatic animals. We hope that doing so in crustaceans will continue to provide crucial insights into the evolutionary history of olfaction, taste, and other chemical senses.

References

Ache BW, Sandeman DC (1980) Olfactory-induced central neural activity in the murray crayfish, *Euastacus armatus*. J Comp Physiol 140:295–301

Altner I, Hatt H, Altner H (1983) Structural properties of bimodal chemo- and mechanosensitive setae on the pereiopod chelae of the crayfish, *Austropotamobius torrentium*. Cell Tissue Res 228:357–374

Anderson PAV, Ache BW (1985) Voltage- and current-clamp recordings of the receptor potential in olfactory receptor cells in situ. Brain Res 338:273–280

Arbas EA, Humphreys CJ, Ache BW (1988) Morphology and physiological properties of interneurons in the olfactory midbrain of the crayfish. J Comp Physiol A 164:231–241

Atema J (1980) Smelling and tasting underwater. Oceanus 23:4–18

Belanger RM, Moore PA (2006) The use of the major chelae by reproductive male crayfish (*Orconectes rusticus*) for discrimination of female odours. Behaviour 143:713–731

Beltz BS, Tlusty MF, Benton JL, Sandeman DC (2007) Omega-3 fatty acids upregulate adult neurogenesis. Neurosci Lett 415:154–158

Bobkov YV, Ache BW (2007) Intrinsically bursting olfactory receptor neurons. J Neurophysiol 97:1052–1057

Brock F (1926) Das Verhalten des Einsiedlerkrebses *Pagurus arrosor* Herbst während der Suche und Aufnahme der Nahrung. Beitrag zu einer Umweltanalyse. Z Morph Ökol Tiere 6:415–552

Caprio J, Derby CD (2008) Aquatic animal models in the study of chemoreception. In: Firestein S, Beauchamp GK (eds) The senses: a comprehensive reference, vol 4, Olfaction & Taste. Academic Press, San Diego, pp 97–133

Carr WES, Gleeson RA, Ache BW, Milstead ML (1986) Olfactory receptors of the spiny lobster: ATP-sensitive cells with similarities to P2-type purinoceptors of vertebrates. J Comp Physiol A 158:331–338

Cate HS, Derby CD (2001) Morphology and distribution of setae on the antennules of the Caribbean spiny lobster *Panulirus argus* reveal new types of bimodal chemo-mechanosensilla. Cell Tissue Res 304:439–454

Cate HS, Derby CD (2002) Ultrastructure and physiology of the hooded sensillum, a bimodal chemo-mechanosensillum of lobsters. J Comp Neurol 442:293–307

Corotto F, Voigt R, Atema J (1992) Spectral tuning of chemoreceptor cells of the third maxilliped of the lobster, *Homarus americanus*. Biol Bull 183:456–462

Derby CD (1982) Structure and function of cuticular sensilla of the lobster *Homarus americanus*. J Crust Biol 2:1–21

Derby CD, Atema J (1982a) Chemosensitivity of walking legs of the lobster *Homarus americanus*: neurophysiological response spectrum and thresholds. J Exp Biol 98:303–315

Derby CD, Atema J (1982b) Narrow-spectrum chemoreceptor cells in the walking legs of the lobster *Homarus americanus*: taste specialists. J Comp Physiol A 146:181–189

Garm A, Hoeg JT (2006) Ultrastructure and functional organization of mouthpart sensory setae of the spiny lobster *Panulirus argus*: new features of putative mechanoreceptors. J Morphol 267:464–476

Garm A, Shabani S, Hoeg JT, Derby CD (2005) Chemosensory neurons in the mouthparts of the spiny lobsters *Panulirus argus* and *Panulirus interruptus* (Crustacea: Decapoda). J Exp Mar Biol Ecol 314:175–186

Goergen EM, Bagay LA, Rehm K, Benton JL, Beltz BS (2002) Circadian control of neurogenesis. J Neurobiol 53:90–95

Grünert U, Ache BW (1988) Ultrastructure of the aesthetasc (olfactory) sensilla of the spiny lobster, *Panulirus argus*. Cell Tissue Res 251:95–103

Hatt H, Bauer U (1982) Electrophysiological properties of pyridine receptors in the crayfish walking leg. J Comp Physiol A 148:221–224

Hindley JPR (1975) The detection, localization and recognition of food by juvenile banana prawns, *Penaeus merguiensis* de Man. Mar Behav Physiol 3:193–210

Horner AJ, Weissburg MJ, Derby CD (2004) Dual antennular pathways can mediate orientation by Caribbean spiny lobsters in naturalistic flow conditions. J Exp Biol 207:3785–3796

Kamiguchi Y (1972) Mating behavior in the freshwater prawn, *Palaemon paucidens*. A study of the sex pheromone and its effects on males. J Fac Sci Hokkaido Univ Ser VI Zool 18:347–355

Kamio M, Matsunaga S, Fusetani N (2002) Copulation pheromone in the crab *Telmessus cheir-agonus* (Brachyura: Decapoda). Mar Ecol Prog Ser 234:183–190

Laverack MS (1968) On the receptors of marine invertebrates. Oceanogr Mar Biol Annu Rev 6:249–324

Laverack MS (1988) The diversity of chemoreceptors. In: Atema J, Fay RR, Popper AN, Tavolga WN (eds) Sensory biology of aquatic animals. Springer, New York, pp 287–312

McKinzie ME, Benton JL, Beltz BS, Mellon D (2003) Parasol cells in the hemiellipsoid body of the crayfish *Procambarus clarkii*: dendritic branching patterns and functional implications. J Comp Neurol 462:168–179

Mellon D (2000) Convergence of multimodal sensory input onto higher-level neurons of the crayfish olfactory pathway. J Neurophysiol 84:3043–3055

Mellon D (2005) Integration of hydrodynamic and odorant inputs by local interneurons of the crayfish deutocerebrum. J Exp Biol 208:3711–3720

Mellon D, Alones V (1993) Cellular organization and growth-related plasticity of the crayfish olfactory midbrain. Microsc Res Tech 24:231–259

Mellon D, Alones V (1995) Identification of three classes of multiglomerular, broad- spectrum neurons in the crayfish olfactory midbrain by correlated patterns of electrical activity and dendritic arborization. J Comp Physiol A 177:55–71

Mellon D, Alones VE (1997) Response properties of higher level neurons in the central olfactory pathway of the crayfish. J Comp Physiol A 181:205–216

Mellon D, Humphrey JAC (2007) Directional asymmetry in responses of local interneurons in the crayfish deutocerebrum to hydrodynamic stimulation of the lateral antennular flagellum. J Exp Biol 210:2961–2968

Mellon D, Munger SD (1990) Nontopographic projection of olfactory sensory neurons in the crayfish brain. J Comp Neurol 296:253–262

Mellon D, Wheeler CJ (1999) Coherent oscillations in membrane potential synchronize impulse bursts in central olfactory neurons of the crayfish. J Neurophysiol 81:1231–1241

Mellon D, Tuten HR, Redick J (1989) Distribution of radioactive leucine following uptake by olfactory sensory neurons in normal and heteromorphic crayfish antennules. J Comp Neurol 280:645–662

Mellon D, Alones V, Lawrence MD (1992) Anatomy and fine structure of neurons in the deutocerebral projection pathway of the crayfish olfactory system. J Comp Neurol 321:93–111

Michel WC, McClintock TS, Ache BW (1991) Inhibition of lobster olfactory receptor cells by an odor- activated potassium conductance. J Neurophysiol 65:446–453

Michel WC, Trapido-Rosenthal HG, Chao ET, Wachowiak M (1993) Stereoselective detection of amino acids by lobster olfactory receptor neurons. J Comp Physiol A 171:705–712

Munger SD, Leinders-Zufall T, Zufall F (2009) Subsystem organization of the mammalian sense of smell. Annu Rev Physiol 71:115–140

Sandeman DC, Denburg JL (1976) The central projections of chemoreceptor axons in the crayfish revealed by axoplasmic transport. Brain Res 115:492–496

Sandeman D, Mellon D (2002) Olfactory centers in the brain of freshwater crayfish. In: Wiese K (ed) The crustacean nervous system. Springer, Berlin, pp 386–404

Sandeman DC, Sandeman RE (1994) Electrical responses and synaptic connections of giant serotonin- immunoreactive neurons in crayfish olfactory and accessory lobes. J Comp Neurol 341:130–144

Sandeman R, Sandeman D (2000) "Impoverished" and "enriched" living conditions influence the proliferation and survival of neurons in crayfish brain. J Neurobiol 45:215–226

Sandeman D, Sandeman R, Derby C, Schmidt M (1992) Morphology of the brain of crayfish, crabs, and spiny lobsters: A common nomenclature for homologous structures. Biol Bull 183:304–326

Sandeman D, Beltz BS, Sandeman R (1995) Crayfish brain interneurons that converge with serotonin giant cells in accessory lobe glomeruli. J Comp Neurol 352:263–279

Schachtner J, Schmidt M, Homberg U (2005) Organization and evolutionary trends of primary olfactory brain centers in Tetraconata (Crustacea+Hexapoda). Arthrop Struct Dev 34:257–299

Schmidt M (2007) The olfactory pathway of decapod crustaceans - an invertebrate model for life-long neurogenesis. Chem Senses 32:365–384

Schmidt M, Ache BW (1992) Antennular projections to the midbrain of the spiny lobster. II. Sensory innervation of the olfactory lobe. J Comp Neurol 318:291–303

Schmidt M, Ache BW (1996a) Processing of antennular input in the brain of the spiny lobster, *Panulirus argus*. I. Non-olfactory chemosensory and mechanosensory pathway of the lateral and median antennular neuropils. J Comp Physiol A 178:579–604

Schmidt M, Ache BW (1996b) Processing of antennular input in the brain of the spiny lobster, *Panulirus argus*. II. The olfactory pathway. J Comp Physiol A 178:605–628

Schmidt M, Derby CD (2005) Non-olfactory chemoreceptors in asymmetric setae activate antennular grooming behavior in the Caribbean spiny lobster *Panulirus argus*. J Exp Biol 208:233–248

Schmidt M, Gnatzy W (1984) Are the funnel-canal organs the "campaniform sensilla" of the shore crab, *Carcinus maenas* (Decapoda, Crustacea)? II. Ultrastructure. Cell Tissue Res 237:81–93

Schmidt M, Gnatzy W (1989) Specificity and response characteristics of gustatory sensilla (funnel-canal organs) on the dactyls of the shore crab, *Carcinus maenas* (Crustacea, Decapoda). J Comp Physiol A 166:227–242

Schmidt M, Van Ekeris L, Ache BW (1992) Antennular projections to the midbrain of the spiny lobster. I. Sensory innervation of the lateral and medial antennular neuropils. J Comp Neurol 318:277–290

Shabani S, Kamio M, Derby CD (2008) Spiny lobsters detect conspecific blood-borne alarm cues exclusively through olfactory sensilla. J Exp Biol 211:2600–2608

Spencer M, Case JF (1984) Exogenous ecdysteroids elicit low-threshold sensory responses in spiny lobsters. J Exp Zool 229:163–166

Spencer M, Linberg KA (1986) Ultrastructure of aesthetasc innervation and external morphology of the lateral antennule setae of the spiny lobster *Panulirus interruptus* (Randall). Cell Tissue Res 245:69–80

Spiegel A (1927) Über die Chemorezeption von *Crangon vulgaris* Fabr. Z Vergl Physiol 6:688–730

Steullet P, Dudar O, Flavus T, Zhou M, Derby CD (2001) Selective ablation of antennular sensilla on the caribbean spiny lobster *Panulirus argus* suggests that dual antennular chemosensory pathways mediate odorant activation of searching and localization of food. J Exp Biol 204:4259–4269

Steullet P, Krützfeldt DR, Hamidani G, Flavus T, Ngo V, Derby CD (2002) Dual antennular chemosensory pathways mediate odor-associative learning and odor discrimination in the Caribbean spiny lobster *Panulirus argus*. J Exp Biol 205:851–867

Sullivan JM, Beltz BS (2001) Neural pathways connecting the deutocerebrum and lateral protocerebrum in the brains of decapod crustaceans. J Comp Neurol 441:9–22

Sullivan JM, Beltz BS (2004) Evolutionary changes in the olfactory projection neuron pathways of eumalacostracan crustaceans. J Comp Neurol 470:25–38

Sullivan JM, Beltz BS (2005a) Integration and segregation of inputs to higher-order neuropils of the crayfish brain. J Comp Neurol 481:118–126

Sullivan JM, Beltz BS (2005b) Newborn cells in the adult crayfish brain differentiate into distinct neuronal types. J Neurobiol 65:157–170

Sullivan JM, Sandeman DC, Benton JL, Beltz BS (2007) Adult neurogenesis and cell cycle regulation in the crustacean olfactory pathway: from glial precursors to differentiated neurons. J Mol Hist 38:527–542

Tautz J, Müller-Tautz R (1983) Antennal neuropile in the brain of the crayfish: morphology of neurons. J Comp Neurol 218:415–425

Thompson H, Ache BW (1980) Threshold determination for olfactory receptors of the spiny lobster. Mar Behav Physiol 7:249–260

Tierney AJ, Voigt R, Atema J (1988) Response properties of chemoreceptors from the medial antennular filament of the lobster *Homarus americanus*. Biol Bull 174:364–372

Voigt R, Atema J (1992) Tuning of chemoreceptor cells of the second antenna of the American lobster (*Homarus americanus*) with a comparison of four of its other chemoreceptor organs. J Comp Physiol A 171:673–683

Voigt R, Weinstein AM, Atema J (1997) Spectral tuning of chemoreceptor cells in the lateral antennules of the American lobster, *Homarus americanus*. Mar Fresh Behav Physiol 30:19–27

Wachowiak M, Ache BW (1994) Morphology and physiology of multiglomerular olfactory projection neurons in the spiny lobster. J Comp Physiol A 175:35–48

Wachowiak M, Ache BW (1998) Multiple inhibitory pathways shape odor-evoked responses in lobster olfactory projection neurons. J Comp Physiol A 182:425–434

Wachowiak M, Diebel CE, Ache BW (1996) Functional organization of olfactory processing in the accessory lobe of the spiny lobster. J Comp Physiol A 178:211–226

Chapter 8
The Neural and Behavioral Basis of Chemical Communication in Terrestrial Crustaceans

Bill S. Hansson, Steffen Harzsch, Markus Knaden, and Marcus Stensmyr

Abstract Within Crustacea, representatives of at least five major taxa have succeeded in the transition from an aquatic to a fully terrestrial lifestyle: Isopoda, Amphipoda, Astacida, Anomura, and Brachyura. Land-living crustaceans are fascinating animals that during a very limited time period at an evolutionary time scale have adapted to a number of diverse terrestrial habitats in which they have become highly successful, and in some case the predominant life forms. Living on land raises new questions regarding the evolution of chemical communication because a transition from sea to land means that molecules need to be detected in gas phase instead of in water solution. The odor stimulus also changes from mainly hydrophilic molecules in aqueous solution to mainly hydrophobic in the gaseous phase. Behavioral studies have provided evidence that some land-living crustaceans, namely terrestrial hermit crabs (Anomura, Coenobitidae) have achieved high efficiency in detecting food from a distance and in responding to airborne odors, in short, that they have evolved an excellent sense of distance olfaction. How do crustaceans on land solve the tasks of odor detection and odor information processing and how have the new selection pressures reshaped the sense of smell? In the present contribution, we review the current knowledge on morphological aspects of the olfactory system of terrestrial crustaceans focusing on representatives of the Anomura and Isopoda. Terrestrial members of the latter taxon have greatly reduced first antennae and seem to have given up their deutocerebral olfactory pathway. Instead, they have shifted gustatory abilities to the second antennae and the tritocerebrum but it remains to be shown if these animals evolved an effective system of aerial olfaction. Within the Anomura, however, terrestrial hermit crabs (Coenobitidae) have greatly inflated those parts of the brain that are responsible for primary olfactory processing, the olfactory lobes. Electro-antennographic detection studies with the well developed first antennae of the giant robber crab *Birgus latro* demonstrated the capacity of this organ to detect volatile chemical information. These experiments point to an olfactory system as

B.S. Hansson (✉)
Department of Evolutionary Neuroethology, Max Planck Institute for Chemical Ecology,
Hans-Knöll-Str. 8, 07745 Jena, Germany
e-mail: hansson@ice.mpg.de

T. Breithaupt and M. Thiel (eds.), *Chemical Communication in Crustaceans*,
DOI 10.1007/978-0-387-77101-4_8, © Springer Science+Business Media, LLC 2011

sensitive as the most sensitive general odor detecting olfactory sensory neurons found in insects and that therefore is well suited to explore the terrestrial olfactory landscape. We also summarize ongoing efforts to explore olfactory-guided behavior of the giant robber crab on Christmas Island, Indian Ocean. We conclude that comparative studies between fully aquatic crustaceans and closely related terrestrial taxa provide a powerful means of investigating the evolution of chemosensory adaptations in these two environments. Future studies must address three main aspects: the behavior towards active odors (both pheromones and other semiochemicals) has to be clearly shown and quantified, the chemical identity of key odor cues and the source of these compounds have to be established, and the sensory base of detection and information processing in the chemical senses need to be further studied.

8.1 Introduction

Most organisms are dependent on semiochemicals to perform vital steps in their reproductive, food-search or predator avoidance behavior. To detect and perceive semiochemicals, animals must possess an olfactory or gustatory system. In general, most known semiochemicals work via the olfactory system, and our description of chemoreception in land-living crustaceans thus concentrate on the sense of smell. How do crustaceans on land solve the tasks of odor detection and odor information processing? Living on land raises new questions regarding the evolution of chemical communication. The fact that crustaceans have lived on land during a very limited time period over an evolutionary time scale raises one of the most interesting questions in this area: how have the dramatic changes in selection pressure not only on the olfactory system, but also on for example pheromone production systems, modified the corresponding bodily functions? In this chapter we will dwell on the present state of knowledge regarding these questions, with a very clear evolutionary twist.

8.2 Classical Work on Land Hermit Crabs in Jena

Through the work of Hans Jürgen Harms (1885–1956) at the Friedrich Schiller University, Jena early on became a focus of research on land hermit crabs. Harms studied zoology and medical sciences in Marburg and also in Manchester, Cambridge and Plymouth (1906). He held positions in Marburg, Münster and Königsberg to get tenure as full professor in Tübingen in 1925 where he stayed until 1935, when he took over the prestigious chair of Zoology at the University of Jena (including the position as director of the Phyletisches Museum), a chair that first was held by Ernst Haeckel from 1865 to 1909. Harms left Jena in 1949 because of political motivations and returned to the University of Marburg. He died in Marburg in 1956 aged 71 years.

During two expeditions to SE Asia, Harms carried out intense studies on the biology of a terrestrial anomuran crab, the giant robber crab *Birgus latro* that resulted in two major publications (Harms 1932, 1937), which today still serves as major sources of information on the morphology, behavior, and physiology of these animals. The collections of the Phyletische Museum hold numerous specimens of *B. latro* that Harms collected during his trips (Fig. 8.1). The two publications on *B. latro* witnessed how closely he observed behavior in the natural environment. Furthermore, numerous ecophysiological experiments were carried out during the expeditions. In the 122 pages of the 1932 paper, Harms also provided a superb anatomical description of *B. latro*. His illustrations of the animal's lungs and circulatory system are still reproduced in zoological textbooks (e.g., Burggren and McMahon 1988, p. 294; Kaestner 1993, p. 945). Of particular interest for the present contribution are Harms' detailed analyses of the *B. latro* sensory organs, specifically the first antennae for which he noted the presence of ca. 780 "Geruchsplatten" (today called aesthetascs) the microscopic architecture of which he described in great depth. In addition he noted the presence of many glands in the pair of first antennae. Harms also carried out a set of lesion experiments under field conditions to study the functions of the visual, tactile and olfactory senses of *B. latro*. In this view, Harms was not only a morphologist, endocrinologist, ecologist, developmental, and behavioral biologist but also a pioneer in the field of crustacean neuroethology.

8.3   The Biology of Terrestrial Crustaceans: General Aspects

Within Crustacea, at least five major lineages have succeeded in the transition from an aquatic to a fully terrestrial life style (Fig. 8.2; reviews Bliss and Mantel 1968; Powers and Bliss 1983; Hartnoll 1988; Greenaway 1988, 1999). These include representatives of the Isopoda, Amphipoda, Astacida, Anomura, and Brachyura (Friend and Richardson 1986; Wägele 1989; Morritt and Spicer 1998; Hartnoll 1988; Richardson and Swain 2000; Greenaway 2003; Richardson 2007). However, we have evidence for olfactory-guided behavior only for members of the terrestrial Isopoda and the Anomura so that only these two taxa will be discussed in the present contribution.

In olfaction, a transition from sea to land means that molecules need to be detected in gas phase instead of water solution. The odor stimulus also changes from mainly hydrophilic molecules in aqueous solution to mainly hydrophobic in the gaseous phase (discussed in Stensmyr et al. 2005). The olfactory system is also, like the rest of the organism, very prone to desiccation and mechanical abrasion in the terrestrial environment. All these new selection pressures take part in reshaping the sense of smell. Behavioral studies have provided evidence that some land-living crustaceans are very effective in detecting food from a distance and in responding to airborne odors, in short, that they have evolved an excellent sense of distance

Fig. 8.1 (a) Phyletisches Museum Jena (with kind permission of Dr. Hans Pohl, Phyletisches Museum Jena). (b) Jürgen Wilhelm Harms (1885–1956), Professor for Zoology at the University of Jena and Director of the Phyletisches Museum Jena from 1935 to 1949. (Picture from the private

olfaction (Rittschof and Sutherland 1986; Vannini and Ferretti 1997; Stensmyr et al. 2005 and references herein).

Within the Anomura, the Coenobitidae are a member of the Paguroidea, a taxon in which the members have evolved the potential to protect the pleon with gastropod shells. They include 15 species of shell-carrying land hermit crabs (the genus *Coenobita*) and the robber or coconut crab *B. latro* (genus *Birgus*), the largest living land arthropod (Harms 1937; Grubb 1971; Greenaway 2003; Fig. 8.3). The early juvenile stages of this impressive creature, which as an adult can weigh >5 kg, carry a gastropod shell, but with subsequent growth suitable shells are not available anymore so that the thorax and pleon harden for protection, as in other Crustacea (Harms 1932, 1937).

The crustacean taxon that was most successful in the colonization of land is the Oniscidea, a subgroup of the Isopoda (Powers and Bliss 1983; Wägele 1989; Kaestner 1993). As in all other peracarid crustaceans, development in the isopods is direct, and fully developed juveniles emerge from the female's brood pouch. The Oniscidea comprises around 3,500 species of which many have established a fully terrestrial life style according to the grade 5 of terrestrialness (Powers and Bliss 1983; Greenaway 1988, 1999). The phylogeny of the Isopoda is a topic of ongoing research (e.g., Wägele 1989; Wetzer 2002; Wägele et al. 2003; Wirkner and Richter 2003), and it is unclear yet if the Oniscidea invaded land several times. Isopoda can be lined up in an ascending order of terrestrialness – *Ligia* – *Oniscus* – *Porcellio* – *Armadillidium* – *Hemilepistus* – to illustrate the evolutionary adaptations to meet the physiological challenges of the terrestrial habitat e.g. by increasingly better protection against loss of water by transpiration, improving capacity for osmoregulation and oxygen uptake, as well as greater reproductive independency from water by sophisticated brood care (Warburg 1968; Powers and Bliss 1983; Kaestner 1993).

8.4   The Olfactory System of Terrestrial Crustaceans – The Substrate of Chemical Communication

In the following, we will provide a very brief overview over the general architecture of the crustacean central olfactory pathway, as this subject is treated much more thoroughly in the chapter by Schmidt and Mellon (Chap. 7), to which we

---

Fig. 8.1 (continued) collection of PD Dr. Wieland Hertel, Institut für Allgemeine Zoologie und Tierphysiologie, Friedrich-Schiller-Universität Jena; with kind permission.). (**c**) A specimen of *Birgus latro* as collected by Harms and displayed in the current exhibition (2008) of the Phyletisches Museum. (**d–f**) Specimens of *B. latro* in the collection of the Phyletisches Museum Jena as acquired by Jürgen Wilhelm Harms during his field trips to the Mollucas, Christmas and Cocos Islands in the 1930s (**c–f**) by Steffen Harzsch with kind permission of Dr. Hans Pohl, Phyletisches Museum Jena)

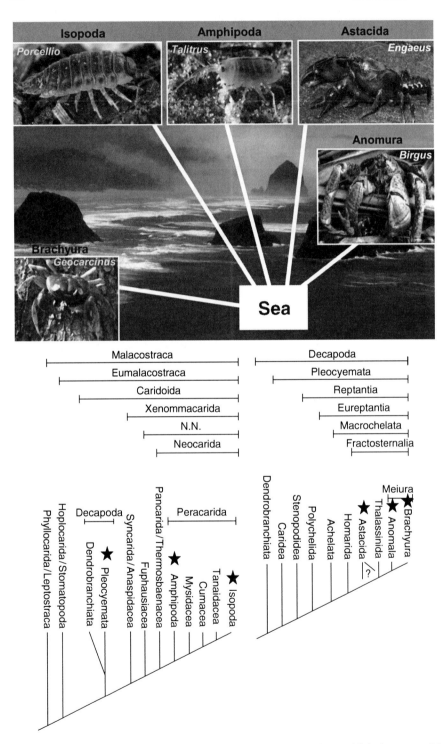

Fig. 8.2 Five lineages of malacostracan crustaceans independently established a terrestrial lifestyle (*upper panel*; photo credits: *Engaeus* – A. Frutiger (crayfishworld.com); *Talitrus* – A. Paul (Creative Commons); *Gecarcinus* – BHNY (Creative Commons); *Porcellio*: A. Karwath

refer the reader. We will then focus on specifics of the olfactory pathway in the various terrestrial crustacean taxa.

In decapod crustaceans, afferent chemosensory input from the olfactory receptor neurons on the paired first antennae is processed in conspicuous deutocerebral neuropil centers, the bilaterally arranged olfactory lobes, which consist of cone-like areas of dense synaptic neuropil, the glomeruli, which are arranged around the periphery of the lobe with the apices pointing to the center of the lobe (Fig. 8.4; Sandeman et al. 1992, 1993; Harzsch et al. in press; and Schmidt and Mellon, Chap. 7). A second-order olfactory neuropil is located in the eyestalks of decapod crustaceans as part of the lateral protocerebrum, which receives a massive input from the olfactory globular tract that originates from the cluster (10) of projection neurons associated with the deutocerebral olfactory and accessory lobes as the major output pathway of these two neuropils. Integration of chemical and hydrodynamic information occurs in the three main areas of the deutocerebrum (the olfactory lobes, the lateral antennular neuropil, and the median antennular neuropil) (Schmidt and Mellon, Chap. 7).

8.4.1 Anomura

Behavioral studies have provided evidence that terrestrial hermit crabs are very effective in detecting food from a distance and in responding to airborne odors, in short, that they have evolved an excellent sense of distance olfaction (Rittschof and Sutherland 1986; Vannini and Ferretti 1997; Stensmyr et al. 2005). The olfactory receptor neurons of crustaceans are associated with specialized structures on the first pair of antennae, the aesthetascs (see Hallberg and Skog, Chap. 6 and references therein). The aesthetascs of Coenobitidae are short and blunt and more similar to those of insects than to those of marine hermit crabs (Fig. 8.5; and compare Ghiradella et al. 1968a, b; Stensmyr et al. 2005). In robber crabs B. latro, they are confined to the ventral side of the primary flagella. Contrary to marine crustaceans, they have an asymmetric profile with the protected side lined with a thick cuticle. The exposed side is covered with a thinner cuticle, a feature that most likely is necessary to enable the passage of odors (Stensmyr et al. 2005). Another clear difference to marine crustaceans is that in the robber crab, the basal bodies and cilia segments are housed well inside the flagellum and are surrounded by a lymph space. Stensmyr et al. (2005) interpreted these morphological features of the aesthetascs as adaptations to terrestrial conditions, more specifically, as mechanisms to minimize water evaporation while maintaining the ability to detect volatile odors from the gaseous phase. Land hermit crabs also show flicking movements of their first antennae to maximize odor sampling, a strategy that is essential for efficient

---

Fig. 8.2 (continued) (Creative Commons); *Birgus*: Hansson and Stensmyr, unpublished). The *lower panel* shows the phylogeny of the Malacostraca according to Scholtz and Richter (1995) and Richter and Scholtz (2001). *Asterisks* indicate those groups with representatives that have conquered land

Fig. 8.3 (a) A population of *B. latro* on Christmas Island (CI). (b) The conspicuous first pair of antennae (*arrows*) in *B. latro*. (c) Hansson Jr. presenting a specimen of *B. latro*.

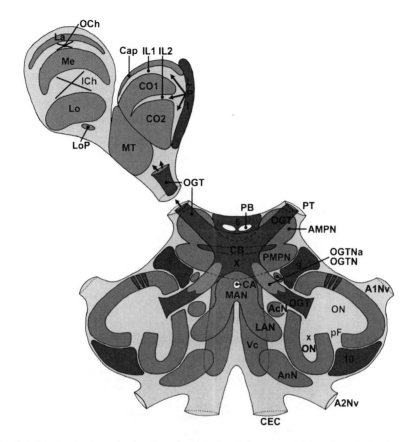

Fig. 8.4 Idealized schematic drawing of the brain of the terrestrial hermit crab *C. clypeatus* (dorsal view; from Harzsch and Hansson 2008; all rights retained by the authors). A1Nv nerve of antenna 1; *A2Nv* nerve of Antenna 2; *AcN* accessory lobe/neuropil; *AMPN* anterior medial protocerebral neuropil; *AnN* antenna 2 neuropil; *CA* cerebral artery; *Cap* cap neuropil of the hemiellipsoid body; *CB* central body; *CEC* circumesophageal connectives; *CO1, CO2* core neuropils 1 and 2 of the hemiellipsoid body; *ICh* inner optic chiasm; *IL1, IL2* intermediate layers 1 and 2 of the hemiellipsoid body; *La* lamina (lamina ganglionaris); *LAN* lateral antenna 1 neuropil; *Lo* lobula (medulla interna); *LoP* lobula "plate"; *LPI* lateral protocerebral interneurons; *MAN* median antenna 1 neuropil; *Me* medulla (medulla externa); *MT* medulla terminalis; *OCh* outer optic chiasm; *OGT* olfactory globular tract; *OGTN* olfactory globular tract neuropil; *OGTNa* accessory olfactory globular tract neuropil; *ON* olfactory lobe/neuropil; *PB* protocerebral bridge; *pF* posterior foramen; *PMPN* posterior medial protocerebral neuropil; *PT* protocerebral tract; *VC* ventral neuropil column; *X* chiasm of the olfactory globular tract

---

Fig. 8.3 (continued) (**d**) An egg-bearing female of *B. latro* heading towards the coast of CI at night. (**e**) One of about 100 females found in a single cave close to the shore of CI, carrying a full load of eggs most likely waiting for the optimal conditions for egg laying (see text for details; (**a–e**) by Bill Hansson). (**f**) An adult specimen of *Coenobita compressus* in the Parque Santa Rosa, Pacific Coast of Costa Rica. (**g**) A juvenile specimen of *Coenobita clypeatus* in the Parque de Cahuita, Atlantic Coast of Costa Rica ((**f, g**) kindly provided by Carsten H. G. Müller)

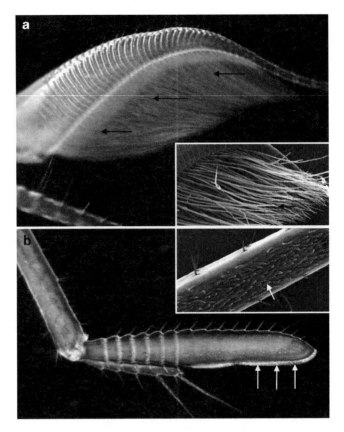

Fig. 8.5 (a) The first antenna of the marine hermit crab *Pagurus bernhardus* bears hundreds of long and slender aesthetascs (*black arrows*; Harzsch, unpublished image). The *inset* shows a scanning electron microscopic image (Hansson and Sjöholm, unpublished). (**b**) The first antenna of the terrestrial hermit crab *Coenobita clypeatus* with short and stout aesthetascs (*white arrows*; Harzsch, unpublished image). The inset shows a scanning electron microscopic image (Hansson and Sjöholm, unpublished)

chemoreception in many aquatic crustaceans (Koehl, Chap. 5 and references therein). In addition, Coenobitidae use their first antennae to touch and sample the ground (Greenaway 2003), which suggests the presence of taste receptors.

The architecture of the central olfactory pathway in land hermit crabs is poorly understood as is the nervous system architecture of Anomura in general (Paul 2003). Previous neuroanatomical studies in concert with the available behavioral reports have indicated the presence of sophisticated olfactory systems in members of the Coenobitidae (Sandeman et al. 1993; Beltz et al. 2003). Thus we decided to explore the brain morphology in this group more closely. We chose to analyze the brain of *Coenobita clypeatus* (Herbst 1791; Anomura, Coenobitidae), a fully terrestrial tropical species that migrates long distances inland, by immunohistochemistry against synaptic proteins, serotonin, FMRFamide-related peptides, and glutamine synthetase (Harzsch and Hansson 2008). We found that the primary olfactory centers in this species dominate the brain, are equipped with a side olfactory lobe, and are composed of many elongate olfactory glomeruli (Fig. 8.4). The secondary olfactory centers, the hemiellipsoid bodies, that receive an input from olfactory projection neurons are almost equally as large as the olfactory lobes and are organized into parallel neuropil lamellae. Our neuroanatomical study confirmed that the sensitive behavioral response to airborne odors as noticed in previous behavioral studies is paralleled by a significant elaboration of brain areas taking part in olfactory processing, as has already been noted by Beltz et al. (2003) who reported that *C. clypeatus* has a fairly high number of elongate olfactory glomeruli compared to other Crustacea. Furthermore, the organization of the visual centers and those areas associated with the second antenna suggest that the visual and mechanosensory skills of *C. clypeatus* are similar to those of their marine relatives (Harzsch and Hansson 2008).

In order to broaden the taxonomic horizon for a comparative analysis of anomuran olfactory systems, we set out to analyze the olfactory neuropils in some additional anomuran taxa of the subgroup Paguroidea which are closely related to the Coenobitidae (Harzsch and Hansson, unpublished data): *Clibanarius erythropus*, *Diogenes pugilator*, and *Calcinus elegans* as members of the Diogenidae and *Pagurus bernhardus* again, as a member of the Paguridae. Furthermore, we have analyzed the giant robber crab *B. latro* (Fig. 8.6; Harzsch et al. 2007) which as a member of the Coenobitidae is most closely related to *C. clypeatus*. From this preliminary study it would appear that the elongate shape of the glomeruli in *C. clypeatus* (at least five times as long as wide) and also in *B. latro* marks one end of the range, whereas *D. pugilator* with glomeruli that are only twice as long as wide marks the other end. *C. erythropus*, *C. elegans*, and *P. bernhardus* fall in between these two extremes (Harzsch and Hansson, unpublished data). *C. clypeatus* has a relatively high number of glomeruli (ca. 800) compared to other Decapoda (Beltz et al. 2003). The need to pack many glomeruli in a radial array and a restricted amount of space may promote the evolution of these elongate glomeruli. A comparison with marine hermit crabs reveals that the existence of an additional side olfactory lobe as shown here for *C. clypeatus* is not a typical feature of other malacostracan crustaceans. However, one of its closest relatives, *B. latro*, has an olfactory neuropil that is even composed of three sublobes (Fig. 8.6; Harzsch et al. 2007). To sum up, in parallel to previous behavioral findings of a good sense of aerial olfaction in *C. clypeatus* and *B. latro*, our results indicate that in fact their central olfactory pathway are most prominent, indicating that olfaction is a major sensory modality that these brains process.

8.4.2 Isopoda

Within the Oniscoidea, a taxon which comprises those isopod species that have developed a fully terrestrial lifestyle, the first antennae are strongly reduced in size (Fig. 8.7) and the second antennae seem to function as major sensory organs (Wägele 1989; Hoese 1989). The tip of the second antennae bears a characteristic

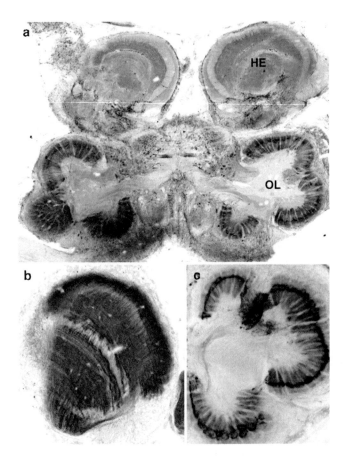

Fig. 8.6 Immunohistochemical localization of the neuropeptide SIFamide in the brain of the giant robber crab *Birgus latro* (Harzsch and Hansson, unpublished data; *black-white* inverted fluorescent images). (**a**) Horizontal vibratome section of the median brain. The olfactory lobes (OLs) and the hemiellipsoid bodies (HE) dominate the brain. (**b**) Detail of the hemiellipsoid body showing its layered structure. (**c**) The OL seems to be composed of three units

apical sensory cone that perceives mechanical and chemical stimuli (see Hoese 1989 and references therein). In the desert isopod *Hemilepistus reaumuri*, this apical organ was shown to respond to both olfactory and gustatory stimuli (Seelinger 1983). Although distance responses to olfactory stimuli have not yet been demonstrated in this species (Seelinger 1983), their mixed olfactory-gustatory organ on the second antennae seems to play a key role in the perception of discriminator substances for social recognition, for family cohesion and for communication (review Linsenmair 2007; see below).

Although some general aspects of brain morphology are known (see e.g., Hanström 1947; Schmitz 1989), little information is available on the central olfactory system in marine species of isopods. Concerning the terrestrial Oniscoidea, it has previously been noted that their primary olfactory centers in the

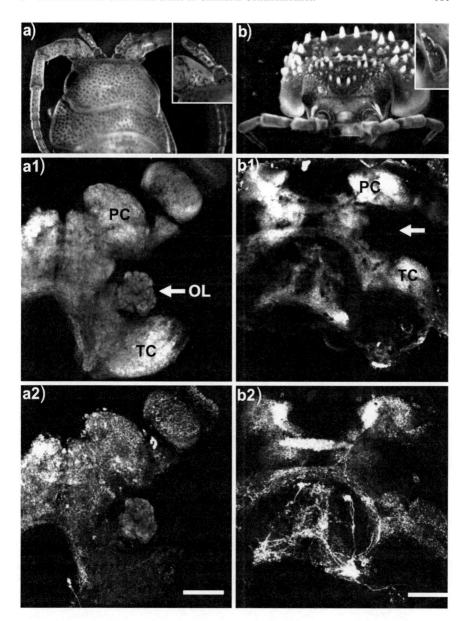

Fig. 8.7 Comparison of the olfactory deutocerebrum in a marine vs. a terrestrial isopod (Harzsch et al., unpublished data). (**a**) The head of *Idotea baltica* (Isopoda, Valvifera) with the pair of large second antennae and the pair of smaller first antennae in-between. *Inset* shows higher magnification of a first antenna. Synapsin-immunoreactivity (**a1**) and the localization of Rfamide-like immunoreactivity (**a2**) shows the presence of a distinct OL composed of spherical glomeruli. The tritocerebrum (TC) receives the afferents from the second antennae and the protocerebrum (PC) receives highly processed visual input. (**b**) The head of the desert isopod *Hemilepistus reaumuri* with a pair of conspicuous second antennae. The *inset* shows the minute first antennae. Synapsin-immunoreactivity (**b1**) and the localization of Rfamide-like immunoreactivity (**b2**) shows that an OL is not present (*arrow*). Scale bars: 100 μm

deutocerebrum, the olfactory lobes, have been reduced in size in response to terrestrialization (Walker 1935; Schmitz 1989; Kacem-Lachkar 2000) which coincides with the minute size of their first pair of antennae (Fig. 8.7). Our own preliminary immunohistochemical studies on the brain architecture of several Isopoda support this idea. The marine isopod *Idotea baltica* is not a representative of the Oniscoidea but of the Valvifera, the members of which are almost exclusively marine (Kaestner 1993). *I. baltica* has a distinct pair of first antennae in addition to the pair of large second antennae (Fig. 8.7a). Immunohistochemical localization of synaptic proteins (synapsins) and the neuropeptide RFamide reveals a distinct olfactory lobe in the deutocerebrum of this species. The olfactory lobe is composed of several dozens of spherical glomeruli. The optic neuropils are very prominent in this species as is the tritocerebrum. As mentioned above, the first pair of antennae is minute in terrestrial Oniscidea (Fig. 8.7b). In all three fully terrestrial species that we analyzed immunohistochemically, *Armadillidium vulgare*, *Porcellio scaber*, and *H. reaumuri*, a distinct olfactory lobe as in the marine species is not present. Their brain is dominated by the optic neuropils and the tritocerebrum. This suggests that in the terrestrial isopods the deutocerebral olfactory pathway does not play a significant role for aerial olfaction. Instead, the tritocerebrum associated with the mixed olfactory-gustatory organ seems to have taken over an important role in processing sensory input. Preliminary data indicate that in this species the tritocerebral neuropil houses hundreds of microglomeruli as a potential neuronal substrate for refined chemosensory processing abilities (Harzsch et al., unpublished data).

8.5   Olfactory-Guided Behavior in Terrestrial Crustaceans

In general, land-living and aquatic animals use very different chemical compounds to communicate. The medium as such raises different demands on the compounds used. In water, molecules have to be more or less water-soluble and stable enough to travel from one individual to another. On land, semiochemical molecules have to be light enough to form a gas in the ambient temperatures where the animals live. The molecules also have to be chemically stable enough to reach the receiving end of the system. In insects pheromone molecules have evolved within quite a restricted group of chemicals, providing the chemical properties listed above (Wyatt 2003; and Baker, Chap. 27). Among marine crustaceans, the presence of sexually active components have been shown in the urine (Ryan 1966), but the chemical structure still remains to be unequivocally determined (but see Hardege and Terschak, Chap. 19). Other types of pheromones are composed of peptides or peptide-like molecules (Rittschof and Cohen 2004). To sum up, regarding chemical communication in land-living crustaceans, our present knowledge is very limited.

8.5.1   *Hermit Crabs of the Genus* Coenobita

The conquest of land has led to different levels of adaptations in terrestrial hermit crabs. *Coenobita caripes*, *C. compressus*, *C. scaevola*, and *C. perlatus* inhabit the shore and need direct access to seawater. *C. rugosus* represents an intermediate form that does not venture further than ~100 m from the shore, but which does not need seawater, except for reproduction. Finally, *C. brevimanus*, *C. clypeatus*, and *C. rubescens* wander several kilometers inland (for a detailed description of habitat preferences see Hartnoll 1988). Hermit crabs have a huge repertoire of orientation mechanisms, including a celestial compass, visual landmarks, and wind direction to orient towards the coastline and to optimize their fleeing direction (e.g., Vannini and Chelazzi 1981).

Much work has been done on the sense of olfaction in terrestrial hermit crabs. It was tested whether these omnivorous crabs are attracted by volatiles emitted by different substances as seawater, well water, distilled water (Vannini and Ferretti 1997), crushed conspecifics or snails (Thacker 1994), fruits, seeds, flowers (Rittschof and Sutherland 1986; Thacker 1996, 1998), and finally even horse feces and human urine (Rittschof and Sutherland 1986). These tests were conducted either as field experiments using pitfalls buried in the ground of the habitat or as lab experiments using T-maze bioassays. Some odors seemed to be attractive in one study, but not in another (e.g., coconut attracted *C. rugosus* in Rittschof's and Sutherland's survey but not in Vannini's and Ferretti's). The contradictory results might be due to different methods used but this does not affect the main outcome of all studies: terrestrial hermit crabs detect volatiles and use them to pinpoint a target. Hence, guided behavior towards volatile odors is not restricted to insects but has evolved also in crustaceans when conquering land. Reaction towards odors may depend on the context and in particular on combinations of different odor stimuli (Thacker 1994, 1996, 1998).

To summarize, olfaction helps terrestrial hermit crabs to localize water, food, and shell resources. It also helps them to distinguish between different resources and to make decisions. Although these capabilities have evolved independently, they are comparable to the odor-guided behavior in insects. It seems to be just a matter of time until further strategies, well established in insects, like e.g. orientation towards sex pheromones, will also be revealed in hermit crabs.

8.5.2   *The Christmas Island Robber Crab – A Case Study*

One of the most striking example of land-living crustaceans on Earth today are the giant robber crabs (*B. latro*; Anomura, Coenobitidae) of the Indian and Pacific Oceans (Fig. 8.3). From earlier being present on most islands in these oceans, robber crabs are today restricted to a small number of refuges, with the largest and densest population found on Christmas Island (CI), an Australian territory situated about 350 km south of Java (review Drew et al. 2010). The robber crab density can exceed 160/ha (Rumpf 1986). These crabs have mainly been known for their size, for voracious omnivorous feeding habits, for their ability to climb tall palm trees in search of coconuts, and subsequently peel and crack open said coconuts, and for an ability to "sniff" out any kind of possible food objects, be it carcasses or left-out garbage.

In 2002 we launched a first expedition to study the olfactory system and smell-driven behavior of the peculiar animals. Following up the local observations that the crabs will find food objects during the pitch black, tropic nights of the area, we set out odor traps baited with attractive objects, such as dead red crabs, coconut pulp, or arenga palm marrow. After baiting, the approach of robber crabs to the traps was observed. These initial, simple experiments clearly showed that robber crabs are odor-guided when looking for food sources during night (Fig. 8.8).

This fact prompted the question how these animals detect odors. By using electrophysiology (see above) we could show that odor molecules were detected on the first antenna, and more specifically on the final segment. We screened a number of synthetic volatile compounds, and noted that the crabs' olfactory system readily could detect and differentiate between these compounds. After establishing the olfactory sensitivity to these synthetic stimuli, we next wanted to establish if they were indeed behaviorally active in themselves, and if

Fig. 8.8 Robber crabs *B. latro* are attracted to potential food sources such as dead red crabs by odor cues. Drawing by Jorge A. Varela Ramos

so, to what degree. In two field trials, the attractivity of gamma-decalactone (coconut) and dimethyl-trisulfide (DMTS; rotten crab) were tested. The DMTS proved to be highly attractive, eliciting orientation behavior in crabs from more than 50 m distance. The gamma-decalactone on the other hand, did not attract any crabs at all, again showing that what smells like coconut to us, does not to the robber crabs.

Upon returning to our home base we analyzed the more detailed structure of the olfactory organs of *B. latro*. The aesthetascs could be divided into at least two types, where one is more prevalent towards the tip of the antenna, and displays what might be a terminal pore. The placement and the possible pore could indicate a function in taste perception. The major part of the aesthetascs was uniformly scale-like without terminal pore. In fact, no pores at all could be observed, not even the minute cuticular pores observed in almost all insect olfactory sensilla. We have so far not been able to elucidate how molecules in gas phase enter the inside of the *B. latro* aesthetascs. Each hair was shown to contain a very large number of olfactory receptor neurons, not only similar to what has been found in other crustaceans, but also in e.g., honeybees and locusts (Hansson et al. 1996; Ochieng et al. 1998).

We also collected brains from robber crabs to examine the structure of the central olfactory system. The brain is surprisingly small (given the enormous size of the animal), hanging in the middle of the large head. The main striking features are two giant eyestalks shooting up from the brain proper. When sectioning the brain, however, the main attention is drawn to the olfactory system, which, compared to other crustaceans, has undergone a significant expansion (see Fig. 8.6 and above). The primary olfactory centers of the brain, the olfactory lobes, have increased in size, and a whole new part has been added. From the olfactory lobes, neurons project to a specific area, known as the hemiellipsoid bodies. Also here a strong expansion has occurred, and a new layered architecture has evolved (Fig. 8.6b). The functional consequences of these expansions, both at the olfactory bulb and the hemiellipsoid body levels, are still unknown. Present experiments will increase our understanding during the coming years.

To get further under the surface of olfactory structure and function, and of population structure in general, a second field expedition was organized in January 2005. During the previous expedition (early December) a mixed-sex population was present in the rain forest area. During the second observation in January, the female proportion had dropped to about 5% in the forest. Since on CI, robber crabs spawn from November to February/March (Drew et al. 2010) our endeavors in search for females were shifted closer to the coastline. While driving along the coastline at nighttime we came across a few females heavily laden with eggs, and purposefully heading for the coast (Fig. 8.3d). Encouraged by these observation we concentrated our further investigations to the rocky coast area close to where we had observed the mobile females. No females were, however, to be found until we by rope ventured into more or less vertical caves close to the shore. Here about 100 females were found in a single cave, all carrying a full load of eggs (Fig. 8.3e). Our interpretation of this phenomenon is that females congregate in the caves to wait for the optimal conditions for egg laying (Hansson et al., unpublished). The caves offer shelter and a very moist environment, both factors highly advantageous to the crabs. From the caves the females can easily reach the rocky beach in a few hours. At these beaches females have earlier been observed dropping their eggs into the ocean, while hanging on the cliffs.

These observations might seem peripheral to discussions regarding chemo-sensation and chemo-communication. However, the fact that the sexes, opposite to e.g., red crabs, do not travel to the oviposition site together, to mate on-site, raises interesting possibilities. It means that the sexes have to communicate within the forest, and most likely at nighttime. Comparing this to other arthropod systems strongly suggests that a sexual communication system built on airborne pheromones could be used. To dissect this possibility is, of course, one of our future goals.

8.5.3   Kin Recognition in the Desert Isopod Hemilepistus reaumuri

Among the crustaceans, the desert isopod *H. reaumuri* definitely exhibits one of the highest levels of terrestrial adaptations (Wägele 1989). Living in the deserts of Northern Africa, the animals' access to water is reduced to dew drops and rare rainfall. Like other terrestrial isopods, *H. reaumuri* is not dependent on water for reproduction. The reason for *H. reaumuri* being a promising subject for chemical ecologists is its highly developed system of kin recognition. *H. reaumuri* is a monogamous species that rears its brood in burrows dug in soil. Both parents care for their young and remain with them within the burrow for up to 200 days. They supply the young with detritus collected during far reaching foraging excur-sions (for a detailed description of the socio-biology of *H. reaumuri* see Linsenmair 2007). In an extensive study including thousands of tests in which members of different families were faced with each other, Linsenmair (2007) could show that the parents are able to distinguish between their own progeny and those belonging to other families. The latter become attacked and in many cases killed and eaten. Also, the young isopods exhibit these discrimination capabilities: any contact to alien conspecifics leads to abrupt escape behavior (Linsenmair 1987). The discrim-ination depends on a set of polar, mainly nonvolatile cuticle compounds produced in epidermal glands. The animals probe each other with the apical sensory cone positioned at the tip of the second antennal pair. The cone has chemo-receptors reacting among others to carbon acid, amines, sugar, fatty acids, and amino acids (Seelinger 1977, 1983). Hering (1981) discovered these types of compounds and some additional esters on the cuticle of the isopods. By manipulating the amino acids he was able to abolish the discrimination capabilities, while the role of the rest of the compounds as discriminators remained unsolved. The composition of compounds produced by an individual seems to be purely genetically determined and, hence, differs for each individual of a family. However, as these compounds are easily transferred during direct body contact (Linsenmair 2007), the mixture of the blends of the whole family forms a family-specific signature. As this family-specific signature changes with each new family member, its identification has to be learned by the individuals. Hence, these small crustaceans are not only able to detect a large set of chemical compounds and discriminate between different mixtures of these compounds but also exhibit startling learning capabilities and therefore provide an interesting model system for chemical ecology and kin recognition.

8.6 Olfactory Physiology in Terrestrial Crustaceans

For any form of olfactory-guided behavior, detection of olfactory information has to be first accomplished. How do terrestrial crustaceans detect and convert volatile chemical information into a neural signal? Unfortunately, our knowledge regarding the physiological properties of the olfactory system of this group of arthropods is rather sketchy.

The only land crab where physiological recordings from the peripheral olfactory system have been successfully obtained is *B. latro* (Stensmyr et al. 2005). In that study, we used so called electro-antennographic (EAG) detection (Schneider 1957) to investigate the physiological properties of the crab's olfactory system. EAG is a technique in which the recording output is thought to represent the summed activity of a large proportion of the olfactory sensory neurons, and where after odor stimulation, the total response of the antenna is recorded as a slow direct current potential. By using a portable setup it was possible for us to obtain electrophysiological recordings on site in the rain forest of CI. EAG recordings from excised terminal antennule segments of *B. latro* clearly demonstrated the capacity of this tissue to detect volatile chemical information. The basal segments of the antennules did not show any activity upon stimulation with odorants. However, it should be noted that other appendages (e.g., legs) also might house chemosensory sensilla capable of detecting volatiles, but this was not investigated.

The terminal antennule segments strongly responded to CO_2. Dose response curves to CO_2 (created by extracting headspace over dry ice) show that the crabs are highly sensitive to this stimulus (Fig. 8.9). During daytime, the animals spend many hours hiding in tight shelters and under rocks so that monitoring CO_2 concentrations may serve an important biological function. Interestingly a very pronounced response to water vapor was also noted. This response was, however, of inverted polarity as compared to all other gaseous stimuli, similar to what is always observed in insects. After these initial observations, we next proceeded to test the odor of a number of attractive substrates: rotting red crab, coconut, arenga palm, banana, and pineapple. These stimulations again revealed some very strong responses, but also showed that the olfactory system was selective, as not all tested substrates elicited responses.

To examine the functional characteristics of the crab's olfactory system in more detail, we proceeded to test a set of odorants representing a number of important

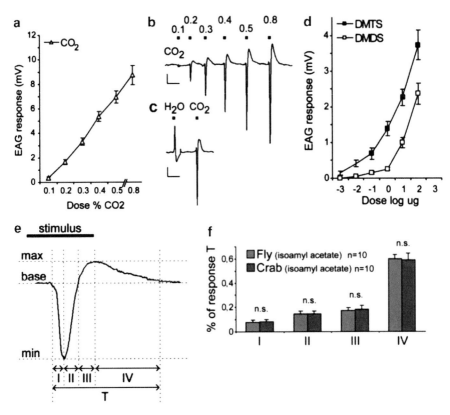

Fig. 8.9 Robber crab *B. latro* olfactory physiology (reprinted from Stensmyr et al. 2005; with permission). (**a, b**) Electroantennogram (EAG) recordings reveal CO_2 as a potent ligand, eliciting responses in a dose-dependent manner (the horizontal scale bars represent 10 s, and the *vertical scale bars* represent 1 mV). (**c**) The antennules are also sensitive to water, which elicits inverted, positive EAGs (the *horizontal scale bars* represent 10 s, and the *vertical scale bars* represent 1 mV). (**d**) EAGs for dimethyltrisulfide (DMTS) and dimethyldisulfide (DMDS) presented in increasing dosages (decadic steps from 1 ng to 100 μg); plotted dose response functions for the two compounds. (**e**) Electroantennogram response kinetics; the EAG responses can be broken up into four distinct phases: (I) initial rapid depolarization, (II) return to baseline, (III) hyperpolarization, followed by (IV), an outdrawn return to baseline. (**f**) Comparing the EAG kinetics from the robber crab with EAG kinetics from an insect (exemplified here by a fruitfly, *D. melanogaster*) shows that the EAG profiles (in response to isoamyl acetate at 1 μg) are indistinguishable from each other

crab resources (as well as different chemical classes) for EAG activity in dose response series (Fig. 8.9). The tested compounds were dimethyl-disulfide and dimethyl-trisulfide (both diagnostic for decaying meat), ethyl-hexanoate (pineapple), isoamyl acetate (banana), phenylacetaldehyde (flower fragrance), and gamma-deca-lactone (coconut). In particular, the tested oligosulfide compounds were detected in minute concentrations, which points to an olfactory system as sensitive as the most sensitive general odor detecting olfactory sensory neurons found in insects (e.g., Hansson et al. 1999). Other volatile compounds, e.g., compounds characteristic for fruit, were also detected at very low concentrations. Interestingly though, the coconut odorants tested were not particularly efficient, in spite of the animal's reported fondness for this resource. Furthermore, the EAG recordings did not only reveal that the system is highly sensitive but also that it is discriminate, as structurally related compounds are readily differentiated (as shown by the dose response curves).

As a comparison to the responses recorded in *B. latro*, we performed exactly the same type of EAG recordings in the fruitfly, *Drosophila melanogaster*. Here, we could observe identical kinetics in the two distant species, indicating the presence of an at least as efficient transduction mechanism in the crab as in the fly. Closer examinations of the EAG response dynamics also hint at a fast and efficient transduction pathway.

These initial experiments indicate that land crabs appear to have a sense of smell that is highly developed and which on the physiological level operates rather similar to that of insects. A question, though, is whether *B. latro* is representative of other land crabs, even other anomurans, or if it is unique with respect to its highly capable "land nose?" The architecture of the peripheral olfactory system of *B. latro* follows closely that of its smaller hermit crab relatives (as outlined in the preceding section), and accordingly it would seem fair to assume that also the operational characteristics are similar. Surprisingly, though, ongoing work in our laboratory indicate that the olfactory system of the terrestrial hermit crab of the genus *Coenobita* appears to function quite differently from that of *B. latro*, responding to other types of chemicals with different response dynamics. Overall, it appears that members of the genus *Coenobita* have a less developed capacity to detect airborne volatiles than *B. latro*, but further work is needed to fully establish this notion.

Isopoda is the only other terrestrial crustacean group where the physiological properties of the olfactory system have been investigated. In two papers, from 1977 to 1983, Seelinger outlines the olfactory response characteristics of the desert isopod *H. reaumuri*. By using electrolytically sharpened platinum-iridium electrodes, Seelinger was able to record neural activity (action potentials) from the axons of chemosensory neurons in the distal antennal nerve, which have their dendritic part innervating the primary chemo- and mechanosensory organ, the apical cone, situated at the tip of the second antennae. The contacted neurons were challenged with a panel of 37 olfactory and gustatory stimuli, including e.g., amines, acids, sugars, and salts. Seelinger predominantly found five groups of chemosensory cells, of which two responded to volatile chemicals. The first group responded stronger to short chain n-fatty acids, such as butyric acid. Seelinger notes, though, that within this group, the response profiles vary gradually from contacted neuron to neuron, and there are no distinct subgroups of neurons. Although there seems to be a fair bit of variation in the responses, examining the presented data in the 1983 paper, a number of distinct subgroups can clearly be seen, suggesting a stereotyped chemosensory gene expression. The second group of cells responded predominantly to amines. The neurons were activated by a number of mono- and diamines, such as butyl-, pentyl- and hexylamine as well as cadaverine and putrescine. As with the fatty acid cells, Seelinger notes that the amine cells do not cluster in distinct subgroups. However, one cell type can clearly be discerned, represented by seven samples (Table 5 of the 1983 paper), which responds selectively and consistently to the above listed amines. All in all, Seelinger's results hint that the desert isopod appears to decode odor information in a manner similar to that of insects.

An interesting point, though, is that none of the contacted neurons ($n = 38$, out of an estimated total of 450) responded significantly to the volatiles benzaldehyde and geraniol, compounds which most, if not all, insects respond to (at a physiological level) at least to some degree. The lack of responses to these compounds could perhaps easiest be explained by the low n value, i.e., sensory neurons tuned to this type of plant volatiles might have been overlooked, but could also hint that the isopods' olfactory system is perhaps not yet fully adapted to the terrestrial odor landscape. All chemicals giving strong responses are decidedly water-soluble and could presumably also have been important sensory cues in the aquatic setting of their ancestors. Accordingly, in the isopods, the shift from sea to land would have changed the architecture of the system, allowing it to operate on land, but perhaps not fundamentally altered the ligand tuning of the chemosensory receptors.

8.7 Why Land-Living Crustaceans?

Land-living crustaceans are truly fascinating animals that during a relatively short time have adapted to a number of highly diverse terrestrial habitats. In these habitats the crustaceans have become highly successful, and in some case the predominant life forms. Comparative studies between fully aquatic crustaceans and closely related terrestrial taxa provide a powerful means of investigating the evolution of chemosensory adaptations in these two environments.

A second very good reason is that these animals form a very interesting case of parallel evolution to the insects. Insects are the most successful arthropods living on land, and their chemical communication systems have been studied in detail. Many pheromones from different types of insects have been identified, the olfactory system has been explored for more than 50 years, and olfactory-guided behavior has been scrutinized meticulously. The terrestrial crustaceans form an ideal group to compare with the insects. Insects have evolved during more than 200 million years of land life, much longer than terrestrial crustaceans. How similar have the different aspects of the systems become?

Overall, what are the most urgent points to study in the close future? The investigations can be divided into three parts: the behavior towards active odors (both pheromones and other semiochemicals) has to be clearly shown and quantified, the chemical identity of key odor cues and the source of these compounds have to be established, and the sensory base of detection and information processing in the chemical senses need to be further studied. When these basic facts regarding terrestrial crustacean life have been established, we are ready to make a new synthesis of the systems involved.

Acknowledgments The writing of this chapter was supported by the Max Planck Society. We gratefully acknowledge the cooperation of the CI branch of National Parks, Australia in our studies of the robber crab. We cordially thank Dr. Hans Pohl (Phyletisches Museum Jena, Friedrich-Schiller-Universität Jena), PD Dr. Wieland Hertel, Institut für Allgemeine Zoologie und Tierphysiologie, Friedrich-Schiller-Universität Jena) and Dr. Carsten H.G. Müller (Universität Rostock) for contributing images. Swetlana Laubrecht kindly assisted compiling the reference list, and Verena Rieger with the immunohistochemical experiments with the isopods.

References

Beltz BS, Kordas K, Lee MM, Long JB, Benton JL, Sandeman DC (2003) Ecological, evolutionary, and functional correlates of sensilla number and glomerular density in the olfactory system of decapod crustaceans. J Comp Neurol 455:260–269

Bliss DE, Mantel LH (1968) Adaptations of crustaceans to land: a summary and analysis of new findings. Am Zool 8:673–685

Burggren WW, McMahon BR (1988) Biology of the land crabs. Cambridge University Press, Cambridge

Drew MM, Harzsch S, Stensmyr M, Erland S, Hansson BS (2010) A review of the biology and ecology of the Robber Crab, *Birgus latro* (Linnaeus, 1758) (Anomura: Coenobitidae). Zool Anz 249:45–67

Friend JA, Richardson AMM (1986) Biology of terrestrial amphipods. Annu Rev Entomol 31:25–48

Ghiradella H, Case J, Cronshow J (1968a) Fine structure of the aesthetesc hairs of *Coenobita compressus* Edwards. J Morphol 124:361–385

Ghiradella H, Cronshow J, Case J (1968b) Fine structure of the aestetasc hairs of *Pagurus hirsutiusculus* in the light and electron microscope. Protoplasma 66:1–20

Greenaway P (1988) Ion and water balance. In: Burggren WW, McMahon BR (eds) Biology of the land crabs. Cambridge University Press, Cambridge, pp 211–248

Greenaway P (1999) Physiological diversity and the colonization of land. In: Schram FR, von Vaupel Klein JC (eds) Proceedings of the fourth international crustacean congress. Brill Academic Publishers, Amsterdam, pp 823–842

Greenaway P (2003) Terrestrial adaptations in Anomura (Crustacea: Decapoda). Mem Mus Vic 60:13–26

Grubb P (1971) Ecology of terrestrial decapod crustaceans on Aldabra. Philos Trans R Soc Lond Ser B 260:411–416

Hansson BS, Ochieng SA, Grosmaitre X, Anton S, Njagi PGN (1996) Physiological responses and central nervous projections of antennal olfactory receptor neurons in the adult desert locust, *Schistocerca gregaria* (Orthoptera: Acrididae). J Comp Physiol A 179:157–167

Hansson BS, Larsson ML, Leal WS (1999) Green leaf volatile detecting olfactory receptor neurones display very high sensitivity and specificity in a scarab beetle. Physiol Entomol 24:121–126

Hanström B (1947) The brain, the sense organs, and the incretory organs in the head in the Crustacea *Malacostraca*. K Fysiogr Sällsk Handl 48:1–45

Harms JW (1932) Die Realisation von Genen und die consecutive adaptation. II. *Birgus latro* L. als Landkrebs und seine Beziehungen zu den Coenobiten. Z Wiss Zool 140:167–290

Harms JW (1937) Lebenslauf und Stammesgeschichte des *Birgus latro* L. von den Weihnachtsinseln. Zeitschr Naturwiss Jena 75:1–34

Hartnoll RG (1988) Evolution, systematics, and geographical distribution. In: Burggren WW, McMahon BR (eds) Biology of the land crabs. Cambridge University Press, Cambridge, pp 6–54

Harzsch S, Hansson BS (2008) Brain architecture in the terrestrial hermit crab *Coenobita clypeatus* (Anomura, Coenobitidae), a crustacean with a good aerial sense of smell. BMC Neurosci 9:58

Harzsch S, Stensmyr M, Hansson BS (2007) Transition from sea to land: adaptations of the central olfactory pathway in the giant robber crab. In: Poster, 7th Göttingen meeting of the German Neuroscience Society, 29.3.-1.4.2007. Online abstracts: http://www.neuro.uni-goettingen.de/NBCsearch/NBC07/nbc07_ab/TS2/TS2-3A.pdf

Harzsch S, Sandeman D, Chaigneau J (in press) Morphology and development of the central nervous system. In: Forest J, von Vaupel Klein JC (eds) Treatise on zoology – crustacea. Koninklijke Brill Academic Publishers, Leiden

Hering W (1981) Zur chemischen Ökologie der tunesischen Wüstenassel *Hemilepistus reaumuri*. Unpublished PhD thesis. Ruprecht-Karl-University of Heidelberg, Heidelberg

Hoese B (1989) Morphological and comparative studies on the second antennae of terrestrial isopods. Monitore Zool Ital (NS) Monogr 4:127–152

Kacem-Lachkar H (2000) Contribution to the study of the nervous system, the accessory glands, and the neurosecretory cells of *Hemilepistus reaumuri* (Audoin, 1826) (Isopoda, Oniscidea). Crustaceana 73:933–948

Kaestner A (1993) Lehrbuch der Speziellen Zoologie. In: Gruner HE, Moritz M, Dunger W (eds) Wirbellose Tiere; 4. Teil: Arthropoda (ohne Insecta). Gustav Fischer Verlag, Jena, pp 1–1279

Linsenmair KE (1987) Kin recognition in subsocial arthropods, in particular in the desert isopod *Hemilepistus reaumuri*. In: Fletcher D, Michener C (eds) Kin recognition in animals. Wiley, Chichester, pp 121–207

Linsenmair KE (2007) Sociobiology of terrestrial isopods. In: Duffy JE, Thiel M (eds) Evolutionary ecology of social and sexual systems – crustaceans as model organisms. Oxford University Press, Oxford, pp 339–364

Morritt D, Spicer JI (1998) The physiological ecology of talitrid amphipods: an update. Can J Zool 76:1965–1982

Ochieng SA, Hallberg E, Hansson BS (1998) Fine structure and distribution of antennal sensilla of the desert locust, *Schistocerca gregaria* (Orthoptera: Acrididae). Cell Tissue Res 291:525–536

Paul DH (2003) Neurobiology of the Anomura: Paguroidea, Galathoidea and Hippoidea. Mem Mus Vic 60:3–11

Powers LW, Bliss DE (1983) Terrestrial adaptations. In: Verneberg FJ, Vernberg WB (eds) The biology of Crustacea Vol. 8: environmental adaptations. Academic, New York, pp 272–333

Richardson AMM (2007) Behavioral ecology of semiterrestrial crayfish. In: Duffy JE, Thiel M (eds) Evolutionary ecology of social and sexual systems: crustaceans as model organisms. Oxford University Press, New York, pp 319–338

Richardson AMM, Swain R (2000) Terrestrial evolution in Crustacea: the talitrid amphipod model. Crustac Issues 12:807–816

Richter S, Scholtz G (2001) Phylogenetic analyses of the Malacostraca (Crustacea). J Zool Systemat Res 39:113–136

Rittschof D, Cohen JH (2004) Crustacean peptide and peptide-like pheromones and kairomones. Peptides 25:1503–1516

Rittschof D, Sutherland JP (1986) Field studies on chemically mediated behavior in land hermit crabs: volatile and nonvolatile odors. J Chem Ecol 12:1273–1284

Rumpf, H. 1986. Freilanduntersuchungen zur Ethologie, Ökologie und Populationsbiologie des Palmendiebes, *Birgus latro* L. (Paguridae, Crustacea, Decapoda), auf Christmas Island (Indischer Ozean). Inaugural-Dissertation zur Erlangung des Doktorgrades der Naturwissenschaften im Fachbereich Biologie der Mathematisch-Naturwissenschaftlichen Fakultät der Westfälischen Wilhems-Universität zu Münster, pp 1–122

Ryan EP (1966) Pheromone: evidence in a decapod crustacean. Science 151(3708):340–341

Sandeman DC, Sandeman RE, Derby C, Schmidt M (1992) Morphology of the brain of crayfish, crabs, and spiny lobsters: a common nomenclature for homologous structures. Biol Bull 183:304–326

Sandeman DC, Scholtz G, Sandeman RE (1993) Brain evolution in decapod Crustacea. J Exp Zool 265:112–133

Schmitz EH (1989) Anatomy of the central nervous system of *Armadillidium vulgare* (Latreille) (Isopoda). J Crust Biol 9:217–227

Schneider DZ (1957) Elektrophysiologische Untersuchungen von Chemo- und Mechanorezeptoren der Antenne des Seidenspinners *Bombyx mori*. Z Vgl Physiol 40:8–41

Scholtz G, Richter S (1995) Phylogenetic systematics of the repatantian Decapoda (Crustacea, Malacostraca). Zool J Linn Soc 113:289–328

Seelinger G (1977) Der Antennenendzapfen der tunesischen Wüstenassel *Hemilepistus reaumuri*, ein kompleyes Sinnesorgan (Crustacea, Isopoda). J Comp Physiol 113:95–103

Seelinger G (1983) Response characteristics and specificity of chemoreceptors in *Hemilepistus reaumuri* (Crustacea, Isopoda). J Comp Physiol 152:219–229

Stensmyr M, Erland S, Hallberg E, Wallén R, Greenaway P, Hansson B (2005) Insect-like olfactory adaptations in the terrestrial giant robber crab. Curr Biol 15:116–121

Thacker RW (1994) Volatile shell-investigation cues of land hermit crabs: effect of shell fit, detection of cues from other hermit crab species, and cue isolation. J Chem Ecol 20:1457–1481

Thacker RW (1996) Food choices of land hermit crabs (*Coenobita compressus* H. Milne Edwards) depend on past experience. J Exp Mar Biol Ecol 199:179–191

Thacker RW (1998) Avoidance of recently eaten foods by land hermit crabs, *Coenobita compressus*. Anim Behav 33:485–496

Vannini M, Chelazzi G (1981) Orientation of *Coenobita rugosus* (Crustacea: Anomura): a field study on Aldabra. Mar Biol 64:135–140

Vannini M, Ferretti J (1997) Chemoreception in two species of terrestrial hermit crabs (Decapoda: Coenobitidae). J Crust Biol 17:33–37

Wägele JW (1989) Evolution und phylogenetisches System der Isopoda. Zoologica 47:1–262

Wägele JW, Holland B, Dreyer H, Hackethal B (2003) Searching factors causing implausible non-monophyly: ssu rDNA phylogeny of Isopoda Asellota (Crustacea: Peracarida) and faster evolution in marine than in freshwater habitats. Mol Phylogenet Evol 28:536–551

Walker G (1935) The central nervous system of *Oniscus* (Isopoda). J Comp Neurol 62:197–237

Warburg MR (1968) Behavioral adaptations of terrestrial isopods. Am Zool 8:545–559

Wetzer R (2002) Mitochondrial genes and isopod phylogeny (Peracarida: Isopoda). J Crust Biol 22:1–14

Wirkner CS, Richter S (2003) The circulatory system in Phreatoicidea: implications for the isopod ground pattern and peracarid phylogeny. Arthropod Struct Dev 32:337–347

Wyatt TD (2003) Pheromones and animal behaviour: communication by smell and taste. Cambridge University Press, Cambridge, 391pp

Part III
Chemical Communication and Behavior

Chapter 9
Chemical Communication Between Copepods: Finding the Mate in a Fluid Environment

Jeannette Yen and Rachel Lasley

Abstract Copepods are small heterosexual aquatic microcrustaceans (1–10 mm, moving at ~1–10 bodylengths/s) that must mate to reproduce. Living in a low Reynolds regime (Re<1 to >1,000), copepods in some families exhibit an unusual mate-seeking behavior using a guidance system not found elsewhere in the animal kingdom. When copepods move through water, they leave a hydrodynamic wake whose structure varies with swimming style. Pheromones, produced specifically by the female or derived generally as a byproduct of metabolic transformations of body contents, are packaged within this hydrodynamic envelope, resulting in a concentration gradient. Chemosensitive male copepods respond to coherent chemical trails with dramatic acceleration along the trail accompanied by precise and accurate trail following, showing little error in staying on the path taken by the female, accompanied by an uncanny ability to retrace their mistaken ways to find and capture his mate. Viscosity-induced attenuation of mixing enables the persistence of small-scale chemical signals and thus, precise mate finding strategies may be a key adaptation for pelagic copepods. By leaving a coherent pheromonal trail, the female increases her signal size by up to 100 times her body size or more, thus multiplicatively improving the probability of encounter with her mate, thereby enhancing reproductive success. Evidence that copepods can follow trails in all directions indicates that guidance of trail-following copepods is solely and uniquely reliant on chemical cues without collimation by other cues, in contrast to so many other organisms that require collimation by fluid flow or gravity to guide them to the source. Thus copepods may provide novel insights into chemically-mediated guidance mechanisms. Small-scale physical gradients in the open ocean may give more definition to the niche of pelagic plankton. The large number of copepod species and the variations in their habitats provide a natural laboratory to examine the significance and specificity of chemical signals in the aquatic environment at the small-scale, and the adaptations microscopic animals have taken to live in the open ocean.

J. Yen (✉)
School of Biology, Georgia Institute of Technology, Atlanta, GA 30332-0230, USA
e-mail: jeannette.yen@biology.gatech.edu

T. Breithaupt and M. Thiel (eds.), *Chemical Communication in Crustaceans*,
DOI 10.1007/978-0-387-77101-4_9, © Springer Science+Business Media, LLC 2011

9.1   Introduction

At the turn of the century, Jeannette Yen (JY) joined the Aquatic Chemical Ecology program at the Georgia Institute of Technology (ACE at GT). There, everybody was studying chemical communication in aquatic systems, from bacteria to fish. Interdisciplinary teams were identifying the signal and source, mapping tracks in 3D and measuring chemo-sensitivity, understanding the role of fluid structure, and organism size. Our laboratory just had discovered that one species of copepod exhibited an unusual response to chemical cues (Doall et al. 1998; Weissburg et al. 1998; Yen et al. 1998). Instead of following a diffusion gradient (Katona 1973), males find and follow tiny filamentous trails that envelop the scent of their female mate and persist in a laminar flow regime. Pheromones, produced specifically by the female or derived generally as a byproduct of metabolic transformations of body contents, are packaged within this hydrodynamic envelope, resulting in a concentration gradient.

Chemosensitive male copepods respond to coherent chemical trails with dramatic acceleration along the trail accompanied by precise and accurate trail following, showing little error in staying on the path taken by the female, accompanied by an uncanny ability to retrace their mistaken ways to find and capture his mate. Living in a low Re regime (<1 to >1,000), these copepods exhibited an unusual mate-seeking behavior using a guidance system not found elsewhere in the animal kingdom. Hence, studies of copepods may provide novel insights into chemically-mediated guidance mechanisms. When Rachel Lasley joined the Yen lab in 2005, she was interested in copepod signaling in an ecological context. Specifically, how do copepods decipher signals of potential mates from other potentially confusing signals in their environment such as those created by other species? The current focus of the Yen lab is aimed toward understanding both the mechanism by which individual copepods detect signals and how they respond differentially to various signals in their environment. We are fascinated by how fast the male copepod could race up the trail to catch up to his mate. How acute does the sensitivity of its sensor array need to be to detect the waterborne cue? Do they integrate multi-modal information from the chemical composition and hydrodynamic structure of the wake? How do the coordinated motions of their multi-oared propulsive system enable them to stay on track? Since then, our understanding of chemical communication between copepods has deepened, revealing unique strategies that demonstrate novel adaptations of copepods to life in the ocean.

Copepods are small (1–10 mm) crustaceans that inhabit marine and freshwater environments. Nearly every conceivable niche (parasitic, free-living, deep-ocean, under-ice) is exploited by these animals. Copepods comprise more than 10,000 species and are arguably the most abundant multicellular organisms on the planet (Humes 1994). By consuming and metabolizing primary production, copepods serve as a key link in the aquatic food web, providing an important food resource to fish. Copepods also are heterosexual and both males and females must find and recognize their mate. Recruitment of copepods depends on the efficiency and success of their mating behavior.

Mating behavior follows a logical sequence: encounter, pursuit, capture, and copulation (Doall et al. 1998). Once close to each other, individual behaviors dominate encounter probability. Research shows that the male copepod performs the mate-searching behavior (Katona 1973; Doall et al. 1998; Tsuda and Miller 1998). Several studies found that males mated more readily with gravid females (Watras 1983; van Leeuwen and Maly 1991). Watras (1983) suggested that since the internal chemistry of a female copepod changes during gametogenesis, the release of secondary metabolites may vary, influencing male copepod behavior. It is not known if these cues are released intentionally or as a by-product where males may be spying on female metabolites.

Chemical communication leads to specificity in mate recognition, and assures species integrity. This specificity can occur at different moments in the mating sequence. First, remote detection of waterborne cues may regulate mate location and identification. Once the male locates the female, he initiates contact by grabbing her with one or several appendages (Blades and Youngbluth 1980). Chemical cues bound to the cuticular surface are critical in mate recognition (see Snell, Chap. 23). The final step occurs when males transfer the spermatophore to the female. Some genera possess genital areas with a key-and-lock fit of coupling plates (Fleminger 1975; Blades and Youngbluth 1980) that may limit the ability of a heterospecific male to correctly place the sperm packet onto the female. However, many copepods lack specialized coupling devices, and it is not known how many copepod species utilize contact chemicals or rely on remotely detected diffusible pheromones for mate recognition.

An examination of their sensors, signals, and sensitivity provides insight on their capabilities to find the mate in a fluid environment. The sensor morphology and a variety of evidence suggest that chemical and mechanosensory cues are the primary signaling modalities for marine copepods (Fields and Weissburg 2005). Chemicals can mediate mating (Snell and Carmona 1994; Yen et al. 1998), obtention of food (Koehl and Stricker 1981; Moore et al. 1999), and predator avoidance (Folt and Goldman 1981) in copepods. Hydrodynamic cues can mediate orientation, prey capture (Yen and Strickler 1996), and escape responses (Fields and Yen 1997) of copepods. We summarize the evidence that both chemical and hydrodynamic cues are important for mating processes in copepods.

9.2   Sensors, Signals, and Sensitivity

9.2.1   Sensory Structures

Many species of calanoid copepods doubled the number of aesthetascs (for description of aesthetascs see Hallberg and Skog, Chap. 6) on the male antennules (Ohtsuka and Huys 2001; Bradford-Grieve 2002) when they colonized the pelagic realm. These chemoreceptors are expressed at the last developmental stage, often accompanied by a reduction in functional mouthparts (Boxshall et al. 1997; Ohtsuka and Huys 2001). This suggests that information about their mates (e.g. location, receptivity, species) may be acquired through chemosensation. Different copepod families show distinct morphological variants in the number of chemosensitive hairs: doublings occur on males only, on both sexes, or on neither sex. In some copepods, the chemoreceptors are well-situated in the proximal region of the antennules, embedded in the high flow region of the anterior locomotory current (see Yen et al. 1998; Lenz and Yen 1993). The morphologically diverse sensory setae have species-specific arrangements and orientation along the paired antennules of copepods (Boxshall et al. 1997; Boxshall and Huys 1998; Fields et al. 2002). Odors can be brought quickly and closely to the aesthetascs in the proximal region of the antennules because of the thinned boundary layer created by the fast flow of the anterior flow field (feeding currents), shortening the diffusion distance (Yen et al. 1998). Studies of feeding current structure indicate that odors precede arrival of the source (Koehl and Strickler 1981), giving the copepod forewarning of the approach of food (Moore et al. 1999). Hence, sensory structures appear to be precisely positioned and acutely sensitive to enhance signal perception in 3D space.

9.2.2 Chemical Communication

Chemical cues are of undisputed significance for mate choice in vertebrates and in invertebrates. Arthropods, mollusks, polychaetes, rotifers, fish, and many others all use chemical cues to locate and identify mates. Studies in insects have been particularly numerous and informative. Most insects use a blend of a small number of compounds (typically 12–18C chain volatile acetates, alcohols or aldehydes) to track and identify mates (Vickers 2000). Divergent mate preferences can be achieved by alterations in blend ratios that are sometimes quite small. This suggests the production of novel chemicals is not necessary, and that specificity may be conferred by minor changes in composition. Studies of marine organisms are not as advanced; few waterborne pheromones have been identified from marine invertebrates (Painter et al. 1998; Hardege and Terschak, Chap. 19; Kamio and Derby, Chap. 20), and only a small number of contact pheromones have been partially characterized. Benthic copepods and rotifers use surface-bound glycoproteins to recognize conspecifics (see Snell, Chap. 23), in both allopatric and sympatric species complexes. A variety of indirect evidence (reviewed in Mauchline 1998) suggests planktonic copepods also respond to diffusible substances released by females.

9.2.3 Hydrodynamic Cues

Copepod species produce distinct fluid mechanical signatures that depend on a variety of factors, including swimming mode and the beat frequency, movement speed, and size of the swimming appendages (Strickler 1984; Yen and Strickler 1996; Yen 2000). Several copepod species (*Cyclops*: Strickler 1998; *Acartia*: Doall, Strickler and Yen unpublished data; Bagøien and Kiørboe 2005a; *Tortanus*: Colin 1995) exhibit the peculiar response of tandem hopping, where the male, when close to a female, hops right after she hops. The millisecond latencies of his hop response matches the response times observed for evoking the capture lunge in response to a microwater jet that was designed to mimic the wake of an escaping copepod prey (Fields and Yen 1997). This illustrates how short-lived fluid deformations provide mechanosensory cues to enable mate finding. Copepods detect mechanosensory cues using highly acute mechanosensory sensilla; velocity thresholds of copepod mechanosensors can be an order of magnitude lower than vertebrate mechanosensory hair cells (Yen et al. 1992). These setae accurately encode the intensity and direction of fluid disturbances (Fields et al. 2002). The amplitude, frequency content, and temporal characteristics of the hydrodynamic wake enable recognition and localization of appropriate potential mates.

9.2.4 Multimodal Strategies

Copepods experience a fluid regime varying from a laminar low Reynolds number (Re) regime (Re < 1–10) through to a quasi-turbulent higher Re regime (Re > 500). A consequence of movement through transitional fluid regimes is that the relative strength of the chemical and fluid-mechanical signals appears to change: the fluid-mechanical signal increases in importance at high Re within the range experienced by copepods. Fast responses resulting in predator avoidance or prey capture tend to be elicited by short-lived fluid-mechanical signals, whereas slow responses resulting in grazing, swarming, and mate-tracking tend to be elicited by persistent chemical signals. Recent behavioral studies have demonstrated that males track female trails over large distances using trail following behavior that strongly implicates they are following a chemical cue (Doall et al. 1998; Tsuda and Miller 1998; Bagøien and Kiørboe 2005b). Small-scale chemically mediated tracking in the open ocean relies on laminar hydrodynamic trails that retain chemical attractants to guide copepods to their mates (Fig. 9.1). Stretching and bending, without breaking of trails, as might occur at mixing intensities with Kolmogorov microscales (for definition see Weissburg, Chap. 4) of less than 1 mm (Lazier and Mann 1989; Yamazaki and Squires 1996), do not necessarily disrupt tracking success. Copepods precisely follow trails indicating that viscous forces indeed dominate and laminar fluid structures persist. In laboratory conditions conducive to mating, the chemically-mediated mate tracking event in *Temora longicornis* occurs in less than a few seconds. This is a smoothly swimming calanoid copepod of 1.2 mm length that lives in coastal marine areas. This copepod does not store sperm, thus giving us the opportunity to observe their repetitive mating events needed for each clutch of eggs. After trail discovery, initiated by the male, he follows the 3D path taken up to 10 s

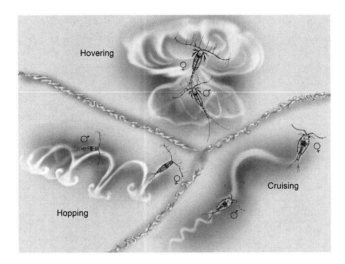

Fig. 9.1 Mating signals of cruising, hopping, and hovering copepods. Smooth swimming creates a thin filamentous wake; erratic swimming, with bursts and pauses, creates intermittent and vortical wakes; and hovering creates a broadly-distributed fluid deformation. Fluid mechanical signals dominate in the wake of hopping copepods, while chemical signals mediate trail following in cruising copepods. Odors are stronger in thinner coherent trails, and weaker when mixed by the vortices of energetic wakes or when diluted by water entrained in the large flow field created by a hovering copepod. At the small scales of copepods, viscous forces damp out fluid mechanical signals quickly, yet these same forces enable chemical signals to persist by limiting mixing to molecular diffusion. Drawing by Jorge A. Varela Ramos

earlier by his female mate. After this amount of time, the fluid mechanical signal generated by a 1 mm copepod swimming 2 mm/s has dissipated below threshold sensitivities. However, the chemical signal remains intact because viscous forces are dominant at this low Re regime. The slow process of molecular diffusion barely changes the chemical trail structure and a coherent gradient is maintained across the trail's long axis. The male makes a final lunge for the female when he is within 1–2 body lengths. It is believed that the initial tracking is mediated by the diffusible pheromone while the final leap is mediated by the hydrodynamic wake generated by the female (Yen et al. 1998).

9.2.5 Copepod Swimming and Mate Finding

Copepod species inherently differ in their characteristic swimming behavior; cruising, hop-and-sink, and hovering. While it is true that most copepod species in fact, exhibit all of these swimming behaviors, what differs is the frequency that copepod species exhibits one behavior more often than another. One can make general conclusions based on the swimming behavior of the species and the resultant signal structure. In copepod species that spend most of their time cruising (i.e. *T. longicornis, Temora stylifera, Centropages hamatus* and *Centropages typicus*) male copepods perform close 3D trail-following of the chemical scent left in the wake of the female (Bagøien and Kiørboe 2005b; Doall and Yen, unpublished; Goetze 2008; Goetze and Kiørboe 2008). Females that exhibit erratic hop-and-sink behavior such as *Oithona davisae* create intermittent high concentration packets along a discontinuous pheromone trail. To find the female, the males circle and rapidly search for the next pheromone packet along her erratic path (Kiørboe 2007). Some species that hop-and-sink seem not to use chemical signals at all. For example, *Acartia tonsa* and *Tortanus* perform tandem hopping, with the male hopping in remarkable synchrony with the female (Colin 1995; confirmed for *Acartia* by Bagøien and Kiørboe 2005a). Finally, females of hovering species such as *Pseudocalanus elongates* create near spherical pheromone clouds with a heterogeneous structure that result from changes in body orientations. These males perform a zigzag swimming behavior, detecting gradients within the cloud (Kiørboe et al. 2005). Comparing species that utilize chemical cues with species that use hydrodynamic cues yields differences in several aspects of their behavior. For example, *T. longicornis* (chemical signaler) and *Acartia tonsa* (hydrodynamic signaler) differ in the following ways: percent swimming (*Temora* swims ca. 100% of the time, *Acartia* spends <10% of its time swimming and instead sinks often); reactive distance (*Temora* follows trails of up to 100 bodylengths (BL); *Acartia* has the shortest reactive distances of 1–2 BL); aggregation structure (*Temora* aggregates loosely when performing its vertical migratory excursions; *Acartia* forms dense swarms where nearest neighbor distance can be within 1–2 BL; Haury and Yamazaki 1995).Trail-following by *Temora* is mediated primarily by chemical cues while tandem hopping by *Acartia* is mediated primarily by the short-lived mechanical cues. In this low Re regime, fluid mechanical signals dissipate $1,000\times$ faster than a chemical molecule diffuses; copepods reliant on fluid mechanical signals must have short perceptive distances while copepods sensitive to odors can detect trails left by females that are more distant, giving a larger perceptive field. Our analysis also suggests that animals utilizing mechanosensory cues form denser aggregates that reduce nearest neighbor distances and promote mating encounters (Mauchline 1998). Thus, mate perception mechanisms are linked to a suite of other behaviors. *C. hamatus* and *C. typicus* spend 50–80% of their time swimming and also rely on trail-following with slightly shorter interaction distances than those of *Temora*. *Tortanus* show mating responses similar to the two species of *Acartia* (*A. hudsonica, A. tonsa*). These data suggest the suite of linked traits (swimming mode/style, aggregation density, dominant sensory modality, reactive distance) occur across a gradient from long-range chemosensory to short-range mechanosensory capabilities.

9.3 Chemically-Mediated Trail Following in Copepods

9.3.1 Following a Laminar Trail

In the following section, we present an independent set of experiments performed in the laboratory at Georgia Tech to illustrate the reproducibility of the effects of diffusible mating pheromones on male copepod tracking behavior, as found elsewhere by others (e.g. Goetze and Kiørboe 2008). We focus on the marine coastal copepod *Temora longicornis* because of its reliance on chemical communication. Using adult stages collected and shipped from the Gulf of Maine, we noted pair formation in the cultures kept in an environmental chamber set at 12°C, matching that of their habitat; this assured us that the mates were receptive. We also made sure that males readily followed the females and successful trackings (Fig. 9.2) were noted when mating pairs

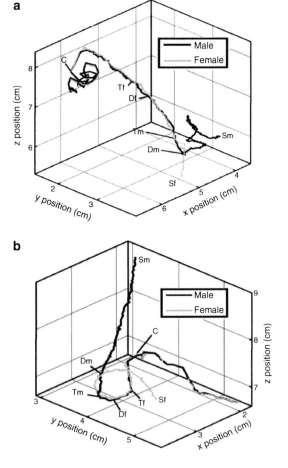

Fig. 9.2 Two representative mating events (**a**, **b**) of the copepod, *Temora longicornis*. Male and female trajectories are shown in *black* and *gray*, respectively. Start positions are labeled as "Sm" (male start position) and "Sf" (female start position). A male exhibits normal cruising behavior until he detects a female's chemical trail (Dm). At the point of signal detection, the female is approximately 1 cm ahead of the male (Df). After the male detects the female, he turns, finds the female's chemical trail and begins tracking (Tm). Female position at initial time of male tracking is denoted "Tf." As the male tracks the female's trail, he speeds up until he catches her and copulation begins (C). These video image sequences of overlapping *x–z* and *y–z* pairs of 2D images were obtained from our three-dimensional Schlieren laser videography system as developed by Strickler and Hwang (1999) and further described by Doall et al. (1998)

were formed. Examination of mating trajectories showed no predominant direction of the turns (Fig. 9.2). The absence of a predominant directionality of the turning angle suggests that no other cue but trail presence, and not light or gravity, guides the behavior. Exploration in random directions also is exhibited when copepods lose the trail, as might occur at a break in the trail caused by a hop by the female which disperses the cue (van Duren et al. 1998; Yen et al. 1998).

We analyzed trail following kinematics of six successful mate tracking events. Before tracking, *T. longicornis* males swam at an average velocity of 0.93 ± 0.18 cm s^{-1} ($n = 6$), which is not significantly different from the females' average swimming speed of 0.88 ± 0.20 cm s^{-1} ($n = 6$). During tracking, *T. longicornis* males swam 1.1–2.3 cm s^{-1}, averaging 1.6 ± 0.5 cm s^{-1}, which is nearly two times faster. Mating pairs of *T. longicornis* swam even faster at 2.8 ± 0.5 cm s^{-1} ($n = 4$; Fig. 9.3a). The fast swimming of the mating pair adds to their conspicuousness due to size, as noted by Jersabek et al. (2007). Like Doall et al. (1998), we also found that males accelerated when tracking females, similar to the behavior of *Centropagus typicus* (Bagøien and Kiørboe 2005b). By plotting the trajectory of the female and the male, we saw that their paths overlapped, with the male following 1–3 s behind the female. As the male closes in on his mate, his speed increases, enabling him to catch up to his mate (Fig. 9.3b). Hence, both the speed and trajectory assured us that the male was competent and that he was interested in mating.

Males began tracking the females many body lengths away, and as noted above, made a directed leap to the female when the separation distance was about 1 BL. We compared the increase in reactive distance gained by using pheromones by calculating chemosensory- vs. mechanosensory-based perceptive distances during mate location. The chemically-mediated perceptive distance was calculated as the linear distance between the male and female at the initiation of mate pursuit behavior. The hydrodynamic perceptive distance was calculated as the linear distance between the male and female when the male lunges to catch the female. The increase in perceptive ability conferred by using chemical perception vs. hydrodynamic perception was characterized as the ratio of chemical perceptive distance to hydrodynamic perceptive distance. Chemically-mediated perceptive distances for *T. longicornis* ranged from 0.45 to 1.72 cm, averaging 0.89 ± 0.55 cm ($n = 6$). Hydrodynamic perception in *T. longicornis* occurred at distances ranging from 0.21 to 0.33 cm, averaging 0.26 ± 0.05 cm ($n = 6$). On average, the ratio of chemical perceptive distance to hydrodynamic perceptive distance in *T. longicornis* was 3.27 ± 1.93 ($n = 6$). Males of *T. longicornis* more than tripled the perceptive distance through which they were able to perceive potential mates by using chemically-mediated perception. Encounter rates can be substantially improved by such increases in the perceptive distance because it depends on distance-squared (Gerritsen and Strickler 1977).

We created an artificial trail to determine whether the diffusible scent of the female was the attractive element in this unusual trail following behavior. We collected female scent by incubating pre-fed adult female copepods in filtered artificial seawater. To test the bio-activity of the scented seawater, we labeled the water with a high molecular compound, dextran, and used Schlieren optics to

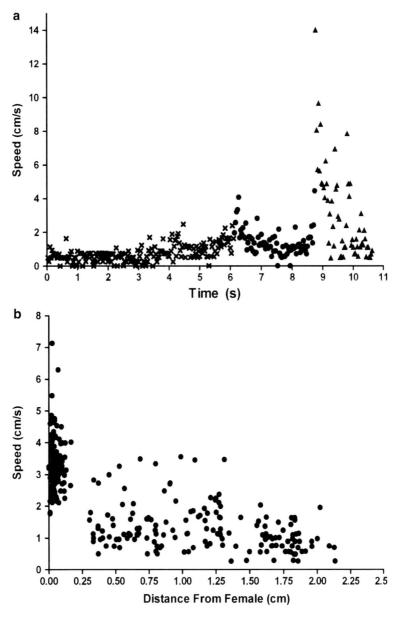

Fig. 9.3 (a) Acceleration and lunging during mate tracking by *T. longicornis*. Speed is calculated from the temporal change in 3D coordinates of the copepod. Swimming speeds before initiation of mate pursuit are marked with x's, speeds during tracking are marked with *circles*, and speeds as a mating pair are marked with *triangles*. *T. longicornis* males accelerate during tracking and lunge immediately before capturing the females. (**b**) Typical mating event for *T. longicornis*. Distance is the straight-line distance between the male–female at the moment the male finds the trail. The male performs an accelerated leap toward the female for the mating capture. The (*x, y, z, t*) data from video image sequences, obtained with the dps velocity 8 (Leitch, Ackworth, GA) interface, were analyzed with Scion Image (Scion Corporation, Frederick, MD)

visualize both the copepods and the trail. Imaging with Schlieren optics relies on differences in refractive indices. Both the copepod and the dextran-labeled scented water have refractive indices that are different than the surrounding water and hence, both the copepod and the trail can be seen as silhouettes when viewed with the appropriate optical path (Doall et al. 1998; Strickler and Hwang 1999). To our delight, when the male copepod intersected the trail, he would race up the trail and we could see how his swimming motions deformed the trail into chevrons (Fig. 9.4). Dual labeling of the trail with the attractant plus a large molecular weight dextran enabled us to see both the deformation of the odor signal and the movement of the male copepod with this behavioral bioassay. An ability to follow a trail mimic, containing only the chemicals exuded by the female with no female present, demonstrates that the trail-following behavior of the male is chemically-mediated.

With our trail-following signal-deforming bioassay, we tested the attractiveness of fresh, 1 day old and frozen scent. We documented the number of trail mimics that

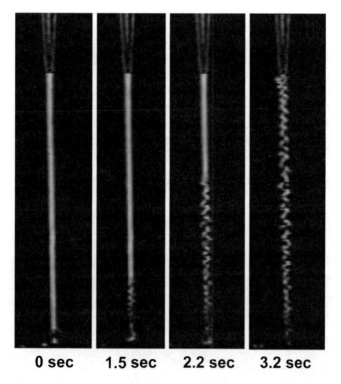

0 sec   1.5 sec   2.2 sec   3.2 sec

Fig. 9.4 Following a chemical trail mimic by *T. longicornis*. Visualized trail and upwardly-directed trail following. A trail mimic was created by releasing seawater, scented with copepod odors. The trail on the *left* is the undisturbed trail. The disturbed chevron-dotted trails on the *right* show how the trail structure changes after the male copepod *T. longicornis* follows odor trail. The copepod is located at upper end of disturbance in trail (Yen et al. 2004). Copyright CRC Press

were followed when encountered. The copepods followed frozen (70% follows for 63 trails intersected) or 1-day old trail-following pheromone (68% follows for 50 trails intersected) just as well as fresh scent (62% follows for 27 trails intersected). This enables future work to collect and chemically analyze frozen scent for bioactive fractions. Continued bio-activity could lead to overstimulation and opens the possibility that copepods might be confused by persistent trails. In the still water of our experimental tanks, molecular diffusive processes appear to erase the trails as surmised by evidence that males rarely followed trails older than 10s. Eddy diffusive processes, which increase with increased input of physical energy (Yamazaki et al. 2002), may cause trails to become more contorted and strained (Visser and Jackson 2004) but copepods are fully capable of following curvy trails (Fig. 9.5). In fact, image analyses showed that, as the male waggles along the trail, the coherent signal becomes deformed into a zigzag-shaped signal (Fig. 9.4). Subsequent trail-following still closely follows the trail; however, as the trail now has a zigzag form, so is the resultant path of the tracking male. This explains the erratic trail following noted in backtracking, when the male, after following the trail the wrong way (away from the female), turns around and retraces his path (Doall et al. 1998). His erratic trajectory matches the deformations he generated in the odor cue. When the male reaches the undisturbed section of the trail, he resumes smooth swimming, confirming the importance of the odor structure in mediating trail following kinematics in copepods. For copepod species with sperm-storing females (Ohtsuka and Huys 2001), male–male competition also places intense intrasexual selective pressure to be the first male to find a receptive female. Disruption of trail structure after discovery (Yen et al. 2004) may prevent other chemosensitive males or predators from using the trail to track down these mating copepods. Todd et al. (2005) confirmed multiple paternities in an ectoparasitic copepod but commented that being the first male can assure single paternity especially

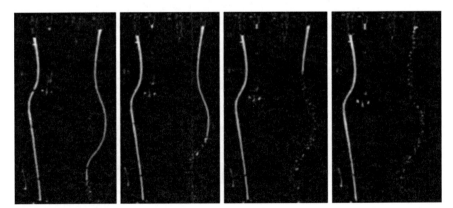

Fig. 9.5 Schlieren image of undisturbed (*left*) and disturbed (*right*) scented trails (four panels). Copepods are perfectly able to follow curvy trails, as illustrated by the signal deformation caused when the male copepod swims up the curved trail on right. Trail on *left* is undisturbed. Small Eppendorf pipette tips which were used to release the female pheromone to form the scent trails can be seen along the top of the panels

for *Lepeophtheirus salmonis*, where the male blocks the copulatory ducts with spermatophore cement.

9.3.2 Sex and Species Specificity of Pheromone Trail

To test for species specificity, trail-mimic bioassays were conducted with 20 reproductively mature *T. longicornis* males in a 3 L tank and chemical trail choices consisted of trails that are likely to be encountered by *T. longicornis* males in the field: *T. longicornis* females, *T. longicornis* males, *Acartia hudsonica* females. Trail following behavior was compared among these treatments and to an unscented control trail containing only filtered seawater and dextran. Male *T. longicornis* were placed in the visualization tank and allowed to acclimate for 30 min after which experiments were run for 2 h. Three chemical trail mimics of the same trail choice then were fed into the tank by an electronic syringe pump via capillary tubes at a slow and constant rate (0.01 mL/min) to control for any hydrodynamic signal. Trail following frequency was recorded for each trail choice. Follows were characterized by a male swimming through a trail and subsequently swimming up or down the trail, disrupting the trail with its antennae and leaving a disturbed trail in his wake. Non-follows were characterized as a male swimming through a trail with no subsequent following behavior.

Our tests for species specificity of the scent (Fig. 9.6) indicate that trail following behavior of *T. longicornis* males is dependent on trail type (i.e. conspecific female, conspecific male, heterospecific female and unscented). Male *T. longicornis* followed conspecific trail-mimics more often than the trail-mimics of heterospecifics or unscented control. However, male *T. longicornis* do not seem to follow conspecific female trails more often than conspecific male trails. This could indicate a sort of mate encounter cue hierarchy, where males use multiple sensory modalities in mate recognition. Males may respond to chemical cues as a proxy for a nearby conspecific individual and respond to additional cues, such as hydrodynamic to decipher between males and females. Previous research by Weissburg et al. (1998) on the kinematics of *T. longicornis* mating behavior revealed that male tracking behavior changes as a result of differences in female swimming style. Close tracking was possible when following a smoothly-swimming copepod, while circling within the pheromone cloud occurred when male copepods were stimulated by a hovering female (documented by Doall et al. 1998 and observed also by Bagøien and Kiørboe 2005b). Therefore, it is likely that swimming style and subsequent hydrodynamic signals play an important role in the male's ability to discriminate between potential mates and nonmates. This would explain why a male would follow another male's species-specific chemical signal in the absence of a sex-specific hydrodynamic cue. It is also possible that males follow the tracks of both males and females with equal frequency in the wild. In other trail following species such as the garter snake *Thamnophis sirtalis parietalis* (Shine et al. 2000), and the snail *Littorina littorea*, mate-searching males follow the chemical trails of

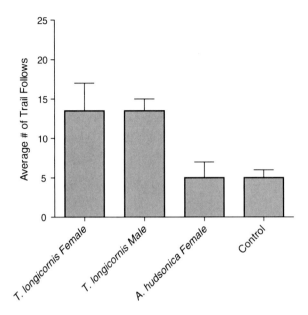

Fig. 9.6 Average number of trail follows executed by *T. longicornis* males for the chemical-trail mimics of *T. longicornis* female, *T. longicornis* male, *Acartia hudsonica* female, and an unscented control. *Error bars* refer to the standard error of mean trail follows out of 25 trail encounters from two experiments. Trail following data were subjected to a chi-square test of associations between trail type and following frequency. Chi Square = 24.79, degrees of freedom = 3, *p*-value < 0.001 (i.e. there is an association between the frequency of follows and the trail type), *n* = 25 (which refers to number of trail encounters per experiment). Statistical analysis was conducted using SigmaStat 3.5 (Port Richmond, CA)

other males almost as frequently as they follow female trails (Saur 1990). This could be an advantageous strategy if the mating encounters are few and the cost of missing a mating opportunity is greater than the cost of making a mistake (wasting a spermatophore on a male).

Recently, the work of Goetze and Kiørboe (2008) indicates that there may be a gradient in the specificity of the pheromonal cue where some species are selective while others will follow any discovered trail. They performed two-way crosses between congeners, demonstrating that *T. stylifera* males would not follow females of *T. longicornis* but *T. longicornis* males would follow *T. stylifera* females. Trails of *T. longicornis* are much thinner than those made by the larger species, *T. stylifera*, and hence the heterospecific signal may be too weak to evoke mate tracking behavior. Also, the distribution of *T. longicornis* shows broad areas dominated by this species, thus reducing the need for species-specificity. In contrast, the population of *T. stylifera* nearly always overlaps with that of *T. longicornis* so that specificity exhibited by *T. stylifera* may have enabled speciation and reproductive isolation via a chemically mediated barrier. The numerous allopatric and sympatric species of copepods with such a variety of unique properties characterizing their mating cue opens up a fascinating opportunity to understand this gradient in mate selective abilities.

9.4 Copepods as Model Organisms in Chemical Communication

The mate tracking behavior of the copepod appears unique in many ways. No other organism has been documented to exhibit the ability to follow a trail in 3D space using only chemical information. The polychaete worm *Nereis succinea* follows a 3D trail of the mate (Ram et al. 2008) but appears to use both chemical and hydrodynamic cues. Catfish use hydrodynamic cues (sensed by the lateral line) to follow the 3D trails of their prey (Pohlmann et al. 2004). Ants (Edelstein-Keshet 1994) and snakes (Schwenk 1994) detect chemical gradients in a trail but these are 2D signals. Crabs live in a benthic 2D environment and can track odors in a turbulent plume. These bottom-dwelling organisms need both the collimating stimuli of water flow and the odor stimulant in order to find the odor source (Weissburg 2000). Moths also rely on the presence of wind to polarize olfactory-guided locomotion in the up-wind direction (Baker, Chap. 27). Shrimp (Hamner and Hamner 1977) and oyster larvae (Tamburri et al. 1996) follow trails downward, where gravity provides a directional signal. The copepod *Calanus marshallae* (Tsuda and Miller 1998) also appears to rely on gravity as males swim downward along the scent trail to find their sinking female. However, we demonstrated that *T. longicornis*, a coastal marine copepod, can follow signals to sources above or below them (Yen et al. 2004). Copepods live in a viscosity-dominated fluid regime, where the little flow remaining is isotropic, providing no information on the direction to the stimulus source. Neither gravity nor flow serves as collimating mechanisms to inform the copepod of the location of the signal source and neither signal is essential in the efficient tracking by copepods. Other organisms that also live in viscosity-dominated realms, such as bacteria (Dusenbery 1998) and nematodes (Pierce-Shimomura et al. 1999), sense odor gradients and respond with specific behaviors to guide them up-gradient, but these organisms are not following trails. Thus, copepods may provide novel insights into chemically-mediated guidance mechanisms. The ability of animals to follow chemical trails is a consequence of viscosity-induced attenuation of mixing that enables the persistence of small-scale chemical signals. Yet, because diffusion does limit the active space to the trail emitted by the female copepod or the phycosphere exuded by a tasty alga, many copepods create currents to advect the odorant to its aesthetacs and facilitate stimulus acquisition. Their aesthetascs are located in the high-velocity region of the flow field so boundary layers are thinner to shorten the diffusion distance. Hence, diffusion is both an ally and a constraint for chemical sensing in copepods.

Spatial detection of stimulus asymmetries has been proposed as a strategy that provides more information than temporal sampling to large aquatic crustaceans when they follow plumes (Weissburg, Chap. 4). Small organisms such as bacteria use temporal comparison to find odor sources (Dusenbery 2001). Copepods use *both* spatial and temporal strategies to track diffusing chemical cues. It appears that the rapid chemically-mediated tracking behavior in copepods shares detection strategies with those employed by animals in both fast and slow flow environments. Fine manipulations of odor properties of turbulent plumes are quite difficult, making copepods and their interactions with filamentous trails an attractive system to examine temporal response properties of the chemosensory system. Now that we can visualize both the trail and the trail-following copepod using Schlieren optics, we can use high-speed high-magnification videography to see the location of the antennules of copepods tracking trails of different width and odor intensities relative to the surrounding fluid. The small size, relative simplicity of the sensory array, and ease of manipulating the signal allow us to examine questions in copepods that are not easily addressed in larger animals, such as the number of sensors needed to code stimulus asymmetry, the time required to detect changes in stimulus intensity, and the reliance of intensity discrimination on absolute concentration or other trail properties. Understanding when animals use temporal and spatial sampling in direct guidance will help evaluate the benefits of a certain orientation strategy under different environmental conditions.

Population densities of these small animals are generally less than 1 individual per liter, so that increases in perceptual distances are required to produce a reasonable chance of encounters between sexes (Gerritsen and Strickler 1977; Yen 1988; Haury and Yamazaki 1995; Kiørboe and Bagøien 2005). Precise mate finding strategies found for pelagic copepods may be a key adaptation, enhancing reproductive success in these open-ocean planktonic populations. In addition to finding mates, the ability to identify them as conspecifics presents another significant challenge for planktonic animals living in the open ocean. If remotely detected signals are not species-specific, then these trails do not save copepods from direct contact for species identification. Species diversity may be promoted because the chemical and hydrodynamic trails contribute to the complexity of small-scale gradients in the physical environment and may give more definition to the niche of open ocean plankton. The large number of copepod species and the variations in their habitats provide a natural laboratory to examine the significance of chemical signals in the aquatic environment at the small scale and the adaptations microscopic animals have taken to live in the open ocean.

9.5 Future Directions

9.5.1 Chemical Composition of Copepod Pheromone(s)

Laboratory and field observations support the hypothesis that copepods use chemical cues for locating mates, and that both diffusible and contact pheromones can be involved (Lonsdale et al. 1998). However, complete structural identification of any intraspecific signaling chemical for copepods is lacking. Glycoproteins are clearly involved in mate recognition and selection in some species of harpacticoid copepods (Snell, Chap. 23). An exuded, diffusible chemical ranging in molecular weight from 100 to 1,000 Da has been detected but not characterized (Lazzaretto et al. 1994). Ingvarsdottir et al. (2002) have provided evidence for lipophilic, diffusible compounds being produced by female copepods to attract males. These compounds are similar to volatile semiochemicals used by terrestrial organisms. It seems likely that these compounds are species-specific, and might also be modified by copepod diet and environment (as found for amphipods by Stanhope et al. 1992). To be highly diffusible, copepod pheromones may be low-molecular weight polar molecules such as small peptides. However, we can envision conditions under which different species use different types of signal molecules. For instance, some species may rely on metabolites associated with reproductive maturity. Others may require chemicals with low diffusion constants (large molecular weight lipophilic compounds; Dusenbery 1992) because the abiotic environment is hydrodynamically unstable, causing rapid trail loss. However, there is a trade-off here: diffusivity of molecules varies with the inverse of their mass to power 1/3, so the copepods get more value for their metabolic investment by producing a large amount of small molecules than a small amount of big molecules. Copepods also have been demonstrated to display feeding responses to particular "cocktails" of dissolved amino acids, while they did not respond to the individual amino acids (Gill and Poulet 1988). A similar response may occur with mate seeking: the individual chemical components are not specific, but the combination and the ratio of the components give information about species and sex. Thus, the molecular identity of the compound may provide insights into potential constraints on the sensory system, such as the need to synchronize mating with reproductive maturity, or a requirement for a particularly cohesive scent trail. To date, there is a wealth of tantalizing observations and ecological data begging a chemical explanation.

9.5.2 Copepod Speciation and Distribution

Marine plankton, which spend their entire life in the water column, present considerable difficulties for explaining speciation via allopatry, since they tend to be widely dispersing. Heterosexuality and speciosity pose opposing constraints: conspecific mates must find each other but unrelated species must remain reproductively isolated to maintain species integrity. Determining the mechanisms for maintaining reproductive isolation and species distinction can improve our understanding of copepod tactics to avoid failed investment of individuals in mating when it does not lead to reproductive success. Organisms have a variety of methods to ensure that they mate only with conspecifics. Such mechanisms often depend on appropriate mates being able to recognize one another, a skill that must be finely honed so that organisms discriminate among potential mates that are morphologically indistinguishable to a human observer. The existence of species that maintain reproductive isolation without detectable morphological change (sibling or cryptic species) strongly suggests that mate recognition by chemical signals can be an important force maintaining species boundaries (Knowlton 1993; Lee 2000; Bucklin et al. 2003; Goetze 2008). This reliance on chemical cues provides an exciting system to study species-specificity in mate-seeking behavior as well as studies of the chemical composition of copepod pheromones and their role in speciation in the aquatic environment. Identifying the signals and their properties therefore will help to understand the evolution of copepod biodiversity and the ecological relationships among species in the plankton.

References

Bagøien E, Kiørboe T (2005a) Blind dating-mate finding in planktonic copepods. III. Hydromechanical communication in Acartia tonsa. Mar Ecol Prog Ser 300:129–133

Bagøien E, Kiørboe T (2005b) Blind dating-mate finding in planktonic copepods. I. Tracking the pheromone trail of Centropages typicus. Mar Ecol Prog Ser 300:105–115

Blades PI, Youngbluth MJ (1980) Morphological, physiological and behavioral aspects of mating in Calanoid copepods. In: Kerfoot WC (ed) Evolution of zooplankton communities. University Press, Hanover, pp 39–51

Boxshall GA, Huys R (1998) The ontogeny and phylogeny of copepod antennules. Philos Trans R Soc Lond B 353:765–786

Boxshall GA, Yen J, Strickler JR (1997) Functional significance of the sexual dimorphism in the array of setation elements along the antennules of Euchaeta rimana Bradford. Bull Mar Sci 61:387–398

Bradford-Grieve JM (2002) Colonization of the pelagic realm by calanoid copepods. Hydrobiologia 485:223–244

Bucklin A, Frost BW, Bradford-Grieve J, Allen LD, Copley NJ (2003) Molecular systematic and phylogenetic assessment of 34 calanoid copepod species of the Calanidae and Clausocalanidae. Mar Biol 142:333–343

Colin SP (1995) A kinematic analysis of trail-following in Temora longicornis and four other copepods: how the male finds his mate. MS thesis, State University of New York at Stony Brook

Doall MH, Colin SP, Strickler JR, Yen J (1998) Locating a male in 3D: the case of Temora longicornis. Philos Trans R Soc B 353:681–689

Dusenbery D (1992) Sensory ecology: how organisms acquire and respond to information. W.H. Freeman & Company, New York

Dusenbery D (1998) Spatial sensing of stimulus gradients can be superior to temporal sensing for free-swimming bacteria. Biophys J 74:2272–2277

Dusenbery D (2001) Performance of basic strategies for following gradients in two dimensions. J Theor Biol 208:345–360

Edelstein-Keshet L (1994) Simple models for trail-following behavior: trunk trails vs. individual foragers. J Math Biol 32:303–328

Fields DM, Weissburg MJ (2005) Evolutionary and ecological significance of mechanosensor morphology: copepods as a model system. Mar Ecol Prog Ser 287:269–274

Fields DM, Yen J (1997) The escape behavior of marine copepods in response to a quantifiable fluid mechanical disturbance. J Plankton Res 19:1289–1304

Fields DM, Shaeffer DS, Weissburg MJ (2002) Mechanical and neural responses from the mechanosensory hairs on the antennule of *Gaussia princeps*. Mar Ecol Prog Ser 227:173–186

Fleminger A (1975) Geographical distribution and morphological divergence in the American coastal-zone planktonic copepods of the genus *Labidocera*. In: Cronin LE (ed) Estuarine research. Academic Press, New York, pp 392–419

Folt C, Goldman CR (1981) Allelopathy between zooplankton: a mechanism for interference competition. Science 213:1133–1135

Gerritsen J, Strickler JR (1977) Encounter probabilities and community structure in zooplankton: a mathematical model. J Fish Res Board Can 34:73–82

Gill CW, Poulet SA (1988) Responses of copepods to dissolved free amino acids. Mar Ecol Prog Ser 43:269–276

Goetze E (2008) Heterospecific mating and partial prezygotic reproductive isolation in the planktonic marine copepods *Centropages typicus* and *Centropages hamatus*. Limnol Oceanogr 53:33–45

Goetze E, Kiørboe T (2008) Heterospecific mating and species recognition in the planktonic marine copepods *Temora stylifera* and *T. longicornis*. Mar Ecol Prog Ser 370:185–198

Hamner P, Hamner WM (1977) Chemosensory tracking of scent trails by planktonic shrimp *Acetes sibogae australis*. Science 195:886–888

Haury LR, Yamazaki H (1995) The dichotomy of scales in the perception and aggregation behavior of zooplankton. J Plankton Res 17:191–197

Humes AG (1994) How many copepods? Hydrobiologia 293:1–7

Ingvarsdottir A, Birkett MA, Duce I, Mordue W, Pickett JA, Wadhams LJ, Mordue AJ (2002) Role of semiochemicals in mate location by parasitic sea louse, *Lepeophtheirus salmonis*. J Chem Ecol 28:2107–2117

Jersabek CD, Luger MS, Schabetsberger R, Grill S, Strickler JR (2007) Hang on or run? Copepod mating versus predation risk in contrasting environments. Oecologia 153:761–773

Katona SK (1973) Evidence for sex pheromones in planktonic copepods. Limnol Oceanogr 18:574–583

Kiørboe T (2007) Mate finding, mating, and population dynamics in a planktonic copepod *Oithona davisae*: there are too few males. Limnol Oceanogr 52:1511–1522

Kiørboe T, Bagøien E (2005) Motility patterns and mate encounter rates in planktonic copepods. Limnol Oceanogr 50:1999–2007

Kiørboe T, Bagøien E, Thygesen U (2005) Blind dating – mate finding in planktonic copepods. II. The pheromone cloud of *Pseudocalanus elongatus*. Mar Ecol Prog Ser 300:117–128

Knowlton N (1993) Sibling species in the sea. Annu Rev Ecol Syst 24:189–216

Koehl MAR, Stricker JR (1981) Copepod feeding currents: food capture at low Reynolds number. Limnol Oceangr 26:1062–1073

Lazier JRN, Mann KH (1989) Turbulence and diffusive layers around small organisms. Deep Sea Res A 36:1721–1733

Lazzaretto I, Franco F, Battaglia B (1994) Reproductive behavior in the harpacticoid copepod *Tigriopus fulvus*. Hydrobiologia 293:229–234

Lee CE (2000) Global phylogeography of a cryptic copepod species complex and reproductive isolation between genetically proximate "populations". Evolution 254:2014–2017

Lenz PH, Yen J (1993) Distal setal mechanoreceptors of the first antennae of marine copepods. Bull Mar Sci 53:170–179

Lonsdale DJ, Frey MA, Snell TW (1998) The role of chemical signals in copepod reproduction. J Mar Syst 15:1–12

Mauchline J (1998) The biology of calanoid copepods. Academic Press, San Diego

Moore PA, Fields DM, Yen J (1999) Physical constraints of chemoreception in foraging copepods. Limnol Oceanogr 44:166–177

Ohtsuka S, Huys R (2001) Sexual dimorphism in calanoid copepods: morphology and function. Hydrobiologia 453:441–466

Painter SD, Clough B, Garden RW, Sweedler JV, Nagle GT (1998) characterization of *Aplysia* attractin, the first water-borne peptide pheromone in invertebrates. Biol Bull 194:120–131

Pierce-Shimomura JT, Morse TM, Lockery SR (1999) The fundamental role of pirouettes in *C. elegans* chemotaxis. J Neurosci 19:9557–9569

Pohlmann K, Atema J, Breithaupt T (2004) The importance of the lateral line in nocturnal predation of piscivorous catfish. J Exp Biol 207:2971–2978

Ram JL, Fei X, Danaher SM, Lu S, Breithaupt T, Hardege JD (2008) Finding females: pheromone-guided reproductive tracking behavior by male *Nereis succinea* in the marine environment. J Exp Biol 211:757–765

Saur M (1990) Mate discrimination in *Littorina littorea* and *L. saxatilis (Olivi)* Mollusca: Prosobranchia. Hydrobiologia 193:261–270

Schwenk K (1994) Why snakes have forked tongues. Science 263:1573–1577

Shine PH, Lemaster MP, Moore IT, Mason RT (2000) The transvestite serpent: why do male garter snakes court (some) other males? Anim Behav 59:349–359

Snell TW, Carmona MJ (1994) Surface glycoproteins in copepods: Potential signals for mate recognition. Hydrobiologia 292–293:255–264

Stanhope MJ, Connelly MM, Hartwick B (1992) Evolution of a crustacean chemical communication channel – behavioral and ecological genetic evidence for a habitat-modified, race-specific pheromone. J Chem Ecol 18:1871–1887

Strickler JR (1984) Sticky water: a selective force in copepod evolution. In: Meyers DG, Strickler JR (eds) Trophic interactions within aquatic ecosystems. Westview Press Inc, Boulder, pp 187–239

Strickler JR (1998) Observing free-swimming copepods mating. Philos Trans R Soc B 353:671–680

Strickler JR, Hwang JS (1999) Matched spatial filters in long working distance microscopy of phase objects. In: Cheng PC, Hwang PP, Wu JL, Wang G, Kim H (eds) Focus on multidimensional microscopy. World Scientific Publishing, River Edge, pp 217–239

Tamburri MN, Finelli CM, Wethey DS, Zimmer-Faust RK (1996) Chemically mediated larval settlement behavior in flow. Biol Bull 191:367–373

Todd CD, Stevenson RJ, Reinardy H, Ritchie MG (2005) Polandry in the ectoparasitic copepod *Lepeophtheirus salmonis* despite complex pre-copulatory and post-copulatory mate-guarding. Mar Ecol Prog Ser 303:225–234

Tsuda A, Miller CB (1998) Mate-finding behavior in *Calanus marshallae*. Philos Trans R Soc B 353:713–720

van Duren LA, Stamhuis EJ, Videler JJ (1998) Reading the copepod personal adds: increasing encounter probability with hydromechanical signals. Philos Trans R Soc B 353:691–700

van Leeuwen HC, Maly EJ (1991) Changes in swimming behavior of male *Diaptomus leptopus* (Copepoda: Calanoida) in response to gravid females. Limnol Oceanogr 36:1188–1195

Vickers N (2000) Mechanisms of navigation in odor plumes. Biol Bull 198:203–212

Visser AW, Jackson GA (2004) Characteristics of the chemical plume behind a sinking particle in a turbulent water column. Mar Ecol Prog Ser 283:55–71

Watras CJ (1983) Mate location by diaptomid copepods. J Plankton Res 5:417–423

Weissburg MJ (2000) The fluid dynamical context of chemosensory behavior. Biol Bull 198:188–202

Weissburg MJ, Doall MH, Yen J (1998) Following the invisible trail: kinematic analysis of mate tracking in the copepod *Temora longicornis*. Philos Trans R Soc B 353:701–712

Yamazaki H, Squires KD (1996) Comparison of oceanic turbulence and copepod swimming. Mar Ecol Prog Ser 144:299–301

Yamazaki H, Mackas DL, Denman KL (2002) Coupling small-scale physical processes with biology. In: Robinson AR, McCarthy JJ, Rothschild BJ (eds) The sea. Wiley, New York, pp 51–112

Yen J (1988) Directionality and swimming speeds in predator-prey and male-female interactions of *Euchaeta rimana*, a subtropical marine copepod. Bull Mar Sci 43:175–193

Yen J (2000) Life in transition: balancing inertial and viscous forces by planktonic copepods. Biol Bull 198:213–224

Yen J, Strickler JR (1996) Advertisement and concealment in the plankton: what makes a copepod hydrodynamically conspicuous? Invertebr Biol 3:191–205

Yen J, Lenz PH, Gassie DV, Hartline DK (1992) Mechanoreception in marine copepods: electrophysiological studies on the first antennae. J Plankton Res 14:495–512

Yen J, Weissburg MJ, Doall MH (1998) The fluid physics of signal perception by a mate-tracking copepod. Philos Trans R Soc B 353:787–804

Yen J, Prusak A, Caun M, Doall MH, Brown J, Strickler JR (2004) Signaling during mating in the pelagic copepod, Temora longicornis. In: Seuront L, Strutton P (eds) Scales in aquatic ecology: measurements, analysis, modelling. CRC Press, New York, pp 149–159

Chapter 10
Chemical Communication in Peracarid Crustaceans

Martin Thiel

Abstract Chemical communication plays an important role during the life of peracarid crustaceans, where the two main taxa, the amphipods and isopods, have representatives in both aquatic and terrestrial habitats. As in other crustaceans, the antennae bear the most important chemosensory structures, which are used for food-finding, predator detection and intraspecific interactions. The chemical nature of peracarid pheromones is unknown, but numerous experimental studies confirm that chemical signals can be soluble/volatile or contact pheromones. Waterborne chemicals mediate mate finding while contact pheromones are mainly involved in mate assessment. Males of some species appear capable to determine the reproductive status of females (closeness to the reproductive molt), which is possibly mediated by chemical compounds. Contrasting with this fine-tuned chemoreception in male–female interactions are other examples that suggest that reproductive isolation between closely related congeneric species is incomplete. The fact that males form precopulatory associations with heterospecific females indicates that chemicals mediating these interactions are not sufficiently specific to permit species discrimination. Gregarious behavior in many species is also guided by chemical cues that lead to aggregation on shared food sources or in communal shelters. Studies on kin recognition in mother–offspring groups have produced ambiguous results – in some species females appear unable to discriminate between their own and unrelated offspring, while females of other species recognize their own juveniles. The best example for kin recognition comes from desert isopods where family-specific chemical signatures allow parents to recognize their offspring. In summary, there is abundant experimental and observational evidence that numerous intra- and interspecific interactions in peracarids are mediated via chemical stimuli, but knowledge about the chemical structure of these compounds is still very limited. Given that all species have direct development and that many species can be easily cultured in the laboratory, peracarid crustaceans are

M. Thiel (✉)
Facultad Ciencias del Mar, Universidad Catolica del Norte,
Larrondo 1281, Coquimbo, Chile
e-mail: thiel@ucn.cl

T. Breithaupt and M. Thiel (eds.), *Chemical Communication in Crustaceans*,
DOI 10.1007/978-0-387-77101-4_10, © Springer Science+Business Media, LLC 2011

proposed as ideal model organisms for studies aiming at the identification of the compounds used in chemical communication.

10.1 Introduction

Some of the most ubiquitous crustaceans on earth are the peracarids, which include the amphipods, isopods, tanaids, and cumaceans among others. Beach hoppers (amphipods) commonly cross our path on the beach, and woodlice (isopods) invade our backyards and sometimes even move into our basements. One of my first scientific experiences with peracarids occurred when I was studying the ecology of nemertean worms. Some nemertean species are dangerous predators of amphipods, one of the main groups of the peracarids. When put together in an aquarium, amphipods swam much more actively above the ground than in control aquaria without nemerteans. Often they would react without touching the nemerteans and when placed in "nemertean" water they would respond with agitated swimming movements. Since at the time I was more interested in learning how the nemerteans managed to find their prey in a dynamic world, I did not dwell more on the question of amphipod chemoreception, but I had gained the strong impression that they might be evaluating their environment via chemical cues. Years later, I started to study parental care behavior in amphipods and isopods. While watching female-offspring groups that lived together for several months, I wondered whether and how the family members would recognize each other, and I searched the literature for information. Some of the most fascinating accounts come from desert isopods where male and female team up to care for the single brood they produce during their life time; family members use chemical signatures to recognize each other (Linsenmair 2007). My search for information also led me to studies on the chemical communication of amphipods during mate search and male–female interactions (e.g., Borowsky 1991). It became evident to me that peracarids do have good chemosensory capabilities to communicate with family members and other conspecifics. In this contribution, I synthesize our present understanding of chemical communication in peracarids. In particular I will report on behavioral observations, which are suggestive of complex chemical interactions, and I hope to convince readers that there are exciting challenges ahead of us.

10.2 Chemoreceptors and Chemicals Perceived by Peracarids

10.2.1 Chemosensory Structures of Peracarid Crustaceans

The most important chemosensory structures of peracarids are located on their two pairs of antennae. The antennules (first antennae) bear numerous sensory hairs, the aesthetascs (e.g., Heimann 1984 for detailed description of peracarid aesthetascs; see

Hallberg and Skog, Chap. 6). The number of aesthetascs varies depending on sex and species. While in the freshwater isopod *Asellus aquaticus,* the total number ranges from 3 to 6 aesthetascs per antennule (Heimann 1984), in some lysianassid amphipods bundles of up to 1,500 individual aesthetascs are found on the basal antennular articles (Kaufmann 1994). Often, the basal articles with these dense bundles of aesthetascs are fused to form the so-called callynophore (Lowry 1986). These callynophores can occur in both sexes, but in many species they are restricted to the males. In general, the antennae, the type and number or presence of sensory structures show a strong sexual dimorphism in many aquatic and terrestrial peracarids (e.g., Sheader 1981; Conlan 1991; Kaufmann 1994; Lefebvre et al. 2000) (Fig. 10.1).

For lysianassid amphipods Kaufmann (1994) suggested that the aesthetascs are involved in perception of waterborne food stimuli. Apparently, the same basic chemoreceptor design employed in food location (aesthetascs) is also used in mate location. While the role of aesthetascs in chemoreception is unquestioned, there are a number of other structures whose chemosensory role remains specula-tive or controversial (e.g., Kaïm-Malka et al. 1999). At the base of the antennae a few amphipod species have cup-shaped structures termed calceoli, which are most common in males. One of the first studies on chemical communication in peracarids suggested that the calceoli on the 2nd antennae of male *Gammarus duebeni* are the

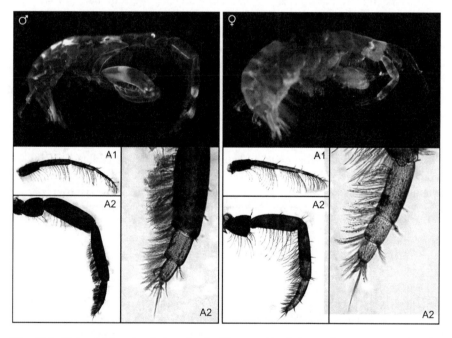

Fig. 10.1 Male and female of the tube-dwelling amphipod *Jassa slatteryi* highlighting the differences in setation of the second antennae. The large numbers of sensory setae on the male antennae are thought to aid in the location and identification of receptive females. A1 refers to first antenna, A2 refers to second antenna (antennule)

structures where female pheromones are perceived (Dahl et al. 1970). Subsequent anatomical (Lincoln 1985) and behavioral studies (Read and Williams 1990) questioned their role in female detection via waterborne pheromones, but Dunn (1998) indicated that calceoli might be used to determine the reproductive status of females via contact chemicals.

The 2nd antennae of many terrestrial isopods terminate in the so-called apical sensory cone (Seelinger 1983). This sensory cone contains many (>400) sensory cells, some of which react to specific chemical substances (Seelinger 1977, 1983). Using electrophysiological techniques, some of the tested chemoreceptor units were shown to respond to volatile compounds, while others reacted to soluble compounds, and they were termed olfactory and gustatory units, respectively (Seelinger 1983). Among the gustatory units of the desert isopod *Hemilepistus reaumuri*, one group responded exclusively to water in which conspecifics had been washed (Seelinger 1977, 1983). More electrophysiological studies are needed to reveal the chemical response spectrum of sensory cells.

Besides chemosensory structures on the antennae, special setae with apical pores have been reported from the gnathopods and the subsequent pereopods of several peracarid species (Holdich 1984; Kaufmann 1994). These setae are often concentrated on the tip or the lateral sides of pereopods, where they come into direct contact with the substratum. They are thought to be important in contact chemoreception, mostly in characterization of food sources (e.g., Kaufmann 1994), but possibly also in the identification of the reproductive status of mates.

When exposed to chemical stimuli, peracarids show several behaviors that are thought to facilitate stimulus transfer. For example, in lysianassid amphipods antennal movements (flicks) increased substantially once a chemical stimulus (e.g., from a food source) is introduced to the water (Kaufmann 1994), probably to enhance water exchange around the antennal chemoreceptors. When walking, desert isopods continuously wave their antennae, briefly (for 40-80 ms) touching the bottom. During these bouts the chemosensory cells in the antennal sensory cone are thought to scan the substratum surface for chemical cues (Seelinger 1983). Holdich (1984) reported that contact chemoreception might be mediated by liquid secretions; some terrestrial isopods were observed to hold the tip of their 2nd antennae to the substratum, releasing a small liquid drop most likely providing an aqueous bridge between antennal chemoreceptors and the substratum.

10.2.2   Food and Predator Cues

Peracarids employ their chemical senses in many aspects of their life, including food search. Some of the best examples for the importance of chemical cues in locating food are scavenging peracarids. Once a piece of carrion reaches the sea bottom, the first peracarids usually appear within <30 min (Premke et al. 2006), and chemical cues released by carrion may attract amphipod scavengers from up to 2 km distance (Sainte-Marie and Hargrave 1987). Search behavior in scavenging amphipods can be induced by amino acid cocktails at concentrations as low as $10^{-5} - 10^{-7}$ mol l^{-1} (Ide et al. 2007). Molecules smaller than 6 kDa are thought to be most efficient in eliciting food-searching behavior (Meador 1989). Different peracarid species react to specific amino acids, which appears to be related to their mobility and potential food sources (Ide et al. 2006). For example, Meador (1989) showed that the mostly crawling amphipod *Orchomene limodes* reacts strongly to tryptophan and taurine, which only occur at very low background concentrations, while Ide et al. (2006) found that the highly mobile swimmer *Scopelocheirus onagawae* responds mainly to glycine and alanine, which are some of the most abundant amino acids in seawater. These authors discussed that more mobile species react to the higher concentrations of common amino acids that rapidly leak out of decaying tissues of larger carrion, while less mobile species orient to low concentrations of rare amino acids, indicating nearby small carrion. The importance of chemical cocktails has also been emphasized (see Hay, Chap. 3), underlining the fact that peracarids may have fine-tuned their reactions to the composition and concentrations of these mixtures to locate preferred (and reachable) food sources.

Detritus-feeding amphipods *Gammarus pulex* were found to move towards leaf and artificial substrata that were conditioned by fungi and bacteria (de Lange et al. 2005). Similarly, terrestrial isopods were attracted to unknown volatile substances probably produced by microorganisms colonizing detritus (Zimmer et al. 1996). In the desert isopod *H. reaumuri*, specific chemoreceptor neurons on the 2nd antennae reacted to extracts of fresh and rotten grass (Seelinger 1977). While soluble or volatile breakdown products of organic matter (animals or plants) attract detritivorous peracarids, a variety of algal secondary metabolites are known to deter herbivorous peracarids (e.g., Hay et al. 1998). Grazing amphipods and isopods react rapidly to defense chemicals produced by attacked algae (Toth and Pavia 2007). How peracarids detect these algal compounds, however, is not known at present. Possibly, taste receptors on the pereopods or in the mouth region are involved.

As already pointed out in the introduction, many peracarid species adjust their behavior and habitat preferences when chemical cues reveal the presence of predators (e.g., Baumgärtner et al. 2003; see Hazlett, Chap. 18). Besides cues from predators themselves, substances leaking from crushed conspecifics also induce antipredator behavior in the amphipod *Gammarus lacustris* (Wisenden et al. 2001). Although less effective, cues from closely related species had similar behavioral effects (Wisenden et al. 1999).

10.2.3 Production and Release of Chemical Signals in Peracarids

There is a variety of substances produced by peracarids themselves, which mediate social interactions. Unfortunately, very little is known about the chemical nature of these substances, which are transmitted via the medium (water or air) or via direct contact. For the amphipod *Microdeutopus gryllotalpa*, Borowsky et al. (1987) showed that waterborne cues contain polar molecules. Some of the chemicals that mediate the mating interactions in aquatic amphipods appear to be released in the urine. Hammoud et al. (1975) demonstrated that urine from receptive females of *G. pulex* could induce males to form a precopulatory pair with nonreceptive females, which they would otherwise ignore. In subsequent experiments Lyes (1979) blocked the urine pores in receptive female amphipods *G. duebeni*, and as a consequence males did not pair up with these females, while they did form precopulatory pairs with unblocked females.

The nature of contact chemicals is also not well known. Since experimental application of the molting hormone ecdysterone to nonreceptive female *G. pulex* induced mate guarding in males, Hammoud et al. (1975) suggested that ecdysterone or related substances are involved in mate recognition. This suggestion was later supported by experiments with the freshwater isopod *Lirceus fontinalis*; when applying a chemical mimic of the molt hormone 20-hydroxyecdysone (20HE), nonreceptive females became attractive to males, which attempted to mate guard them (Sparkes et al. 2000). Recently, Hayden et al. (2007) showed for shore crabs *Carcinus maenas* that 20HE may function as a feeding inhibitor rather than as a sex pheromone; this ensures that mate-guarding males do not cannibalize the recently molted female.

Using various fractionation methods, Takeda (1984) suggested that the aggregation cues of some terrestrial isopods are of alcoholic nature and represent a large molecule, without specifying this further. The aggregation cues of terrestrial isopods are secreted in special glandular areas of the hindgut (*Armadillidium vulgare* and *Porcellionides pruinosus*) or the rectum (*Ligia exotica*) where they are incorporated into the feces (Takeda 1984). Cuticular glands have also been reported from terrestrial isopods, but the composition and function of their secretions is not known (Holdich 1984). For the desert isopod *H. reaumuri*, Linsenmair (1987) showed that the contact signatures are polar and nonvolatile compound mixtures.

10.3   Intraspecific Interactions Mediated by Chemical Stimuli

In most studies on intraspecific interactions that are mediated by chemical substances, the focus is on the receiving individuals. Consequently, in many cases we know little or nothing about the intentions of or benefit to the emitter of the chemicals. Especially in the context of mating interactions it is in the interest of the searching individual to perceive chemical cues indicating the presence of a receptive mate, but it might not necessarily be in the interest of the sought-after individual to reveal its presence or its reproductive status (Manning 1975; Jormalainen 1998).

10.3.1   Mating Interactions

As in many other crustaceans mating in peracarid species follows a particular sequence. In many species males search for receptive females (e.g., Borowsky and Borowsky 1987). Males may find females at variable times before these become receptive. In most peracarid species, females become receptive and mate shortly after the reproductive molt. Thus, females close to their reproductive molt are of high value for males, which guard and defend these females against other males. How do males find and recognize these receptive females?

10.3.1.1   Mate Finding

Numerous studies suggest the existence of pheromone-guided mate finding in amphipods. Early studies by Dahl et al. (1970) had shown that males of the amphipod *G. duebeni* were guided by waterborne substances towards females. The fact that these results could not be replicated by subsequent workers (e.g., Hartnoll and Smith 1980) was ascribed by Dunham and Hurshman (1991) to the experimental conditions. These authors particularly emphasized the importance of flow. Indeed, in simple Y-arrays (also called "olfactometer") the existence of waterborne chemicals have been subsequently confirmed for a number of peracarid species including the free-living amphipods *Gammarus palustris* (Borowsky 1985), *Gammarus lawrencianus* (Dunham and Hurshman 1991), and the tube-/burrow-dwelling amphipods *M. gryllotalpa* (Borowsky 1984, 1991) and *Corophium volutator* (Krång and Baden 2004) (Fig. 10.2). All these authors could show that more males moved towards the branch with odors from receptive females than towards the branch with odors from nonreceptive females or without conspecific odors. Exposure to polluted sediments or waters led to a reduction in male search efficiency (e.g., Lyes 1979; Krång 2007).

Distances over which males moved towards female odors in these Y-arrays with moderate water flow were generally on the order of 20-40 cm. In standing water, Bertin and Cezilly (2005) observed that male isopods *A. aquaticus* only detected receptive females at distances of about 5 cm, but it is unclear whether detection was mediated by chemical cues. In the harem-forming isopod *Paracerceis sculpta*, females were attracted via waterborne chemicals to harems that already contained other females (Shuster 1990).

While there is good evidence that waterborne chemicals play a role as sex attractant in some species, it is not well known how these function in the natural environment. Water flow appears to be important (Dunham and Hurshman 1991). Environments with little or no water flow might not be favorable for the evolution of waterborne attractants. In freshwater amphipods from the genus *Hyalella*, which inhabit small ponds throughout the Americas, encounters between the sexes appear to be random without mediation through waterborne substances (Holmes 1903). The fact that freshwater isopods *Thermosphaeroma thermophilum* do not react to waterborne chemical cues from conspecifics, even in experiments with moderate water flow, seems to support this suggestion (Jormalainen and Shuster 1997). It must also be kept in mind that most behavioral assays with Y-arrays that indicated the existence of waterborne chemical attractants used large aggregations of individuals as lure, usually 30–50 individuals. This might not really reflect a natural

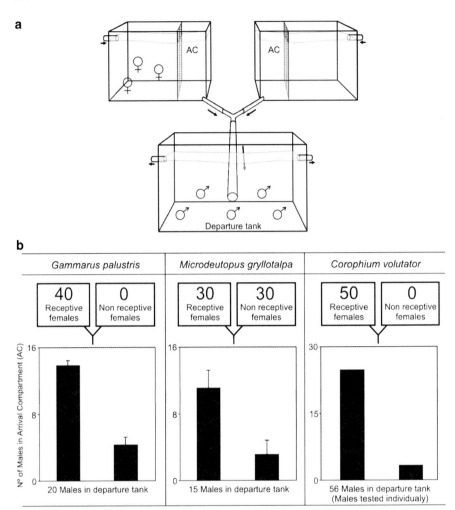

Fig. 10.2 (a) Schematic drawing of Y-apparatus used as "olfactometer" in studies on the role of waterborne chemical cues in aquatic peracarids. Males are placed in the departure tank (individually or several together) and travel through the Y-tubes towards the arrival compartments (AC) of the respective holding tanks with the receptive females or with the control groups (no females or nonreceptive females). Similar set-ups were used to test for the role of airborne chemicals in terrestrial peracarids. (b) Numbers of male amphipods approaching the branch with receptive females and the control branch; modified after Borowsky (1985, 1991) and Krång and Baden (2004)

situation where encounters generally occur between single individuals (Fig. 10.3), which then might form a pair.

Intriguing observations come from studies by Borowsky (1984, 1991) on the tube-dwelling amphipod *M. gryllotalpa*, where males entering the tubes of females often engaged in "intermittent pleopod beating," a behavior never performed by females during these interactions. In most studied cases, the females seem to be

Fig. 10.3 Males of the tube-dwelling amphipod *Corophium volutator* are attracted to chemicals released by the females, but it is not well known whether receptive females actively advertise their presence by flushing water out of the tube (as in this conceptual drawing), or whether males, upon finding a tube, draw water from the tube to identify the sex and reproductive status of the tube inhabitant. Drawing by Jorge A. Varela Ramos

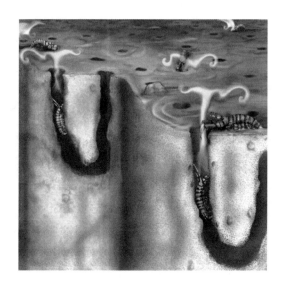

releasing the attractive substances. Possibly, males entering the female's tube draw female water towards them (Borowsky 1984, 1991) to explore the reproductive status of females. Alternatively, the male might flush chemical cues towards the female residents, or a combination of both, resulting in a complex odor exchange between the male and the female, similar to that reported for the American lobster (Atema and Steinbach 2007) and for the blue crab (see Kamio and Derby, Chap. 20). In contrast to the lobster where males monopolize a shelter that is visited by females, in most tube-dwelling peracarids the females are residents in their dwellings and males cruise in search of receptive females (Borowsky 1991). Future studies need to show how males perceive and react to waterborne chemicals released by the females, especially in tube- and burrow-dwelling peracarids.

Distance chemoreception has also been reported for terrestrial peracarids. In the isopod *Venezillo evergladensis*, males appear to perceive the presence of a receptive female via airborne chemical cues (Johnson 1985). Upon introduction of a receptive female, males started running around, holding their antennae high and waving them; this response was never observed with nonreceptive females.

10.3.1.2   Mate Choice

Contact pheromones of unidentified nature have been inferred to guide males in the identification of females and their reproductive status (e.g., Borowsky and Borowsky 1987). After having made first contact with another individual, male amphipods *G. palustris* briefly palpate it (for a few seconds) and then either drop it or, if it is a receptive female, take it in a precopulatory grasp (amplexus). During palpation, males quickly move their antennae, chelae, and mouthparts over the body of the other individual, probably scanning substances on the cuticula. Dunn (1998) confirmed the importance of antennal chemoreceptors as follows: male *Gammarus zaddachi* and *G. duebeni* without the calceoli-bearing 2nd antennae were less efficient in identifying and pairing up with reproductive females than control males (for similar results with *G. pulex* see Hume et al. 2005). While numerous studies indicate that males mainly recognize the reproductive status of females via chemical stimuli produced by females, Dunham et al. (1986) warned that this view might be too simplistic and that other factors, such as female behaviors, might also be involved in these interactions. Borowsky and Borowsky (1987), however, showed that receptive females of *G. palustris* that were freeze-fixed elicited the same responses in males as living females, underlining the importance of chemical cues. Other substances, e.g., algal exudates, might occasionally interfere with female-released pheromones (Borowsky et al. 1987), but this suggestion has not been further explored.

Males are able to distinguish between females that are approaching the reproductive molt (i.e., should be mate guarded) and those that have just molted (i.e., are available for the copula) as shown in behavioral assays with *G. palustris* (Borowsky and Borowsky 1987). Parallel studies demonstrated that gammarid males were even able to estimate the time that females require before their reproductive molt (Hunte et al. 1985; Dick and Elwood 1990). Different methods were used to reveal this astonishing capacity of male amphipods. Hunte et al. (1985) placed precopulatory pairs of *G. lawrencianus* in distilled water and measured the time until the male released the female; this was taken as a measure of his investment. In the congener *G. pulex*, Dick and Elwood (1990) measured the duration of male struggles over a receptive female; here the duration of struggles was taken as a measure of male investment. In both cases there was a strong relationship between the respective measures of male investment and the time until the reproductive molt of the female, which demonstrated that males were able to accurately estimate the value of the female (Fig. 10.4). Touching the female briefly with the antennae appears to be sufficient for the male to detect her reproductive status (Dick and Elwood 1989). While these observations are highly suggestive of chemical communication, other cues (e.g., female behaviors) may also play an important role in male decisions (see e.g., Dunham and Hurshman 1991). The decision to guard a female or not may also depend on the contents of her brood pouch, but whether this is mediated by chemical cues is not well known (Dunham and Hurshman 1991).

Similar observations have been made for the freshwater isopod *A. aquaticus*, where males appear able to distinguish between premolt females that will reproduce and those that will not reproduce, probably based on the presence of a chemical stimulus on the reproductive females (Thompson and Manning 1981). Males of the terrestrial isopod *Helleria brevicornis* were able to distinguish between females approaching the reproductive molt and those approaching a growth molt; when paired females were separated from males and wiped with a moist cotton, males were unable to judge the reproductive status of these females, suggesting that chemicals on the cuticle of the female play an important role in this interaction (Mead and Gabouriaut 1977). Not in all peracarid species males appear capable to recognize the reproductive status of females. For example, males of the isopod

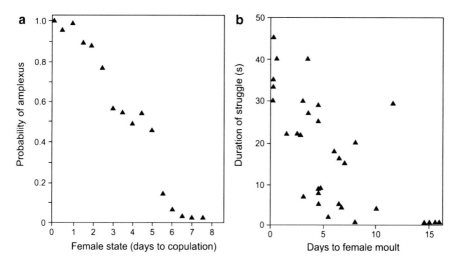

Fig. 10.4 (a) Relationship between female reproductive status and probability of pair-maintenance (amplexus) in the amphipod *Gammarus lawrencianus* (modified after Hunte et al. 1985). (b) Relationship between time until female reproductive molt and duration of male struggles in the amphipod *Gammarus pulex* (modified after Dick and Elwood 1990)

Lirceus fontinalis were unable to distinguish between two types of premolt females, and they not only attempted to form precopulatory pairs with reproductive but also with nonreproductive premolt females (Sparkes et al. 2000). In the symbiotic marine isopod *Jaera hopeana* males guard precocious juveniles (called "manca"), but they cannot distinguish between male and female mancas; consequently many males also guard male mancas, a costly mistake, which males only realize after several days at the first juvenile molt (Franke 1993).

Other studies show variable abilities of males to discriminate against females with parasites. While male *G. pulex* do not distinguish between parasitized and unparasitized females (Poulton and Thompson 1987), male *Gammarus insensibilis* selected against parasitized females (Thomas et al. 1996). Nothing is known about the mechanisms involved, but chemical cues might play a role.

Interestingly, in some cases, males do not seem to be able to distinguish between females of their own and of closely related species. While waterborne chemical cues seem to be species-specific (Borowsky 1985), there are a number of examples where males pair up with females of closely related species. For example, even though male *H. brevicornis* can recognize the reproductive status of females of their own species, they often pair up initially with individuals of another species, *Armadillo officinalis*, albeit only briefly (rudimentary pairing behavior) (Mead and Gabouriaut 1977). In laboratory experiments, males of the brackish-water amphipod *G. zaddachi* frequently form precopulatory pairs with female *Gammarus salinus* and *Gammarus locusta* (Kolding 1986). According to the author, habitat and life cycle segregation reduce the risk of interspecific matings in the field. Stanhope et al. (1992) suggested that habitat-specific characteristics in the food might contribute to premating isolation. They found that in a local population of *Eogammarus confervicolus*, females raised on habitat-specific food algae produced a pheromone that was preferred by males from that same population. However, females raised on other food sources were not recognized by males from their local population. These observations generally confirm the importance of chemical cues for mate identification, yet despite the intriguing nature of those studies and their potential in fostering our understanding of speciation processes and the mechanisms involved, this topic has not been pursued further in peracarids.

10.3.2 Aggregation Behavior

Chemical cues do not only mediate interactions between potential mates, but also with other conspecifics. In the pelagic amphipod species *Parathemisto gaudichaudi*, both sexes are attracted to excretory products from conspecifics, a behavior that is considered important for swarm formation (Sheader 1981). In free-living benthic amphipods (e.g., *G. palustris*), both males and females are attracted to conspecifics, which was interpreted as a behavior facilitating mate finding (Borowsky 1985, 1991). However, conspecific attraction might also be useful in the context of food finding, since aggregations of many individuals might indicate the presence of suitable food items (Fig. 10.5a).

The ubiquitous aggregations of many terrestrial isopods are also mediated by chemical cues (Fig. 10.5b). Early experiments by Kuenen and Nooteboom (1963) showed that some terrestrial isopod species (*A. vulgare*, *Oniscus asellus*, *Porcellio scaber*) are attracted to pheromones released by conspecifics. Later Takeda (1984) reconfirmed for *A. vulgare*, *P. pruinosus*, and *Ligia exotica* that pheromones are species-specific. He showed that isopods quickly aggregate around sites where feces with the active pheromone had been deposited. This behavior is considered advantageous because dehydration of isopods is substantially reduced when they are in aggregations. As suggested above for herbivorous aquatic amphipods, aggregation behavior might also direct terrestrial isopods to potential food sources.

Gregarious behavior might have various advantages ranging from resource exploitation to mate finding or reproductive synchronization. For example in some terrestrial isopods, reproductive activity of females appears to be synchronized through chemical cues transmitted via feces (Mead and Gabouriaut 1988; Caubet et al. 1998), which are common in communal shelters.

10.3.3 Kin Recognition

Kin recognition has not been commonly studied in peracarids. Contrasting reports come from aquatic peracarids, and all of these are between females and their

Fig. 10.5 Aggregations of (**a**) marine amphipods *Parhyalella penai* on their food algae, and of (**b**) terrestrial isopods *Armadillidium vulgare* in their communal shelters. Photos courtesy of Iván A. Hinojosa

offspring brooded in the marsupium. Most studies testing for kin recognition found that females are unable to discriminate between their own and foreign embryos and juveniles (e.g., Shillaker and Moore 1987; Borowsky 1983). However, brooding females of the intertidal amphipod *Apherusa jurinei* recognized their own offspring and preferentially retrieved these into their brood pouches (Patterson et al. 2008). How females are able to recognize their offspring is not known at present, but since females actively manipulated embryos with their antennae before placing them in their brood pouch (kin) or cannibalizing them (unrelated embryos), it appears likely that chemical cues are involved in these interactions.

The best known example of chemical communication comes from desert isopods *H. reaumuri* where the social system and the complex behavioral interactions between family members have been studied over almost 40 years by Karl Eduard Linsenmair

Fig. 10.6 Desert isopods *Hemilepistus reaumuri* at their burrow entrance. The male and the female cooperate in the establishment of the burrow. While one individual is away foraging, the other individual remains at the burrow entrance, defending it against possible intruders. When the mate returns to the burrow, it is carefully examined and only admitted to the burrow if its chemical signature is recognized. Intruders (on the *left*) with unfamiliar chemical "badges" are not admitted to the burrow (Photo courtesy of Karl-Eduard Linsenmair)

(summarized in Linsenmair 2007). In this species, reproduction occurs only once a year, and at the beginning of their single reproductive event, a female and a male form a monogamous pair. They cooperate in digging a deep family burrow and during this time they develop a strong pair bond. In particular, they learn the chemical characteristics of their mate, which is essential in order to recognize each other after periods of separation (Fig. 10.6). Both the female and the male frequently leave the burrow to forage for food in the surroundings, especially while rearing their offspring. In numerous behavioral assays, it was shown that the chemical signature (termed "badge") is a combination of self-produced substances and those acquired from family members (Linsenmair 1987). Each family has a unique chemical badge, which each individual carries on its cuticula. Individuals recognize family members after briefly (<1 s) touching their cuticle with the apical sensory cone on the 2nd antennae. However, even though there is one unique chemical badge, not all individuals within the same family produce the same chemicals. Instead, the badge seems to be a combination of chemicals from all family members (Linsenmair 1987). Due to their nonvolatile nature, the chemicals can only be transferred by directed contact. Recently molted individuals do not have the badge. However, they seem to have a substance that suppresses aggressive behavior from others (possibly 20HE, see above). Just before and after the molt these individuals stay in the burrow, where they reacquire the family badge through contacts with other family members. Only after they are again fully identifiable by their family members, do they reassume foraging excursions or burrow defense tasks. Since an important fraction of the badge is acquired from the other family members, recently molted juveniles can be introduced to new "foster" families. If this happens very shortly after the molt, these juveniles are often accepted by the members of their foster family. Using this knowledge, Karl Eduard Linsenmair conducted numerous experiments to uncover the complex system of kin recognition in these fascinating isopods. The unique chemical badge is considered a crucial asset in the evolution of the highly sophisticated social behavior in these family units (Linsenmair 2007). At present no comparable communication system is confirmed for other crustaceans.

10.4 Peracarid Crustaceans as Model Organisms

The diversity of habitats inhabited by peracarids is unsurpassed among the crustaceans and possibly among the arthropods in general. Some species burrow in aquatic sediments, many graze on algae and plants, others hover above the sea-floor or swim in the water column, some are symbionts on other organisms, and a few species even have conquered deserts or high mountain streams. The variety of selective environments inhabited by peracarids has led to an extraordinary diversity of life-styles and behavioral adaptations. This has also led to a wide diversity in communication systems and peracarids employ visual, tactile, and chemical channels, often simultaneously. The diversity of life-styles and the divergence in basic life history traits (habitat or food preferences) among closely related species (e.g., Wellborn and Cothran 2007) makes peracarids well suited for phylogenetic comparative studies.

Furthermore, peracarids have similar sizes as the well-studied insects and spiders, and consequently they face similar ecological challenges (predation pressure, intra- and interspecific competition) as these arthropods. Aquatic larvae of some insect species are highly sensitive to predator cues and show a high plasticity in their antipredator responses (Ferrari et al. 2008), very similar to what occurs in peracarids. Mate finding and mate competition in peracarids is based on similar mechanisms (chemical cues) as reported for insects (e.g., see Baker, Chap. 27) or spiders (Robinson 1982). However, there are also important differences, one of the most important being the reproductive biology. All peracarids have direct development and offspring emerge from the female's brood pouch as fully developed individuals, which usually have similar needs concerning food and habitat as adults. Thus, adult female peracarids are in direct contact with their own offspring when these hatch out. This situation is rare in most insects and spiders, but occurs in the subsocial insects where parents cohabit with their developing offspring (Costa 2006). Consequently, we can expect that there are parallels between peracarids and subsocial insects in the evolution of parent–offspring interactions and (chemical) communication.

The direct development is also one of the great opportunities that peracarids offer as model organisms for evolutionary ecologists. They are easily raised in laboratory cultures, because adults and offspring have similar requirements (Dunham and Hurshman 1991). For rapidly maturing species it is very feasible to generate specific offspring lines, and to place them in different selective environments. This allows examining a wealth of questions, e.g., how the habitat or social condition affects their communication system.

10.5 Future Directions

More than 30 years ago Dunham (1978) wrote: "The problem has not, therefore, been to demonstrate that chemical interactions occur in Crustacea, but to establish the functional significance of these interactions." Since then we have learned much about chemical interactions in peracarid crustaceans, but our understanding of the adaptive significance of e.g., the chemicals that mediate mating interactions in peracarids remains poor. Are they signals or cues? (Jormalainen 2007). What are the benefits and costs for females of giving away their reproductive state via chemical cues? Can females control the release of these substances? Possibly the answers to these kinds of questions vary over time. There should be a temporal threshold after which the guarding costs for females (e.g., lost feeding opportunities or higher predation risk) are surpassed by their benefits (assuring fertilization at the moment of ovulation) (e.g., Sparkes et al. 2000). Thus, substances perceived by males during the initial phase of mate-guarding might be cues, whereas shortly before the reproductive molt, females may actively release specific chemical signals to inform males about this.

What chemical compounds are involved in these interactions is one of the key questions that needs to be attacked to better understand chemical communication in peracarids and other crustaceans (see Chaps. 19 and 20). Given the relatively easy and economic culture of peracarids and the possibility to select reproductive females for behavioral assays, peracarids could be ideal model organisms to identify the chemicals intervening in the mating interactions. Identifying the chemicals will widely open the door to examine some of the following evolutionary topics:

Species-specificity of chemicals and their role in reproductive isolation: While there are general patterns in the chemical communication systems, there also seem to be species-specific differences. Some amphipod species appear unable to distinguish between mates and individuals of closely related species (Kolding 1986). Also in some isopod species, males pair with females from closely related species (e.g., Hargeby and Erlandsson 2006), which suggests that mate recognition via chemical cues is not (yet) fully developed. Possibly, selection on contact pheromones has been weak, because other life-history traits ensure reproductive isolation under natural conditions (e.g., size, habitat preference).

Parallels between life history and mating (communication) system: Finding a mate is usually one of the first steps in successful reproduction. Organisms that have evolved a sophisticated sensory system for finding food may also utilize this system when searching for mates. For example, lysianassid amphipods are very efficient in locating carrion over long distances. In this group, the males often increase the number or size of chemoreceptors once they become reproductive. This suggests that there might be a functional relationship between food and mate finding via waterborne cues.

10.6 Summary and Conclusions

Peracarids employ chemoreception in a wide variety of contexts, including food selection, predator avoidance, mating interactions, aggregation behavior, and kin recognition. The main chemoreceptors are on the antennae. Some of the chemosensory neurons respond specifically to waterborne or airborne chemical cues and others respond to contact cues. Soluble cues have been found to attract conspecifics of highly mobile, algal-dwelling, amphipods and volatile cues are used by some terrestrial isopods to locate conspecific aggregations. Similar mechanisms function during mate attraction, but at present the water- or airborne attractants have not been identified. In many species, mates are only identified by direct contact. Once contact is made with the receptor-bearing antennae, mate recognition after initial contact can be quick. Males use chemical cues to identify the reproductive status of females and base their guarding decisions on these. Despite overwhelming support of chemical mediation in these mating interactions, one major challenge is confirming the adaptive significance of these chemical cues. The question of whether females release these cues intentionally or whether males eavesdrop on chemicals that the female cannot retain has not yet been resolved. A highly specific communication system has evolved in a desert isopod where parents and offspring cohabit in a soil burrow for many months. The characterization of the involved chemicals remains an important task in this and other crustacean communication systems. Due to their small size and direct development, peracarids are proposed as ideal model organisms for evolutionary ecologists.

Acknowledgements I am grateful to Iván A. Hinojosa for his help with the figures. Thomas Breithaupt, Chuck Derby, Veijo Jormalainen, and Anna-Sara Krång offered many constructive comments, which helped to improve the original draft of this manuscript.

References

Atema J, Steinbach MA (2007) Chemical communication and social behavior of the lobster *Homarus americanus* and other decapod Crustacea. In: Duffy JE, Thiel M (eds) Evolutionary ecology of social and sexual systems – crustaceans as model organisms. Oxford University Press, New York, pp 115–144

Baumgärtner D, Koch U, Rothhaupt KO (2003) Alteration of kairomone-induced antipredator response of the freshwater amphipod *Gammarus roeseli* by sediment type. J Chem Ecol 29:1391–1401

Bertin A, Cezilly F (2005) Density-dependent influence of male characters on mate-locating efficiency and pairing success in the waterlouse *Asellus aquaticus*: an experimental study. J Zool 265:333–338

Borowsky B (1983) Placement of eggs in their brood pouches by females of the amphipod Crustacea *Gammarus palustris* and *Gammarus mucronatus*. Mar Behav Physiol 9:319–325

Borowsky B (1984) Effects of receptive females' secretions on some male reproductive behaviors in the amphipod crustacean *Microdeutopus gryllotalpa*. Mar Biol 84:183–187

Borowsky B (1985) Response of the amphipod crustacean *Gammarus palustris* to waterborne secretions of conspecifics and congeners. J Chem Ecol 11:1545–1552

Borowsky B (1991) Patterns of reproduction of some amphipod crustaceans and insights into the nature of their stimuli. In: Bauer RT, Martin JW (eds) Crustacean sexual biology. Columbia University Press, New York, pp 33–49

Borowsky B, Borowsky R (1987) The reproductive behaviors of the amphipod crustacean *Gammarus palustris* (Bousfield) and some insights into the nature of their stimuli. J Exp Mar Biol Ecol 107:131–144

Borowsky B, Augelli CE, Wilson SR (1987) Towards chemical characterization of waterborne pheromone of amphipod crustacean *Microdeutopus gryllotalpa*. J Chem Ecol 13:1673–1680

Caubet Y, Juchault P, Mocquard JP (1998) Biotic triggers of female reproduction in the terrestrial isopod *Armadillidium vulgare* Latr. (Crustacea Oniscidea). Ethol Ecol Evol 10:209–226

Conlan KE (1991) Precopulatory mating behavior and sexual dimorphism in the amphipod Crustacea. Hydrobiologia 223:255–282

Costa JT (2006) The other insect societies. The Belknap Press of Harvard University Press, Cambridge

Dahl E, Emanuelsson H, von Mecklenburg C (1970) Pheromone transport and reception in an amphipod. Science 170:739–740

De Lange HJ, Lürling M, van den Borne B, Peeters ETHM (2005) Attraction of the amphipod *Gammarus pulex* to waterborne cues of food. Hydrobiologia 544:19–25

Dick JTA, Elwood RW (1989) Assessments and decisions during mate choice in *Gammarus pulex* (Amphipoda). Behaviour 109:235–246

Dick JTA, Elwood RW (1990) Symmetrical assessment of female quality by male *Gammarus pulex* (Amphipoda) during struggles over precopula females. Anim Behav 40:877–883

Dunham PJ (1978) Sex pheromones in Crustacea. Biol Rev 53:555–583

Dunham PJ, Hurshman AM (1991) Precopulatory mate guarding in aquatic Crustacea: *Gammarus lawrencianus* as a model system. In: Bauer RT, Martin JW (eds) Crustacean sexual biology. Columbia University Press, New York, pp 50–66

Dunham PJ, Alexander T, Hurshman AM (1986) Precopulatory mate guarding in an amphipod, *Gammarus lawrencianus* Bousfield. Anim Behav 34:1680–1686

Dunn AM (1998) The role of calceoli in mate assessment and precopula guarding in *Gammarus*. Anim Behav 56:1471–1475

Ferrari MCO, Messier F, Chivers DP (2008) Variable predation risk and the dynamic nature of mosquito antipredator responses to chemical alarm cues. Chemoecology 17:223–229

Franke HD (1993) Mating system of the commensal marine isopod *Jaera hopeana* (Crustacea) I. The male-manca(I) amplexus. Mar Biol 115:65–73

Hammoud W, Comte J, Ducruet J (1975) Recherche d'une substance sexuellement attractive chez les gammares du groupe pulex (Amphipodes, Gammaridea). Crustaceana 28:152–157

Hargeby A, Erlandsson J (2006) Is size-assortative mating important for rapid pigment differentiation in a freshwater isopod? J Evol Biol 19:1911–1919

Hartnoll RG, Smith SM (1980) An experimental study of sex discrimination and pair formation in *Gammarus duebenii* (Amphipoda). Crustaceana 38:253–264

Hay ME, Piel J, Boland W, Schnitzler I (1998) Seaweed sex pheromones and their degradation products frequently suppress amphipod feeding but rarely suppress sea urchin feeding. Chemoecology 8:91–98

Hayden D, Jennings A, Mueller C, Pascoe D, Bublitz R, Webb H, Breithaupt T, Watkins L, Hardege JD (2007) Sex specific mediation of foraging in the shore crab, *Carcinus maenas*. Horm Behav 52:162–168

Heimann P (1984) Fine structure and molting of aesthetasc sense organs on the antennules of the isopod, *Asellus aquaticus* (Crustacea). Cell Tissue Res 235:117–128

Holdich DM (1984) The cuticular surface of woodlice: a search for receptors. Symp Zool Soc Lond 53:9–48

Holmes SJ (1903) Sex recognition among amphipods. Biol Bull 5:288–292

Hume KD, Elwood RW, Dick JTA, Morrison J (2005) Sexual dimorphism in amphipods: the role of male posterior gnathopods revealed in *Gammarus pulex*. Behav Ecol Sociobiol 58:264–269

Hunte W, Myers RA, Doyle RW (1985) Bayesian mating decisions in an amphipod, *Gammarus lawrencianus* Bousfield. Anim Behav 33:366–372

Ide K, Takahashi K, Nakano T, Sato M, Omori M (2006) Chemoreceptive foraging in a shallow-water scavenging lysianassid amphipod: role of amino acids in the location of carrion in *Scopelocheirus onagawae*. Mar Ecol Prog Ser 317:193–202

Ide K, Takahashi K, Omori M (2007) Direct observation of swimming behaviour in a shallow-water scavenging amphipod *Scopelocheirus onagawae* in relation to chemoreceptive foraging. J Exp Mar Biol Ecol 340:70–79

Johnson C (1985) Mating behavior of the terrestrial isopod, *Venezillo evergladensis* (Oniscoidea, Armadillidae). Am Midl Nat 114:216–224

Jormalainen V (1998) Precopulatory mate guarding in crustaceans: male competitive strategy and intersexual conflict. Q Rev Biol 73:275–304

Jormalainen V (2007) Mating strategies in isopods – from mate monopolization to conflicts. In: Duffy JE, Thiel M (eds) Evolutionary ecology of social and sexual systems – crustaceans as model organisms. Oxford University Press, New York, pp 167–190

Jormalainen V, Shuster SM (1997) Microhabitat segregation and cannibalism in an endangered freshwater isopod, *Thermosphaeroma thermophilum*. Oecologia 111:271–279

Kaïm-Malka RA, Maebe S, Macquart-Moulin C, Bezac C (1999) Antennal sense organs of Natatolana borealis (Lilljeborg 1851) (Crustacea: Isopoda). J Nat Hist 33:65–88

Kaufmann RS (1994) Structure and function of chemoreceptors in scavenging lysianassoid amphipods. J Crust Biol 14:54–71

Kolding S (1986) Interspecific competition for mates and habitat selection in five species of *Gammarus* (Amphipoda: Crustacea). Mar Biol 91:491–495

Krång AS (2007) Naphthalene disrupts pheromone induced mate search in the amphipod *Corophium volutator* (Pallas). Aquat Toxicol 85:9–18

Krång AS, Baden SP (2004) The ability of the amphipod *Corophium volutator* (Pallas) to follow chemical signals from con-specifics. J Exp Mar Biol Ecol 310:195–206

Kuenen DJ, Nooteboom HP (1963) Olfactory orientation in some land-isopods (Oniscoidea, Crustacea). Entomol Exp Appl 6:133–142

Lefebvre F, Limousin M, Caubet Y (2000) Sexual dimorphism in the antennae of terrestrial isopods: a result of male contests or scramble competition? Can J Zool 78:1987–1993

Lincoln RJ (1985) Morphology of a calceolus, an antennal receptor of gammaridean Amphipoda (Crustacea). J Nat Hist 19:921–927

Linsenmair KE (1987) Kin recognition in subsocial arthropods, in particular in the desert isopod *Hemilepistus reaumuri*. In: Fletcher DJC, Michener CD (eds) Kin recognition in animals. John Wiley and Sons Ltd, Chichester, pp 121–208

Linsenmair KE (2007) Sociobiology of terrestrial isopods. In: Duffy JE, Thiel M (eds) Evolutionary ecology of social and sexual systems – crustaceans as model organisms. Oxford University Press, New York, pp 339–364

Lowry JK (1986) The callynophore, a eucaridean/peracaridan sensory organ prevalent among the Amphipoda (Crustacea). Zool Scripta 15:333–349

Lyes MC (1979) The reproduction behavior of *Gammarus duebeni* (Lilljeborg), and the inhibitory effect of a surface active agent. Mar Behav Physiol 6:47–55

Manning JT (1975) Male discrimination and investment in *Asellus aquaticus* (L.) and *A. meridianus* Racovitsza (Crustacea: Isopoda). Behaviour 55:1–14

Mead F, Gabouriaut D (1977) Chevauchée nuptiale et accouplement chez l'isopode terrestre *Helleria brevicornis* Ebner (Tylidae). Analyse des facteurs qui contrôlent ces deux phases du comportement sexuel. Behaviour 63:262–280

Mead F, Gabouriaut D (1988) Belated and decreased reproduction in isolated females of *Helleria brevicornis* Ebner (Crustacea, Oniscoidea). Recuperation after the addition of faeces to the female environment. Int J Invertebr Reprod Dev 14:95–104

Meador JP (1989) Chemoreception in a lysianassid amphipod: the chemicals that initiate food-searching behavior. Mar Behav Physiol 14:65–80

Patterson L, Dick JTA, Elwood RW (2008) Embryo retrieval and kin recognition in an amphipod (Crustacea). Anim Behav 76:717–722

Poulton MJ, Thompson DJ (1987) The effects of the acanthocephalan parasite *Pomphorhynchus laevis* on mate choice in *Gammarus pulex*. Anim Behav 35:1577–1579

Premke K, Klages M, Arntz WE (2006) Aggregations of arctic deep-sea scavengers at large food falls: temporal distribution, consumption rates and population structure. Mar Ecol Prog Ser 325:121–135

Read AT, Williams DD (1990) The role of the calceoli in precopulatory behaviour and mate recognition of *Gammarus pseudolimnaeus* Bousfield (Crustacea, Amphipoda). J Nat Hist 24:351–359

Robinson MH (1982) Courtship and mating behavior in spiders. Ann Rev Entomol 27:1–20

Sainte-Marie B, Hargrave BT (1987) Estimation of scavenger abundance and distance of attraction to bait. Mar Biol 94:431–443

Seelinger G (1977) Der Antennenendzapfen der tunesischen Wüstenassel *Hemilepistus reaumuri*, ein komplexes Sinnesorgan. J Comp Physiol 113:95–103

Seelinger G (1983) Response characteristics and specificity of chemoreceptors in *Hemilepistus reaumuri* (Crustacea, Isopoda). J Comp Physiol A 152:219–229

Sheader M (1981) Development and growth in laboratory-maintained and field populations of *Parathemisto gaudichaudi* (Hyperiidea: Amphipoda). J Mar Biol Ass UK 61:769–787

Shillaker RO, Moore PG (1987) The biology of brooding in the amphipds *Lembos websteri* Bate and *Corophium bonnellii* Milne Edwards. J Exp Mar Biol Ecol 110:113–132

Shuster SM (1990) Courtship and female mate selection in a marine isopod crustacean, *Paracerceis sculpta*. Anim Behav 40:390–399

Sparkes TC, Keogh DP, Haskins KE (2000) Female resistance and male preference in a stream-dwelling isopod: effects of female molt characteristics. Behav Ecol Sociobiol 47:145–155

Stanhope MJ, Connelly MM, Hartwick B (1992) Evolution of a crustacean chemical communication channel: behavioral and ecological genetic evidence for a habitat-modified, race-specific pheromone. J Chem Ecol 18:1871–1887

Takeda N (1984) The aggregation phenomenon in terrestrial isopods. Symp Zool Soc Lond 53:381–404

Thomas F, Renaud F, Cezilly F (1996) Assortative pairing by parasitic prevalence in *Gammarus insensibilis* (Amphipoda): patterns and processes. Anim Behav 52:683–690

Thompson DJ, Manning JT (1981) Mate selection by *Asellus* (Crustacea: Isopoda). Behaviour 78:178–186

Toth GB, Pavia H (2007) Induced herbivore resistance in seaweeds: a meta-analysis. J Ecol 95:425–434

Wellborn GA, Cothran RD (2007) Evolution and ecology of mating behavior in freshwater amphipods. In: Duffy JE, Thiel M (eds) Evolutionary ecology of social and sexual systems – crustaceans as model organisms. Oxford University Press, New York, pp 147–166

Wisenden BD, Cline A, Sparkes TC (1999) Survival benefit to antipredator behavior in the amphipod *Gammarus minus* (Crustacea: Amphipoda) in response to injury-released chemical cues from conspecifics and heterospecifics. Ethology 105:407–414

Wisenden BD, Pohlman SG, Watkin EE (2001) Avoidance of conspecific injury-released chemical cues by free-ranging *Gammarus lacustris* (Crustacea: Amphipoda). J Chem Ecol 27:1249–1258

Zimmer M, Kautz G, Topp W (1996) Olfaction in terrestrial isopods (Crustacea: Oniscidea): responses of *Porcellio scaber* to the odour of litter. Eur J Soil Biol 32:141–147

Chapter 11
Mantis Shrimp: Olfactory Apparatus and Chemosensory Behavior

Kristina Mead and Roy Caldwell

Abstract Mantis shrimp (stomatopods) are known to recognize individuals, but this ability varies with species, with reproductive mode, and with the degree of competition for burrow space. In this chapter, we describe the sensory basis, focusing on chemosensory sensilla (aesthetascs) located on the antennules, that allowed the evolution of the intricate communication system found in stomatopods. The efficiency of these chemosensors is supported by self-generated currents, which mantis shrimp employ to both send and receive chemical information. Multiple behavioral experiments, involving paired aggressive contests or studies of mating pairs, highlight the robust abilities of several stomatopod species to both recognize and remember individual opponents and mates. The extent of these capacities can be understood within the context of each species' social and mating systems. Among the main factors selecting for individual recognition, we identified the limited supply of suitable dwellings and the high possibility of repeated encounters among individuals. These, together with powerful weapons (that could inflict lethal damage) and high site fidelity, have led to the evolution of diverse mating systems, some of which (monogamy, for example) facilitate the evolution of individual recognition. Chemical signaling is essential for this, but it is also employed in other contexts: signals indicate sex but not sexual receptiveness, and some species exploit this fact to deceive and to gain access to burrows during mating encounters. Based on our results we suggest that the signaling mechanisms and chemical recognition systems are highly developed in stomatopods, but might also have evolved in other crustaceans that are exposed to similar selective pressures. We end with a projection of future work, focusing on experiments that could improve our understanding of signaling mechanisms, the role of multimodal signaling, and the opportunity to investigate deception and its role in sexual conflict.

K. Mead (✉)
Department of Biology, Denison University,
Granville, OH 43023, USA
e-mail: meadk@denison.edu

T. Breithaupt and M. Thiel (eds.), *Chemical Communication in Crustaceans,*
DOI 10.1007/978-0-387-77101-4_11, © Springer Science+Business Media, LLC 2011

11.1 Introduction: The Value of Learning from Mistakes

Early in Roy Caldwell's (hereafter called "Roy") career working with stomatopods, he was interested in aggression and how these pugnacious creatures communicated with one another. To that end he staged a round robin tournament where each of 16 combatants fought one another over the course of 15 days – a different opponent each day. This would allow him to determine which behaviors were important in establishing dominance and to calculate how much information was being exchanged between individuals. Unfortunately, or actually fortunately, Roy made a book keeping mistake and paired some stomatopods against animals with which they had previously fought. The behaviors resulting from pairings in which the participants had previously encountered one another were dramatically different from those in which contestants had not met previously. In every case of accidental rematch, there was no protracted fight with the animals exchanging threats and blows until dominance by one was established. Rather, one animal immediately fled from the other and did not challenge it. The subordinate acted as if it recognized its opponent and remembered that it had previously lost to it. When Roy discovered his error and checked the previous outcomes of the first pairings, in every case the subordinate animal had also lost the initial fight, suggesting that the loser recognized its opponent. This led Roy to ask how the fighting ability of previous opponents was recognized, and whether or not this involved the recognition of individuals or simply the identification of traits associated with winning. At that time, individual recognition in agonistic interactions was unknown for any invertebrate (although a few studies suggested that mate recognition occurred in some crustaceans and insects). Over the course of the next several years and a few thousand cuticle bruising fights later, he eventually could say with some certainty that some stomatopod species were capable of identifying other individuals that they had previously encountered, and could adjust their agonistic behavior accordingly. Kristina Mead became intrigued by these charismatic crustaceans after seeing their behavior and noticing that their sophisticated social interactions appeared to be mediated by a morphologically simple olfactory apparatus. Here, we first present some background information about stomatopods and their sensory capabilities to lead over to the description of some of the experiments that revealed how and when individual recognition occurs.

11.2 What is a Stomatopod?

Stomatopods, also known as mantis shrimp, are malacostracan crustaceans that typically live in tropical or subtropical oceans, although a few temperate species are known. Mantis shrimp are typically between 2 and 30 cm in length. Their elongate bodies are typically flattened dorsoventrally. Hallmarks include specialized stalked eyes, tripartite antennules that are highly sensitive to a variety of odors, and raptorial appendages capable of very rapid strikes.

Fig. 11.1 Stomatopods in burrows. (**a**) *Neogonodactylus* sp., smasher, in coral rubble cavity, (**c**) *Lysiosquilla tredecimdentata,* spearer, in sandy burrow (photo Nick Hobgood, with permission), (**b**) smasher raptorial appendage, (**d**) spearer raptorial appendage. Most smashers occupy cavities in rock or rubble. These are in limited supply, and so are the subjects of fierce contests between individuals. Most spearers dig their own cavities in soft substrates such as sand or mud. While territory suitable for burrowing is easier to come by, these burrows can still represent a substantial investment. Nonetheless, cavity-dwellers tend to be more aggressive

Most mantis shrimp spend much of their time in burrows (Fig. 11.1a, b), but can emerge to hunt prey with startling swiftness. Their raptorial appendages (derived from enlarged second thoracopods) can be long, spiny spearing appendages, or hardened smashing appendages with extra muscle mass and the ability to generate shock wave-producing cavitation bubbles (Patek and Caldwell 2005), and to break glass aquaria (Caldwell 1987) (Fig. 11.1c, d). Most spearing species make burrows in soft substrata, while most smashing species occupy cavities in coral rubble or other hard substrata.

The burrows or cavities are critical for many aspects of mantis shrimp ecology. They provide safety from predators, a location for processing prey, and a safe haven for mating and for the guarding of eggs and larvae. Many species face strong competition for suitable locations for burrows or for cavities. The combination of the need of a burrow entrance precisely calibrated to the animal's diameter and the increase in body size with each molt means that stomatopods periodically (during growth) require a larger cavity or burrow. Species living in sand or mud can usually expand their burrows without abandoning them, but species living in rock or coral cavities can't easily enlarge their dwellings and must often fight to take control of a new, larger cavity. Furthermore, in nonmonogamous species the animals often face eviction from burrows during mating when individuals leave their burrows to search for a mate. Either the resident mating partner is evicted, or the visitor returns to find its burrow occupied and must fight for it.

This fierce competition for burrow space, combined with the potentially lethal weapons at their disposal, places a premium on rapid, accurate information that can reduce the risk involved in assessing and fighting other stomatopods and other dangerous competitors. As a consequence, the visual and olfactory systems of stomatopods, particularly smashing species living in cavities, are very well developed. Mantis shrimp possess appositional compound eyes with line scanning ability and up to at least 16 types of visual pigments that combine to allow sensitivity to UV and polarized light, 12 channel color vision, and complex spatial vision (Marshall et al. 2007). However, visual information can be compromised by low light levels (e.g., at night or in turbid environments), extraneous light in the environment, or the animal's position in its cavity or burrow. Furthermore, the combinatorial nature of chemical cues provides information among more axes than do visual cues. This includes the ability to recognize the individual odor signature of previous opponents or mates. These factors have led to the extensive use of chemical signals in shelter defense, reproduction, and food acquisition, and the evolution of the equipment and behavior to send and receive such information (Caldwell 1991). Below, we will describe the sensory apparatus that supports individual recognition, and the sending and receiving of chemical signals. We will then discuss the roles of odor cues in individual recognition in the contexts of shelter defense and in reproduction, as well as the roles of asymmetric and deceptive signaling.

11.3   Aesthetascs

As in many other crustaceans, mantis shrimp detect odors from distant sources using specialized chemosensory sensilla (aesthetascs) located on their antennules (Hallberg et al. 1992) (Fig. 11.2a). Stomatopod antennules are triramous, with aesthetascs located on the distal portion of the dorsolateral flagellum (Hallberg et al. 1992; Mead and Weatherby 2002; Derby et al. 2003) (Fig. 11.2b). Mantis shrimp aesthetascs are long, slender structures inserted into the dorsolateral flagellum at an angle of 40–60°. The aesthetasc cuticle is very thin and is permeable to methylene blue (Mead and Weatherby 2002), suggesting that it is also permeable to other small molecules, such as amino acids and simple sugars, which are common components of chemical cues. Aesthetasc length and diameter depend on size and species, varying from 10 μm in diameter and 200 μm length in an 8-mm telson-rostrum length *Gonodactylaceus falcatus* to 32 μm in diameter and 550 μm length

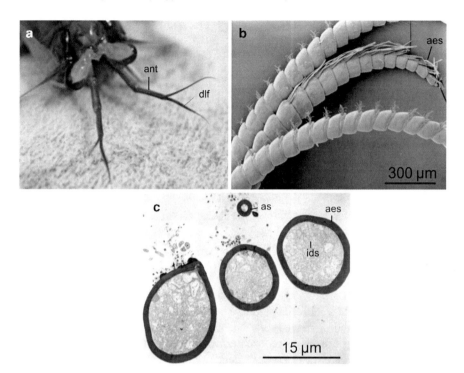

Fig. 11.2 Antennule and aesthetasc structure. (**a**) Head of *Hemisquilla californiensis* showing antennules (ant) and the aesthetasc-bearing dorsolateral flagellum (dlf), which is lying flat against the ventrolateral flagellum so that only two of the three flagella are easily visible. (**b**) Scanning electron micrograph of antennule from *G. falcatus* indicating aesthetascs (aes). (**c**) Transmission electron micrograph of *G. falcatus* aesthetascs in cross-section showing row of three aesthetascs (aes) housing sensory neurons (with inner dendritic segments labeled ids seen in cross-section) and accompanying asymmetric seta (as)

in a 157 mm telson-rostrum length *Hemisquilla californiensis* (Mead, unpublished). The aesthetascs are arranged in rows of three, with one row at the distal margin of each flagellar annulus. Associated with each row of aesthetascs is a small seta (4–5 μm in diameter), possibly analogous to the asymmetric mechanosensory sensilla of the spiny lobster (Gleeson et al. 1993) (Fig. 11.2c). This simple arrangement is in marked contrast to the complex arrays of aesthetascs and associated setae found in some other crustaceans such as lobsters, which can have up to ten antennular setal types (Cate and Derby 2001; Hallberg et al. 1992). There do not appear to be significant morphological differences between male and female aesthetascs, at least in the species studied to date (Mead et al. 1999).

Each aesthetasc is innervated by bipolar olfactory neurons (seen in cross section in Fig. 11.2c) whose cell bodies are located within the dorsolateral flagellum of the antennule (Mead and Weatherby 2002). In *G. falcatus*, there are 12–20 olfactory receptor neurons (ORNs) per aesthetasc. Each of these cells gives rise to multiple dendrites, as in some crabs, mysids, and other crustaceans (see Hallberg and Skog,

Chap. 6). These chemosensory neurons project proximally to the ipsilateral olfactory lobe, which is organized into spherical glomeruli. Derby et al. (2003) have estimated that approximately ten ORNs expressing similar classes of odorant receptor molecules on their membranes converge on each of the 60–80 glomeruli. Nonaesthetasc chemosensory neurons that have mixed chemosensory and mechanosensory function are thought to project proximally to the ipsilateral lateral antennular neuropil. Thus, as in other crustaceans, there appear to be two parallel antennular pathways, one involving aesthetascs, and one for nonaesthetasc sensilla (Derby et al. 2003; see Schmidt and Mellon, Chap. 7). The nonaesthetasc pathway is thought to mediate antennular movements, reflexes such as antennular grooming behavior, and searching for food (Horner et al. 2004), while the aesthetasc pathway appears to be responsible for more complex behaviors, such as courtship behavior in male crabs (Gleeson 1982) and individual recognition in stomatopods (Caldwell 1985, 1992; see also Schmidt and Mellon, Chap. 7 and Aggio and Derby, Chap. 12). The relative roles of the two pathways in mantis shrimp are not yet known, but the dominance of aesthetascs as a sensillar type on antennules suggests that their role is important here as well.

Evidence that the aesthetascs on the distal portion of the dorsolateral antennular flagellum are responsible for individual recognition comes from removal and ablation experiments (Caldwell, unpublished). *Neogonodactylus festae* that had the dorsolateral flagella removed were unable to distinguish known from unknown individuals. In a less drastic procedure, *Neogonodactylus bredini* trained to distinguish between known and unknown residents were treated on the following day by having their dorsolateral flagella aspirated into a pipette containing deionized water for 2 min. This procedure ruptured cellular material within the aesthetascs and possibly other sensilla. As a control, other individuals had their antennules aspirated into a pipette containing seawater. When tested on the third day against resident or stranger water, the animals treated with deionized water were unable to distinguish the two types of odor while controls treated with seawater did. After 5 days, the aesthetascs began to function again and the animals could distinguish between resident and stranger water; this result also highlights the high regenerative capacity of these chemoreceptor cells. Stomatopods with aspirated antennules were not tested with food odors.

11.4 Sending Chemosensory Signals: Maxilliped Pulses and Maxilliped Whirls

In addition to a highly functioning sensory apparatus, mantis shrimp have specific behaviors for sending and receiving olfactory signals. Two behaviors thought to be involved in sending signals are maxilliped pulses and maxilliped whirls (Fig. 11.3a, b). Both behaviors are used by female cavity residents at the entrance of their burrow as a potential intruder approaches (Caldwell 1992). A maxilliped pulse consists of a series of simultaneous anterior extensions of the third, fourth, and fifth pairs of

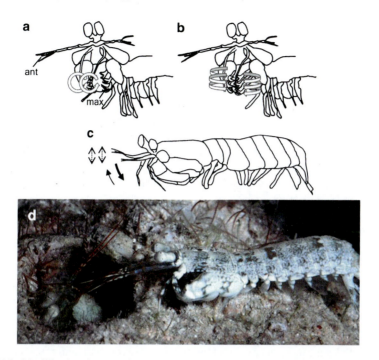

Fig. 11.3 Maxilliped and antennule movements used in the sending and receiving of chemical signal. (**a**) Maxilliped pulsing ant indicates antennules, max indicates maxillipeds. *Grey circular arrows* show orbits of maxilliped dactyls moving in circular strokes up, out, down, and back in 4–8 cycles/s in a vertical plane. (**b**) Maxilliped whirling. *Grey arrows* moving in and out of the page show opposite orbits of maxillipeds in a shallow angled plane. (**c**) Antennulation (*vertical dotted double arrows*) and olfactory flicking (*solid curved arrows* showing fast lateral flick and slower return motion). (**d**) Image of *Haptosquilla trispinosa* antennulating at a burrow entrance

maxillipeds, directed toward the intruder. The appendages move in circular strokes up, out, down, and back in 4–8 cycles/s (Caldwell 1992). Cycle rate, extension distance, and the openness of the maxilliped dactyls all increased the closer the male came to the entrance. Observations of small streams of dye pipetted in front of the female indicate that the maxilliped pulse pushes water from under the female and out of the cavity towards the opponent, with small return currents bringing water from the side and across the female's antennules (Caldwell 1992). Maxilliped pulses, therefore, may function to both sending and receiving signals.

A maxilliped whirl is a more intense type of maxilliped motion, usually taking place for a few seconds in the middle of a bout of maxilliped pulsing. In maxilliped whirls, the maxillipeds are fully extended and the dactyls are open. Rather than pulsing in unison, the appendages of the third, fourth, and fifth pair of maxillipeds rotate 180° out of phase with each other in very small, tight circles at up to 10 cycles/s in a shallow nearly horizontal plane (Caldwell 1992). This behavior is used most often by a resident female in the entrance of her cavity when an approaching male is less than a body length away from the cavity entrance. There have been no recorded instances of male or female intruders using maxilliped whirls (Caldwell 1992). The proximity of the two mantis shrimp and the rapid motion of the maxillipeds made it difficult to quantify dye movement, but the whirling produced a strong churning effect, mixing water in front of the female. Dye was often drawn towards the head of the female, so it is suspected that this behavior also serves to sample the approaching individual. The nature and source of the chemicals distributed by the microcurrents created by both the maxilliped pulses and whirls are not known. Two possibilities include fecal material, as seen in *Haptosquilla* females during burrow defense when they fan their telson pleopods to push jets of water out from the entrance of the cavity (Caldwell, unpublished), or urine, as in lobsters and crayfish (Breithaupt 2001; see Aggio and Derby, Chap. 12 and Breithaupt, 13).

11.5   The Reception of Chemosensory Signals

Mantis shrimp receive chemical information via antennulation and olfactory flicking (Fig. 11.3c), sometimes in combination with maxilliped and/or pleopod movements. Antennulation consists of the forward extension of the paired antennules coupled with short, quick oscillations in the vertical plane (Caldwell 1992) while olfactory flicking can be in any direction and consists of lateral and medial motions (Mead et al. 1999). In the context of mating behavior and individual recognition, antennulation typically occurs as a male is approaching a cavity inhabited by a female. Antennulation at the cavity entrance is the most common behavior by a male during an approach to a cavity (Fig. 11.3d). Antennulation by females, at least in a cavity entrance, and in *N. bredini*, is much less common. Antennulation appears to be a subcategory of olfactory flicking used in many investigatory situations, such as in response to a food odor or in tracking an odor to its source. An example of how maxilliped pulses generated by a female in a cavity might send an odor signal to an antennulating male desirous of entering the burrow is shown in Fig. 11.4.

Fig. 11.4 Artist's sketch of female mantis shrimp in burrow using maxilliped pulses to send an odor signal to a potential male intruder. The male is using antennulation to investigate the signal emanating from the burrow. Drawing by Jorge A. Varela Ramos

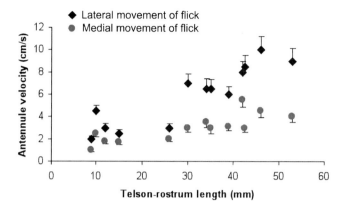

Fig. 11.5 Antennule flicking. Antennule velocity during outward lateral flick (*black diamonds*) and medial return stroke (*grey circles*) increases with body size. Data are means ± standard errors. The outward stroke of the flick is faster than the return stroke of the flick

Although there are no published studies focusing solely on the biomechanics of antennulation, work has been done on olfactory flicking in stomatopods (Mead et al. 1999). Each flick starts with a rapid, outward lateral movement followed by a slower return medial movement (Fig. 11.5a). Experiments studying flow around dynamically scaled physical models suggest that water is pushed into an array of aesthetascs during the rapid outstroke of a flick (Mead et al. 1999; Mead and Koehl 2000). These data inserted into an advection-diffusion model indicate that odor molecules in filaments moving into the sensor array arrive at the surfaces of the aesthetascs within milliseconds (Stacey et al. 2003). In contrast, during the slower return stroke of the flick and during the stationary pause between flicks, water flows around rather than into the array of chemosensory hairs (Mead et al. 1999; Mead and Koehl 2000). Thus, an antennule of a mantis shrimp takes a discrete sample in time and space of its odor environment only during the flick outstroke (Mead and Koehl 2000; Stacey et al. 2003). This pattern of discrete sampling appears to be widespread among crustaceans (see Koehl, Chap. 5).

Olfactory flicking seems to be matched to the animal's olfactory need. For instance, the tropical mantis shrimp *G. falcatus* becomes increasingly involved in sophisticated aggressive and reproductive encounters as it grows. It also hunts faster and more elusive prey. These complex social and foraging behavior patterns require rapid responses. Perhaps to accommodate the need for speed, antennule flicking velocity and frequency increase several fold with body size (Fig. 11.5a, Mead et al. 1999). Flicking reduces the boundary layer so that odors penetrate the array more quickly (Mead and Koehl 2000), and increases the rate at which chemical information is obtained.

In addition to the relatively small scale movements of the antennules and maxillipeds, recent field observations of *Squilla empusa* suggest that vigorous pleopod fanning movements can create relatively large currents in and out of the burrow that may enhance signal reception or dispersion (Mead, unpublished).

This brief overview shows that a variety of morphological adaptations and behavioral mechanisms facilitate the emission and reception of chemical signals during interactions between stomatopods.

11.6   Individual Recognition Using Odor Cues

Individual recognition using odor cues appears to be important both in limiting aggressive contests and thus avoiding injury and in promoting mate recognition during reproduction. The following discussion will start with the less complicated role of individual recognition in aggressive contests and will treat both conspecific and heterospecific recognition. This will be followed by a treatment of the varied roles of individual recognition in both monogamous and polygamous mating systems.

Much of the work investigating individual recognition in stomatopods has been done using paired aggressive contests over artificial shelter cavities (Fig. 11.6a), in the small tropical mantis shrimp species N. festae. In the initial study on chemically based recognition in stomatopods, Caldwell (1979) placed N. festae into artificial cavities. Fifteen minutes after these animals entered their cavities, a second stomatopod matched for size and sex was introduced into the test arena. Invariably the resident and intruder fought for ownership of the cavity and the resident almost always won due to its positional advantage inside the cavity. If the resident was evicted, the pair was tested the next day with their roles reversed. After dominance was established (usually within less than 5 min), both animals were removed and the test apparatus cleaned and refilled with seawater. The cavity was filled with one of three types of water and corked to prevent unwanted diffusion of the odors inside. The three test fluids were: "clean" water that had never contained a stomatopod, "stranger" water from the container of a stomatopod of the same size and sex that the intruder had never encountered, and "known victorious resident" water taken from the previously victorious resident's container removed prior to the encounter between them. After 15 min, the cork was removed and the intruder reintroduced into the far end of the test arena. This testing procedure was repeated over 3 days, each intruder encountering all three odors, with a randomized order of presentation. While intruders all quickly approached the empty cavities, their behavior was markedly different depending on the source of the water. If there was no odor of another stomatopod, the intruders took no defensive action and immediately entered the cavity (Fig. 11.6b). If the cavity contained the odor of another stomatopod with which the intruder had no experience, the intruders entered cautiously, often inserting their armored telson into the entrance as if expecting an attack from the phantom resident (Fig. 11.6c). If the cavity contained the odor of the previously encountered resident that had defeated the intruder earlier, most approached the entrance, sampled the odor diffusing from the cavity with their antennules, and immediately fled the area (Fig. 11.6d). When encountering clean water, 80% of the intruders entered the cavity within 2 min and when exposed to the water containing

Fig. 11.6 Test arena and sample behaviors (**a**). The arena consists of a 16 × 32 × 8 cm *grey* plastic box with a layer of sand on the *bottom*. The water depth is 5 cm. A 30-ml *dark* plastic bottle (the cavity) is screwed into the wall of shallow tank, just above the sand substratum. A 1-cm hole is drilled into the cap, allowing stomatopods to enter the bottle easily. Note the protruding antennules of the resident mantis shrimp. *Neogonodactylus festae* exploring cavities filled with (**b**) clean water, (**c**) water from the container of a stranger stomatopod (matched by species, size, and sex), and (**d**) water from the container of a previously encountered victorious resident stomatopod (matched by species, size, and sex) that had resisted eviction in a previous battle. When encountering clean water, 80% of the intruders entered the cavity within 2 min and when exposed to the water containing the odor of a stranger, 80% entered within 7 min. However, only 41% ever entered the cavity during the course of the 15 min test if it contained the odor of the animal than had previously defeated them

the odor of a stranger, 80% entered within 7 min. However, only 41% ever entered the cavity during the course of the 15 min test if it contained the odor of the animal that had previously defeated them.

While this experiment clearly shows that *N. festae* can quickly learn the odor of an opponent and associate that odor with the outcome of previous agonistic encounters, it does not demonstrate the ability to identify individuals based on their odors. It is possible that the intruders were responding differently to broader classes of animals such as "known" vs. "unknown" opponents. To test for this possibility, a different design was needed where the intruder encountered at least two opponents over a short period of time and then reacted differently to them.

Intruders were first matched against either a resident that was larger and could successfully defend its cavity or against a smaller resident that the intruder could evict. Thirty minutes later the intruder was matched against the other type of resident. The residents were the same sex as the intruder and the order of presentation was randomized. This design let the intruder encounter two stomatopods, one that it defeated and another that defeated it. Thirty minutes later the intruder was tested using the odor of one of the residents it had just fought and 30 min after that it was tested against the odor of the other resident (Fig. 11.7). The results of these rematches were that intruders quickly entered cavities spiked with the odor of an animal that they had evicted (median time = 19 s), but delayed entering cavities with odor from animals that had successfully defended the cavity in an earlier bout (median time = 324 s; Caldwell 1985). This time is shorter than that reported above because the experimental design increased familiarity with the apparatus which led to faster probing of cavities. Since the same intruders failed to respond differentially to odors of larger and smaller unfamiliar conspecifics, it is unlikely that odors providing information on size alone served as the basis for the difference in time to enter the cavity. The response to odor was based on previous experience, suggesting the possibility of individual recognition.

In addition to conspecific individual recognition, the stomatopod *Neogonodactylus zacae* is able to use chemical cues to discriminate between different individuals of *N. bahiahondensis* (Caldwell 1982). The two species are sympatric along the Pacific Coast of Central America, and occupy similar cavities in the same habitat. To test this ability, an individual *N. bahiahondensis* was allowed to occupy the cavity shown in Fig. 11.6a, and a *N. zacae* with 2–3 mm smaller rostrum-telson length was introduced into the tank. The size differential ensured that the occupant would successfully defend the cavity. Later, the same *N. zacae* was returned to the tank with the cavity empty except for odor from either the *N. bahiahondensis* that defeated it or another individual *N. bahiahondensis* with which the subject *N. zacae* had no experience. The median time to entry was much greater when odor from the known victor was in the cavity (552 s) than when the cavity contained unknown odor (68 s). Along with more delay before entering the cavity, the *N. zacae* encountering victor odor were more hesitant once entering the cavity, and were more likely to enter telson-first. This type of entry is a defensive tactic used to reduce the chance of injury since blows to the telson are less likely to cause lasting damage than are blows to the more vulnerable unarmored eyes or antennules. In contrast, all animals entering vials containing unknown odor entered in a more aggressive manner, head-first. Previous experience with an individual, regardless of its species, provides much more reliable information on which to decide to challenge it or not than species-specific cues.

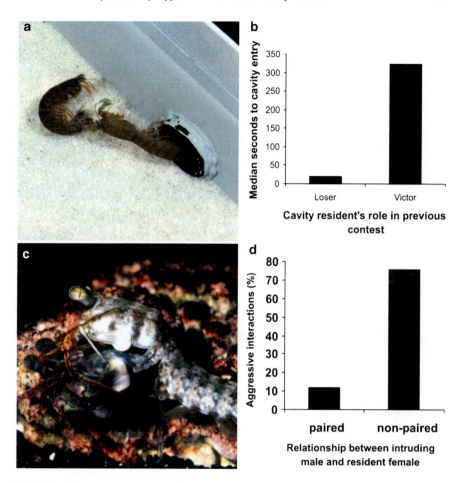

Fig. 11.7 Contests over cavities. (**a**) *N. festae* defending artificial burrow. (**b**) Time to enter cavity held either by animal evicted in previous contest (loser) or held by animal victorious in previous contest (victor). Intruders quickly entered cavities spiked with the odor of an animal that they had evicted (loser; median time = 19 s), but delayed entering cavities with odor from animals that had successfully defended the cavity in an earlier bout (victor; median time = 324 s; Caldwell 1985). (**c**) Photograph of mating behavior in *G*. sp. near cavity. (**d**) Percent of interactions that are aggressive when a male enters a cavity inhabited by a female with whom he previously mated (paired) vs. a female that he has not encountered previously (nonpaired)

11.7 Role of Chemical Communication in Reproduction in Stomatopods

Since cavities and burrows are as essential for reproduction as for shelter from predation, the potential for aggressive contests over mating space and mating partners is high. Reproductive pressure can add impetus to the desire to defend space. For example, 64% of nonbrooding females retained their cavity against slightly larger competitors, while 96% of similarly sized females with eggs successfully defended their cavities against the same intruders, despite the extra bulk of the egg mass (Montgomery and Caldwell 1984).

It is necessary to review stomatopod reproductive behavior to appreciate the spectrum of ways in which odor is implicated in true and deceptive signaling during mating. Sexes are separate, and females use sperm stored in seminal receptacles to fertilize extruded eggs. The eggs are held together with glycoprotein and mucopolysaccharides and form a soft ball which the female carries in her maxillipeds. Mating systems range from life-long monogamy in some lysiosquillids to multiple sequential matings in many neogonodactylids to rampant promiscuity in *Pseudosquilla ciliata* (Caldwell 1987). Mating behavior varies, depending on (1) the searching sex (male, female, or both), (2) whether the male or female burrow is used for mating, (3) the level of potential aggression during mating, (4) the number of mates per brood, (5) the occurrence and behavior associated with sperm storage, and (6) the length of pair bonding (Caldwell 1991). The following examples detail species with three different mating habits ranging from monogamy to polygamy, with no, one, or both partners searching for mates. In each case, the role of individual recognition via odor seems critical, but for different reasons.

Lysiosquillina are spearers that make elaborate (up to 10 m) burrows in soft sediment (Christy and Salmon 1991). Adults of all of the species in this genus that have been studied are always found in pairs. They have reduced armor and are thus very vulnerable outside of their burrow. Typically the burrow entrance is covered with a sand mucus cap with just a small hole for the camouflaged eyes and antennules. Prey such as fish are speared as they pass over the entrance of the burrow. While substratum for excavating burrows is readily available, reducing competitive pressure for space, burrow construction nonetheless represents a considerable investment. One of the most costly components is the large amount of mucus required to stabilize the shifting sands where they live. The biological cost of the mucus prevents adult *L. sulcata* from being able to construct a new burrow if evicted from their old one (Caldwell 1991). Together, the large investment in burrow construction and the vulnerability to predation are thought to have led to a mating system that replaces searching by either or both sexes with long-term monogamy (Christy and Salmon 1991). Monogamy in crustaceans often involves mate recognition based on chemical cues (e.g., *Hymenocera picta*, Wickler 1973; Wickler and Seibt 1981) and this appears to be the case in *Lysiosquillina* (Caldwell unpublished). Thus, in this case, individual recognition facilitates the pair bond.

In contrast, *N. bredini* are smashers that live on shallow reef where they face heavy competition for a limited number of cavities. Their breeding cycles are tightly synchronized. Typically, the male searches for a receptive female and fights to enter her cavity a week before the full moon. They copulate repeatedly and he guards the female and the cavity until the eggs are laid at the time of the full moon. As soon as the female oviposits, the male leaves. The female cares for the developing eggs alone until the eggs hatch 3 weeks later, and then maintains the larvae for another week until they enter the plankton. The presence of the male in the tight confines of the cavity would jeopardize the brood. However, the ejection of the male means that the male must now seek a new cavity since his original cavity has most likely been occupied. Just after the full moon, there are numerous males searching for cavities and trying to evict the residents, even if they contain brooding females. Should a male encounter his former mate while she is still brooding and attempt to evict her, he would be threatening his own offspring. However, if the pair recognized one another and avoided a confrontation, both parties would benefit. Here, individual recognition serves to protect the investment in their offspring.

The ability of *N. bredini* to modulate aggression through chemical recognition was studied by observing 25 mated pairs that produced eggs (Caldwell 1992). Fourteen days after the male left the breeding cavity, Caldwell (1992) compared the interaction between the female (placed in a new cavity) to him and to a stranger male. The males were also tested as intruders against brooding females that they had not previously encountered. An aggressive act (meral spread threat, lunge, strike; see Caldwell 1979 for precise definition) occurred only 3/25 times when the original male was introduced to his brooding mate, but 19/25 times when a stranger male was introduced (Fig. 11.7c, d). Contests between nonpaired animals escalated more rapidly, with the first act being aggressive 87% of the time, vs. 0% of the first acts being aggressive in previously paired animals' contests. The reduced aggression displayed by males that had been paired was not due to their being less likely to act aggressively towards brooding females or the presence of eggs. Males attempted to evict brooding females with which they had not mated and in another experiment, males did not escalate contests when encountering former mates whose eggs had been removed, perhaps because she could be storing his sperm. Other species that experience competition for cavities and high search costs, and guarding of females by males, such as other neogonodactylids and the squillid *Meiosquilla*, are thought to have similar mating systems and thus perhaps may also rely on individual recognition to maintain pair bonding.

P. ciliata are widely distributed stomatopods found in a variety of habitats. Females will copulate at any stage in their reproductive cycle and have been observed to mate with several different males in 1 day. Females are extremely aggressive when pursuing mates, often harassing males until they copulate. Mating occurs in the open and the participants separate immediately after coupling. Since the decreased competition for burrows leads to lower aggression risk, and since there is no long-term sharing of burrows, it would not be surprising if individual recognition of mates did not occur in this species. However something akin to the "Coolidge Effect," in which sexual performance is increased when new receptive potential mates are available (Wilson et al. 1963), seems to be in play for both sexes in this species. Animals that have mated with each other will not remate within the same pair for several hours but they will copulate with a new partner within just a few minutes. This suggests that some form of individual recognition is occurring, this time to avoid the original mate!

Although mantis shrimp are capable of individual recognition, this does not mean that they are immune to the deceptive use of odors. When a male and female stomatopod come into close proximity, the odors that waft from the female to the male signal sex but do not appear to indicate sexual receptivity, here defined as the presence of developed cement glands. Although the most common mating scenario in *N. bredini* is that males seek entrance to female's cavities, females not currently in a pair may go looking for a male when her ovaries become fully ripe, usually toward the full moon. During this part of the lunar cycle, the male usually admits the female without hesitation. She evicts the male 15% of the time if she is reproductive, but 33% of the time if she is nonreproductive (Caldwell 1986). In some cases, the male may even leave the cavity to pursue the female. The female then darts back to the cavity, occupies it, and blocks the male from entering. These males appear to be trading their cavity for the chance to mate. By identifying her sex but not signaling her reproductive status, a *N. bredini* female increases her overall chances (i.e., not just when she is receptive, but throughout her reproductive cycle) of gaining access to a cavity. However, the experiment described above was conducted near the time of the full moon, when roaming adult females are likely to be sexually receptive. When this experiment was repeated during the week of the new moon when few females are receptive, males were significantly less likely to admit a female into their burrow (Caldwell 1986).

11.8   Where Should We Expect Individual Recognition?

Odor-mediated individual recognition has been found in many gonodactylids, and may be a common tool to mediate the aggressive encounters that correlate with their use of a limited number of cavities in rocks and rubble (Caldwell 1979, 1982, 1985). While these observations provide strong evidence for the capacity for individual recognition, not all mantis shrimp seem to possess this ability. The spearer *S. empusa* appears not to have the ability to recognize individuals. This may be because *Squilla* construct their own burrows in soft sediment. In this species, space is more readily available than in rock-dwelling species. In general, there appears to be a correlation between the scarcity of cavities or the energy invested in a burrow and the type and intensity of aggression involved in its defense (Caldwell and Dingle 1976; Dingle 1983). Thus, with burrow space at less of a premium, there may be less selective pressure for chemically mediated individual recognition. It is worth pointing out that some lysiosquillids, despite experiencing less limiting burrow options and less strident competition for burrows, are capable of individual recognition mediated by odor. Their nocturnal nature, coupled with their dull body color, less acute vision, and the deeper depths, greater turbidity, and lower light levels of their preferred habitats may have created their own evolutionary pressure for an effective olfactory system, despite the lower pressure exerted by their particular burrow type.

Social conditions also affect the level of reliance on individual recognition. For instance, if a mantis shrimp lives in a solitary manner, or if it encounters many different individuals a day with no repeats, then individual recognition is unlikely to play a pivotal role for that animal. In contrast, social and environmental factors leading to repeat interactions increase the likelihood of individual recognition playing a role. For example, field observations show that up to eight adult *N. festae* can inhabit individual cavities in the same rock, and that each inhabits the same cavity for at least several consecutive days. Video surveillance shows that they come into frequent contact as they forage over the surface of the coral rubble and interact at territorial boundaries (usually set within a few body lengths of the cavity entrance). Since the most easily defended cavities are size-matched to their inhabitants, animals usually return to the same cavity unless they molt or are evicted. At this point they attempt to find another nearby cavity and try to evict its occupant. This small group setting with moderate permanence suggests that the animals experience repeated contacts among a limited set of individuals, a situation required for the development of individual recognition. Here, too, lysiosquillids are informative: their monogamous mating system may have driven the development of a mate recognition system mediated by odor.

A reasonable question to ask is why individual recognition is mediated by the olfactory system. After all, many gonodactylids are active in the day, are brightly colored, live in clear water in fairly high light areas, and have excellent vision. Certainly, visual signaling may play a large role in these groups. However, their sense of smell is likely to provide cues that may contain more (and possibly more reliable) information than visual cues (see Christy and Rittschof, Chap. 16). Furthermore, for an animal anxious to probe the unseen occupant of a cavity, odor provides a channel of reliable information without requiring dangerous intimate contact. This is important because the cavity resident can easily see intruders in the open and can at least assess the size and vigor of their opponent.

One question we must consider is whether individual recognition is really that unusual in crustaceans. Recent evidence points to odor-mediated individual recognition in lobsters (Johnson and Atema 2005), hermit crabs (Gherardi and Tricarico 2007), hippolytid shrimp (Rufino and Jones 2001), and possibly alpheid shrimp (Boltaña and Thiel 2001; Duffy 2007). In other crustaceans, such as fiddler crabs (Detto and Backwell 2009), individual recognition is most likely mediated by visual cues. It will not be a surprise to find odor-mediated individual recognition in more groups. The most likely places to find individual recognition will be in systems, like stomatopods, where there is site fidelity and a valuable but limited resource that individuals are willing to fight for. Formidable weapons increase the cost of the contest and can drive the evolution of sophisticated recognition systems that could serve to minimize the risk of fighting by providing accurate information. Social and mating systems revolving around a safe dwelling result in repeated contacts among a limited group of individuals, which is expected to increase the likelihood of individual or group recognition.

11.9 Future Directions

Mantis shrimp represent a wonderful model system for investigating individual recognition. We have a basic understanding of the behavioral basis of individual recognition in aggressive and reproductive encounters, but there remain several important gaps in our knowledge. First, the chemical nature of the odor cues remains unknown. We hope that colleagues might take on the isolation and identification of the compounds used in this signaling. Second, we need to improve our understanding of the currents that carry the odors. This could be addressed through high speed video analysis of the maxilliped movements and the resultant water currents. One approach would be to obtain visible currents by injecting the mantis shrimp with fluorescein, as in Breithaupt and Eger (2002). A second approach could use small-scale particle image velocimetry. This latter technique would most likely result in a more mechanistic understanding of maxilliped pulses, whirls and antennulations. It will also be important to supplement laboratory observations with field work. Potential approaches here include the use of SCUVA (self-contained underwater velocimetry apparatus; Katija and Dabiri 2008) to analyze antennule and fluid movement when animals are sending and receiving signals in the field. This will give us the opportunity to compare the relative contributions of antennules and maxilliped movements to burrow currents and ambient flow when looking at the biomechanics of signal dispersion and reception.

Multimodal communication may enhance the information content or signal accuracy (see Hebets and Rundus, Chap. 17), which appears to be especially relevant during interactions with potentially dangerous individuals. Recent work indicates that stomatopods increase their antennular flicking under low light conditions and may place less reliance on visual cues under these circumstances (Cheroske et al. 2009). While the interaction between these sets of stimuli has been examined in *Gonodactylus smithii*, it would be very interesting to see how the relative importance of visual and chemical information varies among species and habitats.

There is also work to be done exploring the role of deception. It would be very interesting to see if males trade their burrows for mating opportunities with the same female more than once. Other forms of sexual conflict may be present in stomatopods. For instance, is the ability to store sperm correlated with mate recognition such that a male is less likely to evict a female that might be carrying his sperm even if there is no brood present? Or, do species in which the females have tightly synchronized receptive periods and thus exhibit more mate searching, show more individual recognition? Further experiments analyzing these phenomena in multiple species and under different environmental pressures may help elucidate broader patterns. Clearly, interdisciplinary approaches will be needed to deepen our understanding about the evolution of chemical communication in mantis shrimp.

References

Boltaña S, Thiel M (2001) Associations between two species of snapping shrimp *Alpheus inca and Alpheopsis chilensis* (Decapoda: Caridea: Alpheidae). J Mar Biol Assoc UK 81:633–638
Breithaupt T (2001) Fan organs of crayfish enhance chemical information flow. Biol Bull 200:150–154
Breithaupt T, Eger P (2002) Urine makes the difference: chemical communication in fighting crayfish made visible. J Exp Biol 205:1221–1231

Caldwell RL (1979) Cavity occupation and defensive behaviour in the mantis shrimp Gonodactylus festae: evidence for chemically mediated individual recognition. Anim Behav 27:194–201

Caldwell RL (1982) Interspecific chemically mediated recognition in two competing stomatopods. Mar Behav Physiol 8:189–197

Caldwell RL (1985) A test of individual recognition in the mantis shrimp *Gonodactylus festae*. Anim Behav 33:101–106

Caldwell RL (1986) Withholding information on sexual condition as a competitive mechanism. In: Drickamer LC (ed) Behavioral ecology and population biology. Privat, Toulouse, pp 83–88

Caldwell RL (1987) Assessment strategies in stomatopods. Bull Mar Sci 41:135–150

Caldwell RL (1991) Variation in reproductive behavior in stomatopod crustacean. In: Bauer RT, Martin JW (eds) Crustacean sexual biology. Columbia University Press, New York, pp 67–90

Caldwell RL (1992) Recognition, signaling, and reduced aggression between former mates in a stomatopod. Anim Behav 44:11–19

Caldwell RL, Dingle H (1976) Stomatopods. Sci Am 234:80–89

Cate HS, Derby CD (2001) Morphology and distribution of setae on the antennules of the Caribbean spiny lobster *Panulirus argus* reveal new types of bimodal chemo-mechanosensilla. Cell Tissue Res 304:349–454

Cheroske AG, Cronin TW, Durham MF, Caldwell RL (2009) Adaptive signaling behavior in stomatopods under varying light conditions. Mar Freshw Behav Physiol 42:219–232

Christy JH, Salmon M (1991) Comparative studies of reproductive behavior in mantis shrimps and fiddler crabs. Am Zool 31:329–337

Derby CD, Fortier JK, Harrison PJH, Cate HS (2003) The peripheral and central antennular pathway of the Caribbean stomatopod crustacean *Neogonodactylus oerstedii*. Arthropod Struct Dev 32:175–188

Detto T, Backwell PRY (2009) Social monogamy in a fiddler crab *Uca capricornus*. J Crust Biol 29:283–289

Dingle H (1983) Strategies of agonistic behavior in Crustacea. In: Rebach S, Dunham DW (eds) Studies in adaptation: the behavior of higher crustacea. New York, Wiley, pp 85–111

Duffy JE (2007) The evolution of eusociality in sponge-dwelling shrimp. In: Duffy JE, Thiel M (eds) Evolutionary ecology of social and sexual systems: crustaceans as model organisms. Oxford University Press, New York, pp 387–409

Gherardi F, Tricarico E (2007) Can hermit crabs recognize social partners by odors? And why? Mar Fresh Behav Physiol 40:201–212

Gleeson RA (1982) Morphological and behavioral identification of the sensory structures mediating pheromone reception in the blue crab, *Callinectes sapidus*. Biol Bull 163:162–171

Gleeson RA, Carr WES, Trapido-Rosenthal HG (1993) Morphological characteristics facilitating stimulus access and removal in the olfactory organ of the spiny lobster. Panulirus argus: insight from the design. Chem Senses 18:67–75

Hallberg E, Johansson KUI, Elofsson R (1992) The aesthetasc concept: structural variations of putative olfactory receptor cell complexes in Crustacea. Micros Res Tech 22:325–335

Horner AJ, Weissburg MJ, Derby CD (2004) Dual antennular chemosensory pathways can mediate orientation by Caribbean spiny lobsters in naturalistic flow conditions. J Exp Biol 207:3785–3796

Johnson ME, Atema J (2005) The olfactory pathway for individual recognition in the American lobster *Homarus americanus*. J Exp Biol 208:2865–2872

Katija K, Dabiri JO (2008) In situ field measurements of aquatic animal-fluid interactions using a self-contained underwater velocimetry apparatus (SCUVA). Limnol Oceanogr Methods 6:162–171

Marshall J, Cronin TW, Kleinlogel S (2007) Stomatopod eye structure and function: a review. Arthropod Struct Dev 36:420–448

Mead K, Koehl MAR (2000) Particle image velocimetry measurements of fluid flow through a model array of mantis shrimp chemosensory sensilla. J Exp Biol 203:3795–3808

Mead KS, Weatherby TM (2002) Morphology of mantis shrimp chemosensory sensilla facilitates fluid sampling. Invert Biol 121:148–157

Mead K, Koehl MAR, O'Donnell MJ (1999) Mantis shrimp sniffing: the scaling of chemosensory sensilla and flicking behavior with body size. J Exp Mar Biol Ecol 241:235–261

Montgomery EL, Caldwell RL (1984) Aggressive brood defense by females in the stomatopod *Gonodactylus bredini*. Behav Ecol Sociobiol 14:247–251

Patek SN, Caldwell RL (2005) Extreme impact and cavitation forces of a biological hammer: strike forces of the peacock mantis shrimp *Odontodactylus scyllarus*. J Exp Biol 208:3655–3664

Rufino MM, Jones DA (2001) Binary individual recognition in *Lysmata debelius*. J Crust Biol 21:388–392

Stacey MT, Mead KS, Koehl MAR (2003) Molecular capture by olfactory antennules: mantis shrimp. J Math Biol 44:1–30

Wickler W (1973) Biology of *Hymenocera picta* Dana. Micronesica 9:225–230

Wickler W, Seibt U (1981) Monogamy in crustacean and man. Z Tierpsychol 57:215–234

Wilson JR, Kuehn RE, Beach FA (1963) Modification in the sexual behavior of male rats produced by changing the stimulus female. J Comp Physiol Psychol 56:636–644

Chapter 12
Chemical Communication in Lobsters

Juan Aggio and Charles D. Derby

Abstract Lobsters are fascinating animals that use chemicals as messages regarding their sexual status, their standing in a social hierarchy, and whether they affiliate with or avoid conspecifics. This, plus their economic importance, makes them important models for the study of intraspecific chemical communication. Our chapter is an overview of these processes, including the types of interactions between lobsters influenced by chemicals, how those interactions are affected by chemicals, and how these chemicals are detected. Since "lobster" refers to a common body plan rather than a taxonomic group and thus includes animals of differing phylogenetic relatedness and lifestyles – most notably clawed lobsters, spiny lobsters, and slipper lobsters, their use of chemicals in intraspecific interactions is diverse. Whenever possible, we compare the different groups of lobsters, though the amount of data available for relevant behaviors varies with the lifestyle of lobsters. Clawed lobsters use urinary chemicals processed by the olfactory pathway to identify previous opponents and maintain a stable social order, which is important because only the most dominant males will mate. After a hierarchy has been established by fighting, subsequent rematches are shorter and less violent, with urinary chemicals playing a key role in this process. Mate choice and mating behavior are also mediated by urinary olfactory cues. These behaviors are disrupted when one of the animals either has a compromised olfactory sense or is not allowed to release urine. Although there is less available data, the picture seems similar in spiny lobsters, with females using urinary chemicals from males as one of the cues in mate selection. Both spiny and slipper lobsters form dominance hierarchies, but little is known about how they are influenced by chemical signals. Conversely, spiny lobsters have been extensively studied regarding the mechanisms of aggregation and avoidance. Aggregation is mediated by urine-borne chemicals and avoidance is mediated by blood-borne chemicals, both processed by the olfactory system. Molecular identification of these compounds will be critical in allowing researchers to study the neural processing of intraspecific chemicals.

J. Aggio (✉)
Neuroscience Institute, Georgia State University,
Atlanta, GA 30302, USA
e-mail: jaggio@gsu.edu

T. Breithaupt and M. Thiel (eds.), *Chemical Communication in Crustaceans,*
DOI 10.1007/978-0-387-77101-4_12, © Springer Science+Business Media, LLC 2011

12.1 Introduction

12.1.1 Why Study Lobsters?

Because of their abundance, size, tastiness, and accessibility, lobsters are appreciated by all who enjoy a good seafood meal, are important in the fisheries of many countries, and are well known by most people throughout the world. Lobsters are also well known to scientists because they have been frequently used as models for the study of, among other things, the chemical senses, including the role of chemicals in mediating intraspecific behavior. Much is known about how lobsters use their chemical senses to find or avoid each other, recognize individuals, mate, and battle. Unfortunately, the chemical identities of most of the compounds driving these behaviors are unknown, which has limited the types of studies that can be done and our current understanding of lobsters' chemical ecology. Nevertheless, chemical communication among lobsters is a fascinating topic that is being investigated in many laboratories worldwide.

In this chapter, we review the ways whereby lobsters chemically communicate with each other and the contexts in which they do it. We make a distinction between different types of chemicals. Based on the terminology of Wyatt (Chap. 2), we use "semiochemicals" as chemicals involved in animal interactions and "pheromones" as a subset of semiochemicals used in intraspecific contexts. We also use "cues" as chemicals that benefit the receiver and not necessarily the sender, with a prime example being alarm cues released in the blood of injured conspecifics.

12.1.2 Is There Such a Thing as a Lobster?

The word "lobster" evokes in most people a precise image: a large, bottom-living, long-tailed marine crustacean. Depending on the geographical location, that image might include big claws. However, to those who study crustacean systematics, lobsters do not exist as a single taxonomic group. Lobster is a polyphyletic group and usually refers to three disparate clades of crustaceans – clawed lobsters, spiny lobsters, and slipper lobsters – although the name lobster is even applied to a group of hermit crabs, the squat lobsters. As seen in Fig. 12.1 (Dixon et al. 2003), clawed lobsters are members of the Homarida and are in reality closely related to crayfish, while spiny lobsters and slipper lobsters are members of the Achelata and are closer to hermit crabs and true crabs. In fact, a common ancestor to both clawed and clawless lobsters is also a common ancestor of, among others, crayfish, hermit crabs, and true crabs.

In spite of this, and because of historical reasons, even in academic publications such as this volume, the word "lobster" is widely used, and we will not be the exception. But be warned: we will use "clawed lobster," "spiny lobster," and

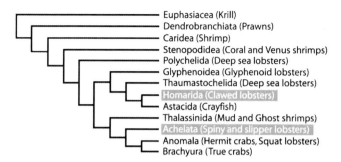

Euphasiacea (Krill)
Dendrobranchiata (Prawns)
Caridea (Shrimp)
Stenopodidea (Coral and Venus shrimps)
Polychelida (Deep sea lobsters)
Glyphenoidea (Glyphenoid lobsters)
Thaumastochelida (Deep sea lobsters)
Homarida (Clawed lobsters)
Astacida (Crayfish)
Thalassinida (Mud and Ghost shrimps)
Achelata (Spiny and slipper lobsters)
Anomala (Hermit crabs, Squat lobsters)
Brachyura (True crabs)

Fig. 12.1 Phylogeny of decapod crustaceans, showing that lobster is not a monophyletic grouping, but rather a body form. The clades representing the lobsters that we discuss in this chapter – clawed, spiny and slipper lobsters – are *shaded* in this figure. Clawed lobsters are more closely related to freshwater crayfishes than spiny and slipper lobsters, which are more closely related to true and hermit crabs, the latter in a group (Anomala) that includes animals also referred to as lobsters. Reproduced with modifications from Dixon et al. (2003). Copyright 2003 by Koninklijke Brill NV, Leiden, The Netherlands

"slipper lobster" when we want to refer specifically to one type, and we will use lobster when we make no distinction.

Significant to our considerations is that the different lobster types vary in their lifestyles, and this has enormous consequences on the types of information they exchange. This results in an unbalanced amount of information about different species, with most of what is known about agonistic and sexual semiochemicals coming from studies of clawed lobsters, and most of what is known about aggregation and sheltering semiochemicals from studies of spiny lobsters. Nevertheless, we will attempt to compare these two groups wherever possible.

12.2   Emission and Reception of Semiochemicals

Urine is often the source of semiochemicals. This is true for sheltering and dominance cues in the spiny lobster *Panulirus argus* (Horner et al. 2006), dominance cues and sex pheromones in the clawed lobster *Homarus americanus* (Atema and Voigt 1995; Bushmann and Atema 1997; Breithaupt and Atema 2000), dominance cues in the clawed lobster *Nephrops norvegicus* (Katoh et al. 2008), and sex pheromones in the spiny lobster *Jasus edwardsii* (Raethke et al. 2004). This seems to be a very general principle, at least within the decapods, with urine as a source of sex pheromones in blue crabs and other true crabs (Gleeson 1980; Kamio 2009) (Hardege and Terschak, Chap. 19; Kamio and Derby, Chap. 20) and crayfish (Breithaupt, Chap. 13). In some publications, the experimental design does not allow for precise identification of the site of origin of the chemical cues, but urine cannot be ruled out as a source. Other sources of semiochemicals are possible. For example, female sex pheromones of freshwater prawns are released from the sternal gland (Kamiguchi 1972a, 1972b), and alarm cues are found in the hemolymph of spiny lobsters (Shabani et al. 2008), crabs (Ferner et al. 2005), and crayfish (Hazlett 1994).

Receptors for semiochemicals of lobsters and many other decapod crustaceans are typically restricted to a very limited set of sensors. These are olfactory sensory neurons contained in specialized setae called aesthetascs, localized in the lateral flagellum of the first antenna or antennule (Fig. 12.2). Ablation experiments show the necessity of aesthetascs for the response of lobsters to different types of semiochemicals (Johnson and Atema 2005; Horner et al. 2008; Shabani et al. 2008). The chemosensory neurons innervating the aesthetascs project their axons exclusively to the olfactory lobes, and thus are considered olfactory, while the other setae contain both chemo- and mechanoreceptors that project to the lateral and medial antennular neuropils (Schmidt and Ache 1996a, 1996b) (Schmidt and Mellon, Chap. 7). Many studies showed that receptor sites for semiochemicals are located in the first antenna without showing them to be in aesthetascs (e.g., Raethke et al. 2004; Skog 2009). Detectors of semiochemicals of crayfish have been

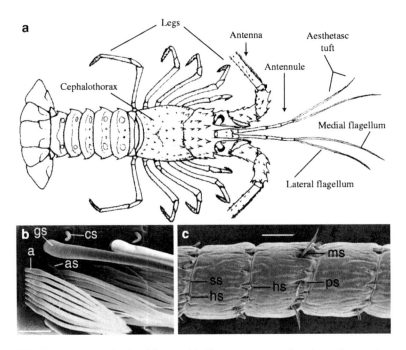

Fig. 12.2 Chemosensors of spiny lobsters. (**a**) Chemosensors are found on all appendages and body surfaces. (**b, c**) The antennules are covered with different types of chemosensilla, including several in the aesthetasc tuft: aesthetasc sensilla (a), guard sensilla (gs), companion sensilla (cs), asymmetric sensilla (as); and those outside the tuft: setuled sensilla (ss), hooded sensilla (hs), plumose (ps), and simple sensilla (ms). The aesthetascs however are the only sensilla that house exclusively chemoreceptors (the others also contain mechanoreceptors). The aesthetascs contain olfactory receptor neurons that mediate responses to intraspecific signals and cues. Reproduced from Steullet et al. (2001) with permission from The Journal of Experimental Biology

located on the first antennae (Tierney et al. 1984; Dunham and Oh 1992) and chelae (Belanger and Moore 2006, 2009).

12.3 Sexual Signals

In sexual communication, the different lifestyles of clawed, spiny, and slipper lobsters have profound effects on the nature of the semiochemicals and the amount of information available to us. Relative to one another, clawed lobsters tend to live solitarily and to not aggregate except when mating, while spiny lobsters tend to have very active social lives year round. Perhaps because of this, the amount of information on sexual communication in spiny lobsters tends to be much less than in clawed ones. Much of our understanding about the role of chemical communication in the sexual behavior of lobsters originated in the laboratory of Jelle Atema and pertains to the American lobster, *H. americanus*. We also have some knowledge of mating signals in the closely related European lobster, *Homarus gammarus*, and also in a spiny lobster, *J. edwardsii*. No information is available on slipper lobsters. Because clawed lobsters are solitary and aggressive towards one another, courtship must include a means to diminish aggression. For this reason, we first describe events leading to and following copulation in clawed lobsters, focusing on what the majority of the available literature supports. Then we briefly discuss data that contradict these results, and lastly we discuss what is currently known in spiny lobsters.

As a broad generalization, mating in *H. americanus* takes place as follows: a female searches for a male, preferably dominant, gains access to his shelter, molts, mates, and eventually leaves his shelter, so that the male may mate again with another female (Cowan and Atema 1990; Atema and Voigt 1995; Atema and Steinbach 2007). Intermolt females also mate, and it is believed that this is caused by lack of enough sperm to fertilize eggs, a situation that may be due to less than ideal mating (Gosselin et al. 2005; Waddy and Aiken 1990). Courtship is strongly influenced by urinary chemical signals, detected by antennular chemoreceptors. Bushmann and Atema (2000) showed that if the release of male urine is blocked, females tend to approach shelters less and, once there, they spend less time attempting to enter them. Because releasing urine from empty shelters does not restore these behaviors, Bushmann and Atema concluded that other sources of chemical information must exist. The response of the male to the female's attempt to enter is highly dependent on his ability to detect the female's urine. If females are prevented from releasing urine, then males still allow females to enter the shelter, but in the face of male aggressive behavior, the number of matings is greatly reduced (Bushmann and Atema 1997).

Two studies on clawed lobsters are inconsistent with the above scenario and with each other (Cowan 1991; Skog 2009). Cowan (1991) reported that the antennules of females, but not males are required for normal mating behavior in *H. americanus*, while Skog (2009) found that male antennules are necessary for normal mating in

H. gammarus. Differences in the findings of these studies might be due to the use of different species, but further studies are needed to resolve this question.

Another contradicting point of view was advanced by Snyder et al. (1993), based on studies of *H. americanus*. These authors reported that blocking urine release had no effect on intra- or intersexual behavior. Once again, although it is possible that there are additional, redundant sources of pheromones such as have been shown in crabs (Bushmann 1999), an alternative explanation is a methodological one – the methods of blocking urine release were different in these studies and the method of Snyder et al. (1993) may not have been completely successful.

In spiny lobsters, which are gregarious, most of the available data do not deal strictly with mating but with aggregation, and we treat this subject elsewhere in this chapter. However, Raethke et al. (2004) addressed the importance of chemical communication in mating of *J. edwardsii* by investigating the role of urine in mate selection by females. Large females normally choose large males, probably to ensure an adequate supply of sperm. In a clever experiment, Raethke and collaborators "reversed" the urines of a small and a large male so that the small male's urine was released close to the big one's nephropores (the site of release of urine) and vice versa. This resulted in the females choosing the large and small males with equal probability, indicating that although urine alone is not sufficient to attract a female, it plays an important role. When they investigated the effect of ablation of the antennules of either males or females on courtship and mating, they found no significant effects, but a tendency to a delayed mating in both experimental conditions, with a slightly greater effect when the ablation was performed in males.

In addition, mating behavior in both spiny and slipper lobsters seems to be simpler and faster than in clawed lobsters, a fact that further hinders our ability to collect data (Lipcius et al. 1983). In their 11-month study, Barshaw and Spanier (1994) failed to observe mating between slipper lobsters.

Overall, results from clawed lobsters paint a picture that urinary chemical cues released by males and detected by females are necessary for the initial location of males by females. Then, when the animals are near each other, females release urinary chemicals that are detected by males, and this drives subsequent behavior. In spiny lobsters, females have been shown to approach tethered males (Raethke et al. 2004), and males have been observed to search for females in the field (Lipcius et al. 1983). More data are needed to establish the precise sequence of events leading to mating in spiny lobsters.

12.4   Individual Recognition and Social Status

The structures of groups of clawed, spiny, and slipper lobsters in the field are, at first sight, very different. However, there are similarities in their use of chemical communication in social interactions: clawed and spiny lobsters establish dominance hierarchies that are maintained, at least partially, via urinary chemical cues.

In clawed lobsters, the hierarchy is very clear and important because normally only the dominant male will mate (Karnofsky and Price 1989). In the wild and in semi-naturalistic aquaria, shelter sharing can occur in juvenile lobsters (Cowan et al. 2001), and networks of relationships can be established and can persist relatively unchanged over time (Karnofsky and Price 1989). The structure of the hierarchy is based on the outcomes of individual fights (Karavanich and Atema 1998a, 1998b). When the size disparity between two animals is large, fights tend to be short, with the smaller animal retreating and assuming the subordinate role (Karnofsky and Price 1989). But when animals are of similar size, fights can be so long and violent that one or both animals may be injured (Karnofsky and Price 1989; Huber and Kravitz 1995). Figure 12.3a–f depicts a battle between two juvenile American lobsters illustrating several of the behavioral patterns displayed. Although the first fight among two individuals may escalate to dangerous levels of violence, subsequent ones are much shorter and milder. For example, Karavanich and Atema (1998a; 1998b) showed that the average duration of two consecutive fights between the same animals drops from 450 to 50 s; concurrently, the behavior of the loser is aggressive on the first fight, but exclusively submissive on the second (Fig. 12.3g). This, of course, explains the short duration of the match.

Thus, when two clawed lobsters that have previously battled meet again, each one knows its place. The question then becomes: How is this accomplished? A series of elegant experiments by Karavanich and Atema (1998a, 1998b) showed that animals are capable of individual recognition that is mediated by the olfactory system detecting compounds in urine. Karavanich and Atema first asked if animals can recognize individuals or an overall dominance status. These two hypotheses have very different predictions, which the authors tested. If animals recognize individuals, then a second fight will only be short and non-violent if the two animals have met before and dominance has been established. If the second hypothesis is true, a "loser" will always retreat from a "winner," even if they have never met before. Karavanich and Atema created "winners" and "losers" by allowing pairs to fight once and then either repeated the same fight (as a control) or paired winners and losers that had not fought each other. As expected, control second fights were short and non-violent. However, when the "loser" had never been exposed to the "winner" in the pair, it did not retreat and actually won in 3 of 10 fights (Karavanich and Atema 1998a). These results support the hypothesis that American lobsters can recognize individuals.

To elucidate the mechanism underlying this recognition, Atema and colleagues (Karavanich and Atema 1998b; Johnson and Atema 2005) staged fights to establish dominance and then re-staged them after manipulating the release of urine or antennular aesthetasc sensilla. They found that ablating the antennular lateral flagella eliminated the dominance that was established in the first fight, and that a similar result was obtained by blocking urine release. Thus, the chemical signals used for individual recognition are in the urine released during the fight and are detected by the olfactory pathway.

The social hierarchy of spiny lobsters in naturalistic conditions is more difficult to observe than in clawed lobsters because many species of spiny lobsters are far

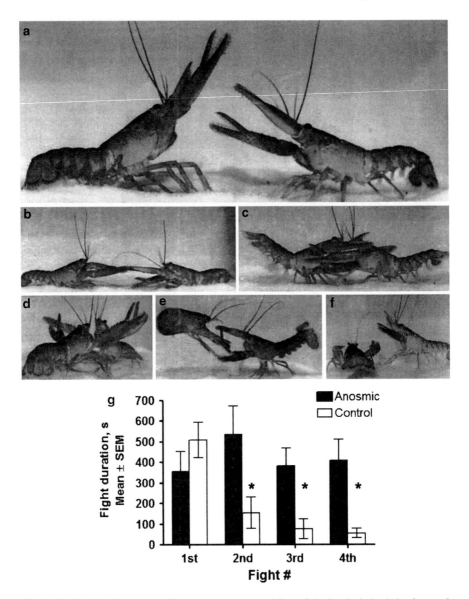

Fig. 12.3 *Top*: Fighting in juvenile American lobsters. After a fight that includes behaviors such as meral spread (**a**), wrestling (**b–d**), and tailflip by one animal (*left* in **e**), the victor (animal on *right* in **e** and **f**) assumes a dominant posture while the vanquished assumes a submissive posture. From Huber and Kravitz (1995) and used with permission. *Bottom*: Successive fights between lobsters become progressively shorter, but only if animals retain their olfactory capabilities. From Karavanich and Atema (1998b). Copyright 1998 by Koninklijke Brill NV, Leiden, The Netherlands

more gregarious and usually share communal shelters. Individual spiny lobsters will compete for food and shelter, displaying aggressive and submissive behaviors (Fielder 1965; Berrill 1975, 1976). When shelters are a limiting factor, they are occupied by dominant individuals that will, if necessary, evict lower-ranked ones (Fielder 1965). In *P. argus*, a highly gregarious species, juveniles introduced into a shelter-containing aquarium will at first exhibit very little aggressive behavior, and consequently shelter density is high; with time, aggressive behavior increases and consequently there is less shelter sharing. Berrill (1975) speculates that stress due to novel surroundings reduces aggression, but that it subsequently resumes as individuals become accustomed and shelter sharing becomes less likely. A similar situation is found in *Panulirus longipes* (Berrill 1976). The solitary species of spiny lobsters are probably more aggressive, but they have received far less attention than the gregarious species (Childress 2007).

Unfortunately, very little is known about chemical communication during social behavior of spiny lobsters, other than aggregation cues (see below). Shabani et al. (2009) showed that in *P. argus* as in clawed lobsters, urine is important in reducing the levels of aggression. Urine release is context- and individual-specific; dominant animals increase urine release when engaged in interactions, while subordinate animals do not (see Fig. 12.4 for an artist's depiction and Fig. 12.5 for original data). Also in *P. argus*, as in clawed lobsters, blocking urine release results in an increase in the number and duration of agonistic encounters. This effect is reversed by reintroducing the urine of one of the combatants into the aquarium. Finally, the aesthetascs are responsible for the responses to urine: solitary spiny lobsters with their olfactory system intact respond to conspecific urine with avoidance behaviors, but when their aesthetascs are ablated they respond with appetitive behaviors, no

Fig. 12.4 Artist's drawing of a pair of Caribbean spiny lobsters engaged in agonistic interaction. The dominant animal (*right*) releases more urine, which in this in drawing is depicted as a white plume, than the subordinate animal (*left*), whose urine is indicated as a smaller black plume. Urine from dominants reduces the level of aggression in subordinates. Drawing by Jorge A. Varela Ramos

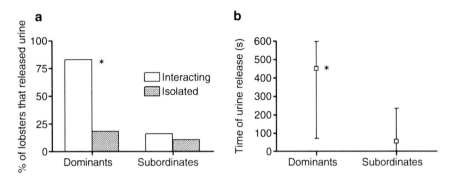

Fig. 12.5 Social interactions and urine release in Caribbean spiny lobster. Dominant animals are more likely to release urine than subordinates (**a**), and dominants release more urine than subordinates (**b**). Adapted from Shabani et al. (2009) with permission from The Journal of Experimental Biology

doubt due to the many other compounds found in urine. It should be noted, however, that spiny lobster urine can also attract conspecifics (Horner et al. 2006), and differences in results may be due to an interaction between the chemical stimulus and the experimental conditions.

Although much less is known about slipper lobsters, one species (the Mediterranean slipper lobster, *Scyllarides latus*) is known to establish and maintain an almost linear dominance hierarchy in semi-naturalistic conditions (Barshaw and Spanier 1994). Interestingly, females were reported to interact more often than males, which were not observed to engage in intrasexual agonistic encounters (Barshaw and Spanier 1994).

Thus, clawed, spiny, and slipper lobsters exhibit similar behaviors; they all engage in agonistic encounters that result in the establishment of a dominance hierarchy, and urine is involved in maintaining it without the need for further, possibly dangerous, combat. Hierarchy maintenance through individual recognition is not restricted to lobsters, as it is also found in mantis shrimp (Mead and Caldwell, Chap. 11) and hermit crabs (Gherardi and Tricarico, Chap. 15).

12.5   Aggregation and Avoidance

As seen above, clawed, spiny, and slipper lobsters have very different lifestyles, and this influences how chemicals are used in their behavioral interactions. Clawed lobsters lead mostly solitary lives and do not cohabitate except when mating, a process controlled by sex pheromones (Atema and Cobb 1980). Some spiny lobsters and slipper lobsters, on the other hand, share communal shelters (Berry 1971; Berrill 1975; Cobb 1981; Zimmer-Faust et al. 1985; Spanier and Almog-Shtayer 1992) (Fig. 12.6a), a behavior that improves survival against predation (Eggleston et al. 1990; Mintz et al. 1994; Weiss et al. 2008). These aggregations are

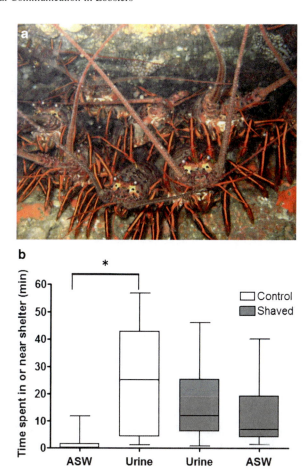

Fig. 12.6 Aggregation and shelter sharing in spiny lobsters. (**a**) California spiny lobsters in a shelter, using their second antennae to protect the entrance to the shelter from predators. From C. White and used with permission. (**b**) Caribbean spiny lobsters are attracted to shelters emanating conspecific urine but only if their aesthetascs are intact. Adapted from Horner et al. (2008) with permission from Springer

chemically mediated, as demonstrated in field and laboratory experiments in which individual animals are given a choice between a control shelter and one from which conspecific chemicals emanate. Conspecific chemicals are attractive, and shelters that emanate them are consistently chosen by higher numbers of animals than predicted by chance alone (Zimmer-Faust et al. 1985; Zimmer-Faust and Spanier 1987, Nevitt et al. 2000).

Although some spiny lobsters and slipper lobsters aggregate, many factors influence this behavior. For example, the introduction of a predator causes *P. argus* individuals to change the preferred shelter size (Eggleston and Lipcius 1992). When there is no predation risk, small spiny lobsters choose either small- or medium-sized shelters whereas medium-sized spiny lobsters prefer medium or big shelters. Upon introduction of a predator, all spiny lobsters prefer medium-sized shelters, presumably because they provide more protection for medium-sized spiny lobsters, while these in turn protect smaller ones (Eggleston and Lipcius 1992). Small (<15 mm carapace length) individuals tend to be solitary, but after they reach 15 mm they begin to aggregate (Marx and Herrnkind 1985; Childress and Herrnkind 1996; Ratchford and Eggleston 1998; Childress 2007). Interestingly, even small individuals release a chemical attractant, but in such small quantities that a relatively high number of animals may be needed to produce enough to have behavioral effects in naturalistic conditions (Ratchford and Eggleston 1998). In *J. edwardsii*, large males do not aggregate year round, but disperse during mating season and try to attract females with which they cohabit exclusively in order to mate (MacDiarmid 1994). This indicates an interaction between aggregation and sexual cues.

Another factor influencing aggregation in spiny lobsters is time of day, because these nocturnal animals forage alone and return to their shelter at dawn (Zimmer-Faust et al. 1985; Weiss et al. 2008). Ratchford and Eggleston (2000) demonstrated that this change is due to the discontinuous production of the chemical cue responsible for aggregation. Having the donor and experimental animals at different subjective times, they were able to prove that the aggregation cue is only released at dawn. The chemical cues mediating aggregation of spiny lobsters are in the urine and detected by aesthetasc chemoreceptors (Horner et al. 2006, 2008) (Fig. 12.6b). Nothing is known about whether chemical cues mediate other types of aggregation, such as the long migratory queues observed in the field.

Finally, once again, studies on slipper lobsters are rare. Barshaw and Spanier (1994) report that *S. latus* aggregates in the daytime, but there is no information regarding what, if any, chemicals influence this behavior. As with spiny lobsters, gregariousness does not seem to be a universal characteristic of this group. Morin and MacDonald (1984) report that in Hawaii the congeneric species *Scyllarides haanii* and *Scyllarides squammosus* have different lifestyles, the former being solitary and the latter gregarious.

For spiny lobsters, as with other animals, avoiding danger and possibly injury and death by staying away from predators is an adaptive strategy (Hay, Chap. 3; Hazlett, Chap. 18). Each stage of an event of predation will result in the release of chemicals that contain information about the event – chemicals from the predator, disturbance or injury-related chemicals from the prey, and finally chemicals released from food during the act of consumption (Wisenden 2000).

Many species of animals, including lobsters, detect and react appropriately to chemicals from predators. For example, Wahle (1992) showed that American lobsters exposed to chemicals from a predatory fish spend significantly more time in their shelters than control animals, and Horner et al. (2006) showed that *P. argus* tends to avoid the scent of a predatory octopus. Hermit crabs are even able to discriminate between predatory and herbivorous crabs by their scent alone (Rosen et al. 2009), although we know of no comparable lobster-specific data.

Disturbance and injury-related chemicals can also influence lobster behavior. Disturbance cues are released by animals that are harassed or disturbed, while the injury-related cues passively leak into the environment due to the damage produced by the predator. Avoidance of injured or freshly dead conspecific has been shown for *Panulirus interruptus* (Zimmer-Faust et al. 1985), *Panulirus cygnus* (Hancock 1974), and *P. argus* (Parsons and Eggleston 2005; Briones-Fourzán et al. 2006). Interestingly, *Panulirus guttatus* does not seem to avoid the scent of injured or freshly killed conspecifics (Briones-Fourzán et al. 2006), providing us with an example of different lifestyles resulting in different modes of chemical communication. Briones-Fourzán et al. (2008) performed a series of experiments specifically designed to compare *P. guttatus* and *P. argus*, which are syntopic species. They confirm that *P. argus* avoids shelters emanating the scent of a freshly killed conspecific and show that they also avoid shelters emanating freshly killed *P. guttatus* scent. In contrast, *P. guttatus* is indifferent to the smell of freshly killed individuals of either species; it selects a den that emanates its scent with the same probability as a control. The authors interpret these results in view of the two species' different lifestyles. They propose that a *P. argus* individual returning to its reef after foraging will gain an advantage by avoiding alarm cues. The fact that the alarm cues may not originate from a member of its own species is not relevant in this scenario because the predator could very well attack the returning spiny lobster too. In the case of *P. guttatus*, there is less advantage in avoiding alarm cues because *P. guttatus* is solitary and only forages very close to its den entrance (Briones-Fourzán et al. 2008). These closely related species also share many predator defense mechanisms, but vary widely in the performance of them, adding support to the idea that their different lifestyles result in different ways of dealing with predation risk (Briones-Fourzán et al. 2006).

Briones-Fourzán (2009) reanalyzed data discussed above by differentiating between lethally and non-lethally injured (one or more autotomized limbs) animals. This analysis shows that *P. argus* only avoids the scent emanating from lethally injured conspecifics, a result that agrees with results from Parsons and Eggleston (2005). For *P. guttatus*, Briones-Fourzán further distinguished between experiments carried out in the spring, which is the peak mating season, and late summer. This new analysis shows that in the spring, when these animals are most gregarious, both intact and slightly-injured animals are attractive to conspecifics.

Hemolymph is the source of alarm cues in *P. argus*; it causes conspecifics to spend more time inside their shelters and counters the effect of food-related chemical cues (Shabani et al. 2008). Alarm cues are detected by aesthetasc chemoreceptors. Spiny lobsters with ablated aesthetascs do not respond to hemolymph with an alarm response; indeed, spiny lobsters without an olfactory sense respond to hemolymph as if it were an appetitive cue and advance towards its source (Shabani et al. 2009). This is due to the fact that hemolymph contains many food-associated compounds that spiny lobsters detect through their non-olfactory antennular chemoreceptors (Steullet et al. 2001, 2002; Schmidt and Mellon, Chap. 7). Interestingly, hemolymph from *P. interruptus* or blue crabs (*Callinectes sapidus*) did not produce the same responses in *P. argus*: they responded with mixed alarm and appetitive behaviors to hemolymph from the California spiny lobster *P. interruptus*, but only with appetitive behaviors to the hemolymph of blue crabs, indicating that phylogeny is more important than sympatry (Shabani et al. 2008). This result is at odds with the one reported by Briones-Fourzán et al. (2008), an effect that might be due to two possible reasons. In the first place, the two studies did not measure the same variable: Briones-Fourzán and her colleagues left the spiny lobsters overnight in a Y-maze and recorded the shelter choice the next morning, whereas Shabani and colleagues evaluated the immediate effect of introducing hemolymph into the aquarium that held a spiny lobster. Another and to us much more interesting explanation rests with the difference in the stimuli used: Shabani et al. (2008) used hemolymph and Briones-Fourzán et al. (2008) used dead animals. Another layer of complexity is added by the fact that injured conspecifics are aversive for a limited time period after injury (Ferner et al. 2005), and so in the case of overnight experiments using dead specimens as stimuli, the actual stimulus may be changing over time.

Our knowledge of avoidance behaviors in clawed lobsters is virtually nonexistent, mainly because there is very little aggregation to begin with. So once again, we find that the extent to which a particular behavior is exhibited (or studied!) depends on an animal's lifestyle. Gregarious spiny lobsters have been extensively used by researchers interested in the nature and mode of action of the chemicals mediating aggregation, while non-gregarious spiny lobsters and clawed lobsters have been largely ignored except in the special and altogether different case in which individuals gather for reproductive purposes. For the same reasons, avoidance and its mechanisms have been studied most extensively in gregarious spiny lobsters. Fortunately, there is an exception to this rule, and Briones-Fourzán and her collaborators (2006) explicitly compared sympatric gregarious and non-gregarious spiny lobsters and found that only the former avoided dead conspecifics.

12.6   Recognition of Diseased Animals

An interesting special case of conspecific avoidance is that *P. argus* is able to detect and avoid conspecifics infected with a lethal virus (Behringer et al. 2006, 2008). When given the choice, these highly social animals avoid shelters occupied by diseased individuals and prefer empty ones, the opposite of what happens when the shelter is occupied by a healthy conspecific. In addition, diseased individuals are avoided before they become infectious, a necessary condition if the avoidance is to reduce the spread of the disease among the population. Although the experimental design does not allow determination if the animals base their decisions on chemical cues, it would be very interesting to evaluate the possibility that they are able to detect the disease from a "safe" distance using their chemical senses, and to elucidate if those signals originate from the host or the virus.

12.7 Summary and Conclusions

We have attempted to give a brief overview of intraspecific chemical communication between lobsters. Lobsters communicate through chemical compounds that are predominantly but not exclusively located in their urine. Other lobsters detect these compounds with olfactory sensory neurons, located in the aesthetasc sensilla on their antennules. As far as is known, this is true for mating in spiny and clawed lobsters, aggregation in spiny lobsters, and social hierarchy establishment and maintenance in spiny and clawed lobsters. Alarm cues, which are in the hemolymph and leak from an injured animal, have some degree of species specificity. These chemical cues can also be detected by the olfactory pathway. One very obvious gap in our knowledge and stumbling block to our progress is that we do not know the chemical identities of these semiochemicals.

12.8 Future Directions

Our review shows that the greatest future contributions to our understanding of the comparative study of chemical communication in lobsters would come from focusing on two areas. The first is to identify the molecules producing these behaviors. With these identified compounds, we will be able to more accurately pinpoint their sources in living animals and perform carefully controlled behavioral experiments. Additionally, this knowledge will allow us to perform physiological experiments to shed light on where and how these important signals are processed. The second area of emphasis is to study the role of chemical communication in mediating the behavior of species that have received less attention. This includes mating, individual recognition, and recognition and avoidance of diseased individuals in spiny lobsters, attraction in clawed lobsters, and just about any aspect of chemical communication in slipper lobsters.

Acknowledgments Funding was provided by NSF grants IBN-0614685.

References

Atema J, Cobb JS (1980) Social behavior. In: Cobb JS, Phillips BR (eds) The biology and management of lobsters: physiology and behavior. Academic, New York, pp 409–450
Atema J, Steinbach MA (2007) Chemical communication and social behavior in the lobster *Homarus americanus* and other decapod Crustacea. In: Duffy JE, Thiel M (eds) Evolutionary ecology of social and sexual systems: crustaceans as model organisms. Oxford University Press, Oxford, pp 115–144
Atema J, Voigt R (1995) Behavior and sensory biology. In: Factor JR (ed) Biology of the lobster *Homarus americanus*. Academic, New York, pp 313–348

Barshaw D, Spanier E (1994) The undiscovered lobster: a first look at the social behaviour of the Mediterranean slipper lobster *Scyllarides latus* (Decapoda, Scyllaridae). Crustaceana 67:187–197

Behringer DC, Butler MJ, Shields JD (2006) Avoidance of disease by social lobsters. Nature 441:421

Behringer DC, Butler MJ, Shields JD (2008) Ecological and physiological effects of Pav1 infection on the Caribbean spiny lobster (*Panulirus argus* Latreille). J Exp Mar Biol Ecol 359:26–33

Belanger RM, Moore PA (2006) The use of the major chelae by reproductive male crayfish (*Orconectes rusticus*) for discrimination of female odours. Behaviour 143:713–731

Belanger RM, Moore PA (2009) The role of the major chelae in the localization and sampling of male odours by male crayfish, *Orconectes rusticus* (Girard, 1852). Crustaceana 82:653–668

Berrill M (1975) Gregarious behavior of juveniles of the spiny lobster, *Panulirus argus* (Crustacea; Decapoda). Bull Mar Sci 25:515–522

Berrill M (1976) Aggressive behaviour of post-puerulus larvae of the Western rock lobster *Panulirus longipes* (Milne-Edwards). Austr J Mar Freshwat Res 27:83–88

Berry PF (1971) The spiny lobsters (Palinuridae) of the east coast of southern Africa. Distribution and ecological notes. Oceanogr Res Inst (Durban) Invest Rep 21:1–16

Breithaupt T, Atema J (2000) The timing of chemical signaling with urine in dominance fights of male lobsters (*Homarus americanus*). Behav Ecol Sociobiol 49:67–78

Briones-Fourzán P (2009) Assessment of predation risk through conspecific alarm odors by spiny lobsters. Commun Integr Biol 2:302–304

Briones-Fourzán P, Pérez-Ortiz M, Lozano-Álvarez E (2006) Defense mechanisms and antipredator behavior in two sympatric species of spiny lobsters, *Panulirus argus* and *P. guttatus*. Mar Biol 149:227239

Briones-Fourzán P, Ramírez-Zaldívar E, Lozano-Álvarez E (2008) Influence of conspecific and heterospecific aggregation cues and alarm odors on shelter choice by syntopic spiny lobsters. Biol Bull 215:182–190

Bushmann PJ (1999) Concurrent signals and behavioral plasticity in blue crab (*Callinectes sapidus* Rathbun) courtship. Biol Bull 197:63–71

Bushmann PJ, Atema J (1997) Shelter sharing and chemical courtship signals in the lobster *Homarus americanus*. Can J Fish Aquat Sci 54:647–654

Bushmann PJ, Atema J (2000) Chemically mediated mate location and evaluation in the lobster, *Homarus americanus*. J Chem Ecol 26:883–899

Childress MJ (2007) Comparative sociobiology of spiny lobsters. In: Duffy JE, Thiel M (eds) Evolutionary ecology of social and sexual systems: crustaceans as model organisms. Oxford University Press, Oxford, pp 271–293

Childress MJ, Herrnkind WF (1996) The ontogeny of social behaviour among juvenile Caribbean spiny lobsters. Anim Behav 51:675–687

Cobb JS (1981) Behaviour of the Western Australian spiny lobster, *Panulirus cygnus* George, in the field and laboratory. Austr J Mar Freshwat Res 31:399–409

Cowan DF (1991) The role of olfaction in courtship behavior of the American lobster *Homarus americanus*. Biol Bull 181:402–407

Cowan DF, Atema J (1990) Moult staggering and serial monogamy in American lobsters, *Homarus americanus*. Anim Behav 39:1199–1206

Cowan DF, Solow AR, Beet A (2001) Patterns in abundance and growth of juvenile lobster, *Homarus americanus*. Mar Freshwat Res 52:1095–1102

Dixon CJ, Ahyong ST, Schram FR (2003) A new hypothesis of decapod phylogeny. Crustaceana 76:935–975

Dunham DW, Oh JW (1992) Chemical sex discrimination in the crayfish, *Procambarus clarkii*: role of antennules. J Chem Ecol 18:2363–2372

Eggleston DB, Lipcius RN (1992) Shelter selection by spiny lobster under variable predation risk, social conditions, and shelter size. Ecology 73:992–1011

Eggleston DB, Lipcius RN, Miller DL, Coba-Cetina L (1990) Shelter scaling regulates survival of juvenile Caribbean spiny lobster *Panulirus argus*. Mar Ecol Prog Ser 62:79–88

Ferner MC, Smee DL, Chang YP (2005) Cannibalistic crabs respond to the scent of injured conspecifics: danger or dinner? Mar Ecol Prog Ser 300:193–200

Fielder DR (1965) A dominance order for shelter in the spiny lobster *Jasus lalandei* (H. Milne-Edwards). Behaviour 24:236–245

Gleeson RA (1980) Pheromone communication in the reproductive behavior of the blue crab, *Callinectes sapidus*. Mar Behav Physiol 7:119–134

Gosselin T, Sainte-Marie B, Bernatchez L (2005) Geographic variation of multiple paternity in the American lobster, *Homarus americanus*. Molec Ecol 14:1517–1525

Hancock DA (1974) Attraction and avoidance in marine invertebrates – their possible role in developing an artificial bait. J Cons Int Explor Mer 35:328–331

Hazlett BA (1994) Alarm responses in the crayfish *Orconectes virilis* and *Orconectes propinquus*. J Chem Ecol 20:1525–1535

Horner AJ, Nickles SP, Weissburg MJ, Derby CD (2006) Source and specificity of chemical cues mediating shelter preference of Caribbean spiny lobsters (*Panulirus argus*). Biol Bull 211:128–139

Horner AJ, Weissburg MJ, Derby CD (2008) The olfactory pathway mediates sheltering behavior of Caribbean spiny lobsters, *Panulirus argus*, to conspecific urine signals. J Comp Physiol A 194:243–253

Huber R, Kravitz EA (1995) A quantitative analysis of agonistic behavior in juvenile American lobsters (*Homarus americanus* L.). Brain Behav Evol 46:72–83

Johnson ME, Atema J (2005) The olfactory pathway for individual recognition in the American lobster *Homarus americanus*. J Exp Biol 208:2865–2872

Kamiguchi Y (1972a) Mating behavior in the freshwater prawn, *Palaemon paucidens*. A study of the sex pheromone and its effect on males. J Fac Sci Hokkaido Univ Ser VI, Zool 18:347–355

Kamiguchi Y (1972b) A histological study of the "sternal gland" in the female freshwater prawn, *Palaemon paucidens*, a possible site of origin of the sex pheromone. J Fac Sci Hokkaido Univ Ser VI, Zool 18:356–365

Kamio M (2009) Towards identifying sex pheromones in blue crabs: using biomarker targeting within the context of evolutionary chemical ecology. Ann NY Acad Sci 1170:456–461

Karavanich C, Atema J (1998a) Individual recognition and memory in lobster dominance. Anim Behav 56:1553–1560

Karavanich C, Atema J (1998b) Olfactory recognition of urine signals in dominance fights between male lobster, *Homarus americanus*. Behaviour 135:719–730

Karnofsky EB, Price HJ (1989) Dominance, territoriality and mating in the lobster, *Homarus americanus*: a mesocosm study. Mar Behav Physiol 15:101–121

Katoh E, Johnson M, Breithaupt T (2008) Fighting behavior and the role of urinary signals in dominance assessment of Norway lobsters, *Nephrops norvegicus*. Behaviour 145:1447–1464

Lipcius RN, Edwards ML, Herrnkind WF, Waterman SA (1983) In situ mating behavior of the spiny lobster *Panulirus argus*. J Crust Biol 3:217–222

MacDiarmid AB (1994) Cohabitation in the spiny lobster *Jasus edwardsii* (Hutton, 1875). Crustaceana 66:341–355

Marx JM, Herrnkind WF (1985) Macroalgae (Rhodophyta: *Laurencia* spp.) as habitat for young juvenile spiny lobsters, *Panulirus argus*. Bull Mar Sci 36:423–431

Mintz JD, Lipcius RN, Eggleston DB, Seebo MS (1994) Survival of juvenile Caribbean spiny lobster: effects of shelter size, geographic location and conspecific abundance. Mar Ecol Prog Ser 112:255–266

Morin TD, MacDonald CD (1984) Occurrence of the slipper lobster *Scyllarides haanii* in the Hawaiian archipelago. Proc Biol Soc Wash 97:404–407

Nevitt G, Pentcheff ND, Lohmann KJ, Zimmer RK (2000) Den selection by the spiny lobster *Panulirus argus*: testing attraction to conspecific odors in the field. Mar Ecol Prog Ser 203:225–231

Parsons DM, Eggleston DB (2005) Indirect effects of recreational fishing on behavior of the spiny lobster *Panulirus argus*. Mar Ecol Prog Ser 303:235–244

Raethke N, MacDiarmid AB, Montgomery JC (2004) The role of olfaction during mating in the southern temperate spiny lobster *Jasus edwardsii*. Horm Behav 46:311–318

Ratchford SG, Eggleston DB (1998) Size- and scale-dependent chemical attraction contribute to an ontogenetic shift in sociality. Anim Behav 56:1027–1034

Ratchford SG, Eggleston DB (2000) Temporal shift in the presence of a chemical cue contributes to a diel shift in sociality. Anim Behav 59:793–799

Rosen E, Schwarz B, Palmer AR (2009) Smelling the difference: hermit crab responses to predatory and nonpredatory crabs. Anim Behav 78:691–695

Schmidt M, Ache BW (1996a) Processing of antennular input in the brain of the spiny lobster, *Panulirus argus*. I. Non-olfactory chemosensory and mechanosensory pathway of the lateral and median antennular neuropils. J Comp Physiol A 178:579–604

Schmidt M, Ache BW (1996b) Processing of antennular input in the brain of the spiny lobster, *Panulirus argus*. II. The olfactory pathway. J Comp Physiol A 178:605–628

Shabani S, Kamio M, Derby CD (2008) Spiny lobsters detect conspecific blood-borne alarm cues exclusively through olfactory sensilla. J Exp Biol 211:2600–2608

Shabani S, Kamio M, Derby CD (2009) Spiny lobsters use urine-borne olfactory signaling and physical aggressive behaviors to influence social status of conspecifics. J Exp Biol 212:2464–2474

Skog M (2009) Male but not female olfaction is crucial for intermolt mating in European lobsters (*Homarus gammarus* L.). Chem Senses 34:159–169

Snyder MJ, Ameyaw-Akumfi C, Chang ES (1993) Sex recognition and the role of urinary cues in the lobster, *Homarus americanus*. Mar Behav Physiol 24:101–116

Spanier E, Almog-Shtayer G (1992) Shelter preferences in the Mediterranean slipper lobster: effects of physical properties. J Exp Mar Biol Ecol 164:103–116

Steullet P, Dudar O, Flavus T, Zhou M, Derby CD (2001) Selective ablation of antennular sensilla on the Caribbean spiny lobster *Panulirus argus* suggests that dual antennular chemosensory pathways mediate odorant activation of searching and localization of food. J Exp Biol 204:4259–4269

Steullet P, Krützfeldt DR, Hamidani G, Flavus T, Ngo V, Derby CD (2002) Dual antennular chemosensory pathways mediate odor-associative learning and odor discrimination in the Caribbean spiny lobster *Panulirus argus*. J Exp Biol 205:851–867

Tierney AJ, Thompson CS, Dunham DW (1984) Site of pheromone reception in the crayfish *Orconectes propinquus* (Decapoda Cambaridae). J Crust Biol 4:554–559

Waddy SL, Aiken DE (1990) Intermolt insemination, an alternative mating strategy for the American lobster (*Homarus americanus*). Can J Fish Aquat Sci 47:2402–2406

Wahle RA (1992) Body-size dependent anti-predator mechanisms of the American lobster. Oikos 65:52–60

Weiss HM, Lozano-Álvarez E, Briones-Fourzán P (2008) Circadian shelter occupancy patterns and predator-prey interactions of juvenile Caribbean spiny lobsters in a reef lagoon. Mar Biol 153:953–963

Wisenden BD (2000) Olfactory assessment of predation risk in the aquatic environment. Phil Trans R Soc Lond B 355:1205–1208

Zimmer-Faust RK, Spanier E (1987) Gregariousness and sociality in spiny lobsters: implications for den habitation. J Exp Mar Biol Ecol 195:57–71

Zimmer-Faust RK, Tyre JE, Case JF (1985) Chemical attraction causing aggregation in the spiny lobster, *Panulirus interruptus* (Randall), and its probable ecological significance. Biol Bull 169:106–118

Chapter 13
Chemical Communication in Crayfish

Thomas Breithaupt

Abstract Crayfish are a species rich group of large decapod crustaceans that inhabit freshwater environments. Having served as important models for the study of the neural and hormonal control of behavior crayfish were among the first crustacean taxa that were reported to use sex pheromones. Decades of research on crayfish chemical communication have, after initial controversies, now generated a comprehensive picture of the role of pheromones in resolving combats and in initiating sexual interactions. Moreover, the structures involved in chemical signal emission and reception have been identified in most cases. Urine, released in the head region, conveys the chemical messages and is directed via water movements such as gill currents or maxilliped generated currents to the receiver. Chemo-receptors on the first antennae were shown – in most cases – to be responsible for pheromone detection. Urinary signals reduce the duration of aggressive interactions and are crucial for the development of a linear dominance hierarchy. The social hierarchy is based on chemical recognition of the dominance status between combatants. Males are more active than females in initiating reproductive interactions. They recognize a female sex pheromone contained in the urine pulses that females release during the initial aggressive bout preceding mating. Female assessment of male quality is multimodal, involving tactile, visual, and in some cases also chemical cues. The recent development of context-specific and less ambiguous bioassays will facilitate the future purification of the molecules that mediate sexual receptivity and social status in crayfish. These pheromones could be valuable for application in the control of alien invasive crayfish species that cause environmental damage.

T. Breithaupt (✉)
Department of Biological Sciences, The University of Hull, Hull HU6 7RX, UK
e-mail: t.breithaupt@hull.ac.uk

T. Breithaupt and M. Thiel (eds.), *Chemical Communication in Crustaceans*,
DOI 10.1007/978-0-387-77101-4_13, © Springer Science+Business Media, LLC 2011

13.1   Introduction

Ever since the first reports of crayfish sex pheromones in the early 1970s there has
been a strong interest in the question how crayfish use chemical signals. My own
interest in the chemical communication of crayfish was initiated after I had spent
2 years in the lab of Jelle Atema in Woods Hole studying the role of urinary signals in
lobsters. In Atema's lab, I learned to be guided by the natural behavior of animals
when investigating the role of the particular sense organs or behavioral mechanisms.
In order to study lobster chemical signaling in natural aggressive interactions I used
catheters to monitor urine release (Breithaupt et al. 1999; Breithaupt and Atema 2000).
This technique allowed – for the first time – to quantify urine output in different
behavioral contexts. The disadvantage of this technique is that the signal is collected in
the catheters and does not arrive at the receiver thereby preventing any natural
response. The solution to this problem was to visualize the chemical signal by
injecting fluorescent dye into the circulatory system so that urine release could be
quantified and still elicit a response. But it took a very gifted student with a lot of
finesse to make the technique work. Petra Eger was trained as a nurse before
she studied biology in Konstanz. She could handle crayfish, syringe, plasticine, and
tape – this is what is needed for dye injection and subsequent sealing of the injection
hole – with only two hands to test the administration of different types of dye for the
visualization of urine release in crayfish. She made the method work and for the first
time allowed us to actually see underwater chemical signaling in action and study
when and how they are sent and received in social interactions (Breithaupt and Eger
2002). This study also gave the unquestionable evidence for chemical communication
in crayfish by showing signal release as well as receiver's response. For me it further
fueled my interest into the field of crustacean chemical communication.

13.2   Crayfish Diversity, Biology and Evidence for Pheromones

Crayfish are the only large decapod crustaceans to occur in freshwater environments
and there are more than 640 species in three different families, Astacidae, Cambaridae and Parastacidae. The Astacidae with 39 species is the smallest family with
species occurring in Europe/Asia (31 species) and North America (8 species).
Cambarid crayfish (>420 species) live in North America while the distribution of
Parastacidae (>170 species) is restricted to the Southern hemisphere including
Madagascar, South America and Australasia (Crandall and Buhay 2008). Crayfish
species have been translocated to other parts of the world mainly for aquaculture and
stocking purposes and have become invasive in the new areas (Gherardi and Holdich
1999). Some of the invasive crayfish species cause considerable ecological damage in
the new areas not least by transmitting the crayfish plague, a fungal disease lethal to
the native crayfish species. One motivation of pheromone research is its potential

application in the control of invasive crayfish species emulating the success of pheromone use in insect pest management (see Baker, Chap. 27).

Earlier studies of crayfish suggested that chemical signals play important roles in various aspects of their life including courtship, agonistic interactions, alarm and brood care (e.g. Ameyaw-Akumfi and Hazlett 1975; Little 1976; Hazlett 1985). Little is known about the chemical communication underlying alarm and brood care and the interested reader is referred to the original literature (Little 1976; Hazlett 1985). The bulk of knowledge about crayfish chemical communication comes from research into aggressive and reproductive interactions. Therefore, in this chapter I will review the evidence that pheromones (chemical signals used in interactions between members of the same species; see Wyatt, Chap. 2) are used for the communication of dominance and sex.

Aggression is a regular part of crayfish social behavior. Crayfish generally live in high population densities and encounters between individuals are frequent (Nystrom 2002). Interactions are often of agonistic nature and escalate into fights (Moore 2007). As a result of repeated fighting, they develop a linear dominance hierarchy (Goessmann et al. 2000). Dominant animals were suggested to have an advantage over other males in gaining better access to shelter, food and mating partners (Bergman and Moore 2003; Herberholz et al. 2007; Moore 2007). Cambarid crayfish males undergo a molt from non-reproductive (Form II) to reproductive morphology (Form I). Form I has larger chelae relative to body size and fully functional reproductive appendages (Reynolds 2002). Courtship generally starts with aggressive behavior of the female toward the male. Mating commences once the male manages to overcome the resistance of the female. The male turns and mounts the female and deposits spermatophores under her abdomen (Mason 1970). Upon release, eggs are fertilized by the stored sperm and attached to the female pleopods under the abdomen. As in most other crustacean species, females carry and ventilate the eggs for extended time periods, thus investing substantially more time and energy into single offspring than males. During this time females are more susceptible to predation because the tail flip mediated escape behavior is severely restricted (for lobsters see Cromarty et al. 1998).

Ameyaw-Akumfi and Hazlett (1975) were the first to report a sex pheromone in crayfish. When tested in an aquarium *Procambarus clarkii* males showed submissive behavior (chelae down, telson curled) in response to female water and displayed aggressive behavior (raised chelae) in response to male odor (Ameyaw-Akumfi and Hazlett 1975). The authors suggested that these differential responses were caused by sex specific pheromones. This study was criticized by Dunham (1978) due to the lack of controls and "blind" observations. Subsequent studies produced contradictory findings and generated a debate about the significance of pheromones in crayfish (for review see Bechler 1995).

Studies over the past 20 years have brought considerable insight into the use of chemical signals in crayfish social behavior. The current review will particularly concentrate on the role of chemical communication in dominance and courtship interactions.

13.3   Chemical Signals: Release, Dispersal, and Detection

More recent studies of crayfish chemical communication suggest that similar to lobsters and crabs (Atema and Steinbach 2007) many chemical signals are contained within the urine. Urine is released anteriorly through a pair of nephropores (Fig. 13.1). The anterior position is advantageous as it facilitates the frontal dispersal of pheromones towards conspecifics (see Fig. 13.2). As in other crustaceans the nephropores are opened and closed by a sphincter muscle (Bushmann and

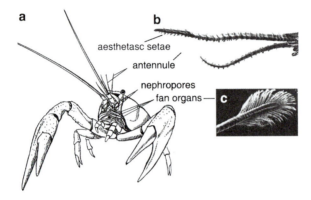

Fig. 13.1 Crayfish appendages involved in chemical communication. (**a**) Position of appendages; (**b**) lateral (*upper*) and medial (*lower*) filament of antennule (Scanning Electron Microscope (SEM) picture). Note the aesthetasc hairs on the lateral filament (**c**) fan organ (SEM picture). Figure modified after Breithaupt (2001)

Fig. 13.2 Photograph showing chemical communication during aggressive interaction of two male *Astacus leptodactylus*. Both males were blindfolded using opaque film wrapped around the eyestalks. Urine signals were visualized using Fluorescein dye injected into the heart (for methodology see Breithaupt and Eger 2002)

Atema 1996) providing the crayfish with the possibility to control the timing of pheromone release. Urine, once released, is drawn into the water currents generated by the animal itself. These are gill currents, expelled from the frontal opening of the gill chambers just underneath the nephropores, and currents generated by fan organs (Fig. 13.1; Breithaupt 2001). Whilst gill currents are always directed forward, fan currents can be directed forward, laterally, upward and backward (Breithaupt 2001). Hence, crayfish have excellent possibilities to control both the timing and the direction of chemical signals comparable to those of lobsters and hermit crabs (Atema and Steinbach 2007).

The fan organs also play a role in the detection of chemical signals. Fan generated currents facilitate acquisition of chemical signals by drawing odor molecules to the antennules, the major chemoreceptors (Denissenko et al. 2007). Fanning may be essential for odor detection in lakes and ponds but less important for detecting odors in a flow environment of a river. This difference in chemical communication between riverine and lacustrine freshwater environments have also been reported from other crustacean species (Thiel, Chap. 10). Multiple studies suggest the aesthetasc hairs on the first antennae (antennules) as the site of phero- mone reception in crustaceans (Fig. 13.1; Hallberg and Skog, Chap. 6). Ablation of antennular flagellae in *P. clarkii* eliminated or reduced the sex-specific responses to female and male water (Ameyaw-Akumfi and Hazlett 1975; Dunham and Oh 1992) or impeded localization of odor from the opposite sex (in *Orconectes propinquus*: Tierney and Dunham 1984; in *P. clarkii*: Giri and Dunham 2000). In contrast, other studies of *P. clarkii* concluded (based on ablation experiments in males and in females) that presence of antennules is not necessary for mating behavior to take place (Corotto et al. 1999). In one study, the major chelae were suggested as alternative sites of sex pheromone receptors (Belanger and Moore 2006). While the significance of aesthetasc receptors for the detection of sex pheromones in crayfish is still unclear, their role in the perception of dominance odors is undis- puted. Blocking of aesthetasc hairs inhibits the maintenance of dominance relation- ships in *P. clarkii* (see below; Horner et al. 2008).

13.4 Fighting Behavior, Social Hierarchies, and the Role of Chemical Signals

The behavioral elements of fighting in crayfish are similar to those of American lobsters (Atema and Voigt 1995) and are described in detail in Moore (2007). Fighting is initiated with one animal approaching the other and includes ritualized non-damaging behaviors (meral spread: a threat displays with animals raising the body and extending the chelipeds; tapping/pushing: tactile displays towards the opponent's body using second antennae or chelipeds; claw locking: claws used to grasp appendages of the opponent). If none of the individuals retreats, fights might escalate to potentially damaging behaviors (claw ripping: a combination of claw

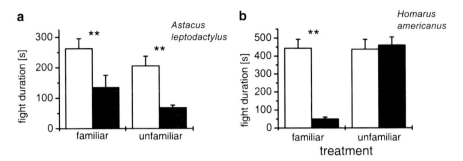

Fig. 13.3 Influence of familiarity on duration of repeated fights between crayfish (**a**) and American lobsters (**b**). "Familiar" denotes that the same pair of size-matched animals fights on two consecutive days: first and second fight. "Unfamiliar" denotes that the loser of the first day fights an unknown winner drawn from a separate fight on the second day. *White bars* denote first fights, *black bars* denote second fights. In crayfish, but not in lobsters, unfamiliar treatments leads to a decrease in fight duration on the second day. This indicates that crayfish recognize the winner-status of their oponent, but not the identity of a familiar oponent. (**a**) Drawn from data in Breithaupt and Eger 2002; (**b**) modified after Karavanich and Atema (1998)

locking and tailflip behavior that can result in removal of the opponent's appendage). Important determinants of winning are body size, claw size, win/lose history and physiological condition (Moore 2007). The closer these conditions are matched between combatants the longer and more intense the fights will be (Schroeder and Huber 2001). First fights of size-matched crayfish last several minutes (Fig. 13.3a; Breithaupt and Eger 2002; see also Horner et al. 2008). Second fights (either on the same or on the subsequent day) are about 50% shorter than first fights, indicating that dominance is maintained over extended time in crayfish (Fig. 13.3a; Breithaupt and Eger 2002; Horner et al. 2008). The previous winner will generally be dominant again (Bergman et al. 2003). In groups of four or five juvenile individuals, fighting activity is high initially, declines after the first hours and drops to low levels after 1 day (Issa et al. 1999). In adults, the gradual decrease in the frequency and mean duration of fights reflect the formation of a stable linear dominance hierarchy with the win/lose ratio being highest for the dominant animal and decreasing towards the most subordinate animals (Goessmann et al. 2000).

13.4.1 The Role of Urine Signals in Dyadic Fights

The significance of urine-borne chemical signals in fighting behavior of crayfish became obvious through blocking and visualization of urine release: Blocking of nephropores leads to a significant increase in fight duration between size-matched males of *Orconectes rusticus* (Zulandt Schneider et al. 2001). This suggests that urine signals are important for crayfish in establishing a dominance relationship with a competitor. Visualization of urine in blindfolded crayfish *Astacus leptodactylus* (Breithaupt and Eger 2002) provided new insight into the use of urine signals in

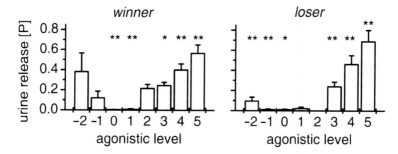

Fig. 13.4 Probabilities (P) of urine release at different agonistic levels in prospective winners (*left*) and prospective losers (*right*). Agonistic levels -2, -1 indicate defensive behaviors and agonistic levels 1–5 indicate aggressive behaviors of increasing intensity (1 = approach, 2 = meral spread, 3 = tapping/pushing, 4 = claw lock, 5 = claw ripping). *Asterisks* denote significant differences with respect to the mean probability over all agonistic levels (*asterisk* $p < 0.05$, *double asterisk* $p < 0.01$; modified after Breithaupt and Eger 2002). Probability of urine release is significantly increased only during physical aggression (levels 3, 4 and 5) while it is decreased during defensive (-2, -1) and neutral behavior (level 0) of losers and during neutral and approach behavior (level 1) of winners

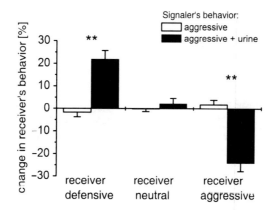

Fig. 13.5 Changes in the receiver response to offensive behavior without urine release (*open bars*) and to offensive behavior accompanied by urine release (*solid bars*). Responses are measured as changes in the relative frequency of occurrence of defensive (levels -2, -1), neutral (level 0) and aggressive (levels 1–5) behaviors. Asterisks denote significant differences between the two conditions (*double asterisk* $p < 0.01$; modified after Breithaupt and Eger 2002)

aggressive interactions: (i) urine is more likely to be released in social interactions than during non-social activities or inactivity, (ii) the eventual winner releases more urine than the eventual loser, (iii) most urine release is coupled to offensive behavior (see Fig. 13.4), and (iv) urine output increases with increasing level of aggression (Fig. 13.4). Since urine is linked to aggressive behaviors we investigated its effect on the receiver by comparing aggressive behaviors that were accompanied by urine release with aggressive behavior not accompanied by urine. Lag sequential analysis revealed that the receiver responds to the combination of offensive behaviors and urinary signals (but not to offensive behavior alone) by decreasing its offensive and increasing defensive behaviors (Fig. 13.5). Together, urine blocking and urine visualization experiments revealed that chemical signals are essential components of fighting behavior in crayfish. Urinary components appear to transfer crucial information in crayfish aggressive interactions (perhaps about resource holding potential; Briffa and Sneddon 2006) promoting faster resolution of the combat.

13.4.2   The Role of Chemical Signals in the Regulation of Social Hierarchies

Chemical signals not only are important in modulating initial encounters between combatants. They also appear to be largely responsible for the maintenance of lasting dominance hierarchies in crayfish. When the olfactory asthetasc sensilla on the antennules were removed in previously fought pairs of red swamp crayfish (*P. clarkii*) the encounter frequency and fight time was significantly altered (Horner et al. 2008). Crayfish lacking olfactory receptors showed similar amounts of fighting within 1 h pairings on three successive days whereas control crayfish decreased the encounter frequency and fight duration in repeated pairings. The authors suggest that dominance is maintained by chemical recognition of the social status of the opponent and that the aesthetascs are a crucial component in this process (Horner et al. 2008). A similar role of the aesthetasc hairs in chemical dominance recognition had previously been shown in American lobsters (Johnson and Atema 2005). Also, in *O. rusticus* previous winners have a higher chance of winning again when paired with a naïve opponent within 20–40 min after the initial fight but have no advantage if antennular receptors are blocked (Bergman et al. 2003). The authors concluded that – in this species – the winner effect lasts less than 60 min and is mediated by antennular receptors. The perception of social odors also appears to have a priming effect (see Wyatt, Chap. 2) on crayfish promoting a prolonged alteration of aggressive behavior. Crayfish (*O. rusticus*), after exposure to odors from a loser, tend to win a subsequent fight while animals exposed to winner chemicals tend to lose (Bergman and Moore 2005). The combined findings from these three studies suggest that olfactory recognition of social status is a key mechanism responsible for the maintenance of social hierarchies in crayfish. The perception of the opponent's odor interacts with the animal's own fight history in influencing the decision to escalate or retreat an ongoing fight.

There is contradictory evidence from different crustacean species as to whether the chemical signals mediate individual identity or social rank of combatants. Experiments on American lobsters and hermit crabs indicate that dominance is maintained by the loser recognizing the individual scent of the familiar opponent it has lost to (Fig. 13.3b; Karavanich and Atema 1998; Gherardi and Tricarico, Chap. 15). Previous losers surrender early on in a fight when paired with a familiar winner. Unfamiliar winners, however, are challenged and these subsequent fights last as long as the first fight (Karavanich and Atema 1998). This is different in the crayfish *A. leptodactylus* (Breithaupt and Eger 2002). When a dominant crayfish who won a fight on a previous day is paired with a subordinate crayfish that lost on the previous day this second fight of both combatants is generally shorter than the first fights (Fig. 13.3a). The decrease in fight duration occurs even in pairings of unfamiliar crayfish, a loser and a winner of previous fights against other opponents. This suggests that *A. leptodactylus* does not depend on individual recognition for the maintenance of dominance relationships. Similar results were obtained in *O. rusticus* (Zulandt Schneider et al. 2001) and *P. acutus* (Gherardi and Daniels 2003). All three studies found no difference in the dynamics of repeated fights between familiar and unfamiliar crayfish pairs rendering individual recognition unlikely as the mechanism for the maintenance of dominance in crayfish. Instead of the individual identity, crayfish appear to recognize the social status (dominance rank) of their opponents as suggested in an earlier study by Copp (1986). In accordance with this result, when tested in a flow-through Y maze, male *P. clarkii* responded more aggressively to dominant male odor than to subordinate odor (Zulandt Schneider et al. 1999).

Contrary to these findings, Seebacher and Wilson (2007) concluded from their study that the Australian crayfish species *Cherax dispar* maintain dominance through individual recognition of familiar opponents. In the highly aggressive *C. dispar*, the closing force of the chela is an important determinant of winning in size matched pairings (Seebacher and Wilson 2006). However, even with disabled chelae previous winners keep winning in subsequent encounters when matched against an unrestrained familiar opponent suggesting that dominance is recognized. Previous winners did not have a higher chance of winning against unfamiliar naïve individuals that had not fought before. Similarly, previous losers were not more likely to lose against unfamiliar naïve individuals (Seebacher and Wilson 2007). Unfortunately, in this study the controls – pairing winners or losers with naïve individuals with no prior fight experience – are not sufficient to exclude the possibility that losers would recognize the status of the claw-restrained dominant rather than its individual identity. This can only be demonstrated by pairing losers with unfamiliar winners. Hence, the current evidence favors status recognition over individual recognition as the mechanisms to maintain dominance relationships in crayfish.

13.4.3 What is Communicated by Urine Signals in Agonistic Interactions?

Urine includes metabolites of current and recent biochemical processes and therefore may open a window for the receiver to the physiological state and resource holding potential of the signaler. Biogenic amines such as serotonin and octopamine have been related to aggressive motivation in crustaceans (Kravitz 2000). In green crabs, high serotonin levels were found in winners of fights (Sneddon et al. 2000).

Injections of serotonin cause *P. clarkii* to adopt a dominance display (meral spread) while octopamine injection triggers a typical subordinate stance (Livingstone et al. 1980) but this study was criticized as it used unnaturally high doses of the biogenic amines. Continuous infusion of low doses of serotonin into subordinate crayfish *A. astacus* caused more subtle changes in the behavior (Huber and Delago 1998). Infusion into the circulatory system was applied throughout the fight using a cannula attached to the carapace. Infused animals showed a reduction in the tendency to retreat during a combat. These neurohormones could also affect the behavior of the opponent if they leak into the environment. Metabolites of serotonin are released with the urine (Huber et al. 1997) and therefore may be a candidate to mediate aggressive motivation of the dominant crayfish to the subordinate receiver. When tested on crayfish with prior social experience, injection of serotonin or octopamine resulted in changes in aggression but did not lead to a permanent inversion of dominance hierarchies as expected (Tricarico and Gherardi 2007). Similarly, when fighting over a resource serotonin-injected American lobsters did not win over sham-injected individuals (Peeke et al. 2000). It still needs to be established whether metabolites of biogenic amines released in the urine act as dominance pheromones in fights or whether other chemical components are involved. Hence, the chemical identity of the pheromones that regulate fighting and social hierarchies in crayfish is still elusive. To date, no bioassay has been established that could serve to guide fractionation of urinary compounds mediating dominance in crayfish (see Hardege and Terschak, Chap. 19).

13.5   The role of Chemical Signals in Reproductive Interactions

13.5.1   Evidence for Female sex pheromones

In some decapod crustaceans including green crabs (see Hardege and Terschak, Chap. 19) and blue crabs (Kamio and Derby, Chap. 20) the existence of female sex pheromones is assured by unambiguous male courtship responses to female odorants (e.g. "cradle carrying of a pheromone labeled object" in green crabs or "courtship stationary paddling" in blue crabs). Male crayfish do not display specific courtship behaviors and their responses to female stimuli are ambiguous, a fact which has hampered research progress into the significance and chemical identity of female chemical signals. In contrast to species such as green crabs, blue crabs and lobsters, where females are mated immediately after having molted, female crayfish mate during the intermolt period, i.e., months after their last molt. Therefore, crayfish females do not profit as much as soft-shelled females from male protection exercised by cradle carrying (in crabs: Hardege and Terschak, Chap. 19) or shelter sharing (in lobsters: Atema and Steinbach 2007) and hence may not actively seek presence of a courting male.

In most crayfish species studied, mating develops from aggressive interactions between the sexes (Gherardi 2002; but see Barki and Karplus 1999). During the agonistic interactions, the male upon recognizing sex and receptivity of the female initiate mating by seizing her rostrum or antennules with the major chelae. He turns the female by rolling her body upside down using chelae and walking legs. The male then mounts the female and deposits spermatophores to the ventral side of her cephalothorax (in Astacidae) or into sperm-receptacles (in Cambaridae) (Reynolds 2002; Stebbing et al. 2003).

Following the controversy about the existence of sex pheromones in crayfish (summarized in Bechler 1995) more recent studies provide new support for female release of and male response to chemical signals during reproductive interactions. Using visualization techniques adopted from Breithaupt and Eger (2002), Simon and Moore (2007) and Berry and Breithaupt (2010) showed that in pairings of receptive females and reproductively active males of the species *O. rusticus* and *P. leniusculus*, respectively, urine is released by both sexes (Fig. 13.6). Males of both species exhibit responses to artificially introduced female stimuli by increasing general activity or by handling the odor source (an aquarium air stone) more readily when conditioned water from receptive females was released but not in response to water from juvenile females (Stebbing et al. 2003) or to male conditioned water (Belanger and Moore 2006). In two cases, males attached spermatophores to the airstone in response to water from receptive females (Stebbing et al. 2003) suggesting a sexual nature of the response. Similarly, male *P. clarkii* appeared to recognize

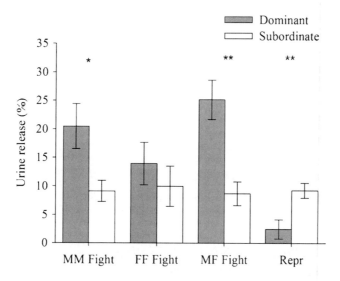

Fig. 13.6 Mean (±SEM) urine release by dominant (*gray bars*) and subordinate (*white bars*) crayfish in male fights (MM Fight), female fights (FF Fight), mixed-sex fights (MF Fight) and reproductive interactions (Repr). In reproductive interactions males are labeled as dominant (*gray bars*) and females as subordinate (*white bars*). *Asterisks* indicate differences between interactants (*asterisk* $p < 0.05$, *double asterisk* $p < 0.01$, paired *t*-test). From Berry and Breithaupt (2010)

a conspecific's gender by chemical stimuli alone. Males showed increased locomotion to both male and female stimuli but responded with aggressive postures to male stimuli only (Aquiloni et al. 2009). Our own initial approaches to bioassay development tested male responses (heart rate and behavioral changes) to a more natural device for odor delivery, a female dummy made of a molt shell (Berry and Breithaupt 2008). Males exposed to female urine showed changes in heart rate as well as significantly increased durations of mounting behavior in comparison to male urine or control water (Berry and Breithaupt 2008). However, other sexual behaviors such as turning and spermatophore deposition were not observed in this study. Perhaps the male needed tactile or visual feedback from a female before continuing mating behavior. Although these studies indicate that males discriminate odor of receptive females from that of unreceptive females or males, the elicited responses (handling of odor source, increased locomotion, mounting of a dummy, changes in heart rate) may still be considered ambiguous as they could occur in reproductive as well as in non-reproductive contexts.

In a recent study we addressed the shortcomings of previous bioassay approaches by studying in *P. leniusculus* the responses of blindfolded males to live females rather than to inanimate odor sources (Berry and Breithaupt 2010). We blocked the natural urine release of the female and compared the effect of artificial urine introduction with introduction of control water. Female urine or water was introduced through a syringe during the initial aggressive bout. In crayfish (*P. leniusculus*), during sexual interactions the male has to overcome resistance of the female and uses a specific sequence of behaviors to seize, turn and mount the female (see above). We scored the presence of any of these behaviors as a sexual response. Males were more likely to show sexual responses toward urine-blocked female when urine from a receptive female was introduced than when control water was introduced (Berry and Breithaupt 2010).

Together, these results provide clear evidence for the existence of female sex pheromones in crayfish. In some species multimodal (chemical and visual) stimulation is required to enable sex recognition by males (see Hebets and Rundus, Chap. 17). In *Austropotamobius pallipes*, a species that mainly inhabits turbulent flow environments, chemical stimuli alone do not elicit male response. Only when accompanied by the sight of a crayfish (male or female), odor was effective in causing increased activity and in reducing the time the male spent under a shelter (Acquistapace et al. 2002). The authors discussed ecological constraints (turbulence causing stronger dispersal of chemical signals) as a cause for the multimodal strategy.

13.5.2   Are Chemical Signals Involved in Female Assessment of Male Quality?

Due to their strong reproductive investment, female crayfish are expected to be highly selective in the choice of the mate. Female quality assessment could include size and dominance status of males, which both affect the resource holding potential of the males. Mate choice based on these characters – if heritable – should be beneficial to the females by increasing her fitness. While the initial aggression of the female to male mating attempts indicates her resistance to mating, there is evidence that females recognize male gender and are able to conduct mate choice.

Sex identification in *P. clarkii* involves both chemical and visual cues (Aquiloni et al. 2009). Females do not show differences in response to male and female conspecifics, if they receive odor alone or if they can see but not smell the other individual. However, when they receive bimodal (chemical and visual) stimuli they display aggressive postures only to females not to the males. How important are chemical stimuli for females to evaluate male quality? To date there is little evidence that chemical stimuli play an important role in mate selection by females. Aquiloni et al. (2008) concluded that female red swamp crayfish (*P. clarkii*) do not show a preference when allowed to select between a dominant and subordinate male of similar size. Only when allowed to watch the fighting interaction between males prior to being exposed to them, female *P. clarkii* display a preference for the dominant male. Hence the previous visual experience was necessary to recognize a dominant male. For the assessment of male size, female *P. clarkii* appeared to require both visual and chemical information. Females approached the larger of two males hidden in plastic containers only when they could see and smell them (Aquiloni and Gherardi 2008b). It is likely that in this experiment, the chemical information was necessary for sex identification while the sight of the males allowed discrimination of size. These results imply that males release some chemicals prior to or during reproductive interactions with females. Urine visualization during reproductive interactions of *P. leniusculus* revealed that males release significantly less urine during sexual interaction than during aggressive interactions (Fig. 13.6; Berry and Breithaupt 2010); 30% of the males did not release any urine when attempting to mate a female. Females, in contrast, readily release urine during their aggressive behavior towards males prior to mating (Fig. 13.6). Figure 13.7 shows an artist illustration of inter- and intrasexual chemical signaling in a crayfish population. Since mating is always preceded by aggressive interactions between the sexes, females get additional tactile information about size and strength of the male by physically interacting with him. Crayfish have very keen tactile senses to explore their environment (McMahon et al. 2005). Due to the nocturnal activity period of most crayfish species tactile information may be even more important than visual information to assess male quality.

Females are also able to perform cryptic mate choice. After mating, females adjust their reproductive effort in relation to male traits. Galeotti and coworkers showed that female crayfish *Austropotamobius italicus* produce larger but fewer eggs for small males with large claws and more numerous but smaller eggs for large-sized, small-clawed males (Galeotti et al. 2006). Similarly, female *P. clarkii* invest in larger eggs after having copulated with large males (Aquiloni and Gherardi 2008a). The authors speculate that crayfish have the ability to reabsorb part of the nutritive substances in the cytoplasm of the oocytes before they spawn. In conclusion, mate choice in crayfish is often cryptic. Chemical and visual stimuli may provide partial information for sex identification but are not always available to the female. Tactile stimuli perceived during physical interaction with the male as well as

Fig. 13.7 Artist impressions of the social interactions between crayfish in a population of high density. Aggressive interactions between males (on the *right*) involve physical contact of the claws (claw lock) and urine release. Reproductive interaction (*bottom*) involve male mating attempt (starting with the male seizing female antennae with left claw) and female aggressive behavior (female claw lock) with urine release. Drawing courtesy of Jorge Andrés Varela Ramos

visual stimuli indicating size of the male appear to be more important than chemical stimuli for females to discriminate quality of males and to display mate choice.

13.6   Crayfish as Model Organisms

In his influential book on "The crayfish," Thomas H. Huxley (Huxley 1879) used the crayfish as a model for the introduction to the study of zoology. This book, due to its integrative approach had a great impact on various zoological disciplines. Particularly the neurosciences have successfully used crayfish preparations to understand the functioning of the nervous system and the link between neural function and behavior ("neuro-ethology") (Edwards et al. 1999). Today, there is an extensive literature on the neurobiology, behavior and ecology of crayfish providing a rich basis for further exploration of their chemical ecology. As freshwater organisms, crayfish are easy to maintain in captivity, have an ideal body size for the study of their social behavior, can be experimentally manipulated (e.g. by blindfolding, dye visualization, urine collection) for laboratory studies and their behavior can be tracked in the field. This makes them stand out as ideal model organism for behavioral studies. In the past 15 years much research effort has been devoted to studies into the neural and behavioral mechanisms of fighting behavior (e.g. Yeh et al. 1997; Herberholz et al. 2001). Fighting behavior in crayfish is easily elicited. Two size-matched crayfish introduced to separated halves of an aquarium will almost always start fighting as soon as the separating divider is lifted. Hence, crayfish may be better suited than many other aquatic organisms including fish and polychaetes to unravel the complex role chemical signals may play in the resolution of animal combats as well as the underlying neural mechanisms. Because neurological methods, e.g. electrophysiological recordings and activity labeling in the brain, are well established (see Schmidt and Mellon, Chap. 7), the identification of crayfish sex or dominance pheromones will enable new insight into the brain function of this highly chemoreceptive group of animals.

Furthermore, crayfish show a high species diversity (>640 species) making them ideal candidates for studies into the evolution of pheromones, a field that so far depended mostly on research of insect pheromones, particularly moths and fruit flies (Wyatt 2003). Crayfish inhabit ecosystems as diverse as wetlands, stagnant ponds and lakes, turbulent creeks and rivers, providing opportunities to disentangle the ecological and phylogenetic constraints of chemical signal evolution.

13.7 Future Directions

While some basic questions about the behavioral significance of sex- and dominance pheromones in crayfish appear to be solved, questions into the exact function of the pheromones are still unanswered and new challenges are waiting to be explored. We now have good evidence that chemical signals are essential for the maintenance of dominance hierarchies in crayfish. Several studies indicate that dominance recognition is mediated by status indicating pheromones and not by individual odors. Status indicating pheromones are expected to consist of a "defined combination of molecules eliciting a particular behavior or response" (Wyatt, Chap. 2) and therefore lend themselves much better for chemical characterization than the highly variable individual odors (signature mixtures; see Wyatt, Chap. 2). Variations in odor composition between individuals such as in lobsters (Atema and Steinbach 2007) render it almost impossible to identify the molecules conveying the signal. Therefore, at the moment, crayfish provide the best opportunity among crustaceans and other aquatic animals to discover the chemical identity of molecules mediating dominance recognition. Once the chemical composition is known, new questions concerning the production and the metabolic costs can be addressed as well as the information content and the possibility of deception by chemical signals (see Christy and Rittschof, Chap. 16).

A better link between endocrinological and pheromone research may lead to new insight about the mechanisms of chemical communication. Barki and coworkers showed that implantation of the androgenic gland into a juvenile female induced male characteristics in the female once it was fully matured (Barki et al. 2003). Normal females accepted these implanted females as males, since they allowed them to mate and deposit spermatophores. This discovery suggests that implanted females may release sex-identifying pheromones typical for males. Further research may allow investigating the potential hormonal origin of sex pheromones. Many fish species have been shown to use hormonal metabolites as pheromones (see Chung-Davidson, Huertas and Li, Chap. 24) but to date there is little evidence for hormonal pheromones in crustaceans (Chang, Chap. 21).

The recent development of context-specific and therefore less ambiguous bioassays (see "evidence for female sex pheromones") will pave the way for research into the chemical purification of female crayfish sex pheromones. Already at an early stage of chemical purification the question can be addressed whether sex pheromones are species specific. As the identity of the molecules unfolds, further questions to their function can be addressed, e.g. whether the molecules specifying sex, receptivity and species are identical such as in fruit flies (Billeter et al. 2009) and how they relate to female reproductive metabolism.

The transport of crayfish species between countries and continents has led to invasions of alien crayfish species with dramatic ecological consequences. Invasive crayfish can cause physical damage (by destroying river banks) as well as ecological damage to the ecosystem including the displacement of the native crayfish species (Hill and Lodge 1999). Protocols including the use of pheromones for species attraction and eradication need to be established to control the invasive crayfish. Here, the knowledge about insect pest management (Baker, Chap. 27) will be important as it may at least in part be transferable to design efficient strategies for the management of invasive crayfish.

13.8   Summary and Conclusions

Decades of research on crayfish chemical communication has, after initial controversies, generated a comprehensive picture of the role of pheromones in resolving combats and in initiating sexual interactions. Moreover, the structures involved in signal emission and reception have been identified in most cases. Urine, released in the head region, conveys the chemical messages and is directed with water currents to the receiver. Aesthetasc receptors on the first antennae were shown – in most cases – to be responsible for pheromone detection. Urinary signals reduce the duration of aggressive interactions and allow the development of a stable linear hierarchy. The social hierarchy is based on the chemical recognition of the dominance status between combatants. Males are more active than females in initiating reproductive interactions. They recognize the female sex pheromones. Female assessment of male quality is multimodal, involving tactile, visual and chemical components. The recent development of context-specific and less ambiguous bioassays will facilitate the future purification of the molecules that constitute the sex and dominance pheromones in crayfish. These pheromones could be valuable for application in the control of alien invasive crayfish species that cause environmental damage.

Acknowledgments I would like to thank my mentors (Prof. Juergen Tautz, Prof. Jelle Atema) and my students who helped me to pursue my interest into this fascinating group of animals. Research was supported by grants from the German research foundation (DFG), the Royal Society and the Natural Environment Research Foundation (NERC). I would like to thank Drs. Martin Thiel, Jelle Atema and two anonymous reviewers for constructive comments.

References

Acquistapace P, Aquiloni L, Hazlett BA, Gherardi F (2002) Multimodal communication in crayfish: sex recognition during mate search by male *Austropotamobius pallipes*. Can J Zool 80:2041–2045

Ameyaw-Akumfi C, Hazlett BA (1975) Sex recognition in the crayfish *Procambarus clarkii*. Science 190:1225–1226

Aquiloni L, Gherardi F (2008a) Evidence of female cryptic choice in crayfish. Biol Lett 4:163–165

Aquiloni L, Gherardi F (2008b) Assessing mate size in the red swamp crayfish *Procambarus clarkii*: effects of visual versus chemical stimuli. Fresh Biol 53:461–469

Aquiloni L, Buric M, Gherardi F (2008) Crayfish females eavesdrop on fighting males before choosing the dominant mate. Curr Biol 18:462–463

Aquiloni L, Massolo A, Gherardi F (2009) Sex identification in female crayfish is bimodal. Naturwissenschaften 96:103–110

Atema J, Steinbach MA (2007) Chemical communication and social behavior of the lobster *Homarus americanus* and other decapod crustacea. In: Duffy JE, Thiel M (eds) Evolutionary ecology of social and sexual systems: crustaceans as model organisms. Oxford University Press, Oxford, pp 115–144

Atema J, Voigt R (1995) Behavior and sensory biology. In: Factor JR (ed) Biology of the lobster *Homarus americanus*. Academic, New York, pp 313–348

Barki A, Karplus I (1999) Mating behavior and a behavioural assay for female receptivity in the red-claw crayfish *Cherax quadricarinatus*. J Crust Biol 19:493–497

Barki A, Karplus I, Khalaila I, Manor R, Sagi A (2003) Male-like behavioral patterns and physiological alterations induced by androgenic gland implantation in female crayfish. J Exp Biol 206:1791–1797

Bechler DL (1995) A review and prospectus of sexual and interspecific pheromonal communication in crayfish. Freshw Crayfish 10:657–667

Belanger RM, Moore PA (2006) The use of the major chelae by reproductive male crayfish (*Orconectes rusticus*) for discrimination of female odours. Behaviour 143:713–731

Bergman DA, Moore PA (2003) Field observations of intraspecific agonistic behavior of two crayfish species, *Orconectes rusticus* and *Orconectes virilis*, in different habitats. Biol Bull 205:26–35

Bergman DA, Moore PA (2005) Prolonged exposure to social odours alters subsequent social interactions in crayfish (*Orconectes rusticus*). Anim Behav 70:311–318

Bergman DA, Kozlowski C, McIntyre JC, Huber R, Daws AG, Moore PA (2003) Temporal dynamics and communication of winner-effects in the crayfish, *Orconectes rusticus*. Behaviour 140:805–825

Berry FC, Breithaupt T (2008) Development of behavioural and physiological assays to assess discrimination of male and female odours in crayfish, *Pacifastacus leniusculus*. Behaviour 145:1427–1446

Berry FC, Breithaupt T (2010) To signal or not to signal? Chemical communication by urine-borne signals mirrors sexual conflict in crayfish. BMC Biol 8:25

Billeter J-C, Atallah J, Krupp JJ, Millar JG, Levine JD (2009) Specialized cells tag sexual and species identity in *Drosophila melanogaster*. Nature 461:987–991

Breithaupt T (2001) Fan organs of crayfish enhance chemical information flow. Biol Bull 200:150–154

Breithaupt T, Atema J (2000) The timing of chemical signaling with urine in dominance fights of male lobsters (Homarus americanus). Behav Ecol Sociobiol 49:67–78

Breithaupt T, Eger P (2002) Urine makes the difference: chemical communication in fighting crayfish made visible. J Exp Biol 205:1221–1231

Breithaupt T, Lindstrom DP, Atema J (1999) Urine release in freely moving catheterised lobsters (Homarus americanus) with reference to feeding and social activities. J Exp Biol 202:837–844

Briffa M, Sneddon LU (2006) Physiological constraints on contest behaviour. Funct Ecol 21:627–637

Bushmann PJ, Atema J (1996) Nephropore rosette glands of the lobster Homarus americanus: possible sources of urine pheromones. J Crust Biol 16:221–231

Copp NH (1986) Dominance hierarchies in the crayfish Procambarus clarkii (Girard, 1852) and the question of learned individual recognition (Decatoda, Astacidea). Crustaceana 51:9–24

Corotto FS, Bonenberger DM, Bounkeo JM, Dukas CC (1999) Antennule ablation, sex discrimination, and mating behavior in the crayfish Procambarus clarkii. J Crust Biol 19:708–712

Crandall KA, Buhay JE (2008) Global diversity of crayfish (Astacidae, Cambaridae, and Parastacidae-Decapoda) in freshwater. Hydrobiologia 595:295–301

Cromarty SI, Mello J, Kass-Simon G (1998) Comparative analysis of escape behavior in male, and gravid and non-gravid, female lobsters. Biol Bull 194:63–71

Denissenko P, Lukaschuk S, Breithaupt T (2007) Flow generated by an active olfactory system of the red swamp crayfish. J Exp Biol 210:4083–4091

Dunham PJ (1978) Sex pheromones in crustacea. Biol Rev 53:555–583

Dunham DW, Oh JW (1992) Chemical sex discrimination in the crayfish Procambarus clarkii. Role of antennules. J Chem Ecol 18:2363–2372

Edwards DH, Heitler WJ, Krasne FB (1999) Fifty years of a command neuron: the neurobiology of escape behavior in the crayfish. Trends Neurosci 22:153–161

Galeotti P, Rubolini D, Fea G, Ghia D, Nardi PA, Gherardi F, Fasola M (2006) Female freshwater crayfish adjust egg and clutch size in relation to multiple male traits. Proc R Soc B 273:1105–1110

Gherardi F (2002) Behaviour. In: Holdich DM (ed) Biology of freshwater crayfish. Blackwell Science, Oxford, pp 258–290

Gherardi F, Daniels WH (2003) Dominance hierarchies and status recognition in the crayfish Procambarus acutus acutus. Can J Zool 81:1269–1281

Gherardi F, Holdich DM (1999) Crayfish in Europe as alien species: how to make the best of a bad situation? A.A. Balkema, Rotterdam

Giri T, Dunham DW (2000) Female crayfish (Procambarus clarkii (Girard, 1852)) use both antennular rami in the localization of male odour. Crustaceana 73:447–458

Goessmann C, Hemelrijk C, Huber R (2000) The formation and maintenance of crayfish hierarchies: behavioral and self-structuring properties. Behav Ecol Sociobiol 48:418–428

Hazlett BA (1985) Disturbance pheromones in the crayfish Orconectes virilis. J Chem Ecol 11:1695–1711

Herberholz J, Issa FA, Edwards DH (2001) Patterns of neural circuit activation and behavior during dominance hierarchy formation in freely behaving crayfish. J Neurosci 21:2759–2767

Herberholz J, McCurdy C, Edwards DH (2007) Direct benefits of social dominance in juvenile crayfish. Biol Bull 213:21–27

Hill AM, Lodge DM (1999) Replacement of resident crayfishes by an exotic crayfish: the roles of competition and predation. Ecol Appl 9:678–690

Horner AJ, Schmidt M, Edwards DH, Derby CD (2008) Role of the olfactory pathway in agonistic behavior of crayfish, Procambarus clarkii. Invert Neurosci 8:11–18

Huber R, Delago A (1998) Serotonin alters decisions to withdraw in fighting crayfish. Astacus astacus: the motivational concept revisited. J Comp Physiol A 182:573–583

Huber R, Orzeszyna M, Pokorny N, Kravitz EA (1997) Biogenic amines and aggression: experimental approaches in crustaceans. Brain Behav Evol 50:60–68

Huxley TH (1879) The crayfish: an introduction to the study of zoology. C. Kegan Paul, London

Issa FA, Adamson DJ, Edwards DH (1999) Dominance hierarchy formation in juvenile crayfish *Procambarus clarkii*. J Exp Biol 202:3497–3506

Johnson ME, Atema J (2005) The olfactory pathway for individual recognition in the American lobster *Homarus americanus*. J Exp Biol 208:2865–2872

Karavanich C, Atema J (1998) Individual recognition and memory in lobster dominance. Anim Behav 56:1553–1560

Kravitz EA (2000) Serotonin and aggression: insights gained from a lobster model system and speculations on the role of amine neurons in a complex behavior. J Comp Physiol A 186:221–238

Little EE (1976) Ontogeny of maternal behavior and brood pheromone in crayfish. J Comp Physiol 112:133–142

Livingstone MS, Harris-Warrick RM, Kravitz EA (1980) Serotonin and octopamine produce opposite postures in lobsters. Science 208:76–79

Mason JC (1970) Copulatory behaviour of the crayfish, *Pacifastacus trowbridgeii* (Stimpson). J Zool 48:969–976

McMahon A, Patullo BW, Macmillan DL (2005) Exploration in a T-maze by the crayfish *Cherax destructor* suggests bilateral comparison of antennal tactile information. Biol Bull 208:183–188

Moore PA (2007) Agonistic behavior in freshwater crayfish: the influence of intrinsic and extrinsic factors on aggressive encounters and dominance. In: Duffy JE, Thiel M (eds) Evolutionary ecology of social and sexual systems: crustaceans as model organisms. Oxford University Press, Oxford, pp 90–114

Nystrom P (2002) Ecology. In: Holdich DM (ed) Biology of freshwater crayfish. Blackwell Science, Oxford, pp 192–235

Peeke HVS, Blank GS, Figler MH, Chang ES (2000) Effects of exogenous serotonin on a motor behavior and shelter competition in juvenile lobsters (*Homarus americanus*). J Comp Physiol A 186:575–582

Reynolds JD (2002) Growth and reproduction. In: Holdich DM (ed) Biology of freshwater crayfish. Blackwell Science, Oxford, pp 152–191

Schroeder L, Huber R (2001) Fight strategies differ with size and allometric growth of claws in crayfish, *Orconectes rusticus*. Behaviour 138:1437–1449

Seebacher F, Wilson RS (2006) Fighting fit: thermal plasticity of metabolic function and fighting success in the crayfish *Cherax destructor*. Funct Ecol 20:1045–1053

Seebacher F, Wilson RS (2007) Individual recognition in crayfish (*Cherax dispar*): the roles of strength and experience in deciding aggressive encounters. Biol Lett 3:471–474

Simon JL, Moore PA (2007) Male-female communication in the crayfish *Orconectes rusticus*: the use of urinary signals in reproductive and non-reproductive pairings. Ethology 113:740–754

Sneddon LU, Taylor AC, Huntingford FA, Watson DG (2000) Agonistic behaviour and biogenic amines in shore crabs *Carcinus maenas*. J Exp Biol 203:537–545

Stebbing PD, Bentley MG, Watson GJ (2003) Mating behaviour and evidence for a female released courtship pheromone in the signal crayfish *Pacifastacus leniusculus*. J Chem Ecol 29:465–475

Tierney AJ, Dunham DW (1984) Behavioral mechanisms of reproductive isolation in crayfishes of the genus *Orconectes*. Am Midl Nat 111:304–310

Tricarico E, Gherardi F (2007) Biogenic amines influence aggressiveness in crayfish but not their force or hierarchical rank. Anim Behav 74:1715–1724

Wyatt TD (2003) Pheromones and animal behaviour: communication by smell and taste. Cambridge University Press, Cambridge

Yeh SR, Musolf BE, Edwards DH (1997) Neuronal adaptations to changes in the social dominance status of crayfish. J Neurosci 17:697–708

Zulandt Schneider RA, Schneider RWS, Moore PA (1999) Recognition of dominance status by chemoreception in the red swamp crayfish, *Procambarus clarkii*. J Chem Ecol 25:781–794

Zulandt Schneider RA, Huber R, Moore PA (2001) Individual and status recognition in the crayfish *Orconectes rusticus*: the effects of urine release on fight dynamics. Behaviour 138:137–153

Chapter 14
Chemical Communication in Decapod Shrimps: The Influence of Mating and Social Systems on the Relative Importance of Olfactory and Contact Pheromones

Raymond T. Bauer

Abstract Interest in chemoreception of decapod shrimps has been stimulated by observations indicative of sex pheromones, such as frenzied male searching and copulatory activity in the presence of premolt or recently postmolt reproductive females. Review of previous studies on shrimp mating behavior led to the formulation of hypotheses about the variation of chemical communication with mating and social systems. Penaeoidean and many caridean species are highly mobile and live in dense aggregations, resulting in frequent contacts among individuals. In such species, males are usually stimulated to copulatory behavior by apparent contact pheromones on the newly molted female's exoskeleton, received by contact with the male's antennal flagella. In species with temporary mate guarding, males search for premolt reproductive females, which release water soluble substances received by olfactory receptors (aesthetascs). Males guard females for some days until the female molt, after which mating occurs, followed by male abandonment to search for other females. In "neighborhoods of dominance" mating systems, it is the premolt parturial female that seeks out a large dominant male, stimulated by his olfactory pheromones. She is then guarded by the male, which will mate with her after her molt. In monogamous mate guarding species, males and females form permanent pairs, with the initial pairing perhaps mediated by sex pheromones emitted and perceived by both sexes. Olfactory sex pheromones are given off by females in many caridean species just after the molt, stimulating nearby males and ensuring mate finding. Recognition of pair partners or social (agonistic) status of an individual is chemically mediated in various decapod shrimps. The exact source and chemical composition of olfactory sex pheromones is still unknown, but both cuticular hydrocarbons and glycoproteins have been implicated as contact sex pheromones. Comparative studies with additional species are required to test these hypotheses about the form of chemical communication in different mating systems. Isolation and chemical identification of sex and individual-recognition

R.T. Bauer (✉)
Department of Biology, University of Louisiana,
Lafayette, LA 70504, USA
e-mail: rtbauer@louisiana.edu

T. Breithaupt and M. Thiel (eds.), *Chemical Communication in Crustaceans*,
DOI 10.1007/978-0-387-77101-4_14, © Springer Science+Business Media, LLC 2011

pheromones is a major avenue of future research. Results of such studies may not only result in a greater understanding of chemoreception per se, but also may lead to commercial applications in shrimp fisheries and aquaculture.

14.1 Introduction

Striking behavioral responses to waterborne odors by decapod shrimps ("natantians") indicate that they, like many other crustaceans, rely on chemical senses in food searching, mate recognition, and social interactions. My initial interest in this subject began in graduate school, stimulated by behavioral observations on small tide-pool shrimps, *Heptacarpus sitchensis*, in aquaria. When food items were placed in the water, shrimps at some distance from the food responded within seconds. Their increased searching and probing of their surroundings indicated that an olfactory sense was at work. However, males did not become noticeably excited when a receptive (recently molted) female was placed at some distance from them in the aquarium. Only when the male wandered idly by and touched the female with one of its chemotactile antennal flagella did it react by frantically turning towards and grasping and copulating with the female. This began a research interest in the olfactory and chemotactile cues used by shrimps in their mating and social behaviors. Chemical communication in decapod shrimps is not surprising because most species are nocturnal or live in turbid environments in which visual cues are limited. However, this research area, although expanding, is still mainly limited to basic behavioral studies on relatively few decapod shrimps from a phylogenetically and ecologically diverse assemblage.

The purpose of this chapter is to briefly review the most important work done on sex pheromones (mate attraction and recognition) in relation to species mating system. Important gaps in our knowledge are presented and hypotheses are proposed that might be tested in future research. The form and role of chemical communication in intraspecific social interactions (dominance relationships and individual recognition) is a relatively unexplored but expanding field of interest in decapod shrimps. Comparisons of chemical communication with other decapods, especially crabs (Sneddon et al. 2003), lobsters (Atema and Steinbach 2007, Aggio and Derby, Chap. 12), and crayfishes (Breithaupt and Eger 2002; Breithaupt, Chap. 13) as well as other taxonomic groups, e.g., the insects (Howard and Bloomquist 2005; Wyatt 2003) will aid in proposing hypotheses for future testing.

Phylogenetic differences in basic ecology and life history impose both constraints and opportunities on reproduction that are reflected in the forms of chemical communication of decapod shrimps. Although "decapod shrimps" refer to an assemblage of taxa ("natantians") with a similar primitive body form ("caridoid facies"), this group is not phylogenetically homogeneous (Bauer 2004; Fransen and De Grave 2009). The carideans (~3,200 spp.) are especially diverse ecologically, with a variety of life styles in marine habitats at all depths and latitudes, and they have invaded freshwater habitats as well. The penaeoideans (over 400 spp.) are schooling shrimps in warm waters of soft-bottom shallow seas, while the sergestoids (~100 spp.) are planktonic. The stenopodideans (~70 spp.) are marine shrimps, often symbiotic with other invertebrates, occurring in male–female pairs. The decapod shrimps, with their diverse ecologies and life histories, show considerable variation in mating systems, and their chemical communication as will be discussed in this chapter.

14.2 Mating Behavior and Mating Systems

The adaptive value of sex pheromones, the most studied form of chemicals mediating communication in decapod shrimps, can be best understood in the context of their basic reproductive biology and mating systems. Penaeoid shrimps have two very different reproductive biologies (Bauer 1991). In the largest family (Penaeidae) "open thelycum" species mate without molting when the female is ready to spawn, at least in the commercially important genus *Litopenaeus* (Yano et al. 1988). "Closed thelycum" species have sperm-storage structures that are filled by the male just after a female molt, which is not closely tied to spawning. In caridean and stenopodidean shrimps, females with vitellogenic gonads mate just after a parturial (prespawning) molt. Spawning takes place soon thereafter, with the female attaching the fertilized eggs (embryos) below the abdomen for an extended incubation (Bauer 2004).

The mating systems of decapod shrimps can be summarized into relatively simple categories (generalizations) based on characteristics such as encounter rate between males and females, the duration of male–female interaction, and sexual dimorphism in body size and weaponry (major chelipeds and third maxillipeds). Bauer (2004) proposed three basic mating systems in carideans, based on those given by Wickler and Seibt (1981). In monogamy ("mate-guarding monogamy"; also considered as "extended mate guarding"), a male and female of similar size live in pairs that are permanent, at least within a single breeding season. The male continuously guards and defends the female from other males to ensure that when she molts, he will be at her side to mate. In temporary female guarding, the usually larger male with enhanced appendage weaponry seeks out females with maturing ovaries and guards her (24 h to a few days; Grafen and Ridley 1983; Ridley 1983) through the mating molt, after which he soon abandons her to seek other females. In pure searching, the usually smaller males guard neither females nor territories prior to mating. These shrimps occur in aggregations or schools, and individual mobility and population density are high. Males frequently encounter other members of the population, checking them for sex and receptivity. Newly molted females are quickly copulated and abandoned. Monogamy and temporary female guarding are characteristic of species with low encounter rates. In such species, males and females do not contact each other frequently because population density is low or their effective mobility is limited (species living in shelters or subject to high predation pressure). Correa and Thiel (2003) described another category of mating system, "neighborhoods of dominance," for shrimps such as *Rhynchocinetes typus* and *Macrobrachium rosenbergii*. In these species, morphologically distinct and aggressively dominant males control areas to which receptive postmolt females are attracted.

The basic form of chemical communication via sex pheromones might be expected to vary with the mating system of the species (Table 14.1 and Fig. 14.1). Females must ultimately advertise their reproductive condition and mating receptivity to males. Receptivity is limited to a brief period (a few hours to a day) after a parturial molt in all carideans, stenopodideans, and many penaeoids shrimps, or it occurs just before spawning in some penaeoids, e.g., *Litopenaeus* spp. (Yano et al. 1988). Females might advertise to attract males before becoming receptive when encounter rates with males are low or male assistance (defense, access to resources) is needed for successful reproduction. On the other hand, when contact with males is frequent as in schooling species (e.g., many pandalids, crangonids, and penaeoids), it might be advantageous for females to limit their advertisement prior to receptivity to prevent harassment from males. Advertisement over a distance is mediated by olfactory pheromones, whereas sexual condition at close range is accomplished by contact (gustatory and chemotactile) pheromones. Where male–female interaction is relatively brief, individual recognition is not predicted to evolve, but in monogamous mating systems it should be adaptive. In the following, I explore hypotheses about the form of chemical communication involved in sex attraction and mating behavior in various categories of decapod shrimp mating systems (Table 14.1), and I present the knowledge leading to these hypotheses.

Table 14.1 Hypotheses on the form of chemical communication in four different categories of decapod shrimp mating systems

| | Mating systems | | | |
	Pure searching	Temporary mate guarding	Monogamy "monogamous mate guarding"	Neighborhoods of dominance
Hypotheses				
Principal sex pheromone	Contact	Olfactory	Olfactory	Olfactory
Timing of pheromone advertisement	Upon female molting or ovulation	$\geq$24 h Prior to female molting	Prior to pairing	Prior to female molting
Sex of pheromone advertises	Female	Female	Male or female	Male
Individual recognition	Absent	Low absent	Present	Low absent

These hypotheses have been derived from information on sex attraction and mating behavior in the published literature which is presented in the chapter

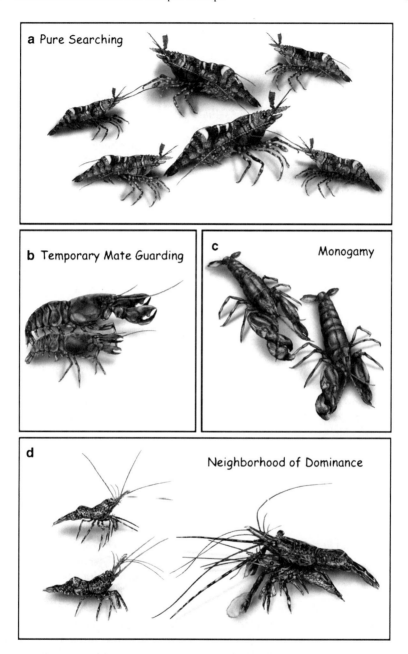

Fig. 14.1 Mating systems in decapod shrimps: sexual dimorphism in body size and weaponry, and pattern of male–female association. (**a**) Several small males searching for a larger receptive female (pure searching); (**b**) large male with hypertrophied chelipeds guarding smaller female (that he has been attracted to) prior to her mating molt (temporary mate guarding); (**c**) male and female of similar body size and weaponry in long-term pairing (monogamy); (**d**) large dominant male with hypertrophied weaponry guarding a receptive female (attracted to him) from smaller subordinate males (neighborhoods of dominance). Drawing by Jorge A. Varela Ramos

14.3   Female Olfactory Sex Pheromones

Behavioral observations on several caridean species (see earlier summary in Table 14.1 in Bauer 1979) as well as a few experimental studies indicate that female olfactory sex pheromones are often emitted *upon* or *after* the mating molt. Such pheromones stimulate obvious male search or courtship behavior in, for example, *Palaemon paucidens* (Kamiguchi 1972), *Hymenocera picta* (Seibt and Wickler 1979), or *M. rosenbergii* (in the small subordinate "sneaker" males; Karplus et al. 2000). On the other hand, careful Y-maze studies by Díaz and Thiel (2004; for Y-maze see Fig. 10.2 in Thiel, Chap. 10) revealed that males of the caridean *R. typus* were unable to locate receptive females using olfactory cues. However, females of this species apparently respond to and locate large dominant males using olfaction. In *Alpheus angulatus*, Mathews (2003) showed with a Y-maze experiment that males, which live paired with females in burrows, were attracted to water exposed to premolt females. However, males did not respond to premolt males and avoided water from intermolt males and females. Attraction of large males to premolt females is also indicated in another alpheid, *Athanas kominatoensis* (Nakashima 1987).

In some shrimp species, males enter into a frenzied search behavior upon detection of a water-soluble postmolt female substance. These males are more likely to respond to a conspecific with some sort of precopulatory behavior. The substance is not related to molting per se, because males do not respond to recently molted (soft exoskeleton) nonparturial females or males. In *H. picta*, the female olfactory pheromone is effective in attracting males for 3–5 h after her molt (Seibt and Wickler 1979). Males are induced to court any female when parturial female molt water is introduced into an aquarium (Seibt 1973). Kamiguchi (1972) showed with careful observations that a water-soluble substance was only effective in stimulating males of *P. paucidens* to search for ~30 min. However, males could recognize receptive females upon contact for a few hours after the female's parturial molt.

A postmolt rather than premolt release of pheromone may be viewed as adaptive in species living at high density aggregations with pure searching mating systems (e.g., many *Pandalus, Palaemon, Palaemonetes,* and *Lysmata* spp.), where advertisement by females to males long before the molt might to lead to their harassment and injury by males. The effect of the postmolt female pheromone has been shown in a rigorous experimental setup using *P. paucidens* as a model species by Kamiguchi (1972). Isolated males initiated searching behavior when water from an aquarium with a recently molted parturial female was pipetted into the male aquarium. Mating experiments (Fig. 14.2) showed that males primarily responded to postmolt parturial females. However, some males attempted to copulate with females in other reproductive states in the presence of postmolt female water, which presumably contained olfactory sex pheromone.

In the caridean *Heptacarpus paludicola*, Bauer (1979) was unable to detect obvious searching behavior by males for a receptive female in the same aquarium.

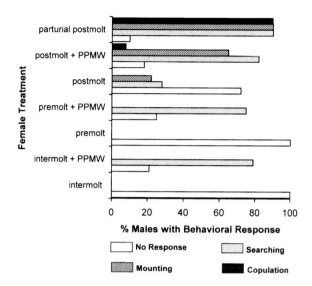

Fig. 14.2 Results of experiments by Kamiguchi (1972) showing effect of postmolt sexual pheromone given off by females of *Palaemon paucidens*. Males were exposed to females in different reproductive states, with or without "postparturial molt water" (PPMW), i.e., water from a tank containing a recently molted parturial female. Male responses (none, searching, mounting, and copulation) were recorded. Parturial postmolt: a recently molted female with mature ovaries; postmolt: a recently molted female without mature ovaries; premolt: female with mature ovaries nearing a parturial molt; intermolt: a female neither near molting nor spawning (from Bauer 2004)

To test for a more subtle response to such females, an "olfactometer" was devised. Males were kept downstream of an inlet chamber containing either an intermolt female incubating embryos (not sexually receptive) or a newly molted parturial (receptive female). Male response (swimming behavior) was recorded. First, however, the olfactometer was tested using a known stimulatory food item (positive control) and a nonstimulatory small rock (negative control) and found satisfactory to measure an olfactory response (Fig. 14.3a). Male response to receptive and nonreceptive females was then tested, resulting in a higher quantitative response toward the former (Fig. 14.3b). However, a striking qualitative response by males was not apparent, indicating that olfactory detection of females may not be important in this species. Postmolt contact pheromones are the primary means of sex attraction and recognition in this and other aggregated species. Perhaps females are not advertising with a pheromone but rather hiding their reproductive condition to avoid harassment, but they are not completely successful in doing so. Males, on the other hand, have evolved a method, albeit a relatively poor one, of detecting the upcoming female molt. Thus, a conflict of sexual interests between males and females may be occurring in such species.

A *premolt* "awareness" of a soon-to-molt female has been reported in some caridean species, perhaps indicative of a female olfactory pheromone. However, the period is rather short, approximately an hour or less before the molt. In *Palaemonetes*

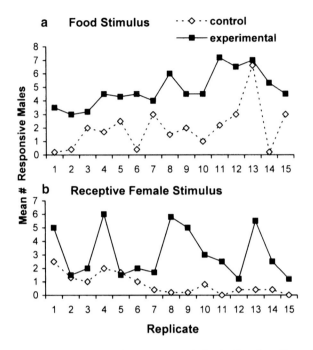

Fig. 14.3 Olfactometer experiments in *Heptacarpus paludicola* (data from Bauer 1979). In each replicate, waterborne stimuli were presented to a group of ten males located downstream of the stimuli. Five observations on the number of active males were made per replicate. (**a**) Experimental stimulus = food item, control = rock; (**b**) experimental stimulus = a newly molted parturial (receptive) female, control = an intermolt (nonreceptive) female (reproduced from Bauer 2004)

pugio, males made more frequent contact with females in the minutes just before the female's molt (Bauer and Abdalla 2001; Fig. 14.4). This does not result in a pairing with the female in *P. pugio* and might be interpreted as an increase in male searching behavior, enhancing the probability of encountering the soon-to-be molted female. In *Lysmata wurdemanni*, a protandric simultaneous hermaphrodite, male-phase individuals approach and remain in contact with a prespawning female-phase individual about an hour before the latter's molt, when mating occurs (Bauer 2002; Zhang and Lin 2006). In both *P. pugio* and *L. wurdemanni*, it may be that the male detects the female by a water-soluble substance. In both species, it is difficult to separate response to an olfactory vs. a contact pheromone. These studies on mating behavior were conducted within a small arena where physical contact (especially with the long antennal flagella) was probably frequent.

Bauer and Abdalla (2001) proposed an alternative hypothesis to explain the apparent male "awareness" within a short period of an upcoming female parturial molt in aggregating species. Females of such species are actually hiding their reproductive condition until the molt to prevent harassment from males. Males simply may be sensing a female metabolite, indicating her reproductive state that she can no longer control as the actual molt approaches.

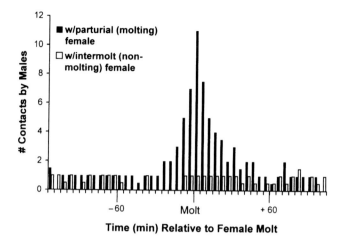

Fig. 14.4 Responses of males to parturial and intermolt females in *Palaemonetes pugio* (Bauer and Abdalla 2001). The graph shows the number of contacts per 5-min intervals by two males with a parturial female and a nonparturial (intermolt) female before and after the molt of the parturial female. Note that males recognized the parturial female (made significantly more contacts) starting about 25 min prior to the molt (reproduced from Bauer 2004)

Penaeoid shrimps characteristically live in aggregations ("schools") and have the expected pure search mating system of such species. Observations on mating behavior in penaeoidean species are limited, but suggest male responses to females similar to that of carideans with a similar aggregated life style (e.g., Hudinaga 1942; Bauer 1996). In *Litopenaeus vannamei*, Yano et al. (1988) cited unpublished observations, suggesting that males are stimulated to precopulatory chasing of other individuals (not necessarily females) when "female water" from a reproductively mature female is introduced into their environment. Thus, in the usually schooling penaeid species, as in similarly aggregated caridean species, males may be using water-soluble substances inadvertently released by females, which become attractive as her physiological state approaches receptivity.

Olfactory sex pheromones might be found in caridean species with mating systems involving temporary or extended (monogamous) male guarding of females before the parturial molt. In some carideans, males guard females by remaining near them (attendance) and defending them from the advances of other males. For example, Nakashima (1987) worked on the alpheid *A. kominatoensis*, which lives in association with sea urchins. In this nonmonogamous alpheid species, a large combative male guards an isolated parturial female prior to her molt but abandons her after mating. It would be advantageous for males to detect, find, and monopolize such females before other males do. Similar to the American lobster *Homarus americanus* (Atema and Steinbach 2007) and the shrimp *R. typus* (Díaz and Thiel 2004), premolt parturial females of *M. rosenbergii* (Karplus et al. 2000) and *M. australiense* (Lee and Fielder 1982) seek out dominant males. The male apparently attracts the female by regularly or continuously releasing his water-soluble pheromone. The dominant male may then recognize the female (and vice versa) by an interchange of olfactory and/or contact pheromone cues before allowing her into the mating nest.

14.4 Contact Sex Pheromones

In some caridean species, males are not noticeably stimulated at a distance by parturial (prespawning) females (e.g., *Palaemonetes vulgaris*, Burkenroad 1947; *H. sitchensis, H. paludicola*, Bauer 1976, 1979; *P. pugio*, Bauer and Abdalla 2001). It is only when males touch a newly molted parturial female with the long antennal flagellum that they react dramatically, seizing the female and copulating with her (Fig. 14.5). Burkenroad (1947) first proposed in *P. vulgaris* that a "nondiffusible" substance on the female exoskeleton was perceived by the male, which stimulated it to copulate. The duration of attractiveness was several hours, similar to that found by Caskcy and Bauer (2005) for *P. pugio* (8 h; Fig. 14.6) in uninseminated newly molted females. As soon as spawning takes place (2–3 h after mating), the female becomes unattractive to males. The period of attractiveness of postmolt

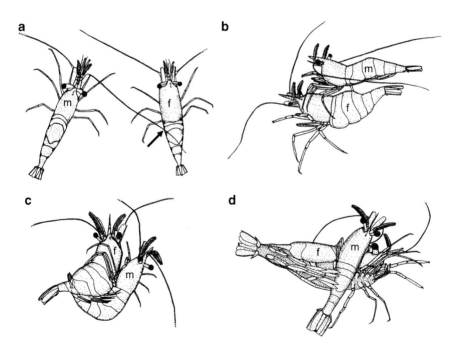

Fig. 14.5 Mating behavior of the caridean *Heptacarpus sitchensis* (Bauer 1976), a species with a pure searching mating system. (**a**) Male makes contact with female via antennal flagellum (*arrow*); (**b**) male mounts female; (**c, d**) copulation (male dips abdomen below that of female, which lowers pleopods to allow spermatophore deposition) (adapted from Bauer 2004)

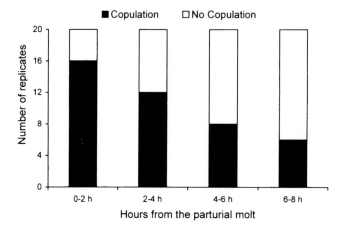

Fig. 14.6 Duration of female attractiveness to males after the female molt in the caridean *Palaemonetes pugio*. Females at varying intervals after the parturial molt were presented to males, and the presence or absence of copulation ("attractiveness") was recorded ($n = 20$ replicates per time interval) (reproduced from Caskey and Bauer 2005)

uninseminated females may be longer, up to 24 h (*Heptacarpus* spp., Bauer 1976, 1979; *R. typus*, Díaz and Thiel 2004).

Caskey and Bauer (2005) proposed that the "nondiffusible substance" is a contact sex pheromone perceived by receptors on setae borne by the antennal flagella and/or pereopods. The sexual condition of the female is thus perceived by "taste" rather than "smell," adaptive in the context of an aggregated species with a pure searching male mating strategy. The substances involved in chemotactile sex attraction and recognition must have little or no solubility and probably occur in relatively high concentrations on the exoskeleton compared with olfactory sex pheromones. Because the male receptors must make actual or very near contact with the surface bearing such substances, they can be termed "contact pheromones" (see Wyatt, Chap. 2; Snell, Chap. 23).

Experiments using individuals in different sexual and molting states have eliminated textural cues such as a soft postmolt cuticle or visual cues in sex attraction and recognition. For example, Bauer (1979: *H. sitchensis*) and Caskey and Bauer (2005: *P. pugio*) performed experiments in which males showed no mating response toward recently molted (soft) males and only a slight or no response to recently molted nonparturial females. Copulatory response was elicited only by newly molted females and only upon antennal or pereopod contact by the male.

In some high density species, visual cues may attract males to a receptive female once one or more males has contacted such a female and is attempting copulation. Díaz and Thiel (2004) used the term "tumult" to describe this situation which they documented with careful experiments. Once a male contacts a newly molted female, the jumping around and chasing of the female attracts the attention of other nearby males, presumably by vision. Similar tumults have also been observed in *L. wurdemanni* (pers. obs.). Response to tumults may be common in aggregating or schooling shrimps. It also occurs in response to other stimuli, e.g., discovery of large food items (pers. obs.). The rapidity with which these tumults develop underlines the importance for the female to hide its reproductive status in order to maintain some control of mating interactions and avoid male harassment (e.g., tumults) which might cause injury or death.

14.5   Chemically Mediated Recognition of Individual, Sexual, and Social Status

Recognition of particular or unique individuals may also be adaptive, e.g., identification of a male or female partner in pair-living species, or perhaps of individuals frequently or previously encountered in an individual's environment (winners or losers of previous agonistic encounters). Many stenopodidean shrimps live in pairs associated with other invertebrates, and there are numerous species of carideans, especially in the families Alpheidae, Palaemonidae (Pontoniinae), Hymenoceridae, and Hippolytidae that live in pairs. In such species, identification of the opposite sex and specific individuals (pair partners) at a distance might be important. In other carideans in which mating territories or colony homes are defended (e.g., *Macrobrachium* with large-bodied, large-clawed males; eusocial *Synalpheus* spp.), social or family group recognition is adaptive. In the eusocial *Synalpheus* spp. (Duffy 2007), chemical cues may mediate sterility of colony workers as well as communal colony defense, as in many of the social insects (Wyatt 2003).

Johnson (1969, 1977) used courtship and agonistic behaviors in the pair-living shrimp *Stenopus hispidus* to test the hypotheses that individuals could identify (a) members of the opposite sex (possible pair partners) and (b) a previous pair partner from among other individuals of the same sex. Members of the opposite sex could be identified by "touch" (presumably contact chemoreception) as shown by reduced agonistic interactions and courtship behaviors between previously unpaired males and females after a short period of physical contact (Johnson 1969). Individuals of the same sex, however, remained aggressive, often fighting to the death and showed no courtship behaviors (Johnson 1969). Olfactory cues allowed individuals to detect the presence of a conspecific but not their sex. In another set of experiments (Johnson 1977), "mates" (previously paired male and female) showed low levels of courtship when introduced to each other after a period of separation (Fig. 14.7a). On the other hand, individuals of the opposite sex that had never met before ("strangers") showed more intense courtship behavior (Fig. 14.7b). This demonstrated that previous mating partners recognized and "remembered" each other. They did not need to expend as much effort on courtship behaviors with previous mates as with strangers. Recognition of previous mates was possible after separations up to 6 days (a complete molt cycle). Contact chemoreception was considered the primary sensory mode involved in individual recognition.

In the pair-living caridean *H. picta*, the role of olfactory pheromones in individual recognition has been shown convincingly by simple behavioral observations and

Fig. 14.7 Recognition of individual mating pair partners in *Stenopus hispidus* after a separation of 2 days (data from Johnson 1977). (**a**) Previous mating partners ("mates") exhibited low levels of courtship behavior when reintroduced compared to (**b**) individuals of the opposite sex which had never met ("strangers")

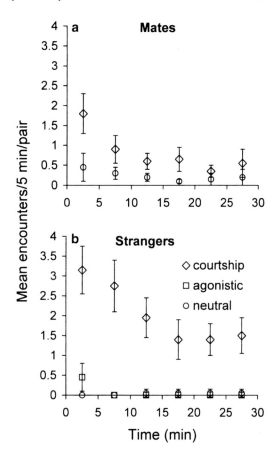

experiments by Seibt (1973). Variations in natural color markings allowed observers to distinguish between individuals of pairs in an aquarium microcosm. Male and female pair partners remained together for the period of observation. In one experiment, males were separated from their female partners, which were hidden from visual and physical contact by nets. After separation, males found and remained with their original partner from among several other similarly hidden females. Y-maze olfactometer experiments showed that *H. picta* could identify and choose water emanating from food vs. no food and from a conspecific vs. some other crustacean. Using olfactory cues, males could distinguish between males and females; a male could also identify its original female partner from other females. These tests were made on intermolt females so that the pre- and postmolt olfactory pheromones of parturial females mentioned earlier were not involved. Although members of this species can recognize other conspecifics visually, sex recognition from a distance can occur in the dark when visual cues are eliminated, and Y-maze tests indicated that olfaction was the sensory mode involved.

Dominance relationships among snapping shrimps (*Alpheus heterochaelis*) have been the focus of several studies. These shrimps live in burrows in female–male pairs. Both male and female defend the burrow and each other with the "snapping claw" which shoots a powerful jet of water in front of the animal (Bauer 2004). With same sex, Y-maze experiments and measures of aggressive activity, both Ward et al. (2004) and Obermeier and Schmitz (2003a) demonstrated that individuals can recognize dominants (individuals against which the focal individual has lost a previous agonistic encounter) from other individuals without previous fighting experience. The latter authors also showed that individuals do not distinguish (behave differently) between familiar and unfamiliar dominants. In experiments with *A. heterochaelis* on olfactory cues, Schein (1975) demonstrated with a Y-maze study that males can distinguish males from intermolt females, but Hughes (1996) found no such ability. However, Mathews (2003) found with similar techniques that males do recognize and are attracted to premolt females with olfactory cues. Rahman et al. (2001), measuring degree of aggressive behavior, found that both sexes can distinguish former mates from strangers. Recognition of mating partners, important in a socially monogamous species, has evolved in *A. heterochaelis*, as it has in *H. picta*, and perhaps will be found to be a common feature of pair-living species. Certainly, it is important for individuals of either sex to distinguish and avoid dominant individuals, which have previously defeated them in agonistic interactions. If individuals can detect and recognize conspecifics, either individually or by group (male vs. female, partner vs. nonpartner, and dominants vs. subordinates), a great deal of potentially injurious fighting can be avoided.

14.6 Transmission and Reception of Pheromones

The actual source and composition of olfactory sex pheromones in decapod shrimps is unknown. One can only speculate by comparison with other decapods (e.g., Atema and Steinbach 2007) that olfactory sex pheromones are contained in and emitted in the urine. However, the simple experiment of plugging nephrophores (antennal gland openings) has not been done in decapod shrimps, probably because of the small body size of most species. Given that the most striking attraction of males by female olfactory substances occurs after and not before the female mating molt in caridean shrimps, there is a good possibility that substances emanating from other sources (e.g., the newly molted exoskeleton) might also be involved.

Transmission of olfactory sex pheromones and odors involved in individual or group recognition is likely to be similar to those found in lobsters and crayfishes. In these decapods, gill and "fan-organ" currents are generated by the excurrent respiratory stream and its modification by beating of the maxillipedal exopods. Herberholz and Schmitz (2001) used plastic microparticles and ink to demonstrate similar currents in the shrimp *A. heterochaelis* and their role in intraspecific interactions. A rather slow "normal" anteriorly directed current is generated by the beating of the scaphognathites (gill bailers), which produces the respiratory currents. The anterior current can be directed laterally, either right or left, by the unilateral beating of the exopods of the second and third maxillipeds. It can also be accelerated anteriorly ("fast anterior gill current"), perhaps by faster beating of the scaphognathites. A posteriorly directed pleopod current produced by in-place beating of the abdominal swimmerets (pleopods) was not shown to be important in social interactions. The anterior fast current was implicated in carrying chemical information from one individual to an agonistic opponent and was produced most frequently by winners of agonistic encounters. The lateral current is a mechanism to remove an individual's own odor from in front of the olfactory antennules so that the individual can perceive the odor of its opponent.

Reception of olfactory cues in shrimps appears similar to that by other decapods, i.e., with specialized setae (aesthetascs) on the outer antennular (first antenna) flagella. This has been demonstrated by ablation of these flagella, a somewhat crude and traumatic technique but effective when coupled with similar trauma in control treatments, i.e., ablation of the medial (inner) flagella (Obermeier and Schmitz 2003b) and/or second antennae (Zhang and Lin 2006).

Behavioral observations (Burkenroad 1947; Bauer 1976, 1979; Caskey and Bauer 2005; Zhang and Lin 2006) indicate that setae on the antennal (second antenna) flagella, as well as those on the third maxillipeds and anterior pereopods, are the site of reception of contact pheromones. Aside from aesthetascs, the study of shrimp antennular (Obermeier and Schmitz 2004) and antennal setae (Bauer and Caskey 2006) is still in the morphological and ultrastructural stages with more to be done; the physiology of these setae is poorly known. The latter statement also applies to possible chemotactile ("taste" or gustatory) setae on the maxillipeds and pereopods. "Antennulation," i.e., mutual touching of antennular and antennal flagella during physical interactions between conspecifics indicates a complex interplay between olfactory and contact chemical cues.

14.7 Chemical Composition of Pheromones

Olfactory sex pheromones of decapod shrimps have not yet been identified. Water-soluble compounds involved in sexual or recognition behaviors are most probably metabolites excreted in the urine or by the gills. Contact pheromones, given their probable higher concentration and distinct localization on the exoskeletal surface, offer good possibilities for identification. Long-chained cuticular hydrocarbons (CH), a rich source of identified contact sex pheromones in insects (Howard and Bloomquist 2005), have been proposed as possible candidates in decapod shrimps (Caskey and Bauer 2005; Caskey et al. 2009a). In insects, these compounds evolved primarily to prevent water loss through the cuticle and have secondarily been used as contact sex pheromones. Although they are much less abundant in aquatic crustaceans, these and similar compounds can be extracted by hexane and other solvents used in gas chromatography/mass spectrometry (GC-MS). Using this technique in the shrimp *P. pugio*, Caskey et al. (2009a) found a variety or "blend" of CHs that differed significantly among parturial postmolt (PPM, recently molted, sexually attractive and receptive) females and intermolt females, as well as from that of postmolt nonparturial females and males. The particular CH mix on the surface of PPM females potentially could serve as a contact pheromone to stimulate male mating. A suitable bioassay needs to be devised to test this hypothesis. Substances on the surface of PPM females in a variety of caridean species lose their attractiveness within hours to a day of molting, indicated that the substances are modified or made unavailable to male receptors by the chemical changes that accompany cuticular hardening after the molt.

Alternatively, other classes of compounds incorporated into the exoskeleton or secreted onto its surface might serve as contact pheromones, e.g., glycoproteins (Kelly and Snell, 1998; "arthropodin hypothesis" of Dunham, 1988; Snell, Chap. 23). Using one technique from Kelly and Snell (1998), Caskey et al. (2009b) exposed PPM females of *P. pugio* to males in seawater with a 50 mM concentration of glucosamine, a glycan component of N-acetyl glucosamine glycoproteins in crustaceans. Mating in these pairings was significantly lower than those arranged in aquaria with 50 mM glucose or natural seawater. In other experiments, significantly greater increases in Ca^{2+} activity were recorded when isolated male antennae were touched to carapaces of anesthetized PPM females or exposed to glucosamine, but not when touched to carapaces of postmolt nonparturial and intermolt females or postmolt males, or when exposed to glucose. Other observations indicated that neurons within male antennal setae were stimulated by glucosamine, as they might be when touching the exoskeleton of a PPM female with such a surface moiety. The reduction in male copulatory behavior observed in the glucosamine treatment of mating experiments might be explained by the binding of glucosamine dissolved in the water to male antennal receptors. Thus, males with blocked or habituated receptors could not perceive a glucosamine-containing glycoprotein on the female exoskeleton, blocking the copulatory response. In sum, this study implicates surface glycoproteins as contact pheromones in *P. pugio*. On the other hand, Zhang et al. (2010), using the lectin-binding method by Kelly and Snell (1998), suggested that surface glycoproteins do not mediate male mating response. Thus, the possible role of glycoproteins, as well as surface CH, as contact pheromones remains unresolved and open to further study.

If contact sex pheromones in shrimps are CHs, changes in activity of surface tegumental glands in the hours after the molt might account for the decrease in female attractiveness. Similarly, sex-specific cuticular glycoproteins might become less stimulatory because of changes associated with quinone tanning during postmolt sclerotization of the exoskeleton. Males of at least some shrimp species lose interest in PPM females immediately after mating and spermatophore transfer (e.g., *P. pugio*, Bauer and Abdalla 2001; *L. wurdemanni*, pers. obs.). The rapid decline in female attractiveness after spermatophore transfer might be due to a number of factors such as change in emission or form of the contact pheromone, a change in female behavior, or perhaps even a male-repulsive substance released near or even deposited on the female by the mating male, e.g., contained within deposited spermatophores.

14.8   Future Directions and Applied Aspects

Future work on sex pheromones should be conducted and interpreted within the context of the species' mating and social systems. The hypotheses in Table 14.1 about the form of chemical communication in different mating systems are based on relatively few species. Future observations and experimental studies, both field and laboratory, on more species in each category are needed to adequately test these hypotheses. The caridean genus *Lysmata* would be a good test group. It contains both species in aggregations ("crowd" species) with pure searching mating system and apparently monogamous ("paired") species (Bauer 2000). Behavioral testing for suspected pheromones should follow as closely as possible the guidelines so clearly stated by Dunham (1978), i.e., use of negative and positive controls, blind observation procedures, and precise response definitions. Nephropore blockage experiments will help to identify whether urine contains possible pheromones. Precise sampling of urine and water generated from the respiratory stream, both possible sources of olfactory pheromones, is a necessary prelude to their chemical description. Isolation and chemical identification of contact sex pheromones, e.g., CHs, glycoproteins, or other substances, is just in the beginning phase. The cause of the very rapid decline in female attractiveness to males after mating also requires investigation. Identification of the setal and neuronal receptors of both olfactory and contact pheromones needs to go beyond the relatively crude technique of ablation experiments and on to more sophisticated physiological techniques.

Chemical communication in shrimps is of applied interest as well. Manipulation of male and female behavior with olfactory and contact pheromones to trap or attract a particular set of individuals (e.g., large-sized males or females, reproductive females) in fisheries or aquaculture might be accomplished with sex pheromones, once identified and synthesized (Barki, Jones, and Karplus, Chap. 25). Matings between preferred genotypes might be promoted using sex pheromones to increase attractiveness and stimulate matings of selected individuals with desired traits.

14.9   Summary and Conclusions

Olfactory and contact chemoreception is the principal means of communication during sexual and social interactions in decapod shrimps. The relative importance of one or the other type of pheromone appears to vary with the mating and social system of the species (Table 14.1). In all carideans and in the "closed thelycum" penaeoids, reproductive females are sexually receptive just after molting. In these aggregated or schooling species, female contact-sex pheromones, possibly cuticular CHs or glycoproteins, are perceived by the antennal flagella, stimulating the male to copulate. In some of these species, a possible olfactory pheromone may be released from the female during or just after the molt which stimulates nearby males to searching behavior. In other aggregated caridean species, males become aware of the female's reproductive state for only a short time prior to her molt. In such species, release of male-stimulating substances appears to be inadvertent and metabolically unavoidable on the part of the female, which may be hiding its reproductive status prior to molting to prevent harassment from the many nearby males. Olfactory pheromones, released by urine or in the anterior respiratory outflow, together with subsequent contact chemoreception, appear to mediate individual mate recognition in pair-living species and agonistic interactions in a number of species. The scant knowledge about the chemistry of decapod shrimp pheromones can be greatly expanded by the development of techniques for behavioral assays with rigorous protocols, leading to isolation and chemical identification of putative olfactory and contact pheromones. Sex pheromones, once characterized chemically, might be used to manipulate mating and social interactions in commercially important shrimp species to increase catch in fisheries, as well as to increase productivity and genetic quality in aquaculture species.

Acknowledgments The author wishes to acknowledge support on shrimp pheromone research from NOAA Louisiana Sea Grant R/SA-03. This is contribution no. 123 of the University of Louisiana at Lafayette Laboratory for Crustacean Research. My sincere thanks to Thomas Breithaupt and Martin Thiel for their careful editorial work and for the invitation to write this chapter on such an intriguing topic. I am grateful to outside reviewers for their helpful comments and suggestions.

References

Atema J, Steinbach MA (2007) Chemical communication and social behavior of the lobster *Homarus americanus* and other decapod crustacea. In: Duffy JE, Thiel M (eds) Evolutionary ecology of social and sexual systems. Oxford University Press, New York, pp 115–144

Bauer RT (1976) Mating behaviour and spermatophore transfer in the shrimp *Heptacarpus pictus* (Stimpson) (Decapoda: Caridea: Hippolytidae). J Nat Hist 10:315–440

Bauer RT (1979) Sex attraction and recognition in the caridean shrimp *Heptacarpus paludicola* Holmes (Decapoda: Hippolytidae). Mar Behav Physiol 6:157–174

Bauer RT (1991) Sperm transfer and storage structures in penaeoid shrimps: a functional and phylogenetic perspective. In: Bauer RT, Martin JW (eds) Crustacean sexual biology. Columbia University Press, New York, pp 183–207

Bauer RT (1996) A test of hypotheses on male mating systems and female molting in decapod shrimp, using *Sicyonia dorsalis* (Decapoda: Penaeoidea). J Crust Res 16:429–436

Bauer RT (2000) Simultaneous hermaphroditism in caridean shrimps: a unique and puzzling sexual system in the Decapoda. J Crust Biol 20(spec no 2):116–128

Bauer RT (2002) Tests of hypotheses on the adaptive value of an extended male phase in the hermaphroditic shrimp *Lysmata wurdemanni* (Caridea: Hippolytidae). Biol Bull 203:347–357

Bauer RT (2004) Remarkable shrimps: adaptations and natural history of the Carideans. University of Oklahoma Press, Norman

Bauer RT, Abdalla JA (2001) Male mating tactics in the shrimp *Palaemonetes pugio* (Decapoda, Caridea): precopulatory mate guarding vs. pure searching. Ethology 107:185–199

Bauer RT, Caskey JL (2006) Flagellar setae of the second antennae in decapod shrimps: sexual dimorphism and possible role in detection of contact sex pheromones. Invert Reprod Dev 49:51–60

Breithaupt T, Eger P (2002) Urine makes the difference: chemical communication in fighting crayfish made visible. J Exp Biol 205:1221–1231

Burkenroad MD (1947) Reproductive activities of decapod Crustacea. Am Nat 81:392–398

Caskey JL, Bauer RT (2005) Behavioral tests for a possible contact sex pheromone in the caridean shrimp *Palaemonetes pugio*. J Crust Biol 25:571–576

Caskey JL, Hasenstein KH, Bauer RT (2009a) Studies on contact pheromones of the caridean shrimp *Palaemonetes pugio*: I. Cuticular hydrocarbons associated with mate recognition. Invert Reprod Dev 53:93–103

Caskey JL, Watson GM, Bauer RT (2009b) Studies on contact pheromones of the caridean shrimp *Palaemonetes pugio*: II. The role of glucosamine in mate recognition. Invert Reprod Dev 53:105–116

Correa C, Thiel M (2003) Mating systems in caridean shrimp (Decapoda: Caridea) and their evolutionary consequences for sexual dimorphism and reproductive biology. Rev Chil Hist Nat 76:187–203

Díaz ER, Thiel M (2004) Chemical and visual communication during mate searching in rock shrimp. Biol Bull 206:134–143

Duffy JE (2007) Ecology and evolution of eusociality in sponge-dwelling shrimp. In: Duffy JE, Thiel M (eds) Evolutionary ecology of social and sexual systems. Oxford University Press, New York, pp 387–409

Dunham PJ (1978) Sex pheromones in Crustacea. Biol Rev 53:555–583

Dunham PJ (1988) Pheromones and behavior in Crustacea. In: Laufer H, Downer RGH (eds) Endocrinology of selected invertebrate types. Alan R. Liss, New York, pp 375–392

Fransen HJM, De Grave S (2009) Evolution and radiation of shrimp-like decapods: An overview. In: Martin JW, Crandall KA, Felder DL (eds) Decapod crustacean phylogenetics. CRC, Boca Raton, pp 245–259

Grafen A, Ridley M (1983) A model of mate guarding. J Theor Biol 102:549–567

Herberholz J, Schmitz B (2001) Signaling *via* water currents in behavioral interactions of snapping shrimp (*Alpheus heterochaelis*). Biol Bull 201:6–16

Howard RW, Bloomquist GJ (2005) Ecological, behavioral, and biochemical aspects of insect hydrocarbons. Annu Rev Entomol 50:371–393

Hudinaga M (1942) Reproduction, development, and rearing of *Penaeus japonicus*. Jpn J Zool 10:305–393

Hughes M (1996) The function of concurrent signals: visual and chemical communication in snapping shrimp. Anim Behav 52:247–257

Johnson VR (1969) Behavior associated with pair formation in banded shrimp *Stenopus hispidus* (Olivier). Pac Sci 23:40–50

Johnson VR Jr (1977) Individual recognition in the banded shrimp *Stenopus hispidus* (Olivier). Anim Behav 25:418–428

Kamiguchi Y (1972) Mating behavior in the freshwater prawn. *Palaemon paucidens*. A study of the sex pheromone and its effect on males. J Fac Sci Hokkaido Univ Ser VI Zool 18:347–355

Karplus I, Malecha SR, Sagi A (2000) The biology and management of size variation. In: New MB, Valenti WC (eds) Freshwater prawn culture: the farming of *Macrobrachium rosenbergii*. Blackwell Science, Malden, pp 259–289

Kelly LS, Snell TW (1998) Role of surface glycoproteins in mate-guarding of the marine harpacticoid copepod *Tigriopus japonicus*. Mar Biol 130:605–612

Lee CL, Fielder DR (1982) Maintenance and reproductive behaviour in the freshwater prawn *Macrobrachium australiense* Holthuis (Crustacea: Decapoda: Palaemonidae). Aust J Mar Freshw Res 33:629–646

Mathews LM (2003) Tests of the mate-guarding hypothesis for social monogamy: male snapping shrimp prefer to associate with high-value females. Behav Ecol 14:63–67

Nakashima Y (1987) Reproductive strategies in a partially protandric shrimp, *Athanas kominatoensis* (Decapoda: Alpheidae): sex change as the best of a bad situation for subordinates. J Ethol 5:145–159

Obermeier M, Schmitz B (2003a) Recognition of dominance in the big-clawed shrimp (*Alpheus heterochaelis* Say 1818). Part I. Individual or group recognition? Mar Fresh Behav Physiol 36:1–16

Obermeier M, Schmitz B (2003b) Recognition of dominance in the big-clawed shrimp (*Alpheus heterochaelis* Say 1818). Part II. Analysis of signal modality. Mar Fresh Behav Physiol 36:17–29

Obermeier M, Schmitz B (2004) The modality of the dominance signal in snapping shrimp (*Alpheus heterochaelis*) and the corresponding setal types on the antennules. Mar Fresh Behav Physiol 37:109–126

Rahman N, Dunham DW, Govind CK (2001) Mate recognition and pairing in the big-clawed shrimp, *Alpheus heterochelis*. Mar Fresh Physiol 34:213–226

Ridley M (1983) The explanation of organic diversity: the comparative method and adaptations for mating. Clarendon, Oxford

Schein H (1975) Aspects of the aggressive and sexual behavior of *Alpheus heterochaelis* Say. Mar Behav Physiol 3:83–96

Seibt U (1973) Sense of smell and pair bond in *Hymeoncera picta* Dana. Micronesica 9:231–236

Seibt U, Wickler W (1979) The biological significance of the pair-bond in the shrimp *Hymenocera picta*. Z Tierpsychol 50:166–179

Sneddon LU, Huntingford FA, Taylor AC, Clare AS (2003) Female sex pheromone-mediated effects and competition in the shore crab (*Carcinus maenas*). J Chem Ecol 29:55–70

Yano I, Kanna RA, Oyama RN, Wyban JA (1988) Mating behaviour in the penaeid shrimp *Penaeus vannamei*. Mar Biol 97:171–175

Wickler W, Seibt U (1981) Monogamy in crustacea and man. Z Tierpsychol 57:215–234

Ward J, Saleh N, Dunham DW, Rahman N (2004) Individual discrimination in the big-clawed snapping shrimp, *Alpheus heterochelis*. Mar Fresh Behav Physiol 37:35–42

Wyatt TD (2003) Pheromones and animal behavior: communication by smell and taste. Cambridge University Press, Cambridge

Zhang D, Lin J (2006) Mate recognition in a simultaneous hermaphroditic shrimp, *Lysmata wurdemanni* (Caridea: Hippolytidae). Anim Behav 71:1191–1196

Zhang D, Zhu J, Lin J, Hardege JD (2010) Surface glycoproteins are not the contact pheromones in the *Lysmata* shrimp. Mar Biol 157:171–176

Chapter 15
Chemical Ecology and Social Behavior of Anomura

Francesca Gherardi and Elena Tricarico

Abstract Anomura is an extremely diverse assemblage of crustacean decapods with a large variety of forms and lifestyles, which makes the taxon ideal for comparative studies of social systems and means of communication. Notwithstanding this, our knowledge of Anomura is mostly restricted to the crab–shell relationships in hermit crabs, whereas promising fields of research, such as chemical ecology, are still in their infancy. This review will analyze the role played by chemical signals and cues in the social life of this taxon. Although our knowledge of anomuran chemical ecology is biased towards hermit crabs, we will illustrate the high potential of this taxon for complex chemosensory abilities. Case studies will be provided, showing that waterborne chemicals are used, for instance, to identify the sex of a potential mate in *Pagurus geminus*, to assess females' reproductive state and fecundity by *Pagurus filholi* males, to select the higher-quality males in the lithodid *Hapalogaster dentate*, to attract *Clibanarius vittatus*, *Pagurus longicarpus*, and *P. pollicaris* to shell-recruitment sites, and to individually recognize conspecifics in *P. longicarpus*. Finally, we will discuss the directions to follow in the near future. Particularly, more scientific attention should be paid to unravel the identity of the substances involved in the transmission of information and to extend ecological studies to the still underexplored groups of lithodids, galatheids, and porcellanids.

15.1 Introduction

The infraorder Anomura encompasses an astonishingly diverse assemblage of crustacean decapods that includes hermit crabs (Paguroidea), king crabs (Lithodoidea), sand and mole crabs (Hippoidea), squat lobsters and porcelain crabs (Galatheoidea),

F. Gherardi (✉)
Dipartimento di Biologia Evoluzionistica "Leo Pardi", Università degli Studi di Firenze,
Via Romana 17, 50125, Firenze, Italy
e-mail: francesca.gherardi@unifi.it

T. Breithaupt and M. Thiel (eds.), *Chemical Communication in Crustaceans*,
DOI 10.1007/978-0-387-77101-4_15, © Springer Science+Business Media, LLC 2011

the freshwater Aegloidea, hairy stone crabs (Lomisoidea), and the least known superfamily of Kiwaoidea (McLaughlin et al. 2007). This variety of forms reflects an extremely wide diversity of lifestyle, which makes the taxon ideal for a comparative analysis of social systems and means of communication. Notwithstanding this, today's knowledge of Anomura is biased towards hermit crabs: of the 389 scientific articles published in 1998–2007 (after Web of Science), over 50% focused on Paguroidea. Anomurans of great importance to fisheries, such as the Lithodoidea, have been seldom studied (only 15% of articles), in spite of the collapse of the natural stocks of some king crab species (Dawson 1989) and the alarming expansion of species with strong invasive potentials (e.g., *Paralithodes camtschaticus*; Jørgensen and Primicerio 2007).

A relatively large fraction of studies on Paguroidea dealt with their ecology (33% compared to 9% of behavioral studies), being mostly focused on crab–shell interactions. This is not surprising; the lack of calcification of the abdominal exoskeleton in hermit crabs dictates their occupancy of exogenous shelters, i.e., empty gastropod shells (but see Gherardi 1996). These housings afford protection against broad environmental conditions, but their supply in the habitat is limited and hermit crabs have not evolved the ability to steal them from live mollusks (Hazlett 1981). Growth, survival, fecundity, reproductive success, and ultimately fitness all depend on the adequate size and shape of the inhabited shell (Gherardi 2006). Because of their shell-orientated life, hermit crabs have been privileged models for the study of resource assessment in invertebrates (Hazlett 1978; Elwood and Neil 1992).

The diversified strategies adopted by hermit crabs to acquire shells were the subjects of Francesca Gherardi's research at the beginning of her scientific career. Under the stimulus of her mentor, Marco Vannini, in 1986 she started to investigate the adaptive value of the bizarre behavior of forming aggregations (or clusters) during low tides by *Calcinus laevimanus* and *Clibanarius virescens* on the shore of a small gulf in Somalia, Awadaxan. After a few years, the civil war erupted in Somalia and the Italian researchers were obliged to move to the muddy and mosquito-infested mangrove forests of the neighboring Kenya; there, the phenomenon of clustering was even more impressive: up to 5,000 individuals of *Clibanarius laevimanus* crowd together in clusters at low tide near mangrove roots and disperse as the flood rises, marching in line to feeding areas to return at ebb tide to the aggregation grounds following the same route. Starting from the study of this amazing behavior, Francesca Gherardi increased her interest in the sociality of hermit crabs and extended her analysis to more and more model organisms and ecological systems; together with Elena Tricarico, at that time a PhD student, her attention was soon attracted to the chemical ecology of hermit crabs.

Indeed, a major breakthrough for the understanding of the anomurans' chemical ecology dates back over 30 years, when McLean (1974) described the ability of hermit crabs to find gastropod predation sites and suggested its chemical mediation. A few years later, in a series of pioneering field experiments, Rittschof (1980a, b) demonstrated that small peptides released from prey gastropod tissues by salivary enzymes of the predatory snail convey information to hermit crabs about the availability of empty shells (Rittschof et al. 1990; Rittschof and Cohen 2004)

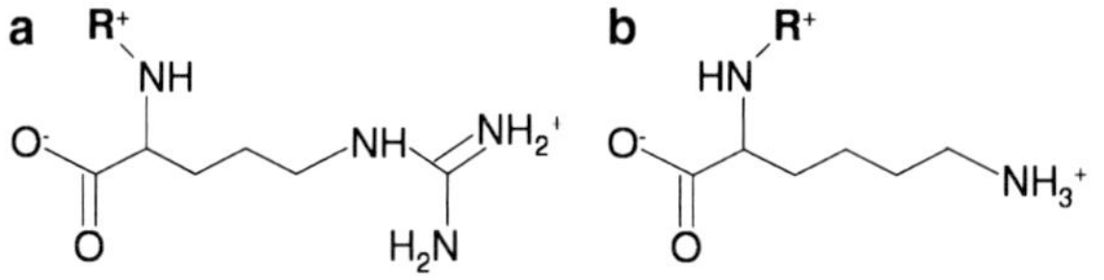

Fig. 15.1 The general form of trypsin-like serine protease generated peptides: (a) R-Arg; (b) R-Lys. R can be any number of amino acids attached through peptide bonds. The peptides are shown in the form that they would occur in sea water at pH 7.8

(Fig. 15.1). Rittschof's results had the merit of catalyzing highly productive research programs on this phenomenon. Subsequent studies showed, for instance, that chemical attraction depends on the shell species (e.g., Gilchrist 1984; Rittschof and Sutherland 1986) and mostly acts on individuals inhabiting badly fitting shells (e.g., Kratt and Rittschof 1991). There is, however, still little published information on how the high level of specificity in shell cues is accomplished; a "working hypothesis" is that it is in part based on the dominance of Arginine or Lysine carboxyl terminals in the natural blends of peptides (Rittschof and Cohen 2004). Regardless of this lasting gap of knowledge, the above listed studies gave a fundamental impulse to the further development of research on the chemical ecology of aquatic invertebrates.

This review will analyze the role played by chemical signals and cues (as defined by Wyatt, Chap. 2) in the social life of Anomura. Three general aspects will emerge from this chapter: (1) the potential of this taxon for complex chemosensory abilities, (2) the multiple – but highly taxon-biased – behavioral evidence of chemically mediated exchange of information among individuals, and (3) the lack of knowledge on the identity of the substances involved.

15.2 Sex Identification and Mate Selection

After the first record in the crab *Portunus sanguinolentus* (Ryan 1966), many attempts have been made to identify sex pheromones in other decapods (e.g., Hazlett 1996a). Among Anomura, *Pagurus geminus* was used as a model organism by Imafuku (1986) in Japan. Males of this species usually "jump" onto any encountered conspecific and apparently start "sniffing" and "tasting" it by beating their third maxillipeds and flicking their antennules and by inserting their chelipeds into its shell. A few minutes of exploration are enough to allow them to identify the sex of the conspecific: males are rejected whereas females are soon accepted. The exploring males grasp the edge of the female's shell with their left minor cheliped and then start to drag her ("pinching" behavior) as a prelude of the subsequent mate-guarding and copulation. Imafuku (1986) used pinching as a behavioral assay to show that waterborne chemicals allow for sex identification in *P. geminus* and to

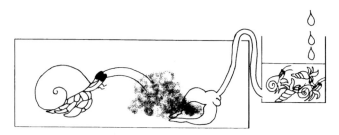

Fig. 15.2 The method used by Imafuku (1986) to analyze the response of *Pagurus geminus* males to putative sex pheromones produced by females. In a cylindrical chamber, a test male was offered with an empty *Crona margariticola* shell which exuded "female water" or "male water." The exuded water was derived at a controlled velocity through a silicon tube from a test vessel in which five females or five males were confined. Modified by G. Mazza after Imafuku (1986)

confirm that the antennules are involved in chemoreception. Males display a positive reaction to empty shells from which "female water" exuded (Fig. 15.2), but do not show any response after removal of their antennules. However, as Imafuku puts it, the use of distant chemical cues is only the first step of mating in hermit crabs: contact chemical cues (Hazlett 1970) and/or tactile stimuli (Hazlett 1966) provide additional information to the pairing individuals. However, notwithstanding that Imafuku's results opened the way for potentially fruitful research, an integrated study examining the diverse modalities of sex identification in hermit crabs is still missing.

Similarly, the use of chemical cues in the context of mate selection is still largely unexplored. Few studies addressed the possibility that not only sex recognition but also additional information is mediated by chemical cues during female–male interactions (Contreras-Garduño and Córdoba-Aguilar 2006). Among these studies, Goshima et al. (1998) analyzed whether *Pagurus filholi* males can assess both reproductive state and fecundity in females. The authors showed that males of this species use chemical cues to choose the females in an advanced reproductive state, i.e., close to spawning. In this way, males decrease the risks of wasting time with females that eventually will not spawn (during mate-guarding, in fact, females cease to feed: starved females usually do not spawn). They may also reduce the time spent mate-guarding and, on the contrary, increase the likelihood of mating with other females in the same reproductive season, thus maximizing their overall reproductive fitness.

Finally, in hermit crabs, and in other invertebrate taxa as well, the role of females during reproduction has been little studied, even though it is often the female that determines whether mating takes place (Hazlett 1966, 1989, 1996a). Many scattered observations pinpointed the key role played by females in hermit crabs' reproduction, but more focused studies are obviously needed. Similarly to other decapods, anomuran females might be able to assess males' ability to offer resources to them and/or to provide indirect benefits to their offspring (e.g., good genes) (Contreras-Garduño and Córdoba-Aguilar 2006). Males of large size, for instance, are usually preferred because they more likely win intrasexual contests than smaller males, furnish good genes, and pass larger ejaculates to the females. Conversely, males with depleted sperm reserves are typically avoided.

The need for females to make correct choices is particularly crucial when (1) they mate only once during a particular reproductive season, (2) fertilization success is constrained by sperm supply, (3) males engage in multiple matings, and (4) sperm reserves require time to be reconstituted. This is the case of the stone crab, the lithodid *Hapalogaster dentata* (Sato and Goshima 2007): females of this species cannot undergo multiple matings because of the long and firm postcopulatory mate-guarding by the grasping male; small males provide insufficient sperm to fertilize the entire clutch and males of all sizes require more than 10 days to replenish their sperm reserves after a previous mating. Sato and Goshima (2007) brilliantly demonstrated that female stone crabs, when allowed choosing between two different males, use chemical cues to select the higher-quality male, i.e., larger individuals over smaller ones and unmated males over depleted ones.

15.3 Aggregation and Alarm

Empty shells are such important and limited resources that a number of behavioral tactics have evolved to enable obligate shell users to obtain them (Tricarico and Gherardi 2006). Hermit crabs are attracted to gastropod predation sites, usually when a large predatory snail is consuming a prey snail (McLean 1974). These sites function as "markets" where shells may be directly acquired by aggression and/or negotiation between the attracted crabs; alternatively, the attendants may passively benefit from a cascade of shell exchanges resulting from a vacancy chain process (Briffa and Austin 2009) (Fig. 15.3). As a consequence, even if only a single empty shell is made available at these sites, the likelihood of obtaining better shells is relatively high for all attendants, thus favoring the evolution of refined chemosensory abilities to detect the odor of partially digested gastropod flesh (see above).

Apart from snail odors, attractants that might enable efficient location of available shells are the odors released by dead or dying con- or heterospecifics. This possibility was investigated by Rittschof et al. (1992) and Hazlett (1996b) in marine species and by Small and Thacker (1994) in land hermit crabs. In the marine *Clibanarius vittatus*, *Pagurus longicarpus*, and *P. pollicaris*, the chemicals released by crushed conspecifics induced similar responses as the solutions of dead gastropod flesh. Hermit crabs heightened their activity, did not attempt to feed upon the carcass, and actively investigated all shells in the area; the attracted hermit crabs formed aggregations in which cascades of shell switches occurred. The terrestrial *Coenobita compressus* and *C. perlatus* shared a similar behavior: in the presence of the odors of dead conspecifics, their shell-acquisition behaviors were intensified, and their responses were up to 10 times stronger than those elicited by food odors

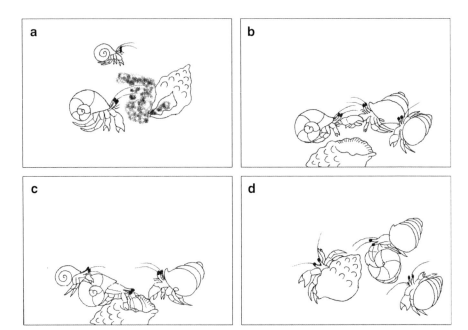

Fig. 15.3 Chemicals released from partly digested snail tissues attract hermit crabs (**a**, **b**); gastropod predation sites function as "markets" where shells may be acquired by aggression or negotiation between hermit crabs (**c**); the other attendants may benefit occupying shells left vacant (**d**). Drawn by G. Mazza

(Small and Thacker 1994). So, the ability to detect chemicals released by dead conspecifics and to associate them with the potential for a new shell has been conserved in the transition from sea to land: the attractant is contained in the crab hemolymph, its power being displayed immediately upon the hemolymph removal without the need of any proteases, as required with snail odors. A recent gas chromatography–mass spectrometry study (Schmidt et al. 2009) showed that it is the volatile alcohol 3-decanol found in the hemolymph of *Clibanarius vittatus* that stimulates shell investigation behavior in conspecifics.

Interestingly, the intensity and type of responses to these substances reflect the shell fit of the responding crabs (Rittschof et al. 1992). Rittschof et al. (1992) showed that the number of hermit crabs attracted to concealed sources of the odors from dead conspecifics varied as a function of their carapace length/shell weight ratio and of the type of shell, in the same fashion as observed for their attraction to snail odors (Fig. 15.4). Similarly, in the land crab *Coenobita* spp., experimental switches of hermit crabs from worse to better shells led to a decreased number of crabs investigating the shells of conspecifics (Thacker 1994). In the marine *Clibanarius vittatus*, the number of grasps to the shells of conspecifics was significantly higher for crabs with a greater shell deficit, at least in the first day of observation (Hazlett 1996b). This effect faded during the subsequent six days of

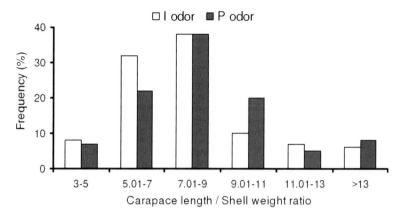

Fig. 15.4 Frequency distributions of carapace length/shell weight ratios for *Pagurus longicarpus* attracted to *Ilyanassa obsoleta* (I) or to crushed conspecifics (P). Low ratios indicate relatively large shells for crab size. Modified after Rittschof et al. (1992)

experimentation in which the crabs had been starved, thus suggesting that the relative attractive capacity of conspecific body odors may vary as a function of some internal factors, such as hunger level.

Attraction to dead conspecifics may also vary between species: *Clibanarius infraspinatus*, but not *C. vittatus*, increased both locomotion and the frequency of shell grasping in response to the conspecific and heterospecific hemolymph (Hazlett 1996b). Finally, the responses to conspecific hemolymph may vary among populations of the same species on the basis of their average shell fit, as found for terrestrial *Coenobita* sp. in Panama (Thacker 1994). Interestingly, even the meaning given to body odors may change among populations of the same species, possibly as a reflection of the different ecological pressures they are subject to. Individuals of a Cape Cod (Massachusetts, USA) population of *P. longicarpus*, either in well- or in badly fitting shells, maintained a complete immobility (a typical antipredator behavior) in the presence of the odor released by dead conspecifics (Gherardi and Atema 2005a). On the contrary, the North Carolina population studied by Rittschof et al. (1992) always showed attraction. These different results might indicate that odors emitted by dying or freshly dead conspecifics act as alarm substances (Hazlett, Chap. 18) in areas where predation is strong, but they also indicate availability of shells from dead conspecifics in areas where it is low.

Hermit crabs are always faced with multiple sensory inputs in their natural environments and seem to be capable to integrate multiple sources of chemical information (Fig. 15.5). Some intrinsic factors of the receiver, such as its shell status (but also its hunger level, reproductive state, etc.), may determine the appropriate responses for a given environmental context; these responses may also depend on extrinsic factors, such as resource availability and variability. In essence, the hermit crabs' responses to aggregation/alarm cues appear to be particularly complex, as a reflection of the behavioral flexibility that makes the

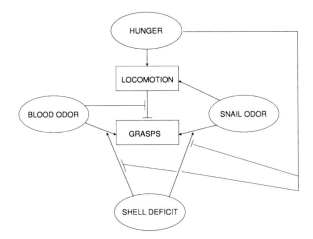

Fig. 15.5 Flow chart of integration among multiple chemical sources. Lines with arrows indicate an increase in activity while simple lines with perpendicular ends indicate inhibitory effects. Modified after Hazlett (1996c)

Paguroidea extremely well adapted to the temporal and spatial variability of their environment (Hazlett 1988).

15.4   Dominance and Individual Recognition

Hermit crabs, particularly when aggregating at shell-supplying sites, interact in pairs in agonistic encounters over ownership of the gastropod shells at stake. The two contestants often play two well-defined roles. The "attacker" initiates a fight by first contacting the conspecific's shell with its chelipeds and then grasping it with its walking legs, whereas the defender withdraws tightly into its shell. The two perform a range of behavioral patterns that include chela displays and grappling by both opponents, and "shell rapping" by the attacker (Briffa and Elwood 2000). Shell rapping consists in the attacker using its abdominal muscles and walking legs to swing the two shells together in a series of bouts, separated by brief pauses. After several bouts of rapping, the fight ends either by the defender giving up, in which case it allows the attacker to evict it from its shell, or by the attacker giving up without evicting the defender. During shell fights, a wide range of visual and tactile information is exchanged between the two contestants to advertise their fighting ability or "resource holding potential" (RHP). Within the framework of evolutionary game theory, a plethora of studies has been carried out to understand how the attacker assesses the quality of the shell at stake (Elwood and Neil 1992) and how its ability and motivation influence its decision (e.g., Elwood et al. 1998; Briffa and Elwood 2000; Gherardi 2006).

Only recently was the hypothesis raised that chemical cues produced by the contestants may also be used during shell fights (Gherardi and Tiedemann 2004a). Indeed, there is evidence that hermit crabs are sensitive to odors of live conspecifics. In the laboratory, *P. longicarpus* (Gherardi and Atema 2005a) executes more numerous investigatory acts on empty shells in the presence of water conditioned by the body odor of live conspecifics than in plain seawater. Similarly, in the field the odor of live conspecifics is as efficacious as the odor of dead conspecifics in attracting *P. longicarpus* to simulated shell-supplying sites (Tricarico and Gherardi 2006).

During shell fights, the more similar the size of the contestants, the more crucial is the acquisition of information about the fighting experience of the opponent. By signaling status level, fight intensity and duration may be reduced, thereby lowering the risk of injury. To explore the potential of hermit crabs to use chemical cues during shell fights, Briffa and Williams (2006) exposed focal *P. bernhardus* crabs to seawater containing three types of chemical cue: (1) plain seawater, and cues from (2) fighting and (3) non-fighting hermit crabs. Focal individuals were monitored for their activity and for the execution of behavioral patterns such as withdrawal into the shell and walking across the substratum. At the end of the observation, a startle response was induced to the same hermit crabs by presenting them with a novel stimulus (a weighted cardboard square) that usually causes them to withdraw into the shell; the time taken by hermit crabs to reemerge was used as an index of their motivational state for exploration. The results clearly show that, in the presence of the fighting cue, focal crabs spend more time withdrawn into their shell and less time in locomotion than those exposed to cues from non-fighting hermit crabs or to plain seawater (Fig. 15.6). Those exposed to the fighting cue took longer to recover when presented with the cardboard square than the hermit crabs in the other groups, indicating that the motivation of the former was significantly lower.

Although Briffa and Williams (2006) provided evidence that fighting hermit crabs release chemicals, there is no clear indication of the message they convey. Chemicals might be some physiological byproducts of demanding activities and may simply indicate "disturbance": for instance, when exposed to water conditioned by stressed conspecifics, individuals of *Calcinus laevimanus* showed a significant increase in locomotion (Hazlett 1990).

In other decapod models, the individuals engaged in agonistic contests may inform the opponent about their dominance status and/or their identity. On one hand, they may convey generic information about their agonistic abilities and experience. On the other, they may display some individual "badges": in this case, the receiver detects an individual-specific cue released by a given animal and associates this cue with information that might be translated into the generic "I have already met this individual" (in the case of a "binary individual recognition") or into the more specific "I have already met it and lost to it" (in the case of the more refined "true individual recognition"). Later, the receiver uses these cues to identify that individual and makes decisions on how to behave with it (retreat, attack, ignore).

Surprisingly, dominance status signals have been never investigated in anomurans, as, on the contrary, abundantly done in crayfish (Breithaupt, Chap. 13).

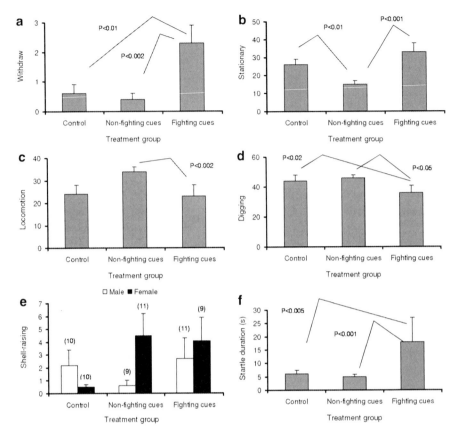

Fig. 15.6 Effects of treatment group (Control, Non-fighting cues, and Fighting cues) on the exploratory behavior of isolated *Pagurus bernhardus*. The recorded parameters were: percent time of (**a**) withdrawn into shell, (**b**) stationary, (**c**) locomotion, (**d**) digging in the substrate, and (**e**) shell-raising (sample size in brackets). The total time taken to recover from a startle response induced at the end of the observation period is reported in (**f**). The figure gives the raw data with the standard errors bars and *P*-values from post-hoc Fisher's PLSD. Modified after Briffa and Williams (2006)

Conversely, in spite of the limited knowledge of the phenomenon in invertebrates (suggested to occur in a few species among crustaceans and insects; Tibbetts and Dale 2007), there is plenty of evidence that hermit crabs are capable of a relatively refined form of individual recognition (Hazlett 1969; Gherardi and Tiedemann 2004b) and that this is mediated by chemical substances (Gherardi and Tiedemann 2004a; Gherardi et al. 2005).

To understand whether chemicals are used by hermit crabs in signaling identity, it was first necessary to define a simple, noninvasive bioassay able to compare the behavior of the tested individuals in different contexts. These characteristics were found in the investigatory behavior, i.e., the response that hermit crabs display in the presence of an empty gastropod shell (Gherardi and Atema 2005a). Using this bioassay, Gherardi and Tiedemann (2004a) confirmed the ability of *P. longicarpus* to discriminate between familiar and unfamiliar individuals (i.e., conspecifics that they had or had not met previously; Gherardi and Tiedemann 2004b), and also showed that crabs respond stronger in the presence of unfamiliar rather than of familiar stimuli when they were allowed to smell a conspecific.

Chemical recognition in *P. longicarpus* is, however, more refined than a binary identification (Gherardi and Atema 2005b). Hermit crabs are in fact still able to classify a conspecific as familiar after having experienced 1-day interactions with other individuals, indicating that memory of a former rival could not be erased, at least after one day of separation, by experiences with different crabs (and with different odors, all familiar). However, a more accurate test for a true individual recognition would involve different responses to two, equally-known individuals. In a study with *P. longicarpus* (Gherardi et al. 2005), we examined the investigatory behavior by "test" crabs towards a well-fitting empty shell (target shell) in the presence of odors of different proveniences: test crab, unfamiliar conspecifics, and familiar conspecifics that either won or lost previous fights with the test crab, i.e., were dominant or subordinate. We found that they can chemically discriminate (1) between themselves and others and (2) between at least two familiar individuals, the winner and the loser. On the contrary, test crabs were not able to distinguish among unfamiliar size-matched crabs in a similar shell and with the same winning/ losing experience, which suggests that the ability to chemically recognize at least two different familiar individuals was not due to the putative odors of rank, size class, or shell type, but to the experience that test crabs have had with previously encountered individuals. At least in hermit crabs, individual recognition seems to be different from a typical associative learning, explaining changes in the behavior of an individual as the results of its exposure to a correlation or contingency between two events. In fact, first, hermit crabs do not need to be trained over repeated trials and can recognize a conspecific after only a brief exposure to stimulus animals (less than 30 min) (Gherardi and Atema 2005b). Second, they remember a conspecific without external, experimenter-provided reinforcement or punishment, and third, memory can be acquired in a few minutes and lasts for about five days without further exposure to the stimulus animal (Gherardi and Atema 2005b).

In a second experiment, Gherardi et al. (2005) tested the hypothesis that the receiver crab associates the odor from a social partner with one of the relevant attributes of the transmitter, such as its rank and size or the quality of its shell (or with a combination of them). Triplets of crabs that differed for a single characteristic, the others being equal, were left to interact for 24 h in the familiarization phase. In this phase, hermit crabs had the time to both make acquaintance of each other and assess the relative quality of the conspecifics' shells by visual and tactile means. Then, the crabs with the intermediate attribute were subjected to three treatments in random succession: they were offered with a target shell as in the previous experiment in the presence of the odor of two familiar crabs, which had (1) the same size and shell quality as the receiver but a different rank (being either dominant or subordinate), (2) a different size (being either larger or smaller) but the same shell quality, and (3) the

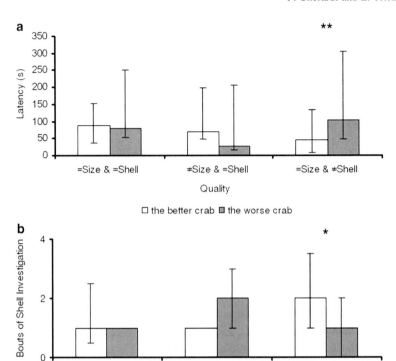

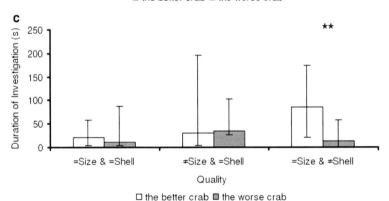

Fig. 15.7 Association between the odor of a familiar conspecific and one of three attributes (rank, body size, and shell quality) in *Pagurus longicarpus* (after Gherardi et al. 2005). The authors observed the response of a test crab to a well-fitting empty shell in the presence of seawater conditioned by the odor of two familiar conspecifics, analyzed separately for 5 min for each odor in random order. Three series of experiments were conducted: same body size and same shell quality (1), different body size and same shell quality (2), and same body size and different shell quality (3). The better crab (and the worse crab) was the dominant (and subordinate) crab in (1),

same size but a shell of either higher or lower quality than the receiver's. The results showed that receivers responded in the same fashion to the target shell when the attributes of the senders differed for their rank only and/or for their size. On the contrary, they showed a quicker response to the target shell (Fig. 15.7a) and investigated it more often (Fig. 15.7b) and for a longer time (Fig. 15.7c) when the odor was emitted by crabs occupying shells of higher rather than of lower quality. This might indicate that the odor of a hermit crab is used by the opponent as a "label" of the quality of the inhabited shell. If, based on the previous experience with that crab, this label denotes high quality, the target shell "merits" to be investigated, so its investigation will be intense; otherwise investigation will be scarce or absent. Hermit crabs can also change the association between individual odor and shell quality. Once the opponent has been switched to a shell of a different quality, the receiver, exposed to its odor, modulates its response to the offered shell accordingly. So, investigation of the target shell by the receiver will increase (and decrease) when exposed to the odor of a crab that had been switched from a low- to a high-quality shell (and *vice-versa*).

In synthesis, *P. longicarpus* seems to have a "concept" of other individuals based on the quality of its shell and behaves accordingly in an adaptive fashion. In the turbid environments in which this species often lives, the odor of a familiar conspecific as a label of shell quality might be a more reliable cue than its sight; moreover, the exclusive use of tactile information from the shell would require the consumption of time and energy in repeated investigatory acts. Indeed, *P. longicarpus* quickly learns the chemical identity of a social partner (Gherardi and Atema 2005b), is inaccurate in discriminating shells by sight (Gherardi and Tiedemann 2004a), and often switches shells without prior investigation (Scully 1986). Obviously, any shell exchange breaks the link between a given hermit crab and a given shell – and the aggregations of hermit crabs around gastropod predation sites are characterized by a cascade of shell switches. Therefore, the plastic response to the cues associated with high-quality shells is a key factor to optimize shell acquisition and to reduce errors.

Chemical recognition of social partners has the potential to be a thrilling subject of study also in other anomurans. For instance, some symbiotic Porcellanidae species, such as *Allopetrolisthes spinifrons*, known to defend their sea anemone hosts from intruders may transmit information on their RHP not only through visual displays (as suggested by Baeza et al. 2001) but also through chemicals. Indeed, porcellanid crabs have refined chemosensory abilities that allow some of them, such as *Porcellana sayana*, to find potential hosts (Brooks and Rittschof 1995).

◀———————————————————————————————————

Fig. 15.7 (continued) the bigger (and the smaller) crab in (2), and the crab occupying a higher-quality (and a lower-quality) shell in (3). The figure gives median values (and interquartile ranges) of (**a**) latency (time until first shell investigation; when the test crab never investigated the shell, we arbitrarily assigned a time equal to 305 s), (**b**) bouts of investigation of the target shell, and (**c**) duration of investigation. * and ** denote $P < 0.05$ and $P < 0.01$, respectively, after a Friedman two-way analysis of variance. Modified after Gherardi and Tricarico (2007)

15.5   The Way Ahead

In spite of the growing evidence that Anomura have complex chemosensory systems, there are still many gaps in our understanding of the ways they use them. Along with the already lamented lack of studies on the identity of the chemical signals transmitted, chemical communication in the majority of Anomura, particularly in lithodids, galatheids, and porcellanids, remains still understudied. This is unfortunate because this taxon might provide a fruitful ground for evolutionary studies, particularly when the interest is directed to the transition from sea to land (see, e.g., Hansson et al., Chap. 8) or from sea to fresh water. Closely related taxa of anomurans may be found from the lower to the higher intertidal zone, which may help compare chemical communication systems along a gradient of ecological pressures (Hazlett et al. 2000). Furthermore, several aspects of the social communication of Aeglidae, the only freshwater anomuran family, are worth investigating, particularly in the context of mating (Almerao et al. 2010) and of mother–juvenile relationships (López Greco et al. 2004).

Even in the case of the more extensively studied taxon of Paguroidea, the use of chemicals in several social contexts, such as breeding, has been underappreciated. Yet, hermit crabs have a set of unique characteristics, which make them suitable for these studies: they are relatively easy animals to manipulate and to observe because of their relatively large size (larger than many insects), some species can be found in large numbers, and above all, male and the female effects can be easily separated (Contreras-Garduño and Córdoba-Aguilar 2006).

A second relevant issue where research is needed (and where breakthroughs are also expected) is the chemically mediated recognition of conspecifics, particularly in the context of aggression. Several questions remain unanswered. First, is it really "good" for hermit crabs to be "different" (Tibbetts and Dale 2007)? Does a crab get a benefit from being identified by a conspecific? Based on our studies in *P. longicarpus*, the answer seems to be negative: by revealing its identity a hermit crab might increase the risks of having its shell stolen. If this is true, under which selective pressure have chemical badges evolved? Do they primarily serve other functions? Particularly in the context of individual recognition, the knowledge of the involved molecules might be of great help to answer these crucial questions. Second, is individual recognition used in the "real world" of the species (Tricarico and Gherardi 2006)? Has this ability been inherited by some of their ancestors? Possibly, individual recognition has evolved as "a matter of life or death" (Gherardi and Tricarico 2007): even if a crab uses it seldom in its life and (thanks to it) succeeds in occupying a high-quality shell, both its life expectancy and individual fitness might score such a sharp increase to make the evolution of this ability highly adaptive.

This avalanche of questions denotes that studies on chemical communication in Anomura are just beginning to gather momentum. The way ahead may be long and harsh, but the promises of an exciting progress towards a better understanding of animal communication make it worth being followed.

References

Almerao M, Bond-Buckup G, Mendonca MD (2010) Mating behavior of *Aegla platensis* (Crustacea, Anomura, Aeglidae) under laboratory conditions. J Ethol 28:87–94

Baeza JA, Stotz W, Thiel M (2001) Agonistic behaviour and development of territoriality during ontogeny of the sea anemone dwelling crab *Allopetrolisthes spinifrons* (H. Milne Edwards, 1837) (Decapoda: Anomura: Porcellanidae). Mar Fresh Behav Physiol 35:189–202

Briffa M, Austin M (2009) Effects of predation threat on the structure and benefits from vacancy chains in the hermit crab *Pagurus bernhardus*. Ethology 115:1029–1035

Briffa M, Elwood RW (2000) Analysis of the finescale timing of repeated signals: does shell rapping in hermit crabs signal stamina? Anim Behav 59:159–165

Briffa M, Williams R (2006) Use of chemical cues during shell fights in the hermit crab *Pagurus bernhardus*. Behaviour 143:1281–1290

Brooks WR, Rittschof D (1995) Chemical detection and host selection by the symbiotic crab *Porcellana sayana*. Invertebr Biol 114:180–185

Contreras-Garduño J, Córdoba-Aguilar A (2006) Sexual selection in hermit crabs: a review and outlines of future research. J Zool 270:595–605

Dawson EW (1989) King crabs of the world (Crustacea: Lithodidae) and their fisheries: a comprehensive bibliography. Misc. Publ. 101. New Zealand Oceanogr. Inst., Div. Water Sci., Wellington, Australia

Elwood RW, Neil SJ (1992) Assessment and decisions: a study of information gathering by hermit crabs. Chapman and Hall, London

Elwood RW, Wood KE, Gallagher MB, Dick JTA (1998) Probing motivational state during agonistic encounters in animals. Nature 393:66–68

Gherardi F (1996) Non-conventional hermit crabs: pros and cons of sessile, tube-dwelling life in *Discorsopagurus schmitti* (Stevens). J Exp Mar Biol Ecol 202:119–136

Gherardi F (2006) Fighting behavior in hermit crabs: the combined effect of resource-holding potential and resource value in *Pagurus longicarpus*. Behav Ecol Sociobiol 59:500–510

Gherardi F, Atema J (2005a) Effects of chemical context on shell investigation behavior in hermit crabs. J Exp Mar Biol Ecol 320:1–7

Gherardi F, Atema J (2005b) Memory of social partners in hermit crab dominance. Ethology 111:271–285

Gherardi F, Tiedemann J (2004a) Chemical cues and binary individual recognition in the hermit crab *Pagurus longicarpus*. J Zool 263:23–29

Gherardi F, Tiedemann J (2004b) Binary individual recognition in hermit crabs. Behav Ecol Sociobiol 55:524–530

Gherardi F, Tricarico E (2007) Can hermit crabs recognize social partners by odor? And why? Mar Fresh Behav Physiol 40:201–212

Gherardi F, Tricarico E, Atema J (2005) Unraveling the nature of individual recognition by odor in hermit crabs. J Chem Ecol 31:2877–2896

Gilchrist S (1984) Specificity of hermit crab attraction to gastropod predation sites. J Chem Ecol 10:569–582

Goshima S, Kawashima T, Wada S (1998) Mate choice by males of the hermit crabs *Pagurus filholi*: do males assess ripeness and/or fecundity of females? Ecol Res 13:151–161

Hazlett BA (1966) Social behavior of the Paguridae and Diogenidae of Curaçao. Stud Fauna Curaçao 23:1–143

Hazlett BA (1969) 'Individual' recognition and agonistic behaviour in *Pagurus bernhardus*. Nature 222:268–269

Hazlett BA (1970) Tactile stimuli in the social behavior of *Pagurus bernhardus* (Decapoda, Paguridae). Behaviour 36:20–40

Hazlett BA (1978) Shell exchanges in hermit crabs: aggression, negotiation or both? Anim Behav 26:1278–1279

Hazlett BA (1981) The behavioral ecology of hermit crabs. Ann Rev Ecol Syst 12:1–22

Hazlett BA (1988) Behavioural plasticity as an adaptation to a variable environment. In: Chelazzi
   G, Vannini M (eds) Behavioral adaptation to intertidal life. Plenum, New York, pp 317–332
Hazlett BA (1989) Mating success of male hermit crabs in shell generalist and shell specialist
   species. Behav Ecol Sociobiol 25:119–128
Hazlett BA (1990) Disturbance pheromone in the hermit-crab *Calcinus laevimanus* (Randall,
   1840). Crustaceana 58:314–316
Hazlett BA (1996a) Reproductive behavior of the hermit crab *Clibanarius vittatus* (Bosc, 1802).
   Bull Mar Sci 58:668–674
Hazlett BA (1996b) Comparative study of hermit crab responses to shell-related chemical cues.
   J Chem Ecol 22:2317–2329
Hazlett BA (1996c) Organisation of hermit crab behaviour: responses to multiple chemical inputs.
   Behaviour 133:619–642
Hazlett BA, Bach CE, McLay C, Thacker RW (2000) A comparative study of the defense
   syndromes of some New Zealand marine Crustacea. Crustaceana 73:899–912
Imafuku M (1986) Sexual discrimination in the hermit crab *Pagurus geminus*. J Ethol 4:39–47
Jørgensen LL, Primicerio R (2007) Impact scenario for the invasive red king crab *Paralithodes
   camtschaticus* (Tilesius, 1815) (Reptantia, Lithodidae) on Norwegian, native, epibenthic prey.
   Hydrobiologia 590:47–54
Kratt CM, Rittschof D (1991) Peptide attraction of hermit crabs *Clibanarius vittatus* Bosc: roles of
   enzymes and substrates. J Chem Ecol 17:2347–2365
López Greco LS, Viau V, Lavolpe M, Bond-Buckup G, Rodriguez EM (2004) Juvenile hatching
   and maternal care in *Aegla uruguayana* (Anomura, Aeglidae). J Crustacean Biol 24:309–313
McLaughlin PA, Lemaitre R, Sorhannus U (2007) Hermit crab phylogeny: a reappraisal and its
   "fall-out". J Crustacean Biol 27:97–115
McLean RB (1974) Direct shell acquisition by hermit crabs from gastropods. Experientia 30:206–208
Rittschof D (1980a) Chemical attraction of hermit crabs and other attendants to gastropod
   predation sites. J Chem Ecol 6:103–118
Rittschof D (1980b) Enzymatic production of small molecules attracting hermit crabs to simulated
   predation sites. J Chem Ecol 6:665–676
Rittschof D, Cohen JH (2004) Crustacean peptide and peptide-like pheromones and kairomones.
   Peptides 25:1503–1516
Rittschof D, Kratt CM, Clare AS (1990) Gastropod predation sites: the role of predator and prey in
   chemical attraction of the hermit crab *Clibanarius vittatus*. J Mar Biol Assoc UK 70:583–596
Rittschof D, Sutherland JP (1986) Field studies on chemically mediated behavior in land hermit
   crabs: Volatile and nonvolatile odors. J Chem Ecol 12:1273–1284
Rittschof D, Tsai DW, Massey PG, Blanco L, Kueber GL Jr, Haas RJ Jr (1992) Chemical
   mediation of behavior in hermit crabs: alarm and aggregation cues. J Chem Ecol 18:959–984
Ryan EP (1966) Pheromone: evidence in a decapod crustacean. Science 151:340–341
Sato T, Goshima S (2007) Female choice in response to risk of sperm limitation by the stone crab,
   Hapalogaster dentata. Anim Behav 73:331–338
Schmidt G, Rittschof D, Lutostanski K, Batchelder A, Harder T (2009) 3-Decanol in the haemo-
   lymph of the hermit crab *Clibanarius vittatus* signals shell availability to conspecifics. J Exp
   Mar Biol Ecol 382:47–53
Scully EP (1986) Shell investigation behavior of the intertidal hermit crab *Pagurus longicarpus*
   Say. J Crust Biol 6:749–756
Small MP, Thacker RW (1994) Land hermit crabs use odors of dead conspecifics to locate shells.
   J Exp Mar Biol Ecol 182:169–182
Thacker RW (1994) Volatile shell-investigation cues of land hermit crabs: effect of shell fit,
   detection of cues from other hermit crab species, and cue isolation. J Chem Ecol 20:1457–1482
Tibbetts E, Dale J (2007) Individual recognition: it is good to be different. Trends Ecol Evol
   22:529–537
Tricarico E, Gherardi F (2006) Shell acquisition by hermit crabs: which tactic is more efficient?
   Behav Ecol Sociobiol 60:492–500

Chapter 16
Deception in Visual and Chemical Communication in Crustaceans

John H. Christy and Dan Rittschof

Abstract Deception in animal communication occurs when one animal causes another to respond to a condition that does not exist or to fail to respond to one that does. Signals that bluff produce the first kind of error, behavior that hides causes the second, and both are common in visual communication by crustaceans. In contrast, crustacean chemical communication may usually be honest because the communicative chemicals typically are byproducts of the biochemical mechanisms by which crustaceans are built and operate. These cues, which often are released in urine, reliably reveal an individual's identity, sex, reproductive state, and condition. There are, however, opportunities for deception by bluffing and by hiding in the chemical channel. Bluffing may occur when dominance relationships are learned and individuals recognize each other. Subordinates may avoid known dominants even after the condition of the dominant has declined and it is no longer able to win a fight with the subordinate. Frequent probing by subordinates should check such bluffing. Hiding in the chemical channel may occur in escalated fights in which one animal fails to chemically announce its intent to strike and wins by delivering a blind-side punch to an unprepared opponent. Receptive female crabs and lobsters may also withhold cues of their receptivity to avoid courtship by some males, yet direct the same cues to preferred potential mates. In species with multiple male morphs, we speculate that subordinate males may hide from dominants by withholding male odors or mimicking female odors. In species with internal fertilization we also suggest that male seminal fluids may contain chemicals that affect female reproductive processes and bias the rate the male's sperm fertilize the female's eggs. Detecting deception in chemical communication will be very challenging, but we encourage crustacean researchers to keep this possibility in mind when examining signaling behavior via chemicals.

J.H. Christy (✉)
Smithsonian Tropical Research Institute, Balboa,
Republic of Panama
e-mail: christyj@si.edu

T. Breithaupt and M. Thiel (eds.), *Chemical Communication in Crustaceans*,
DOI 10.1007/978-0-387-77101-4_16, © Springer Science+Business Media, LLC 2011

16.1   Introduction

Deception is common between species (e.g., mimicry), but, primates excepted, it seems to be rare between individuals of the same species even though the interests of signalers and receivers usually differ (Searcy and Nowicki 2005). Ironically, I (JC) first encountered apparently deceptive behavior in an animal after I stopped studying howler monkeys and began observing fiddler crabs. Over the past 3 decades, I have discovered several cases of visual deception in fiddler crab communication perhaps because the profound differences in our nervous systems make it easy for me to detect deception that goes unnoticed by the crabs. These discoveries are not unique; visual deception appears to be more common in crustaceans than in any other nonprimate taxon. In contrast, the second author's (DR) studies suggest that crustacean chemical communication usually is honest because the chemicals are often byproducts of basic physiological processes and cannot be faked. Here we combine our different research backgrounds and perspectives to explore how crustaceans might deceive using chemicals. We review visual deception by bluffing, hiding, and mimicry, speculate freely on the possible existence of these modes of deception in the chemical channel, and offer a few tentative examples. First, we define some terms.

16.2   Definitions

Although consensus definitions for common terms and concepts used in animal communication research are close at hand (Searcy and Nowicki 2005), disagreement remains.

16.2.1   Signals and Cues

A signal is a feature of an individual's phenotype that evolves because it elicits a response from other organisms that increase the signaler's fitness (after Hasson 1994). Maynard Smith and Harper (2003) and Scott-Phillips' (2008) definitions of signaling require that the response also evolves because it increases the receiver's fitness. Unfortunately, this requirement disqualifies as signals all deceptive traits that elicit costly responses that decrease the receiver's fitness (Hasson 1994; Wiley 1994) and classifies them as coercion instead (e.g., Scott-Phillips 2008). We also do not require that deceptive signals be variants of signals that are honest on average (Johnstone and Grafen 1993), though sometimes they are (see Table 16.1). In order to accommodate deception by hiding and mimicry, we find it necessary to allow some signals to elicit costly responses that do not benefit receivers (see also Hebets and Papaj 2005).

Table 16.1 Deception in crustacean visual communication

Context	Species	Signal/behavior	Kind of deceit	Mechanism	References
Aggression	*Alpheus heterochaelis* snapping shrimp, males	Display large snapping claw	Bluffing RHP	Claw size increases with body size which determines RHP. Small males with relatively large claws bluff RHP when they signal and they bluff most often in encounters they are sure to lose if they escalate to fighting	Hughes (1996, 2000)
	Pagurus bernhardus, hermit crab, males	Present or extend larger (right) claw	Bluffing aggressive intent	When males approach each other, one or both extend their chelae toward the other crab. Larger (body size) males usually win fights for shells. Both small and large males with relatively large claws tend to display more often. About 3% of the time chela displays are not backed up with future aggression. Claw size residuals do not predict winners in escalated fights for shells	Laidre (2009) and Arnott and Elwood (2010)
	Cherax dispar Australian slender crayfish, males	Display large chelae	Bluffing strength	Males with the more powerful chelae win fights, but most encounters are resolved without fighting. Winners usually have larger but not stronger chelae; hence they bluff strength. Male large chelae deliver forces about half those of females and they vary widely in strength	Wilson et al. (2007, 2009)
	Uca annulipes, *Uca mjobergi* and related species, fiddler crab, males	Wave and threaten with large claw	Bluffing strength	Large males with relatively large claws win fights. Males assess opponents visually and avoid larger males. When males lose their large claws they grow (regenerate) new ones that are long,	Jennions and Backwell (1996), Backwell et al. (2000), and Lailvaux et al. (2009)

(continued)

Table 16.1 (continued)

Context	Species	Signal/behavior	Kind of deceit	Mechanism	References
				thin, and weak compared to their original claws. Opponents avoid fighting males with long weak claws as often as those with long strong ones. Hence, regenerated claws bluff strength.	
	Neogonodactylus bredini, mantis shrimp, both sexes	Display merus of large maxilliped	Bluffing strength	Just before molting, individuals frequently challenge their neighbors using the meral spread display. Immediately after molting they are too weak to fight, but nevertheless use the display and bluff their ability to strike	Caldwell (1986) and Adams and Caldwell (1990)
	Uca pugilator, Uca beebei, Uca rapax fiddler crabs, both sexes	Furtive approach and motionless crouch	Hiding (perceptual) of self	A crab without a burrow crouches low and motionless several centimeters from a resident's burrow and the resident stops interacting with it. When the resident moves away from the burrow the crouching crab dashes into the burrow before the resident can return to defend it	Christy (1980) and B. Greenspan, personal communication
	Uca arcuata, U. beebei; Uca capricornis, Uca formosensis, fiddler crabs, both sexes but typically females	Chimneys	Hiding (physical) of the burrow entrance	Crabs build a cylinder of mud centered on their burrow entrance. Burrows with mud chimneys are less frequently found by crabs seeking new burrows than are burrows without chimneys. In U. beebei chimneys are taller than the eyes of intruders thus making the burrow opening impossible to see. Chimneys built by male U.	Wada and Murata (2000), J. Christy unpublished, Slatyer et al. 2008, and Shih et al. 2005

	Species	Signal	Deception type	Description	Reference
				formosensis may hide the male from rivals as he expands the burrow after attracting a mate	
Courtship	*Uca annulipes* (and possibly its sister species, *U. lactea*, *U. perplexa*, *U. mjobergi*) fiddler crab, males	Wave large claw	Bluffing condition	Males with long, thin, weak regenerated claws lose fights, but they court females by claw waving as do males with robust original-form claws that win fights. Females are as attracted to males with regenerated claws as they are to males with original claws	Backwell et al. (2000)
	Uca terpsichores, U. beebei and perhaps other structure-building species; fiddler crabs, males	Sand hoods, mud pillars	Mimicry, sensory trap	Crabs without burrows orient to and hide against objects on the surface to reduce their predation risk. Females leave their burrows and search for mates by visiting several courting males at their burrows before they choose one. Males sometimes build structures at the entrances to their burrows. Females are more attracted to burrows with than without structures and this preference increases with perceived predation risk. The preference reduces females' predation risk and incidentally biases mate choice in favor of males with structures	Christy (2007) (review)
	U. terpsichores, U. beebei, U. pugilator, fiddler crabs, males	Startling male displays	Mimicry, sensory trap	As a mate searching female leaves the burrow of a male she has visited but not mated, the male, with claw held low, moves quickly away, stops, raises his claw high and runs back to	Christy and Salmon (1991) and J. Christy unpublished

(continued)

Table 16.1 (continued)

Context	Species	Signal/behavior	Kind of deceit	Mechanism	References
				his burrow. This may startle the female into the burrow whereupon the male follows and the pair may mate. *U. terpsichores* males with sand hoods may instead move behind them, climb over the top and jump down in front of the burrow startling the female into the burrow. These rapid, elevated male movements present stimuli that are similar to those usually associated with predatory birds perhaps explaining why they startle females	

A cue also is a feature of an individual's phenotype (or the environment) to which other organisms respond. Receivers benefit when they respond to cues and their response can be either beneficial or harmful to the cue bearer. For example, tracks and odors left by a passing social group of prey could be used by lost group members to relocate the group or by predators to find a meal. Unlike signals, cues do not evolve because the responses to them increase the fitness of the cue barer (Hasson 1994). Cues can be traits with noncommunicative functions, byproducts of those traits, or nonfunctional features of the phenotype (e.g., tracks in the above example). However, once cues elicit responses that affect the fitness of the cue barer, they may be elaborated as signals if they are beneficial or they may be lost from the phenotype if they are costly and these costs are not balanced by benefits in other contexts.

16.2.2 Honesty and Deception

Maynard Smith and Harper's (2003) definition of signaling (see above and Wyatt, Chap. 2) requires that receivers benefit from their response in the signaling context and reliable delivery of such benefits constitutes honesty. There has been much debate about what ensures signal reliability (Harper 2006). It is now generally agreed that the costs of signal development, production, or maintenance and those imposed by skeptical receivers when they probe to detect deception ensure signal reliability (Searcy and Nowicki 2005). However, receivers can make errors in honest signaling systems when they have problems either detecting signals due to noise (Johnstone and Grafen 1992; Wiley 1994) or interpreting them when signalers vary in both the costs and benefits of signaling at a given level (Johnstone and Grafen 1993). The former errors are mistakes caused by the environment and are not the result of deception. The later errors provide the opportunity for deception as illustrated by some crustacean visual threat displays (e.g., snapping shrimp and hermit crabs, Table 16.1).

Although deception is familiar to us all, there is no widely accepted definition of deceiving by humans (Mahon 2007). The difficulties include specifying the intentions of the deceiver, what both parties do and do not believe, and how they come to hold those beliefs. Fortunately, a biological definition of deception is easier to write because the intentions and beliefs of animals are unknowable and are excluded from the definition. The following definition combines elements of those by Mitchell (1986) and Mahon (2007).

> To deceive is to act or appear in a way that causes another organism to respond appropriate to condition x, when x is not the case, or to fail to respond appropriate to condition x, when x is the case.

Thus, there are two broad classes of deception that differ according to the kind of error a receiver makes. Receivers make an error of the first kind when they falsely respond to signals that bluff or mimic and they make an error of the second kind when they fail to respond to behavior that hides a trait of interest; we present cases of each below. These two kinds of errors are analogous to statistical type one (false alarms) and type two (missed opportunities) errors (Hasson 1994; Wiley 1994). The first results in a cost to the receiver because its response is inappropriate and the second because its lack of response results in lost potential benefits. We exclude from deception cases in which an animal forcefully interferes with the ability of another to respond; an animal that wins a fight by blinding its opponent is using coercion, not deception. Next, we present examples of visual deception by bluffing, mimicry, and hiding by crustaceans during fighting and courtship, the two social contexts in which the interests of sender and receiver seldom coincide.

16.3 Visual Deception

16.3.1 Aggression

Equipped with hard exoskeletons and multiple appendages allowing specialization, crustaceans have evolved a diversity of powerful weapons (Emlen 2008), many of which can deliver fatal blows. When death is a possible outcome of a fight, it is in the mutual interest of opponents to use displays to assess their relative RHP (Resource Holding Potential) without exchanging blows; the weaker withdraws and avoids injury and the stronger wastes little time in an interaction it is sure to win. Not surprisingly, aggression between crustaceans typically begins with displays allowing assessment of weapons. These displays can include pushing and rubbing and other relatively nonforceful tactile components (chemical cues of sex or status are often present as well). The honesty of such threat displays should be maintained by escalated fights in which individuals that bluff RHP by displaying large but weak weapons suffer costly losses. Nevertheless, there are four examples of apparent bluffing of RHP or strength by crustaceans via weapon display and one example of bluffing aggressive intent (Table 16.1, Fig. 16.1).

In an especially perspicacious study of aggressive displays and fighting in snapping shrimp, Hughes (1996, 2000) (Table 16.1, Fig. 16.1a) showed that individuals with signals (chelae) that are larger than average for their competitive ability (determined by body size) will appear to be better competitors than they are. Some receivers may avoid fighting an individual with a large signal residual even though they might win. As predicted, Hughes (2000) found that smaller individuals with positive signal residuals, those with the most to gain by exaggerating their body size, were more likely to display aggressively, effectively bluffing their competitive ability, yet they were not more likely to win when receivers probed the deception and contests escalated to fights. This approach to detecting deception in an otherwise honest signaling system is useful only when signal residuals do not affect fighting ability (e.g., Briffa 2006). It has been used to detect bluffing of aggressive intent in hermit crabs (Arnott and Elwood 2010) and the evidence

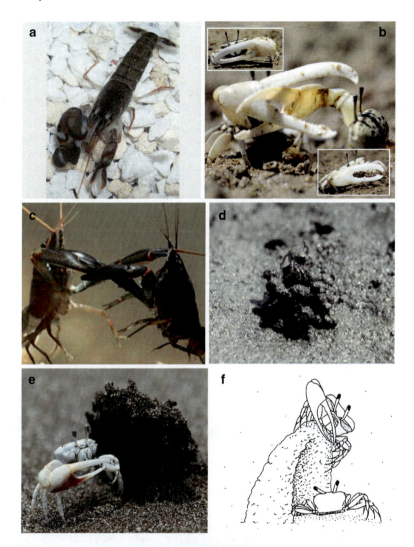

Fig. 16.1 Visual deception in crustaceans. Bluffing RHP wherein the size of the weapon may not reliably indicate strength: (**a**) *Alpheus heterochaelis*; (**b**) *Uca perplexa*; (**c**) *Cherax dispar*. Hiding burrow entrance and self: (**d**) female *Uca beebei* and her chimney. Sensory traps based on responses selected by predation: (**e**) *Uca terpsichores* male and his hood; (**f**) *U. terpsichores* male running over his hood and startling a female into his burrow (Image credits: (**a**) Melissa Hughes; (**b**, **d**, and **e**) John Christy; (**c**) Anthony O'Toole courtesy of Robbie Wilson; (**f**) drawn by Donna Conlon)

generally supports low levels of deception in this context (Laidre 2009). However, in many crabs, clawed lobsters, and crayfish, individuals with relatively large claws are better competitors (Table 16.1, Fig. 16.1b, c), in which case Hughes' signal residual analysis is inappropriate. Examples in Table 16.1 ("bluffing strength") show how threat signals that are honest on average can be corrupted when variation in the traits (muscle mass and cuticle strength) likely to affect RHP are not apparent to the viewer. Such variation may be caused by individual differences in nutrition, molt stage (Fig. 16.2), or even history of weapon loss and regeneration as in fiddler crabs (Fig. 16.1b). The crustacean exoskeleton coupled with the visual display and use of lethal weapons may predispose these animals to bluffing RHP.

There are two cases of deception by hiding during fiddler crab aggression, one by use of a furtive behavior, the other by concealment. I (JC) first saw furtive behavior (July 1974, Table 16.1) in male sand fiddler crabs *Uca pugilator* (Christy 1980) and subsequently in both sexes of several other species, most notably *Uca beebei*. Most remarkable is the ability of the crouching crab to determine when the resident is sufficiently far from its burrow to make the dash worthwhile and to orient to the burrow entrance it may not see (Zeil and Layne 2002).

The second example shows how fiddler crabs that are burrow residents sometimes build chimneys around their burrows that conceal both the burrow entrance and themselves (Table 16.1, Fig. 16.1d). Chimney building is known in at least 22 of the 97 recognized species of fiddler crabs (Rosenberg 2001; 21 species listed by Shih et al. 2005, plus *U. beebei*; J. Christy, unpublished). There is considerable variation in the sex, reproductive status, and sizes of crabs that build chimneys, how they are built, their shape, and how often and when they are built. Chimneys may have additional, and perhaps, multiple functions (e.g., for defense; Salmon 1987).

Fig. 16.2 Stomatopods *Neogonodactylus bredini* recognize each other by odor. They may bluff their dominance over another individual when they molt or when other physiological changes not reflected in their individual odor decrease their RHP (Image by Roy Caldwell)

16.3.2   Courtship

Male decapods use their weapons to signal during courtship. Berglund et al. (1996) pointed out that the crucible of combat should ensure that weapons honestly indicate RHP and that RHP should reliably indicate male condition and viability. Hence, females that choose mates by assessing their weapons may mate with males in good condition. However, when males bluff RHP with large but weak weapons, they may also bluff their mate quality. Indeed, this appears to be the case in *U. annulipes* (Table 16.1).

Two sensory traps, a mode of signaling based on mimicry (Christy 1995), have been discovered in the courtship of the fiddler crab *Uca terpsichores* and a few other species (Table 16.1). A sensory trap operates during courtship when a signal or display elicits a response from a receiver that increases the chance the signaler will mate, but that has not evolved for mate choice (Christy 1995). The first case, the subject of over a decade of field research, was summarized recently by Christy (2007). The female response to the male signal, a structure the male builds at the opening of his burrow (Fig. 16.1e), is selected by predation, increases in strength with predation risk (Kim et al. 2007, 2009), and incidentally results in mate choice. The female's response results in a direct immediate benefit (reduced risk of predation), but it also may be costly if it results in choice of a mate that is different from her choice had the sensory trap signal not been used (Bradbury and Vehrencamp 2000). However, males that are in better condition are more likely to build structures (Backwell et al. 1995; Kim et al. 2010). Hence, females that mate with structure builders may benefit provided that female fitness increases with the condition of their mate. If so, then females that are caught in this sensory trap may doubly benefit and we leave it to the reader to decide whether this "white lie" qualifies as deception (Christy 1997).

The second sensory trap in fiddler crab courtship was discovered in *U. pugilator* (Christy 1980; Christy and Salmon 1991) and has subsequently been seen in *U. terpsichores* and *U. beebei*. When a female that is searching for a mate stops at a male's burrow and begins to leave, the male usually courts the female with claw waving. However, just as the female begins to depart, males sometimes execute two displays that include rapid movement of their large claw or entire body above the female's visual horizon (Table 16.1, Fig. 16.1f). Such looming stimuli elicit a startle response (Layne 1998); the females dash back to the burrow as if they have detected an approaching predator. Hence, the deception relies on females' mistaken responses to displaying males as if they were predators.

16.4   Chemical Communication with Cues, not Signals

As chapters by Duffy and Thiel (2007) attest, crustaceans respond to chemicals that originate from conspecifics during hatching, settlement, maternal care, and family interactions, while seeking and interacting with mates and when competing for food and shelter. Some of these chemicals have not been modified for communication and therefore fit the definition of a cue rather than a signal (see below). Few chemicals used in social interactions have been identified. In contrast, studies of chemical communication by crustaceans have produced detailed descriptions of signaling behavior and the structures used to deliver and receive chemicals.

Peptides mediate hatching, settlement and, in hermit crabs, shell-seeking behavior. Research on these peptides spans 5 decades and has produced over 100 publications (summarized in Rittschof and Cohen 2004). The peptides result from hydrolysis of structural proteins by trypsin-like serine proteases. These proteases are endogenous to the source or are produced by bacteria that feed on the proteins or by the digestive enzymes of predators of gastropod shell occupants, or as part of coagulation processes (Matsumura et al. 2000; Dickinson et al. 2009). Hence, the production of the active chemical is routinely a predictable byproduct of an important activity and is not under direct control of the source as would be expected of most signals. As is typical of cues but not signals, the peptides and proteins are active at very low concentrations (e.g., 10^{-7} to $\ll 10^{-11}$ M for natural cues from eggs that stimulate larval release behavior in females) and they do not stimulate greater responses at higher concentrations. Many features of peptide-based chemical interactions in crustaceans and other phyla support the view that the active chemicals are cues of interest to responders, that they evolved before the responses to them, and that they have not been modified as signals. Substituted amino sugars mediate behavioral interactions in some crustaceans (Rittschof and Cohen 2004); since these molecules too are byproducts of the basic chemical pathways by which crustaceans and other aquatic organisms are built and manage hygiene, they are cues.

It is not known whether chemical social signaling in crustaceans uses cues or chemicals that are specifically modified as signals. The recent identification of the first crustacean sex pheromone (Hardege and Terschak, Chap. 19) supports the former view. The chemical is uridine diphosphate (UDP) which is a byproduct of the final step in chitin synthesis. Interestingly, as expected of a conserved cue but not a highly evolved signal, UDP elicits male sexual behavior in several unrelated crabs (Bublitz et al. 2008). Lacking evidence to the contrary, we therefore speculate that chemical signaling in crustaceans may typically be accomplished by modulated production, release, and delivery of chemical cues. If this is correct, then one way that deception in this channel, if it exists, could be achieved is through changes in the behaviors used to release and direct otherwise genuine cues to receivers.

16.4.1 Bluffing Dominance

Many benthic crustaceans live at high densities in self-made or natural refuges for which they compete intensely. Dominance relationships are established through fighting and may often be maintained through individual recognition (IR) based on chemical cues (Table 16.2). IR occurs when receivers perceive and recognize others

Table 16.2 Possible mechanisms of deception in crustacean chemical communication

Context	Kind of deceit	Mechanism	Possible examples
Aggression	Bluffing individual dominant status	Individuals recognize each other by odor. When an animal fights and looses, it associates the odor of the winner with its dominant status. When the individuals again meet, the subordinate recognizes the dominant by odor and defers to it even if the fighting ability of the dominant has declined making it likely to lose if the animals fought again	Gonodactylid stomatopods, Caldwell (1986) and Adams and Caldwell (1990) *Homarus americanus*, Karavanich and Atema (1998) *Pagurus bernhardus*, Hazlett (1969) *Pagurus longicarpus*, Gherardi et al. (2005)
	Hiding RHP to deliver a "blind side" winning punch	Dominant males tend to release more urine more often in fights and urine release tends to be coincident with aggressive acts. However, when fights escalate, the variance in urine release by dominants increases in *H. americanus*, fights are just as likely to reach the highest levels whether or not males release urine in *Astacus leptodactylus* and, at the highest levels of fights, dominant *Orconectes rusticus* cease directing urine toward opponents. Hence, when fights escalate dominant males may deliver winning forceful aggressive acts without producing urine or directing it toward opponents and chemically signaling their aggressive intent	*H. americanus*, Breithaupt and Atema (2000) *A. leptodactylus*, Breithaupt and Eger (2002) *O. rusticus*, Bergman et al. (2005)
	Hiding sexual receptivity but not sex	Male stomatopods *N. bredini* can detect females by odor, but not their state of receptivity. Around the time of the full moon, female *N. bredini* that are not sexually receptive exploit the tendency of males to admit females to their cavities for mating. Such females gain access to males' cavities and prevent the males from re-entering about twice as often as do sexually receptive females	*N. bredini*, Caldwell (1986), Mead and Caldwell, Chap. 11
Courtship	Hiding sexual receptivity	Sexually active males detect females and their state of	*H. americanus*, Atema and Steinbach (2007)

(continued)

Table 16.2 (continued)

Context	Kind of deceit	Mechanism	Possible examples
		sexual receptivity by odors in their urine. During mate sampling and assessment females control the release of their urine and its transmission in currents toward or away from males. In this manner they can hide from and thus reject some males and reveal their sex and status to others they prefer as mates	*Carcinus maenas*, Bamber and Naylor (1996) *Callinectes sapidus*, Jivoff and Hines (1998)

by their uniquely distinctive cues and respond in a manner appropriate to the identified individual (Caldwell 1986; Sherman et al. 1997). IR mediates dominance when an individual that has fought and lost learns to associate unique cues with the winner. Although the response "defer to dominant" may be a class-level response, it is elicited because the dominant is identified individually (Mead and Caldwell, Chap. 11; Aggio and Derby, Chap. 12; Wyatt, Chap. 2).

We suggest that dominance relationships that are mediated by IR are vulnerable to bluffing because the cue used for IR and the property of the individual that allows it to dominate another are not immutably linked (Caldwell 1986). Presumably, IR cues are individually distinctive body odors that are unrelated to condition or other traits that affect RHP. Indeed, lack of such relationships would be advantageous because IR could mediate a variety of interactions based on other learned qualities and physiological states. Thus, an individual's identity will remain constant, while some of the factors that affect its transitory physiological or condition-dependent qualities vary (However, a change in body odor when an individual molts would reveal a temporary reduction in RHP). If an individual's RHP declines after the individual has established its dominance over another, the weakened individual can use its IR cues to bluff dominance until the subordinate calls the bluff. This situation is analogous to bluffing RHP with weapons made of exoskeleton that do not decrease in size with a loss in condition or muscle mass. Experiments in which an individual's RHP is decreased by forced exercise or starvation could be used to explore bluffing dominance by IR. Under natural conditions, the decay in the relationship between IR and dominance, as measured by the extinction of the subordinate's submissive deference when the two meet, should relate to the average reduction in individuals' RHP with time and, consequently, a reduction in the costs to the subordinate of probing the relationship. The bluff should be effective and used for a time determined by the rate of decay of condition and the decrease in the costs of probing. If bluffing dominance by IR is common, subordinates should "forget" their status sooner rather than later or use other cues to detect a reduction in RHP (e.g., coincident with molting). Under experimental laboratory conditions, American lobsters ceased to recognize individuals that beat them in fights after about 1 week, even though when they challenged the dominant they lost again (7 cases,

Karavanich and Atema 1998). Subordinates may "forget" and probe sooner than dominants lose their relative RHP. We emphasize that the use of IR to bluff may be possible when the relationship between IR cues and class membership is arbitrary, learned, and ephemeral. Hence, bluffing may occur in other relationships based on IR chemical cues and learned temporary states that affect social interactions.

16.4.2 Hiding RHP

Chemicals that mediate social interactions in decapod crustaceans often are present in urine, with highly controlled directed release (Atema and Steinbach 2007). Urine contains metabolites that are potential cues of an individual's identity, membership in a social or biological class, physiological or motivational status, and diet or future behavior (Atema and Steinbach 2007). Here we explore urinary signaling during fighting in lobsters and crayfish and during sexual interactions in lobsters and crabs. Our intent is to identify situations in which it might be to the sender's advantage to withhold the chemical. We then present possible evidence of chemical "hiding" (Table 16.2).

Homarid (clawed) lobsters and crayfish that are not familiar with each other interact aggressively when they meet and these interactions often escalate to fights in which individuals may be injured or (rarely) killed. Fights typically escalate following a patterned progression of exchange of visual, tactile, and chemical cues. The relationships between fight outcome and the release of urine have been explored on fine temporal scales during symmetrical interactions between male *Homarus americanus* (Breithaupt and Atema 2000), *Astacus leptodactylus* (Breithaupt and Eger 2002), and *Orconectes rusticus* (Bergman et al. 2005). Lobsters and the crayfish *O. rusticus* release urine as opponents approach and meet, and winners are more likely to release urine than are losers. In *A. leptodactylus* winners also tend to release urine more often than do losers, but significant urine release occurs only after the animals begin to fight. In all species, urine release is associated with offensive behaviors and it increases significantly, especially by dominants, as interactions escalate. In lobsters the probability that a male releases urine at a given level in a fight is strongly positively associated with the probability that it has released urine up to 30 s before. Hence, urine release both precedes and accompanies offensive acts as one would expect of a threat. However, at the highest level of aggression (claw ripping) in lobsters there is no significant increase in urine release compared to the level at which they begin to interact but cannot touch each other (Breithaupt and Atema 2000). Examination of the data (Fig. 16.3a) suggests this is a consequence of high variance in urine release at the highest level of escalation. Similarly, male *A. leptodactylus* are just as likely to escalate to a physical fight following a threat at a distance whether or not they released urine when they gave the threat (Breithaupt and Eger 2002 (Fig. 16.3b). Yet, in escalated fights receivers decrease their subsequent aggression and respond with defensive acts only if an opponent's offensive behavior is accompanied by the release of urine (see

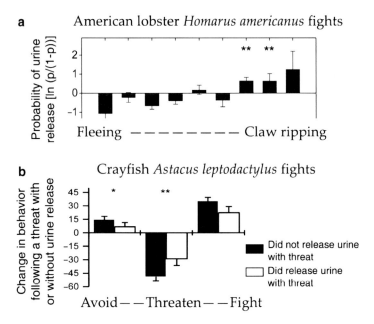

Fig. 16.3 Lobsters and crayfish may hide chemical cues of aggression in escalated fights: (**a**) the high standard error of the mean probability (log odds ± SEM) of urine release by male lobsters *Homarus americanus* at the highest level of combat suggests winners sometimes stopped releasing urine which otherwise reliably predicts aggressive intent (urine release both immediately precedes and accompanies aggressive acts); probability is measured as ln (p/(1-p)); categories of aggression range from defense (fleeing) to forceful offensive acts (claw ripping) (**b**) fighting male crayfish *Astacus leptodactylus* are more likely to exhibit defensive behavior if their opponent releases urine when it acts aggressively (Breithaupt and Eger 2002). However, about half the time, fights escalate to the highest levels of aggression following threats in which males do not release urine and this may reduce the chance opponents will mount effective defenses. (**a**) Adapted with permission from Journal of Experimental Biology after Breithaupt and Eger (2002), and (**b**) modified after Breithaupt and Atema (2000) with kind permission from T. Breithaupt and Springer Science + Business Media

Fig. 13.5 in Breithaupt, Chap. 13; Breithaupt and Eger 2002). Finally, when crayfish grab each others' claws, dominant *O. rusticus* tend to produce less urine and to generate posterior currents sending the urine away from opponents just before the end of a fight. The temporal patterns of fighting and urine release in these crustaceans suggest that males may sometimes attempt to catch opponents off guard and deliver a "blind-side punch" by withholding urine or directing it away before they deliver a highly aggressive and winning act. The advantage of such cheating, measured by an increase in the probability of a win, comes from the failure of the opponent to respond effectively to the blow due to the lack of the threat that usually precedes it. Such cheating would persist only if it was attempted infrequently.

Fights may also be won by withholding cues not of RHP, but of sexual receptivity in the stomatopod *Neogonodactylus bredini* (Table 16.2; Mead and Caldwell, Chap. 11). During peak mating periods males come out of their cavities (where these mantis shrimp mate) and allow females to enter whether or not the female is receptive. Evidently, males chemically detect the sex, but not the reproductive status of the female. Once in the cavity, a nonreceptive female can use her positional advantage to exclude the male. However, it is not clear whether nonreceptive females more often seek cavities during peak mating periods as would be expected if there is selection to deceive males into relinquishing their burrows.

16.4.3   Hiding Sex

Clawed lobsters and many species of crabs and shrimps usually mate a short time after the female molts. Depending on species, either or both sexes may seek the other by following chemical cues (Table 16.2). For example, female *H. americanus* approach dominant males in dens and these males make their presence, dominance, and probably individual identity known by broadcasting cues in urine. Once at a preferred male's den, both sexes direct jets of urine at each other, presumably for mutual mate assessment (Atema and Steinbach 2007). In the blue crab *Callinectes sapidus* females also locate males using chemical cues (Bushmann 1999; Kamio et al. 2008). Sexually active males take up a relatively fixed position and dribble seminal fluid. They direct cues toward females by paddling their last walking legs which they hold high above the posterior edge of their carapace (Jivoff and Hines 1998; Kamio et al. 2008; Kamio and Derby, Chap. 20). Receptive premolt females move until they come close to a dribbling male whereupon they release urine which contains cues that inhibit feeding and elicit cradling by the male (D. Rittschof, unpublished) and guarding (Jivoff and Hines 1998). Using restrained males in aquaria, Bamber and Naylor (1997) showed that sexually receptive female green crabs *C. maenas* play an active role in courtship and mate choice by soliciting guarding or copulation. These observations strongly suggest that female lobsters and crabs seek out and discriminate among potential mates using chemical and perhaps other cues and that they indicate their selection by sending chemical cues preferentially to their future partner. Females may thereby increase the net benefits of being guarded by minimizing the costs of overly long guarding by typically always eager males (Jivoff and Hines 1998). Indeed, males are far less selective than are females: in the lab they court, guard, and attempt to mate with sponges (Kamio et al. 2000), stones (Bublitz et al. 2008), and even other males (Hardege et al. 2002) that are treated with chemical cues from sexually attractive females. In the field male blue crabs *C. sapidus* also sometimes carry smaller males (D. Rittschof, unpublished). Hence, by withholding chemical cues of sexual receptivity, female crabs and lobsters can avoid certain males as mates. Their lack of signaling can be interpreted as a form of perceptual hiding, analogous to the motionless crouch of a fiddler crab that prevents its visual detection by a burrow resident.

We close with two highly speculative suggestions for other modes of chemical deception and manipulation in crustaceans. First, in species with polymorphic males of graded social status, such as the shrimps *Rhynchocinetes typus* (Correa et al. 2000) and *Macrobrachium rosenbergii* (Kuris et al. 1987), sexual access to females and fertilization success usually is strongly biased toward the dominant morph (Ra'anan and Sagi 1985; Correa et al. 2003). In this context selection may favor female-like morphology, behavior, and chemical cues in subordinate classes. By mimicking females in several sensory modalities, subordinates may be able to remain near to dominants and gain occasional uncontested access to females that the dominant male defends. Chemical mimicry of females by subordinate males is not without precedent in arthropods; the passive winged morphs of males of the ant *Cardiocondyla obscurior* use sexual mimicry to gain sexual access to emerging reproductive females without having to fight aggressive wingless males (Cremer et al. 2008).

Second, in *Drosophila* and a number of other insects (Arnqvist and Rowe 2005), male seminal fluid is a cocktail of chemicals that induces a large range of physiological and behavioral responses in females that increase male reproductive success. Decapods typically pass sperm and seminal products to females as external spermatophores or as a viscous fluid stored in the spermathecae (Sainte-Marie 2007). Females may store seminal fluid briefly, or until they oviposit following a molt, or for life depending on species and life history. Storage of seminal fluids in the female creates the opportunity for selection to favor the incorporation of chemicals in seminal fluid that manipulate female sexual receptivity. As emphasized by Sainte-Marie (2007), there is a dearth of information on male gametic strategies in decapods (and other crustaceans).

16.5 Looking Forward

Deception in crustacean chemical communication has not yet been detected. This mode of communication may most often involve the transfer of cues that are byproducts of essential biochemical processes and that cannot be faked. Yet, we see no reason, based on first principals, why deception via the withholding and directed release of such cues should be rare. But we also acknowledge that detecting such deception will be very difficult. Techniques to monitor and visualize the production and release of chemicals by senders and the detection of and responses to them by receivers under both seminatural and experimental conditions have just begun to be used to study signaling in aggressive contexts in lobsters and crayfish.

Acknowledgments We thank Martin Thiel and Thomas Breithaupt for their guidance, helpful criticism, and especially their patience. J. Christy thanks Pat Backwell for freely sharing ideas and for her expert assistance in field studies of visual deception in fiddler crabs. D. Rittschof thanks Sarah McCall, Zach Darnell, Kelly Darnell, and Ruth McDowell for the quality time in the field with sexually receptive blue crabs.

References

Adams ES, Caldwell RL (1990) Deceptive communication in asymmetric fights of the stomatopod crustacean *Gonodactylus bredini*. Anim Behav 39:706–716

Arnott G, Elwood RW (2010) Signal residuals and hermit crab displays: flaunt it if you have it! Anim Behav 79:137–143

Arnqvist G, Rowe L (2005) Sexual conflict. Princeton University Press, Princeton

Atema J, Steinbach MA (2007) Chemical communication and social behavior of the lobster *Homarus americanus* and other decapod crustacea. In: Duffy JE, Thiel M (eds) Evolutionary ecology of social and sexual systems. Oxford University Press, New York, pp 115–144

Backwell PRY, Christy JH, Telford SR, Jennions MD, Passmore NI (2000) Dishonest signaling in a fiddler crab. Proc R Soc Lond B 267:719–724

Backwell PRY, Jennions MD, Christy JH, Schober UM (1995) Pillar building in the fiddler crab *Uca beebei*: evidence for a condition-dependent ornament. Behav Ecol Sociobiol 36:185–192

Bamber SD, Naylor E (1996) Chemical communication and behavioural interaction between sexually mature male and female shore crabs (*Carcinus maenas*). J Mar Biol Assoc UK 76:691–699

Bamber SD, Naylor E (1997) Sites of release of putative sex pheromone and sexual behaviour in female *Carcinus maenas* (Crustacea: Decapoda). Estuar Coast Shelf Sci 44:195–202

Berglund A, Bisazza A, Pilastro A (1996) Armaments and ornaments: an evolutionary explanation of traits of dual utility. Biol J Linn Soc 58:385–399

Bergman DA, Martin AL, Moore PA (2005) Control of information flow through the influence of mechanical and chemical signals during agonistic encounters by the crayfish, *Orconectes rusticus*. Anim Behav 70:485–496

Bradbury JW, Vehrencamp SL (2000) Economic models of animal communication. Anim Behav 59:259–268

Breithaupt T, Atema J (2000) The timing of chemical signaling with urine in dominance fights of male lobsters (*Homarus americanus*). Behav Ecol Sociobiol 49:67–78

Breithaupt T, Eger P (2002) Urine makes the difference: chemical communication in fighting crayfish made visible. J Exp Biol 205:1221–1231

Briffa M (2006) Signal residuals during shell fighting in hermit crabs: can costly signals be used deceptively? Behav Ecol 17:510–514

Bushmann PJ (1999) Concurrent signals and behavioral plasticity in Blue crab (*Callinectes sapidus* Rathbun) courtship. Biol Bull 197:63–71

Bublitz R, Sainte-Marie B, Newcomb-Hodgetts C, Fletcher N, Smith M, Hardege J (2008) Interspecific activity of the sex pheromone of the European shore crab (*Carcinus maenas*). Behaviour 145:1465–1478

Caldwell RL (1986) The deceptive use of reputation by stomatopods. In: Mitchell RW, Thompson NS (eds) Deception, perspectives on human and nonhuman deceit. State University of New York Press, Albany, pp 129–145

Cremer S, D'Ettorre P, Drijfhout FP, Sledge MF, Truillazzi S, Heinze J (2008) Imperfect chemical female mimicry in males of the ant *Cardiocondyla obscurior*. Naturwissenschaften 95:1101–1105

Christy JH (1980) The mating system of the sand fiddler crab *Uca pugilator*. Ph.D. thesis, Cornell University

Christy JH (1995) Mimicry, mate choice and the sensory trap hypothesis. Am Nat 146:171–181

Christy JH (1997) Deception: the correct path to enlightenment? Trends Ecol Evol 12:160

Christy JH (2007) Predation and the reproductive behavior of fiddler crabs (genus *Uca*). In: Thiel M, Duffy JE (eds) Evolution of social behavior of crustaceans. Oxford University Press, Oxford, pp 211–231

Christy JH, Salmon M (1991) Comparative studies of reproductive behavior in mantis shrimps and fiddler crabs. Amer Zool 31:329–337

Correa C, Baeza JA, Dupré E, Hinojosa IA, Thiel M (2000) Mating behavior and fertilization success of three ontogenetic stages of male rock shrimp *Rhynchocinetes typus* (Decapoda: Caridea). J Crust Biol 20:628–640

Correa C, Baeza JA, Hinojosa IA, Thiel M (2003) Male dominance hierarchy and mating tactics in the rock shrimp *Rhynchocinetes typus* (Decapoda: Caridea). J Crust Biol 23:33–45

Dickinson GH, Vega IE, Wahl KJ, Orihuela B, Beyley V, Rodriguez EN, Everett RK, Bonaventura J, Rittschof D (2009) Barnacle cement: a polymerization model based on evolutionary concepts. J Exp Biol 212:3499–3510

Duffy JE, Thiel M (2007) Evolutionary ecology of social and sexual systems. Oxford University Press, Oxford

Emlen DJ (2008) The evolution of animal weapons. An Rev Ecol System 39:387–413

Gherardi F, Tricarico E, Atema J (2005) Unraveling the nature of individual recognition by odor in hermit crabs. J Chem Ecol 31:2877–2896

Hardege JD, Jennings A, Hayden D, Müller CT, Pascoe D, Bentley MG, Clare AS (2002) Novel behavioural assay and partial purification of a female-derived sex pheromone in *Carcinus maenas*. Mar Ecol Prog Ser 244:179–189

Harper DGC (2006) Maynard Smith: amplifying the reasons for signal reliability. J Theor Biol 239:203–209

Hasson O (1994) Cheating signals. J Theor Biol 167:223–238

Hazlett BA (1969) "Individual" recognition and agonistic behavior in *Pagurus bernhardus*. Nature 222:268–269

Hebets EA, Papaj DR (2005) Complex signal function: developing a framework for testable hypotheses. Behav Ecol Sociobiol 57:197–214

Hughes M (1996) Size assessment via a visual signal in snapping shrimp. Behav Ecol Sociobiol 38:51–57

Hughes M (2000) Deception with honest signals: signal residuals and signal function in snapping shrimp. Behav Ecol 11:614–623

Jennions MD, Backwell PRY (1996) Residency and size affect fight duration and outcome in the fiddler crab *Uca annulipes*. Biol J Linn Soc 57:293–306

Jivoff P, Hines AH (1998) Effect of female molt stage and sex ratio on courtship behavior of the blue crab *Callinectes sapidus*. Mar Biol 131:533–542

Johnstone RA, Grafen M (1992) Error-prone signaling. Proc Royal Soc Lond B 248:229–233

Johnstone RA, Grafen M (1993) Dishonesty and the handicap principle. Anim Behav 46:759–764

Kamio M, Matsunaga S, Fusetani N (2000) Studies on sex pheromones of the helmet crab, *Telmessus cheiragonus* 1. An assay based on precopulatory mate-guarding. Zool Sci 17:731–733

Kamio M, Reidenbach MA, Derby CD (2008) To paddle or not: context dependent courtship display by male blue crabs, *Callinectes sapidus*. J Exp Biol 211:1243–1248

Karavanich C, Atema J (1998) Individual recognition and memory in lobster dominance. Anim Behav 56:1553–1560

Kim TW, Christy JH, Choe JC (2007) A preference for a sexual signal keeps females safe. PLoS ONE 2(5):e422

Kim TW, Christy JH, Dennenmoser S, Choe J (2009) The strength of a female mate preference increases with predation risk. Proc Royal Soc Lond B 276:775–780

Kim TW, Christy JH, Rissanen JR, Ribeiro PD, Choe J (2010) The effect of food addition on the reproductive intensity and timing of both sexes of an intertidal crab. Mar Ecol Prog Ser 401:183–194

Kuris AM, Ra'anan Z, Sagi A, Cohen D (1987) Morphotypic differentiation of male Malyasian giant prawns, *Macrobrachium rosenbergii*. J Crust Biol 7:219–237

Laidre ME (2009) How often do animals lie about their intentions? An experimental test. Am Nat 173:337–346

Lailvaux SP, Reaney LT, Backwell PRY (2009) Dishonest signaling of fighting ability and multiple performance traits in the fiddler crab *Uca mjoebergi*. Funct Ecol 23:359–366

Layne JE (1998) Retinal location is the key to identifying predators in fiddler crabs (*Uca pugilator*). J Exp Biol 201:2253–2261

Mahon JE (2007) A definition of deceiving. Int J Appl Phil 21:181–194

Matsumura K, Hills JM, Thomason PO, Thomason JC, Clare AS (2000) Discrimination at settlement in barnacles: laboratory and field experiments on settlement behaviour in response to settlement-inducing protein complexes. Biofouling 16:181–190

Maynard Smith J, Harper D (2003) Animal signals. Oxford University Press, New York

Mitchell RW (1986) A framework for discussion deception. In: Mitchell RW, Thompson NS (eds) Deception, perspectives on human and nonhuman deceit. State University of New York Press, Albany, pp 3–40

Ra'anan Z, Sagi A (1985) Alternative mating strategies in male morphotypes of the freshwater prawn *Macrobrachium rosenbergii*. Biol Bull 169:592–601

Rittschof D, Cohen JH (2004) Crustacean peptide and peptide-like pheromones and kairomones. Peptides 25:1503–1516

Rosenberg MS (2001) The systematics and taxonomy of fiddler crabs: a phylogeny of the genus *Uca*. J Crust Biol 21:839–869

Salmon M (1987) On the reproductive behavior of the fiddler crab *Uca thayeri*, with comparisons to *U. pugilator* and *U. vocans*: evidence for behavioral convergence. J Crust Biol 7:25–44

Sainte-Marie B (2007) Sperm demand and allocation in decapods crustaceans. In: Duffy JE, Thiel M (eds) Evolutionary ecology of social and sexual systems. Oxford University Press, New York, pp 191–210

Scott-Phillips TC (2008) Defining biological communication. J Evol Biol 21:387–395

Searcy WA, Nowicki S (2005) The evolution of animal communication, reliability and deception in signaling systems. Princeton University Press, Princeton

Sherman PW, Reeve HK, Pfennig DW (1997) Recognition systems. In: Krebs JR, Davies NB (eds) Behavioural ecology: an evolutionary approach. Blackwell Scientific, Cambridge, pp 69–96

Shih H, Mok H, Chang H (2005) Chimney building by male *Uca formosensis* Rathbun, 1921 (Crustacea: Decapoda: Ocypodidae) after pairing: a new hypothesis for chimney function. Zool Studies 44:242–251

Slatyer RA, Fok ESY, Hocking R, Backwell PRY (2008) Why do fiddler crabs build chimneys? Biol Letters 4:616–618

Wada K, Murata I (2000) Chimney building in the fiddler crab *Uca arcuata*. J Crust Biol 20:505–509

Wiley RH (1994) Errors, exaggerations and deception in animal communication. In: Real LA (ed) Behavioural mechanisms in evolutionary ecology. University of Chicago Press, Chicago, pp 157–189

Wilson RS, Angilletta MJ Jr, James RS, Navas C, Seebacher F (2007) Dishonest signals of strength in male slender crayfish (*Cherax dispar*) during agonistic encounters. Am Nat 170:284–291

Wilson RS, James RS, Bywater C, Seebacher F (2009) Costs and benefits of increased weapon size differ between sexes of the slender crayfish, *Cherax dispar*. J Exp Biol 212:853–858

Zeil J, Layne JE (2002) Path integration in fiddler crabs and its relation to habitat and social life. In: Wiese K (ed) Crustacean experimental systems in neurobiology. Springer, Berlin, pp 227–246

Chapter 17
Chemical Communication in a Multimodal Context

Eileen A. Hebets and Aaron Rundus

Abstract All animals are equipped with multiple sensory systems (e.g., visual, chemical, acoustic, tactile, electrical, thermal), and signals perceived via these sensory systems facilitate communication. Such communication often involves displays that incorporate more than one signal from more than one sensory modality, resulting in multimodal signaling. The number of empirical and theoretical studies addressing issues of multimodal signaling is ever-increasing and this chapter highlights why crustaceans, as a taxonomic group, are ideal for advancing such studies. Early classifications of multimodal signaling sought to categorize signal components as either redundant or nonredundant, while more recent classifications lay out specific hypotheses relating to multimodal signal function. Two common empirical approaches used in studying multimodal signaling involve signal isolation and signal playback designs – both of which are extremely amenable to crustaceans.

Chemical communication is considered the oldest and most widespread channel for communication, and as such, it is not surprising that numerous crustaceans incorporate chemical signals into multimodal displays. In this chapter, we review multimodal signaling in crustaceans with a focus on those displays that incorporate a chemical component. Specifically, we highlight examples of taxa that combine chemical and hydrodynamic as well as chemical and visual cues. We conclude that despite the plethora of excellent studies examining crustacean responses to isolated signal components, relatively few studies are couched in a communication framework – ultimately limiting the conclusions that can currently be drawn with respect to multimodal signal evolution and function in crustaceans. We suggest that future studies using a hypothesis-testing framework of multimodal signal function could greatly advance our understanding of multimodal signaling in this group. Furthermore, studies involving signal manipulations and correlations between signaler attributes and variation in signal form could be extremely informative. These avenues are wide open for crustacean biologists. We argue that several aspect of

E.A. Hebets (✉)
School of Biological Sciences, University of Nebraska,
Lincoln, NE 68588-0118, USA
e-mail: ehebets2@unl.edu

T. Breithaupt and M. Thiel (eds.), *Chemical Communication in Crustaceans*,
DOI 10.1007/978-0-387-77101-4_17, © Springer Science+Business Media, LLC 2011

crustacean biology (e.g., their abundance, the ease with which they can be manipulated, the ease with which their environment can be manipulated, their morphological diversity, the diversity of habitats in which they live, etc.) make them ideal for studying multimodal signaling!

17.1 Introduction

All animals are equipped with multiple sensory systems (e.g., visual, chemical, acoustic, tactile, electrical, thermal) and signals perceived via these sensory systems facilitate communication. It has long been the goal of scientists to understand the selection pressures that influence signal form and function. Yet, despite the significant progress that has been made, it is becoming increasingly clear that animal displays are often complex, incorporating more than one signal or component, frequently from more than one sensory modality (i.e., multimodal Rowe 1999) into a single display. This recently acknowledged complexity of animal displays has increased the difficulty for conducting empirical studies of signal function, yet it has simultaneously given birth to exciting new theory.

Multimodal signals comprise a fascinating and widespread category of complex signaling in which components from more than one sensory modality are combined into a single multimodal display. The addition of signaling modalities presumably carries with it several costs such as increasing a signaler's energy expenditure, increasing its conspicuousness, and/or adding new avenues available for eavesdropping or aggressive mimicry (Hebets and Papaj 2005; Partan and Marler 2005). Why then is multimodal signaling so common? This question can be asked in several ways. For example, from an evolutionary viewpoint, we might ask whether general patterns regarding the way in which sensory modalities are combined can shed light on selection pressures driving the evolution of multimodal signaling. From an ecological point of view, we might ask whether ecological constraints can select for multimodal signaling. If the signaling environment is variable, and signal efficacy is tightly tied to environmental characteristics, simultaneously signaling in multiple modalities might prove advantageous. From a functional point of view, we might ask whether signals in different modalities, or in different combinations of modalities, are better at effectively eliciting the desired behavioral response(s) from signal targets. This latter line of questioning can then lead to more proximate examinations of how an animal's neural architecture influences the reception and processing of multisensory input. For those of us interested in integrative approaches to understanding trait evolution, multimodal signaling provides an excellent opportunity to simultaneously explore the evolution and function of a trait at multiple levels of analysis.

Although certainly not at the forefront of our minds, when pressed, most of us can relate well to the importance of multimodal signaling. Being part Sicilian for example, I (EAH) find that I rely heavily on gestures and body language when telling a story, teaching a class, etc. I typically incorporate gestures to emphasize a point or to more clearly convey my verbal information. As a result, I am much more effective at conveying information in person, for example, than over the phone. Similarly, my now 3 ½-year-old daughter became proficient in multimodal communication while learning verbal language. My husband, Jay Storz, and I taught Jessie sign language well before she could speak. She was quick to pick up the signs she was taught and soon realized that she could further increase the efficacy of her communication by incorporating additional visual signals in the form of facial expressions – resulting in multicomponent visual signaling. Once her verbal vocabulary started to surface, Jessie began combining her visual signals with her verbal words and her language became truly multimodal. Depending upon both the situation and the information she wished to convey, Jessie's multimodal signaling (visual and acoustic) functioned to convey the same information (for example, because I could not figure out what she was trying to tell me using only one component, e.g., redundant signal hypothesis) or to convey multiple bits of information (e.g., multiple messages). Regardless of reason, as a receiver, I have no doubt that I was more likely to understand what my daughter was trying to tell me and to respond appropriately if I could both see and hear her!

As is often the case with new trends in research topics, much of the functional work on multimodal signaling has thus far focused on a few taxonomic groups (e.g., birds, spiders). Throughout this chapter, we emphasize why crustaceans should not be overlooked for their potential contribution to this field! We will call attention to several aspects of crustacean biology that make them excellent candidates for research on multimodal signaling, many of which are unique to the group (Fig. 17.1). Furthermore, crustaceans rely heavily on chemical communication, a sensory modality deserving more attention in multimodal signaling. However, before delving deeper into crustacean multimodal communication, we will first lay the conceptual groundwork by describing the framework and general approaches used, as well as by providing a few well-studied examples from other taxonomic groups.

Fig. 17.1 Head region of *Ovalipes trimaculatus* showing the antennae (which perceive chemical and hydrodynamic stimuli) and the stalked eyes (which perceive visual stimuli). Photograph by Iván A. Hinojosa

17.2   General Aspects of Multimodal Signaling

17.2.1   Conceptual Framework of Multimodal Signaling

The study of multimodal signaling is rich with theory, as evidenced by the plethora of recent reviews of the subject (Candolin 2003; Hebets and Papaj 2005; Partan and Marler 2005). Early classifications sought to place multimodal signals into categories of redundant versus nonredundant signals based upon receiver responses (Partan and Marler 1999), while more recent classifications attempt to lay out more specific hypotheses relating to complex signal function (Candolin 2003; Hebets and Papaj 2005). Much of the theoretic framework addresses the ultimate question - "why" – why use multimodal signals? This question is most commonly addressed from a content-based approach, with studies focusing on the characteristics of signaling individuals and exploring how these characteristics are, or are not, reflected in the form of their signal. In other words, content-based approaches focus upon the purpose of the signal and the putative information it conveys. Using this approach, the "multiple messages" and "redundant," or "back-up," signals hypotheses are the most commonly tested functions of multimodal signaling (Møller and Pomiankowski 1993; Johnstone 1996). The "multiple messages" hypotheses simply propose that different signals reflect different information or content (Møller and Pomiankowski 1993; Johnstone 1996). For example, one modality may correlate with foraging history and another with parasite resistance. In contrast, the "redundant signal" hypothesis suggests that multiple signals provide the same information, with some error, and thus enable receivers to more accurately assess signal content (Møller and Pomiankowski 1993; Johnstone 1996). Several studies across numerous taxonomic groups have provided evidence for a multiple messages function of complex signaling, while fewer studies have found support for a role of backup, or redundant, signals (see Candolin 2003; Hebets and Papaj 2005).

Regardless of the information content of a signal, it is only effective if it can be produced, transmitted, perceived, and processed effectively. As such, signals are subject to what is referred to as efficacy-based selection in addition to content-based selection (Guilford and Dawkins 1991). An example of how efficacy-based selection can drive the evolution of multimodal signaling is well-articulated by Candolin (2003) in her "multiple sensory environments" hypothesis of multiple cues used in mate choice. Stated simply, the transmission and detection of signals and cues is influenced by environmental conditions and thus variability in the signaling environment may select for multimodal signals independent of information content (see Candolin 2003; Hebets and Papaj 2005). In addition to environmental variability, there likely exists variability among receivers in terms of signal detection and processing, which may also exert selection pressure on signal form. Additionally, signals need not be independent and may interact such that one alters either the production, reception, or processing of another (Intersignal interaction hypotheses – Hebets and Papaj 2005), making some multimodal signals functional units upon which selection can act. In summary, there is a wealth of ideas regarding how multimodal signals function and why they might exist and many of these hypotheses remain untested – making multimodal signaling an exciting and timely focus of research (see Candolin 2003; Hebets and Papaj 2005).

While progress has been made regarding multimodal signal function within select taxonomic groups, there are countless questions that remain unanswered. For instance, published examples of multimodal signaling most often include displays typified by the simultaneous production and transmission of signals and/or components from more than one sensory modality (see Candolin 2003; Hebets and Papaj 2005), yet it is likely that some (even many) multimodal signals function in a sequential fashion. The importance of simultaneous versus sequential production, transmission, and perception of multimodal signals is a relatively untouched topic in theoretic discussions of multimodal signal function and exemplifies the work that remains to be done in this field. Interestingly, the sequential arrival of information to receivers has previously received attention from crustacean biologists (e.g., Hazlett and McLay 2000), setting a precedent for similar future studies.

17.2.2 Approaches to Studying Multimodal Signaling

Depending upon the questions of interest, approaches to studying multimodal signaling may include morphological studies, ecological studies, behavioral manipulations and assays, psychological experiments, histology, electrophysiology, phylogenetic analyses, and computational studies. Generally, however, studies of multimodal signaling fall within the fields of behavior and/or neurobiology. The "how" questions – how is simultaneous input from multiple sensory systems received and potentially integrated in the nervous system – are commonly addressed with neurobiological techniques such as histology and electrophysiology. In contrast, the "why" questions – why do animals use multiple modalities when communicating – are addressed with a whole suite of complementary behavioral techniques and are rife with theoretical hypotheses (see above).

An approach championed by Partan and Marler (1999, 2005) to address the "why" questions of multimodal signaling involves comparing receiver responses to signal components both in isolation and jointly in an attempt to categorize the components as redundant versus nonredundant. In many taxonomic groups, this presents a difficult, frequently insurmountable challenge, yet as we will demonstrate subsequently, crustacean researchers can readily overcome this challenge. Currently, the two most common ways to assess receiver responses are using (1) signal isolation and/or (2) signal playback experiments (for discussion of approaches see Uetz and Roberts 2002; Hebets 2008). Signal isolation experiments involve manipulating signal production, signal transmission, or receiver reception such that a receiver's response to a single signal in isolation can be assessed. Such experiments require animals for which such manipulations are feasible – such as crustaceans. Signal playback experiments rely on technology such as video/acoustic playback and/or robotics, again, requiring taxonomic groups amenable to such techniques – again, such as crustaceans. Below, we will use our own study systems to highlight two noncrustaceans examples of studies utilizing multiple techniques to address questions of multimodal signal evolution and function.

17.2.3   Example I: A Comparative Approach to Multimodal Signal Function in Spiders

Male *Schizocosa* wolf spiders exhibit tremendous variation in courtship displays with respect to their use of visual and seismic (substratum-coupled vibrations) signaling. While all *Schizocosa* males incorporate some form of seismic courtship signaling, a subset also incorporates visual signals (Stratton 2005). Among the visual signaling species, some possess pigmentation or large brushes of hair on their forelegs, which are waved during courtship, while others lack foreleg ornamentation (Stratton 2005; Framenau and Hebets 2007). The observed variation in multimodal courtship signaling among members of the genus allows for comparisons between closely related species and facilitates an understanding of the evolution and function of multimodal signaling in this group.

Using a combination of signal isolation and video playback experiments, recent studies have explored multimodal signal function in two closely related *Schizocosa* species – *S. uetzi* and *S. stridulans*. Males of both species possess ornamentation on their forelegs and produce multimodal courtship displays (seismic plus visual). In order to understand the relative importance of each signaling modality, researchers designed mating arenas that prevented the transmission of each signaling modality independently. For example, to remove the seismic signal, males and females were placed on separate substrata, thereby removing the effective transmission of their substratum-coupled vibrations. Similarly, in order to remove the visual signal, an opaque barrier was placed in between a courting male and female, removing the successful transmission of visual signals. Using such signal isolation techniques, *S. uetzi* females were found to be more receptive to seismic signals as compared to visual signals, while *S. stridulans* females responded equally to signals in both modalities (Hebets and Uetz 2000). These results suggested that the seismic and visual courtship signals are redundant for *S. stridulans*, but nonredundant for *S. uetzi*. However, follow-up experiments examining actual copulation frequency in the presence versus absence of seismic and visual signals demonstrated that for both species, seismic signaling was crucial for mating success, while the visual signal had no impact on mating frequency – suggesting nonredundancy (Hebets and Uetz 2000; Hebets 2005, 2008).

Results of video playback experiments add an interesting twist to the story. For both *S. uetzi* and *S. stridulans*, courtship sequences were digitized and modified into three separate video loops in which the male foreleg ornamentation was modified in the following way: (1) brushes of hair were added (*S. uetzi*) or enlarged (*S. stridulans*) on the male's foreleg tibiae – "brushes" video, (2) no changes were

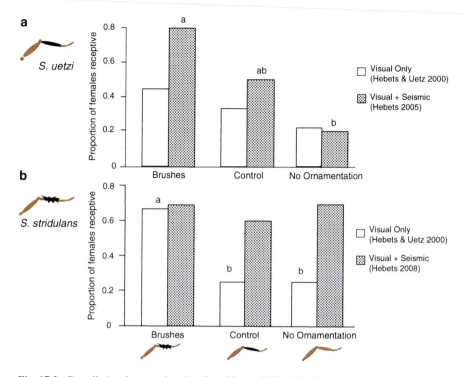

Fig. 17.2 Compilation figure (using data from Hebets 2005, 2008; Hebets and Uetz 2000) depicting female receptivity responses to video playbacks of courting conspecific males in the presence (*black bars*) and absence (*white bars*) of a seismic courtship signal for (**a**) *Schizocosa uetzi* and (**b**) *S. stridulans*

made to the male – "control" video, and (3) all pigmentation and brushes were removed from the foreleg tibiae – "no ornamentation" video. Conspecific female receptivity was then assessed to playbacks of these courtship sequences both in the presence and absence of a seismic courtship signal. When the seismic signal was absent, *S. uetzi* females did not distinguish among the visual courtship sequences. However, in the *presence* of a seismic courtship signal, females were more likely to display receptivity to the "brushes" video versus the "no-ornamentation" video (Hebets 2005; Fig. 17.2a). These results, in combination with others, suggest that the seismic signal of *S. uetzi* functions to alter a female's visual attention (Hebets 2005). In contrast, results from *S. stridulans* show an opposite pattern. In the *absence* of the seismic signal, *S. stridulans* females are more likely to display receptivity to more ornamented males (i.e., "brushes" video). However, in the *presence* of the seismic signal, female receptivity is independent of the visual stimulus (Hebets 2008) (Fig. 17.2b). The results from *S. stridulans* suggest that the seismic signal is dominant to the visual courtship signal (Hebets 2008). Thus, in two closely related species (see Stratton 2005; Hebets and Vink 2007), multimodal courtship signaling appears to function in very different ways: in *S. uetzi*, the signals interact such that the seismic signal appears to alter a female's visual attention, and in *S. stridulans*, the seismic signal appears dominant to the visual signal.

The above examples were chosen to demonstrate several things: (1) studying multimodal signaling can be extremely complicated and often requires multiple approaches and techniques, (2) multimodal signal function may vary greatly even among closely related species, (3) taxonomic groups for which manipulations of both the signaling environment and the signals themselves are feasible make ideal organisms for studying multimodal signaling, and (4) the comparative approach can be extremely valuable in understanding the evolution of multimodal signaling. We highlight these points here because we feel that these same points are relevant and applicable to crustaceans. Crustaceans not only encompass tremendous diversity in terms of communication systems, lifestyles, and habitat use (to name a few), but they are also extremely tractable, making them an excellent taxonomic group for comparative work on multimodal signal form and function!

17.2.4   Example II: Sensory Systems Lend Insight into Multimodal Signaling in Squirrels

Before finally turning our focus to crustaceans, we wish to provide another brief example of multimodal signaling, this time borrowing from the vertebrate literature. We chose this example because it highlights the importance of natural history information and knowledge of an organism's basic anatomy and physiology – knowledge that appears to be abundant among crustacean biologists! Specifically, our example illustrates how selection for sensory integration in a foraging context has potentially facilitated the evolution of multimodal signaling in a communication context. Our example comes from interactions between California ground squirrels, *Spermophilus beecheyi*, and their rattlesnake predators (*Crotalus oreganus*).

Although rattlesnakes are sensitive to visual stimuli at close range, they also possess pit organs, a highly specialized sensory system that enables them to detect infrared radiation or "radiated heat" at a distance. Bimodal neurons in the rattlesnake's optic tectum respond to both infrared and visual stimuli (Hartline et al. 1978; Newman and Hartline 1981), creating a visual and thermal representation of the snake's environment and enabling the integration of visual and thermal stimuli. These rattlesnakes will often feed on pups of California ground squirrels. In defense of their vulnerable young, adult ground squirrels vigorously confront predatory snakes with a suit of antipredator behaviors, including a visual signal called tail flagging in which they wave their piloerected tails from side-to-side (Owings and Coss 1977). Recently, a closer examination of this tail flagging display revealed that when the signal is directed towards a rattlesnake, the squirrels increase the temperature of their tails by 2–3°C (Fig. 17.3a). Interestingly, the squirrels do not increase their tail temperature when tail flagging to infrared insensitive gopher snakes, *Pituophis melanoleucus*, (Fig. 17.3b; Rundus et al. 2007). Thus, when encountering an infrared sensitive

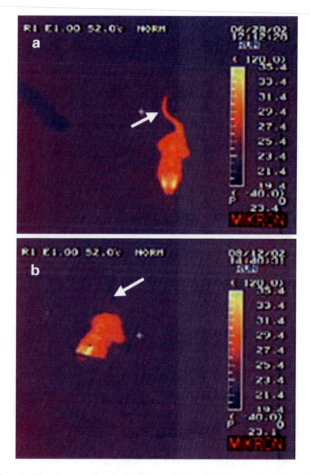

Fig. 17.3 Infrared video frames of a squirrel interacting with (**a**) a rattlesnake and (**b**) a gopher snake during experimental trials. Lighter pixel color corresponds to warmer object surface temperature. Note that the tail regions of the squirrel, referenced by the white arrows, are considerably warmer than background during the rattlesnake trial, but not the gopher snake trial. Figure adapted from Rundus et al. (2007)

snake, California ground squirrels produce a multimodal (thermal and visual) signal, which capitalizes on the multimodal sensory specializations of their signal target. The addition of the infrared component to the visual tail flagging shifts rattlesnakes from predatory to defensive behavior to a greater degree than the visual component alone (Rundus et al. 2007), suggesting an enhancing effect of this additional component. Again, we highlight this example because it clearly demonstrates how an understanding of an animal's sensory system might help direct our studies of communication, enabling the discovery of signals or components that may lie beyond our human perceptual capabilities. Given the wealth of knowledge regarding crustacean sensory systems, a sensory systems approach might prove valuable for understanding the evolution of crustacean multimodal signaling.

17.3 Chemical Communication and Multimodal Signaling

Chemical signaling is considered to be the oldest and most widespread channel for communication (Bradbury and Vehrencamp 1998; Johansson and Jones 2007). Not surprisingly then, examples of animals that incorporate chemical signals into multimodal displays are numerous (e.g., >50% of provided multimodal signaling examples in a recent review involve a chemical component: see Tables 2 and A1 in Partan and Marler 2005). Interestingly, of those reviewed studies that included chemical signals and examined the redundancy of signal components, >60% of the taxa used were invertebrates (Partan and Marler 2005). We point this out here to stress the value of invertebrate taxa, such as crustaceans, in advancing our understanding of multimodal signaling.

Despite the prevalence of chemical components in multimodal displays, many of the theoretical and functional studies of multimodal signaling involve audio-visual displays (see Candolin 2003; Hebets and Papaj 2005). This likely reflects a greater difficulty in identifying and characterizing chemical signals as well as an effect of observer bias. For example, relating a receiver's response to the presence/absence of a chemical stimulus is relatively straightforward, yet it is far less intuitive to correlate variation in a chemical stimulus to variation in signaler attributes. Studying multimodal signals that incorporate a chemical component can face an additional challenge – discontinuity of components in space and time. A majority of studies of multimodal signal function have focused on signal components that are closely linked in either space (e.g., infrared and thermal, visual and mechanical) or time (e.g., visual and acoustic). Given the wide area across which chemical signals often disperse and the often extended time frames for their emission, interpreting relationships between components, particularly intersignal interactions, can become quite challenging. In fact, as mentioned earlier, little is known about the role of synchrony of multimodal signal components (but see Narins et al. 2005) and complex displays that contain "immediate" or "fast acting" components (e.g., visual, acoustic) paired with more "tonic" components (chemical) would be ideal candidates for further exploration.

17.4 Crustaceans and Multimodal Signaling

17.4.1 Chemical and Hydrodynamic Multimodal Signaling

The sensory system of any animal has presumably evolved under sources of selection generated by its environment. In an aquatic environment, for example, one expects animals to be able to sense fluid dynamics, since chemical cues important for survival are necessarily embedded in these movements (Mellon 2007). Not surprisingly then, aquatic crustaceans possess multiple fluid-flow detectors, chemoreceptors, and even bimodal chemo-mechanoreceptors (Hallberg and Skog, Chap. 6). A wealth of literature exists on all of these sensory structures and the bimodal chemo-mechanosensory sensilla specifically have been examined extensively in the spiny lobster *Panulirus argus* (Cate and Derby 2001, 2002). The widespread existence of these bimodal sensilla is intriguing and the significance of combining these two modalities into a single sensillum has received much attention from neurobiologists (Schmidt and Mellon, Chap. 7). From a communication perspective, given that individuals are capable of such sensory integration at the peripheral sensory level, one might expect at least some signalers to capitalize on this in a similar way that the California ground squirrels capitalize on the bimodal (visual plus thermal) sensory perception of rattlesnakes ("sensory bias for multisensory integration" hypothesis of Sensory Constraints – Hebets and Papaj 2005).

In addition to the above-highlighted capacity to receive and process bimodal cues (chemical plus hydrodynamic), aquatic crustaceans must often couple chemical signal production with the generation of water currents in order to transmit these signals – making many chemical signals obligately tied to hydrodynamic cues. For example, in lobsters, urine containing chemical signals is "carried by anteriorly directed gill currents" (Atema 1985). The fan organs of crayfish have also been shown to enhance the flow of chemical information (Breithaupt 2001) and forward-directed gill currents were proposed as the means of transporting urine signals during male–male contests in the crayfish *Astacus leptodactylus* (Breithaupt and Eger 2002). Similar fanning behavior is observed in terrestrial animals as a means to disperse chemical signals (e.g., male sac-winged bats, *Saccopteryx bilineata*; Voigt and von Helversen 1999). Despite the often obligatory tie between chemical and hydrodynamic cues in many crustaceans, some receivers still focus mostly on a single modality. For example, the mating behavior in copepods seems to fall into two separate strategies – species in which males follow chemical trails left by females, and those for which males perform a tandem hopping behavior, mediated primarily by hydrodynamic cues left by a female (see Yen and Lasley, Chap. 9). Ultimately, in both groups, the final leap of the male (just before capturing the female) is thought to be mediated by hydrodynamic cues (Yen et al. 1998).

Given that (1) receivers can simultaneously detect both chemical and hydrodynamic cues and (2) hydrodynamic cues are often obligately tied to chemical signal transmission, it seems reasonable to expect that in some circumstances, selection has acted on the hydrodynamic cues (i.e., current production) for a communication function. Indeed, hydrodynamic signals are used in agonistic displays of hermit crabs, though here the evidence of their use in a multimodal signal is only via a temporal coupling with a visual display (Barron and Hazlett 1989). Multiple studies using various crustacean groups have demonstrated both the production of several distinct types of water currents during social interactions (Fig. 17.4) and the context-dependent production of such water currents (e.g., Herberholz and Schmitz 2001; Bergman and Moore 2005a; Simon and Moore 2007), suggesting a function in signaling. Some of these studies have confirmed that hydrodynamic currents act in conjunction with chemical signals for multimodal communication (e.g., Bergman and Moore 2005b; Simon and Moore 2007). Despite our knowledge of the existence of these multimodal displays, however, their function remains unknown. For example, while Simon and Moore (2007) suggest that the water currents generated by the

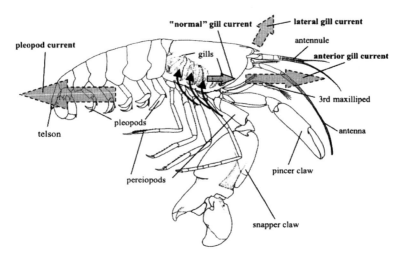

Fig. 17.4 Drawing (lateral view) of a snapping shrimp (originally modified after Kim and Abele 1988) showing four different water currents (*gray arrows*): the "normal" gill current, the lateral gill current, the anterior gill current, and the pleopod current. *Black arrows* show the direction of water entering the gill chamber (from Herberholz and Schmitz 2001)

crayfish *Orconectes rusticus* during male–female pairings do not provide information but instead simply facilitate the transmission of the information-carrying chemical signal, this hypothesis has not been explicitly tested. In summary, much work remains regarding understanding the function of combining chemical plus hydrodynamic stimuli, and the prevalence of these combined modalities among crustaceans begs for such future work.

17.4.2 Chemical and Visual Multimodal Signaling

While chemical and hydrodynamic signals may be functionally connected in aquatic environments, the addition of visual signals makes intuitive sense for animals that live in terrestrial or shallow-water environments. Not surprisingly, numerous studies have documented an interaction between chemical and visual cues in crustaceans across a variety of contexts (e.g., foraging, navigation, antipredator behavior, intraspecific interactions, etc.). For example, in a study examining individual recognition, Gherardi and Tiedemann (2004) used a signal isolation design to examine the behavior of the hermit crab, *Pagurus longicarpus*, towards familiar or unfamiliar conspecifics. They examined individual behavior in the presence of visual only, chemical only, or visual plus chemical cues. Although they found olfaction to be the dominant modality used for individual recognition, the combination of visual and olfactory cues resulted in an enhancement of the receiver's response (Gherardi and

Tiedemann 2004; Table 17.1), suggesting some benefit to bimodal (chemical plus visual) sensory acquisition. In a similar study examining individual recognition in the crayfish *Cherax destructor,* Crook et al. (2004) were able to demonstrate that individuals can discriminate familiar from unfamiliar opponents using either chemical only or visual only cues. Their results suggest that chemical and visual cues may act as backups to each other – potentially providing redundant information (Crook et al. 2004). Finally, using similar cue/signal isolation designs, numerous studies have also demonstrated that the presence of an odor influences an animal's orientation response to a visual stimulus, again suggesting an interaction between chemical and visual stimuli (Chiussi et al. 2001; Diaz et al. 2001; Chiussi and Diaz 2002; Huang et al. 2005; Table 17.1).

Several studies have also examined the importance of chemical and visual stimuli in the context of mate choice. A recent study paired male and female red swamp crayfish, *Procambarus clarkii,* in the presence of visual only, chemical only, or visual plus chemical cues and examined the influence of bimodal information on mate choice (Aquiloni and Gherardi 2008). The authors found that females chose larger males only in the presence of both stimuli and that, when given only one cue, females often acted aggressively (Aquiloni and Gherardi 2008; Table 17.1). In contrast, a male's choice of larger females was based solely on chemical cues, but they engaged in mating behavior most when both stimuli were present (Aquiloni and Gherardi 2008). Results of this study suggest that males and females may use bimodal information differently. A second example is provided by Acquistapace et al. (2002) in which the authors demonstrate that male crayfish, *Austropotamobius pallipes,* react to the combination of chemical and visual cues from conspecific females more than they do to chemical cues in isolation. Although this latter study examined male responsiveness across four treatments (female odor alone, male odor

Table 17.1 Evidence for visual and chemical sensory integration in crustaceans

Species	Purpose/context	Reference
Uca cumulanta (Fiddler crab)	Predator-avoidance	Chiussi and Diaz (2002)
Synalpheus demani (Snapping shrimp)	Shelter-seeking	Huang et al. (2005)
Clibanarius antillensis (Hermit crab)	Shelter-seeking/predator-avoidance	Chiussi et al. (2001)
Austropotamobius pallipes (Crayfish)	Sex recognition	Acquistapace et al. (2002)
Alpheus heterochaelisi (Snapping shrimp)	Mate choice/male competition	Hughes (1996)
Callinectes sapidus (Crab)	Mate choice	Kamio et al. (2008)[a]
Procambarus clarkii (Crayfish)	Mate choice	Aquiloni and Gherardi (2008)
Cherax destructor (Crayfish)	Individual recognition	Crook et al. (2004)
Pagurus longicarpus (Hermit crab)	Individual recognition	Gherardi and Tiedemann (2004)

[a]In addition to visual and chemical signals, the authors also suggest hydrodynamic signals

alone, female odor plus visual cues, and male odor plus visual cues), the authors did not test male responses to visual cues in isolation, making it difficult to interpret their results in terms of multimodal signal function (Acquistapace et al. 2002).

Despite the fact that many of the above-mentioned studies clearly demonstrate an impact of bimodal (visual plus chemical) sensory stimuli on an individual's behavior, they do not necessarily reflect communication and thus may not represent examples of multimodal signaling per se. For example, while the presence versus absence of an individual is clearly a manipulation of visual stimuli, it is unlikely that the individual itself constitutes a signal (i.e., has been selected for a communicative function; see chapter of Wyatt, Chap. 2). Rather, the presence of an individual might be more appropriately treated as a cue (Bradbury and Vehrencamp 1998). Similarly, the chemical stimuli used in many of the studies may or may not reflect past selection. The distinction between signals and cues is often difficult, especially with respect to complex signaling – where a multimodal display can be a functional unit upon which selection can act, regardless of whether its components are signals or cues (Hebets and Papaj 2005). In many of the above-mentioned cases, further investigation of both chemical and visual stimuli will likely confirm them as true examples of multimodal signaling.

One crustacean species in which chemical plus visual multimodal signaling appears to be important is the snapping shrimp *Alpheus heterochaelis*. Using signal isolation techniques, Hughes (1996) examined male responses to chemical signals from males and females in isolation and found that male responses were independent of the chemical signal. Next, she presented chemical signals jointly with the visual signal of an open chela display – a visual signal produced by both sexes. She found that the response of the male shrimp to the visual display was dependent on the chemical signal (Hughes 1996). Males responded more to the open chela display in the presence of a male versus female chemical signal (Figs. 17.5 and 17.6). In addition, the male's response was dependent on the chela size only in the presence of the female chemical signal (Hughes 1996). These experiments clearly show that the chemical signal (male vs. female) alters the male's response to the visual, open chela, display. Although this study preceded many of the theoretical reviews of the topic, it appears to represent an example of multimodal signaling in which one signal provides a context in which a perceiver can interpret and respond to a second (Hebets and Papaj 2005).

Another recent study documents a complex courtship display that combines visual, chemical, *and* hydrodynamic signals into a trimodal signal! Prior studies established that male blue crabs, *Callinectes sapidus*, have complex chemical signals, likely incorporating both urine and nonurine sources (Bushmann 1999). During courtship, these males sometimes adopt a stationary paddling display in which they paddle their swimming legs while standing in an elevated position with their chelae open (Kamio et al. 2008). The paddling behavior described above increases the volume of water males pump out of their gill chambers, suggesting a potential hydrodynamic signal (Gleeson 1991). In addition, sexually receptive females have previously been shown to respond to models of male crabs in this stationary posture and the exposed blue coloration on the chelae during the spread

Fig. 17.5 Drawing depicting how male *Alpheus heterochaelis* (*left*) respond more to an open chela display in the presence of male (*top right*) versus female (*bottom right*) chemical signals. Greater response is artistically represented by male orientation. Drawing by Jorge A. Varela Ramos

Fig. 17.6 Actual data representing male responses to open chela display (as indicated by the number of open chela displays) in the presence of male versus female chemical signals. (Data modified from Hughes (1996), Figure 5).

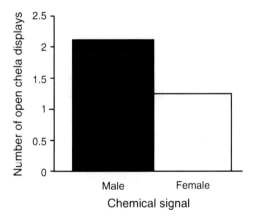

display suggests a visual signal (Gleeson 1991). Recently, Kamio and colleagues examined the context in which this stationary paddle display was performed. Specifically, they manipulated the male's signaling environment into treatments where females were accessible versus inaccessible (putatively mimicking low visibility environments in which females may be recessed in a refuge) and assessed their use of stationary paddling. Results demonstrated that males were more likely to engage in stationary paddling in environments where females were inaccessible versus accessible (see Fig. 17.7). In addition, the authors were able to verify that a larger volume of water and a larger velocity of flow were generated during the

Fig. 17.7 Courtship-related behaviors shown by male blue crabs in the presence of an accessible female and in the presence of an inaccessible female. While there was no significant difference in total courtship behavior (Carry cradle + Courtship paddling) between the two testing conditions, the number of males exhibiting courtship stationary paddling was significantly higher in the inaccessible condition. (From Kamio et al. 2008)

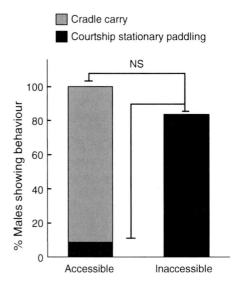

stationary paddling display than when males were not performing this display (Kamio et al. 2008). Taken together, the authors suggest that the trimodal aspect of male courtship display is context-dependent and functions as an efficacy-backup in the presence of environmental variability ("Multiple sensory environments" – *sensu* Candolin 2003; Hebets and Papaj 2005). Few studies have documented context- or environment-dependent use of multimodal signals (for discussions of multimodal signaling dependent on social context see Leger 1993; Partan and Marler 2005) and thus this study adds a tremendous amount to our understanding of complex signal function.

17.5 Future Directions

To date, numerous researchers and numerous studies have thoroughly examined the effects of different sensory stimuli on crustacean behavior; however, many of these studies are difficult to interpret within the framework of multimodal signaling. For example, many of the experiments discussed previously were not designed from a communication or signaling perspective – many of the experiments did not incorporate a full factorial design; many did not allow for potential intersignal interactions; most did not address the distinction between signals and cues; and none examined signal form in relation to signaler quality, identity, etc. Unfortunately, as a result of the different perspectives from which these studies were conducted, we are currently limited in the conclusions that can be drawn regarding multimodal communication in crustaceans. Regardless, the studies do highlight the importance of bimodal, and even trimodal, stimuli on crustacean behavior and suggest that multimodal signaling, specifically chemical plus hydrodynamic and visual, is likely prevalent among crustaceans. These studies set the stage for well-designed future work couched in a solid conceptual and theoretical communication framework.

In addition to the various design issues briefly discussed above, most crustacean researchers have not yet explicitly tested hypotheses of multimodal signal function (but see Kamio et al. 2008). To date, most studies have simply involved signal isolation designs in which responses of receivers are assessed to isolated cues/signals. However, many hypotheses of multimodal signal function require an examination of correlations between signaler and signal form and/or require further manipulations of either the signal itself or the signaling environment. Elegant manipulations of crustacean signals have already been conducted in other contexts, demonstrating that crustaceans are indeed amenable to the types of experimental approaches and techniques necessary for addressing questions of multimodal signaling function. For example, signal ablation studies have used reversible blindfolding to occlude visual signals (Breithaupt and Eger 2002) and more recently, numerous creative techniques have been used to manipulate visual signals for playback-style studies. Mirrors have been incorporated into studies to magnify claw size in waving fiddler crabs while controlling for waving rate and waving motion (McLain and Pratt 2007). In addition, robotic male crabs have been implemented in a study aimed at exploring female preference for synchrony (Reaney et al. 2008). The incorporation of these new technologies into studies examining multimodal signal function will surely provide exciting results!

Given the diverse environments in which crustaceans live, they also offer a unique opportunity to study the influence of signaling environment on the evolution of multimodal signaling. Specifically, they offer the possibility to compare and contrast the multimodal communication systems of closely related aquatic versus terrestrial organisms. Along these lines, prior comparative studies examining the reproductive behavior of mantis shrimp and fiddler crabs have already suggested several factors hypothesized to have been important in the evolution of their respective sexual signals (Christy and Salmon 1991). In comparing multimodal signaling between aquatic and terrestrial crustaceans, *a priori*, we might expect a prevalence of chemical plus hydrodynamic or visual signals in aquatic crustaceans while more terrestrial species might pair visual and acoustic signals. Anecdotally, on a recent trip to Costa Rica, I (EAH) was struck by the multimodal displays of a common beach-side fiddler crab. Upon approach, individuals waved their brightly colored chela in the air while simultaneously producing a sound – resulting in a multimodal display (visual + acoustic) that presumably functions to warn off predators. Acoustic communication is well known among different groups of crustaceans (Müller 1989; Popper et al. 2001), and the combination of visual and acoustic signals has already been reported in mating contexts (e.g., singing and dancing in the ghost crab *Ocypode platytarsus* – (Clayton 2008)). Although our focus in this chapter has been limited to multimodal signals incorporating chemical components, crustaceans likely encompass numerous examples of multimodal signals that combine other sensory modalities.

17.6   Summary and Conclusions

Despite the plethora of both theoretical and empirical studies emerging on the topic of multimodal signal evolution and function, innumerable questions remained unanswered. For example, what is the importance of simultaneous versus sequential production, transmission, and perception of signals from multiple modalities on multimodal signal function? How important are various selection pressures such as signaling environment (e.g., aquatic vs. terrestrial), information content or signal purpose, receiver sensory systems, etc. on multimodal signal evolution? Why are certain modalities paired together over others? For many of the reasons highlighted throughout this chapter, we believe that crustaceans are ideal organisms to use for filling in these gaps in our knowledge of multimodal signal evolution and function. As such, we hope that this chapter will inspire future work on crustacean multimodality, as we believe that such work will undoubtedly lead to important contributions to our broader understanding of complex signal evolution.

Acknowledgments We would like to extend a special thanks to both Martin Thiel and Thomas Breithaupt for the invitation to participate in this book (despite the fact that neither of us knew much of anything at the onset about crustaceans)! We have both learned a tremendous amount and have thoroughly enjoyed getting to know some of the arachnid's arthropod relatives! We would also like to thank Martin and Thomas for steering us towards relevant literature and for their excellent editorial comments. In addition, we would like to thank John Christy for pointing us in the right direction for acoustical references. We would also like to thank three anonymous reviewers for incredibly insightful and helpful comments. Finally, we would like to thank all of the crustacean researchers for providing stimulating reading!

References

Acquistapace P, Aquiloni L, Hazlett BA, Gherardi F (2002) Multimodal communication in crayfish: sex recognition during mate search by male *Austropotamobius pallipes*. Can J Zool 80:2041–2045

Aquiloni L, Gherardi F (2008) Assessing mate size in the red swamp crayfish *Procambarus clarkii*: effects of visual versus chemical stimuli. Freshwater Biol 53:461–469

Atema J (1985) Chemoreception in the sea: adaptations of chemoreceptors and behaviour to aquatic stimulus conditions. Soc Ex Biol Symp 39:386–423

Barron LC, Hazlett BA (1989) Directed currents: a hydrodynamic display in hermit crabs. Mar Behav Physiol 15:83–87

Bergman DA, Moore PA (2005a) Prolonged exposure to social odours alters subsequent social interactions in crayfish (*Orconectes rusticus*). Anim Behav 70:311–318

Bergman DA, Moore PA (2005b) The role of chemical signals in the social behavior of crayfish. Chem Senses 30:i305–i306

Bradbury JW, Vehrencamp SL (1998) Principles of animal communication. Sinauer Associates, Massachusetts

Breithaupt T (2001) Fan organs of crayfish enhance chemical information flow. Biol Bull 200:150–154

Breithaupt T, Eger P (2002) Urine makes the difference: chemical communication in fighting crayfish made visible. J Exp Biol 205:1221–1231

Bushmann PJ (1999) Concurrent signals and behavioral plasticity in blue crab (*Callinectes sapiduss> Rathbun*) courtship. Biol Bull 197:63–71

Candolin U (2003) The use of multiple cues in mate choice. Biol Rev 78:575–595

Cate HS, Derby CD (2001) Morphology and distribution of setae on the antennules of the Caribbean spiny lobster *Panulirus argus* reveal new types of bimodal chemo-mechanosensilla. Cell Tiss Res 304:439–454

Cate HS, Derby CD (2002) Hooded sensilla homologues: structural variations of a widely distributed bimodal chemomechanosensillum. J Comp Neurol 444:345–357

Chiussi R, Diaz H (2002) Orientation of the fiddler crab, *Uca cumulanta*: responses to chemical and visual cues. J Chem Ecol 28:1787–1796

Chiussi R, Diaz H, Rittschof D, Forward RB (2001) Orientation of the hermit crab *Clibanarius antillensis*: effects of visual and chemical cues. J Crust Biol 21:593–605

Christy JH, Salmon M (1991) Comparative-studies of reproductive-behavior in mantis shrimps and fiddler-crabs. Am Zool 31:329–337

Clayton D (2008) Singing and dancing in the ghost crab *Ocypode platytarsus* (Crustacea, Decapoda, Ocypodidae). J Nat Hist 42:141–155

Crook R, Patullo BW, MacMillan DL (2004) Multimodal individual recognition in the crayfish *Cherax destructor*. Mar Freshwater Behav Physiol 37:271–285

Diaz H, Orihuela B, Forward RB, Rittschof D (2001) Effects of chemical cues on visual orientation of juvenile blue crabs, *Callinectes sapidus* (Rathbun). J Exp Mar Biol Ecol 266:1–15

Framenau VW, Hebets EA (2007) A review of leg ornamentation in male wolf spiders, with the description of a new species from Australia, *Artoria schizocoides* (Araneae, Lycosidae). J Arachnol 35:89–101

Gherardi F, Tiedemann J (2004) Chemical cues and binary individual recognition in the hermit crab *Pagurus longicarpus*. J Zool 263:23–29

Gleeson RA (1991) Intrinsic factors mediating pheromone communication in the blue crab, *Callinectes sapidus*. In: Martin JW, Bauer RT (eds) Crustacean sexual biology. Columbia University Press, New York

Guilford T, Dawkins MS (1991) Receiver Psychology and the Evolution of Animal Signals. Anim Behav 42:1–14

Hartline PH, Kass L, Loop MS (1978) Merging of modalities in the optic tectum: infrared and visual integration in rattlesnakes. Science 199:1225–1229

Hazlett BA, McLay C (2000) Contingencies in the behaviour of the crab *Heterozius rotundifrons*. Anim Behav 59:965–974

Hebets EA (2005) Attention-altering interaction in the multimodal courtship display of the wolf spider *Schizocosa uetzi*. Behav Ecol 16:75–82

Hebets EA (2008) Seismic signal dominance in the multimodal courtship display of the wolf spider *Schizocosa stridulans* Stratton 1991. Behav Ecol 19:1250–1257

Hebets EA, Papaj DR (2005) Complex signal function: developing a framework of testable hypotheses. Behav Ecol Sociobiol 57:197–214

Hebets EA, Uetz GW (2000) Female responses to isolated signals from multimodal male courtship displays in the wolf spider genus *Schizocosa* (Araneae: Lycosidae). Anim Behav 57:865–872

Hebets EA, Vink CJ (2007) Experience leads to preference: experienced females prefer brush-legged males in a population of syntopic wolf spiders. Behav Ecol 18:1010–1020

Herberholz J, Schmitz B (2001) Signaling via water currents in behavioral interactions of snapping shrimp (*Alpheus heterochaelis*). Biol Bull 201:6–16

Huang D, Rittschof D, Jeng M (2005) Visual orientation of the symbiotic snapping shrimp *Synalpheus demani*. J Exp Mar Biol Ecol 326:56–66

Hughes M (1996) The function of concurrent signals: visual and chemical communication in snapping shrimp. Anim Behav 52:247–257

Johansson BG, Jones TM (2007) The role of chemical communication in mate choice. Biol Rev 82:265–289

Johnstone RA (1996) Multiple displays in animal communication: 'backup signals' and 'multiple messages'. Phil Trans Roy Soc Lond Ser B 351:329–338

Kamio M, Reidenbach MA, Derby CD (2008) To paddle or not: context dependent courtship display by male blue crabs, *Callinectes sapidus*. J Exp Biol 211:1243–1248

Leger DW (1993) Contextual sources of information and responses to animal communication signals. Psychol Bull 113:295–304

McLain DK, Pratt AE (2007) Approach of females to magnified reflections indicates that claw size of waving fiddler crabs correlates with signaling effectivess. J Exp Mar Biol Ecol 343:227–238

Mellon D (2007) Combining dissimilar senses: central processing of hydrodynamic and chemo-sensory inputs in aquatic crustaceans. Biol Bull 213:1–11

Møller AP, Pomiankowski A (1993) Why have birds got multiple sexual ornaments. Behav Ecol Sociobiol 32:167–176

Müller W (1989) Untersuchungen zur akustisch-vibratorischen Kommunikation und Ökologie tropischer und subtropischer Winkerkrabben. Zool Jb Abt Syst Ökol Geogr Tiere 116:47–114

Narins PM, Grabul DS, Soma KK, Gaucher P, Hodl W (2005) Cross-modal integration in a dart-poison frog. Proc Nat Acad Sci USA 102:2425–2429

Newman EA, Hartline PH (1981) Integration of visual and infrared information in bimodal neurons of the rattlesnake optic tectum. Science 213:789–791

Owings DH, Coss RG (1977) Snake mobbing by California ground squirrels: adaptive variation and ontogeny. Behavior 62:50–69

Partan S, Marler P (1999) Behavior - Communication goes multimodal. Science 283:1272–1273

Partan SR, Marler P (2005) Issues in the classification of multimodal communication signals. Am Nat 166:231–245

Popper AN, Salmon M, Horch KW (2001) Acoustic detection and communication by decapod crustaceans. J Comp Phys A 187:83–89

Reaney LT, Sims RA, Sims SWM, Jennions MD, Backwell PRY (2008) Experiments with robots explain synchronized courtship in fiddler crabs. Curr Biol 18:R62–R63

Rowe C (1999) Receiver psychology and the evolution of multicomponent signals. Anim Behav 58:921–931

Rundus AS, Owings DH, Joshi SS, Chinn E, Giannini N (2007) Ground squirrels use an infrared signal to deter rattlesnake predation. Proc Nat Acad Sci USA 104:14372–14376

Simon JL, Moore PA (2007) Male-female communication in the crayfish *Orconectes rusticus*: the use of urinary signals in reproductive and non-reproductive pairings. Ethology 113:740–754

Stratton GE (2005) Evolution of ornamentation and courtship behavior in *Schizocosa*: insights from a phylogeny based on morphology (Araneae, Lycosidae). J Arachnol 33:347–376

Uetz GW, Roberts JA (2002) Multisensory cues and multimodal communication in spiders: insights from video/audio playback studies. Brain Behav Evol 59:222–230

Voigt CC, von Helversen O (1999) Storage and display of odour by male *Saccopteryx bilineata* (Chiroptera, Emballonuridae). Behav Ecol Sociobiol 47:29–40

Yen J, Weissburg MJ, Doall MH (1998) The fluid physics of signal perception by mate-tracking copepods. Phil Trans Roy Soc Ser B 353:787–804

Chapter 18
Chemical Cues and Reducing the Risk of Predation

Brian A. Hazlett

Abstract The use of chemical cues to recognize elevated risk of predation is widespread in crustaceans and the responses to chemical cues can decrease the risk of predation. There are multiple sources of such cues: odors from damaged conspecifics, damaged heterospecifics, conspecifics digested by predators, predator odors, and odors associated with predation risk by learning. The most common responses to such odors are behaviors such as decreased movement or movement away from the source of cues, but in small planktonic species development of defensive morphologies such as spines also occurs. When faced with combinations of chemical cues, such as danger cues and food cues, most crustaceans respond primarily to the danger cue. Starvation can eliminate the dominance of danger cues over food cues. In the crab *Heterozius rotundifrons*, there is no response to chemical cues indicating increased predation risk unless tactile cues are also detected. Learned associations can result in crustaceans showing responses to cues that in themselves may be weak indicators of elevated predation risk. More field work is needed to document the extent to which patterns reported from the laboratory are important in nature.

18.1 Introduction

Avoiding predation is clearly of overriding importance in determining the fitness of individuals (Lima and Dill 1990). Missing one feeding or mating opportunity is insignificant compared to failure of avoiding predation. Thus, we should expect evolution to favor phenotypes that are as effective as possible in detecting elevated predation risk and behaving in such a way as to reduce successful predation.

B.A. Hazlett (✉)
Department of Ecology and Evolutionary Biology, University of Michigan,
830 North University, Ann Arbor, MI 48109-1048, USA
e-mail: bhazlett@umich.edu

T. Breithaupt and M. Thiel (eds.), *Chemical Communication in Crustaceans*, 355
DOI 10.1007/978-0-387-77101-4_18, © Springer Science+Business Media, LLC 2011

Of course, animals must also find food, mates, shelter, etc. (Sih 1992), but if they are eaten, the other activities are eliminated.

The detection of elevated predation risk can occur via input from any sensory channel, but in aquatic environments, the primary medium for the majority of crustaceans, chemical cues are of particular importance (Rittschof 1992; Chivers and Smith 1998; Wisenden 2000; Moore and Crimaldi 2004). In this review, I will examine some aspects of the use of chemical cues that indicate elevated predation risk to crustaceans. I will consider the benefits of behavioral patterns shown, the sources of odors, the types of responses shown, the effects of other factors on the responses to chemical cues, and the role of past experience in the utilization of cues.

However, it should be noted that as important as predation avoidance is, perhaps this chapter is not appropriate for a book with "communication" in the title. The strict definition of communication includes sending and receiving signals with the result that both sender and receiver benefit from the exchange (see Wyatt, Chap. 2). And the strong implication of the use of the word is that signals have evolved as information delivery systems, i.e., are ritualized over evolutionary time (Hazlett 1972). Indeed the release of alarm substance from specialized cells in ostariophysan fish (Smith 1992) has been cited as an example of communication. I remember being impressed with the highly stereotyped responses of fish to alarm odor when F. Jan Smith showed me the system in 1988. Both senders and receivers can benefit from this chemical communication system (see Sect. 18.2). However, there is no evidence for crustaceans that the source of chemicals that can indicate elevated predation risk has evolved for communication purposes, and recent work (Chivers et al. 2007; Carreau-Green et al. 2008) indicates that the ostariophysan fish example is also questionable as a good example of communication. Certainly crustaceans use chemicals to evaluate the level of predation risk, but those chemicals have evolved for other functions, not for communication. Thus, in this chapter I will imagine quotation marks around the word "communication" and avoid labeling chemicals as alarm substances (see Aggio and Derby, Chap. 12). I admit to slipping in some of my own papers and have used the term as shorthand for chemicals indicating elevated risk of predation. I have worked on crustacean behavior since undergraduate days, including responses to chemicals (Hazlett and Winn 1962).

18.2 Benefits of Predator Avoidance

There have been numerous descriptions of behavioral patterns elicited by chemical cues that logically should decrease predation risk (see Sect. 18.4). However, there have been just a few studies actually documenting the benefits of antipredation behaviors in crustaceans elicited by chemical cues and the best studies have been with other taxa. Work with several fish species (Mathis and Smith 1993; Mirza and Chivers 2001, 2003), the toad *Bufo boreas* (Hews 1988) and the wolf spider *Pardosa milvina* (Persons et al. 2001), has shown increased survival rates by prey that showed antipredation behavioral patterns following detection of chemical cues

Fig. 18.1 The limb extension posture of the crab *Heterozius rotundifrons* is assumed for a longer time when the odor of crushed conspecifics is combined with tactile input (from Hazlett and McLay 2000)

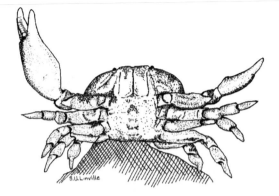

indicating increased predation risk. And while release of alarm odor by an injured animal might not seem advantageous for that individual, Chivers et al. (1996) showed for the minnow *Pimephales promelas* that attraction of a second predator to the area by the released prey alarm odor can lead to interference of the predation attempt. And this in turn leads to an increased probability of escape by the injured minnow.

The responses of the amphipod *Gammarus minus* to the odor of injured conspecifics resulted in an increase in the time until first attack by a fish predator (Wisenden et al. 1999). This delay should enhance the chances of prey to escape. The induction of a morphological change, spine development, in *Daphnia* by detection of predator odor results in increased chance of escape once attacked (Havel and Dodson 1984). The brachyuran crab *H. rotundifrons* responds to cues indicating increased predation risk by assuming an immobile, appendage-extended posture (Fig. 18.1). This posture mechanically inhibits predation by fish (Hazlett and McLay 2000) and exposure to the odor of crushed conspecifics increases the duration of this defensive posture.

18.3   Sources of Odors

The chemicals that indicate an increased risk of predation can come from several sources. Perhaps the most widespread are chemicals released by physical damage to a conspecific crustacean. Almost every species tested has shown an increase in behavior patterns that can be related to predation avoidance when odors of "crushed" conspecifics are presented. There is no indication of crustaceans having specialized cells such as the epidermal club cells (the previously purported source of alarm odor in the narrow sense) found in ostariophysan fish (Smith 1989, 1992). But rather some chemical or chemicals, very probably including peptides, found in the hemolymph seem implicated as the cue in crustaceans (Rittschof et al. 1992; Rittschof 1993; Rittschof and Cohen 2004; Acquistapace et al. 2005), as well as in fish (Wisenden et al. 2009). Chemicals from physically damaged conspecifics would seem to be a very reliable cue of increased danger of predation.

In some cases, crustaceans also respond to the odor of crushed heterospecific individuals. The brachyuran crab *H. rotundifrons* responded to the odors from other brachyuran species (*Cyclograpsus lavauxi* and *Hemigrapsus sexdentatus*) just as strongly as to crushed conspecifics, but when confronted with crushed anomuran crabs (*Petrolisthes elongates* and *Pagurus novizealandiae*) it responded less strongly (Hazlett and McLay 2005b). In three pairs of crayfish species it was found that individuals of invasive species responded to heterospecific cues more strongly than did native species: (1) the invasive *Orconectes rusticus* vs. the native North American species *O. propinquus* (Hazlett 2000b), (2) *Cherax tenuimanus* vs. *C. albidus* in Australia (Gherardi et al. 2002), and (3) *Procambarus clarkii* vs. the Italian native *Austropotamobius pallipes* (Hazlett et al. 2003). This stronger response by invasive crayfish species may be related to successful invaders evolving in regions with higher species richness, thus exposing them to a greater variety of cues indicating elevated predation risk on crayfish (Hazlett 2000b). Use of a greater variety of danger cues may help when invasives are in new environments where they are exposed to new predators.

A second major category of chemical cues indicating increased predation risk are chemicals resulting from digestion of conspecifics by predators. For example, *Daphnia galeata* developed protective spines after detecting odors of conspecific individuals that have been digested by predators, but not just crushed conspecifics (Stabell et al. 2003). Because growth of spines involves more energetic costs than a momentary change in behavior, the chemical cue apparently has to be a very accurate indication of high risk of predation.

A third category of cues I will mention are predator odors. These are chemicals from a predator that has not necessarily been feeding on conspecifics. Juvenile blue crabs, *Callinectes sapidus*, respond to both water in which a predator (the mud crab *Panopeus herbstii*) was held for one hour as well as water containing crushed conspecifics (Diaz et al. 2003). Naïve individuals of the crayfish *Orconectes virilis* and *O. rusticus* do not initially respond to snapping turtle (*Chrysemys picta*) odor, but do so after simultaneous exposure to snapping turtle odor and the odor of crushed conspecifics (Hazlett and Schoolmaster 1998). Associative learning can lead to other odors being treated as danger cues and the sources of those odors can be diverse, as will be discussed below. The work of Cohen and Forward (2005) suggests that the copepod *Calanopia americana* responds to modified amino sugars in the mucus of predators as the cue associated with increased predation risk. Aminosugar disaccharides were also implicated as the cue from predators affecting the larvae of the estuarine crab *Rhithropanopeus harrisii* (Cohen and Forward 2003). While not strictly a response to elevated predation risk, the crayfish *O. virilis* (Hazlett 1985) and *P. clarkii* (Zulandt Schneider and Moore 2000) respond to odors released from stressed conspecifics by showing behavior patterns as if on low-level alert. In both species, urine released from stressed (but not physically damaged) animals has been implicated (Hazlett 1990a; Zulandt Schneider and Moore 2000). Disturbance odor was also reported in the hermit crab *Calcinus laevimanus* (Hazlett 1990b) and two species of crayfish from Australia, *Cherax destructor* and *Euastacus armatus* (Hazlett et al. 2007).

Table 18.1 Types of responses shown by crustaceans to chemical cues indicating elevated risk of predation

Type of response	Species	Reference
Decrease in movement		
	Pacifastacus leniusculus	Blake and Hart (1993)
	Procambarus acutus	Acquistapace et al. (2004)
	Cambarus bartonii	Ciruna et al. (1995)
	Astacus astacus	Hirvonen et al. (2007)
	Orconectes virilis	Hazlett (1994)
	Diogenes avarus	Hazlett (1997)
	Petrolisthes elongates	Hazlett (2000a)
	Halicarcinus innominatus	Hazlett (2000a)
	Notomithrax ursus	Hazlett (2000a)
	Gammarus pseudolimnaeus	Williams and Moore (1985)
Increase in locomotion		
	Clibanarius vittatus	Hazlett (1996a)
Movement away from source of odor		
	Gammarus minus	Wisenden et al. (1999)
	Gammarus pseudominaeus	Williams and Moore (1985)
	Callinectes sapidus juveniles	Diaz et al. (2003)
	Dardanus venosus	Brooks (1991)
	Pagurus pollicaris	Brooks (1991)
	Cancer magister megalopae	Banks and Dinnel (2000)
	Homarus americanus postlarvae	Boudreau et al. (1993)
	Daphnia magma	Van de Meutter et al. (2004)
	Calanopia americana	Cohen and Forward (2005)
Increase defensive displays		
	Paranephrops zealandicus	Shave et al. (1994)
	Synalpheus hemphilli	Hazlett and Winn (1962)
Increase preference for protective shells		
	Pagurus filholi	Mima et al. (2003)
Cease reproductive activity		
	Female, but not male, *C. vittatus*	Hazlett (1996a, b)
Develop defensive spines		
	Daphnia galeata	Stabell et al. (2003)
	Daphnia pulex	Spitze (1992)
Delay metamorphosis		
	C. sapidus megalopae	Forward et al. (2001)

18.4   Responses to Chemical Cues of Elevated Predation Risk

Perhaps the most commonly reported response to cues indicating elevated predation risk is a decrease in movement (Table 18.1). Given that predators often detect prey by movement, it is logical that many crustacean species respond to the odor of crushed conspecifics by reducing locomotion and other movements such as feeding activity. Table 18.1 lists some of the species that show this response. Some species, though, respond in an opposite manner to these cues. For example, the hermit crab *Clibanarius vittatus* showed an increase in locomotion when exposed to the odor of stone crabs (*Menippe mercenaria*), a predator that detects prey via chemical cues rather than visual cues (Hazlett 1996a). Similarly, the hermit crab *Pagurus granosimanus* decreased the time spent withdrawn in its shell and moved faster when detecting the odor of the predatory crab *Cancer productus*, which can crush shells (Rosen et al. 2009). Possibly, these hermit crabs try to move out of the activity radius of these shell-crushing predators.

A second category of behaviors is movement in a particular direction (away from the source of a predator-related odor) or to a particular place (where the predator is less likely to occur) (Table 18.1). The amphipod *G. minus* responded to the odor of injured conspecifics by moving to the bottom as well as reducing activity (Wisenden et al. 1999); other algal-dwelling amphipods, e.g., *G. pseudolimnaeus* (Williams and Moore 1985), showed similar reactions. Megalopae of *C. sapidus* decrease upstream swimming when exposed to predator odor, thus passively moving away from the source of the danger cue (Forward et al. 2003). When presented with the odor of predatory damselfly larvae, *Daphnia magna* move horizontally from macrophyte-rich regions to more openwater habitats (Vande Meutter et al. 2004). Detection of chemicals associated with predators by the copepod *C. americana* altered the photoreponses involved in diel vertical migration, resulting in reduced ascent at sunset (Cohen and Forward 2005). Spiny lobsters, *Panulirus argus*, avoid shelters that contain the odor of injured conspecifics and heterospecifics (*P. guttatus*) (Briones-Fourzán et al. 2008). In *P. argus*, the cues are detected by the aesthetasc sensillae (Shabani et al. 2008). Blue crabs, *C. sapidus*, avoid food-baited traps containing injured conspecifics (Ferner et al. 2005).

Another category of behaviors shown by some crustaceans is an increase in defensive chela display, shown by the crayfish *Paranephrops zealandicus* when exposed to chemicals from the skin mucus of predatory eels (Shave et al. 1994) and by the snapping shrimp *Synalpheus hemphilli* when exposed to the odor of crushed conspecifics (Hazlett and Winn 1962). A more specialized response was shown by the hermit crab *Pagurus filholi* when exposed to a predatory crab: individuals increased their preference for a species of shell that provides superior predation protection (Mima et al. 2003). In contrast, the hermit crab *C. vittatus* did not alter shell selection when exposed to conspecific hemolymph (Hazlett 1995). How hermit crabs respond to hemolymph is strongly affected by shell fit (Katz and Rittschof 1993). When exposed to the odor of a predatory crab, female *C. vittatus* stopped showing precopulatory behavior and no longer released sex pheromone, but males were unaffected and continued courtship behaviors (Hazlett 1996b).

A final category of responses to cues indicating increased predation risk is changes in morphology or development (Harvell 1990). Stabell et al. (2003) showed that defensive spine induction in *Daphnia* follows detection of odors that result from digestion of conspecifics by predators, while the odor of a predator (the insect *Chaoborus americanus*) induces spine formation in *Daphnia pulex* (Spitze 1992). The megalopae of *C. sapidus* delay metamorphosis to the crab stage when they detect odors from benthic predators (Forward et al. 2001), thus staying in the water column longer and delaying potential predation on the substratum where the crabs and benthic predators live.

18.5   Effects of Other Factors on Responses to Chemical Cues

How crustaceans respond to chemical cues of elevated predation risk can be influenced by a variety of factors (see Fig. 18.2). Animals are always detecting multiple stimuli in a number of sensory modalities and those inputs can clearly alter responses to chemical danger cues. When the crayfish *Pacifastacus leniusculus* was exposed to both visual and chemical cues associated with predation risk, the reduction in locomotion and increase in shelter use was greater than when either stimulus was presented alone (Blake and Hart 1993). Detection of multiple cues related to increased predation risk clearly can be an indication of greater probability of predatory activity.

The fiddler crab *Uca cumulanta* oriented to visual cues indicating favorable environmental locations more strongly when predator odors were detected (Chiussi

Fig. 18.2   What to do? How a crayfish responds to predator odor depends on hunger level, strength of odor cues, and recent experiences with odors

and Diaz 2002) as did the megalopae of the blue crab *C. sapidus* (Diaz et al. 1999) and the mangrove crab *Aratus pisonii* (Chiussi 2003). The threshold of light decrease needed to elicit a shadow response (descent in the water column upon sudden decrease in irradiance) was decreased by exposure to predator odor in the larvae of *R. harrisii* (Cohen and Forward 2003).

Because feeding often involves movement that could attract a predator, it is not surprising that exposure to the odor of crushed conspecifics almost eliminates responses to food odors in a variety of species: the crayfish *O. virilis* and *O. rusticus* (Hazlett 2003a) and *P. zealandicus* (Hazlett 2000a), the anomuran crab *Petrolisthes elongatus* (Hazlett 2000a), and the brachyuran crab *Halicarcinus innominatus* (Hazlett 2000a). Consistent with reduced feeding responses to food odor in presence of odor from crushed conspecifics was the increase in time taken to find food by the crayfish *O. virilis* (Tomba et al. 2001). Predator odor greatly reduced the responses of the hermit crab *C. vittatus* to both food cues and shell cues (Rittschof and Hazlett 1997). However, exposure of *C. vittatus* to a combination of predator odor and the odor of crushed conspecifics did not increase the magnitude of responses compared to predator odor alone (Hazlett 1996a). The responses shown by individuals of *C. vittatus* to conspecific hemolymph depended strongly on the fit of the shell they occupy. Crabs in shells that fit well show increased locomotion following exposure to hemolymph, while crabs in shells that are too small tend to grasp and investigate any shell around at an increased rate (Rittschof et al. 1992), apparently treating a cue of a conspecific being damaged as a shell-might-be-available cue (Fig. 18.3). Clearly, how and how long animals respond to a particular environmental cue can depend upon a number of motivational states. In addition,

Fig. 18.3 How hermit crabs react to the odor of crushed conspecifics depends on shell fit. Crabs in shells close to the ideal size flee while crabs in shells that are too small increase the rate of grasping other shells (from Rittschof 1993)

Fig. 18.4 When the odor of crushed conspecifics is presented, the responses to food odor are very strongly reduced in the crayfish *Orconectes virilis*. Starvation significantly reduces this effect (data from Hazlett 2003a). Drawing of *O. virilis* courtesy of Aleta Karstad

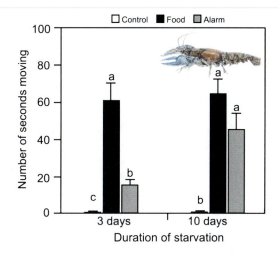

physical factors can influence responses. Smee et al. (2008) have shown that turbulence leads to a decrease of the distance at which prey (clams) respond to chemical cues from predators (blue crabs); environmental conditions may also mediate cue dispersion and the responses of crustaceans to their predators.

The reduction in responses to food cues by exposure to the odor of crushed conspecifics significantly diminished after 10 days of starvation in the crayfish *O. virilis* (Hazlett 2003a, Fig. 18.4). The shore crab *Gaetice depressus* responded differently to predation risk cues depending upon the time of day (Sakamoto et al. 2006). *Octopus* odor provoked no change in behavior during the day, but suppressed activity at night. Because octopi are primarily nocturnal hunters, it makes sense to respond more strongly when the risk of predation is higher. The odor of crushed conspecifics resulted in an increase in activity during the day, but a decrease at night (Sakamoto et al. 2006). The crayfish *Orconectes propinquus* responded more strongly to a tactile cue when the odor of crushed conspecifics was present, but only at night (Bouwma and Hazlett 2001) when tactile cues gain importance over visual cues.

The situation is more complex in the brachyuran crab *H. rotundifrons*: crabs just exposed to the odor of crushed conspecifics showed no overt change in behavior. They responded to food odors similarly if exposed to the food odor alone or in combination with the odor of crushed conspecifics. However, if a crab received tactile input, as if handled by a predator, it assumed an appendage-extended immobile posture for a number of seconds or even minutes (Fig. 18.1). And the duration of this posture was longer if the odor of crushed conspecifics was present either before or during the assumption of the posture (Hazlett and McLay 2000). Thus, the chemical danger cue is detected, but a change of behavior was manifested only when a second danger cue is detected. Interestingly, if a third danger cue was added, e.g., shadows passing over the crab, the crabs switched strategies and remained in the immobile posture for a shorter time even though shadows alone

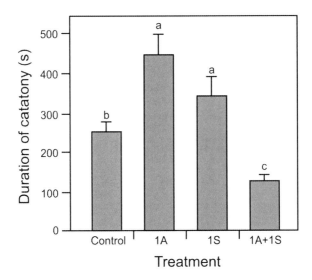

Fig. 18.5 Duration of the antipredator limb-spread posture by individuals of *H. rotundifrons* depending upon treatment prior to the tactile induction of the posture. Mean (±SE) in seconds under control conditions, addition of alarm odor, passage of shadows, and the combination of alarm odor and shadows (data from Hazlett and McLay 2005a)

caused an increase in the duration of the tactile-induced posture when the chemical cue was not added (Hazlett and McLay 2005a). It would appear that when the indications of predation risk are high enough (indicated by cues detected in three sensory modalities), the animals switched to a "cut and run" strategy rather than waiting out the predator (Fig. 18.5).

Cues indicating increased predation risk have been used to study the "structure" of behavior, i.e., how animals respond to combinations of inputs. By exposing animals to a cue related to gastropod shell acquisition and differing strengths of predator odor, the hermit crab *Diogenes avarus* responded in an all-or-none fashion showing responses typical of either one or the other cue (Hazlett 1997). On the other hand, the crayfish *O. virilis* responded to differing combinations of food and predator cues in a gradual way, reducing but not eliminating feeding when predation risk cues were less than full strength (Hazlett 1999). From the results of similar tests on a series of New Zealand species (Hazlett 2000a), I proposed that animals can show the gradation pattern when they have in their repertoire an alternative response to predation in addition to the reduction in movement. The tail flip of crayfish and limb autotomy of the crab *P. elongatus*, both behaviors induced mainly by tactile input, are examples of alternative predation-reducing behaviors. When the option of limb autotomy was eliminated, by inducing the autotomy of both chelipeds prior to testing, individuals of *P. elongatus* show a complete cessation of feeding when exposed to the odor of crushed conspecifics rather than a partial reduction (Hazlett 2004, Fig. 18.6).

Fig. 18.6 Removal of the
option to autotomize
chelipeds results in a switch
from an intermediate
inhibition of feeding
behaviors by the odor of
crushed conspecifics to a
complete inhibition in the
anomuran crab *Petrolisthes
elongatus* (data from Hazlett
2004). Photographs by Colin
McLay and Erasmo Macaya

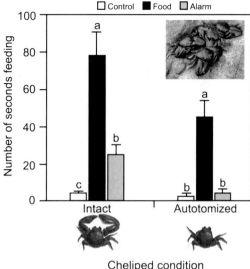

18.6   Role of Experience in Recognizing Danger Cues

The responses of animals to the odor of damaged conspecifics do not appear to
depend on past experience. However, in a number of cases it has been shown that
responses to predator odor are dependent on the past experience of simultaneous
detection of predator odor and the odor of damaged conspecifics. For example,
individuals of the crayfish *O. virilis* with no past association with snapping turtles
(*C. picta*) showed no response to the introduction of snapping turtle odor. Following simultaneous exposure to turtle odor and the odor of crushed conspecifics, turtle
odor elicited antipredator responses (Hazlett and Schoolmaster 1998). These
responses were shown by both very young crayfish raised in the laboratory and
adults from a pond with no snapping turtles. The latter result was an accident in that
I was just testing animals collected from a convenient habitat to further investigate
responses to predatory odor. When they failed to respond (as had conspecifics from
other habitats containing snapping turtles), a new line of research resulted.

Classical conditioning can lead to crayfish treating the odor of nonpredators as
danger cues. Following the method of Chivers and Smith (1998), we have used the
odor of goldfish (*Carassius aurtus*), clearly not a predator on crayfish, to explore a
number of aspects of learning of predator odor in crayfish. Individuals of invasive
species remember a learned association longer than native species (the invasive *O.
rusticus* vs. *O. virilis* and the invasive *P. clarkii* vs. the Italian native *A. pallipes*;
Hazlett et al. 2002). Formation of a learned association can be blocked by either
latent inhibition (repeated exposure to the predator odor alone prior to the simultaneous exposure of predator odor and the odor of crushed conspecifics)

(Acquistapace et al. 2003) or by learned irrelevance (random exposure to the two odors prior to simultaneous exposure) (Hazlett 2003b).

In fact, the crayfish *O. rusticus* can show second-order conditioning (Hazlett 2007). When an odor cue that became a danger cue after pairing with the odor of crushed conspecifics was paired with a third (formerly neutral) odor, crayfish treated that third odor as a danger cue. Thus, prey can show antipredation behaviors when exposed to a chemical cue that was never directly linked to an accurate danger cue. It has been suggested that this could occur in part because the cost of such behavior (lost foraging time) is small compared to the benefit of avoiding predation (Koops 2004).

Conditioning can alter the response a crustacean shows to a particular chemical mixture even when that mixture elicited a different response prior to conditioning. Spiny lobsters (*P. argus*) showed aversive predator avoidance behaviors after conditioning to a mixture that originally elicited feeding behaviors (Fine-Levy et al. 1988).

18.7   Conclusions and Future Directions

While numerous studies have revealed many interesting aspects of how crustaceans use chemical cues to reduce their chances of predation, major areas require more attention. We clearly need additional studies addressing the benefits and costs of antipredation responses. But perhaps the area of greatest demand is field studies of what the patterns of exposure to chemicals are and how do animals behave in the field. For example, the potential of the role of learning can be studied in the laboratory, but the actual effectiveness of past experience can only be elucidated if we can follow animals in the field and know what their chemical experiences are. This of course first dictates further work on the identification of the chemicals of importance for a given system. Then, combining knowledge of plume structure in the field situation (Keller et al. 2003) and techniques that allow following particular molecules (Moore et al. 1992) has the potential for telling us what is actually going on in the chemical lives of our favorite crustaceans.

Certainly, crustaceans play an important role in aquatic food webs, frequently are of economical importance, and are usually easy to maintain in the laboratory. Combined with the wealth of behaviors they exhibit, they can be considered to be a model group for the study of behavioral and chemical ecology.

References

Acquistapace P, Calamai L, Hazlett BA, Gherardi F (2005) Source of alarm substances in crayfish and their preliminary chemical characterization. Can J Zool 83:1624–1630

Acquistapace P, Daniels WH, Gherardi F (2004) Behavioral responses to 'alarm odors' in potentially invasive and non-invasive crayfish species from aquaculture ponds. Behaviour 141:691–702

Acquistapace P, Hazlett BA, Gherardi F (2003) Unsuccessful predation and learning of predator cues by crayfish. J Crustacean Biol 23:364–370

Banks J, Dinnel P (2000) Settlement behavior of Dungeness crab (*Cancer magister* Dana, 1852) megalopae in the presence of the shore crab *Hemigrapsus* (Decapoda, Brachyura). Crustaceana 73:223–234

Blake MA, Hart PJB (1993) The behavioural responses of juvenile signal crayfish *Pacifastacus leniusculus* to stimuli from perch and eels. Freshwater Biol 29:89–97

Briones-Fourzán P, Ramírez-Zaldívar E, Lozano-Álvarez E (2008) Influence of conspecific and heterospecific aggregation cues and alarm odors on shelter choice by syntopic spiny lobsters. Biol Bull 215:182–190

Brooks WR (1991) Chemical recognition by hermit crabs of their symbiotic sea anemones and a predatory octopus. Hydrobiologia 216(217):291–295

Boudreau B, Bourget E, Simard Y (1993) Behavioural responses of competent lobster postlarvae to odor plumes. Mar Biol 117:63–69

Bouwma P, Hazlett BA (2001) Integration of multiple predator cues by the crayfish *Orconectes propinquus*. Anim Behav 61:771–776

Carreau-Green ND, Mirza RS, Martínez NL, Pyle GG (2008) The ontogeny of chemically mediated antipredator responses of fathead minnows *Pimephales promelas*. J Fish Biol 73:2390–2401

Chiussi R (2003) Orientation and shape discrimination in juveniles and adults of the mangrove crab *Aratus pisonii* (H. Milne Edwards, 1837): effect of predator and chemical cues. Mar Freshw Behav Phy 36:41–50

Chiussi R, Diaz H (2002) Orientation of the fiddler crab, *Uca cumulanta*: responses to chemical and visual cues. J Chem Ecol 28:1787–1796

Chivers DP, Smith RJF (1998) Chemical signaling in aquatic predator-prey systems: a review and prospectus. Ecoscience 5:338–352

Chivers DP, Brown GE, Smith RJF (1996) The evolution of chemical alarm signals: attracting predators benefits alarm signal senders. Am Nat 148:649–659

Chivers DP, Wisenden BD, Hindman CJ, Michalak TA, Kusch RC, Kaminskyj SGW, Jack KL, Ferrari MCO, Pollock RJ, Halbgewachs CF, Pollock MS, Alemadi S, James CT, Savaloja RK, Goater CP, Corwin A, Mirza RS, Kiesecker JM, Brown GE, Adrian JC Jr, Krone PH, Blaustein AR, Mathis A (2007) Epidermal 'alarm substance' cells of fishes are maintained by non-alarm functions: possible defence against pathogens, parasites and UVB radiation. Proc Biol Sci 274:2611–2620

Ciruna KA, Dunham DW, Harvey HH (1995) Detection and response to food versus conspecific tissue in the crayfish *Cambarus bartonii* (Fabricius, 1798) (Decapoda, Cambaridae). Crustaceana 68:782–788

Cohen JH, Forward RB Jr (2003) Ctenophore kairomones and modified aminosugar disaccharides alter the shadow response in a larval crab. J Plankton Res 25:203–213

Cohen JB, Forward RB Jr (2005) Photobehavior as an inducible defense in the marine copepod *Calanopia americana*. Limnol Oceanogr 50:1269–1277

Diaz H, Orihuela B, Forward RB Jr, Rittschof D (1999) Orientation of blue crab, *Callinectes sapidus* (Rathbun), megalopae: responses to visual and chemical cues. J Exp Mar Biol Ecol 233:25–40

Diaz H, Orihuela B, Forward RB Jr, Rittschof D (2003) Orientation of juvenile blue crabs, *Callinectes sapidus* Rathbun, to currents, chemicals, and visual cues. J Crustacean Biol 23:15–22

Ferner MC, Smee DL, Chang YP (2005) Cannibalistic crabs respond to the scent of injured conspecifics: danger or dinner? Mar Ecol Prog Ser 300:193–200

Fine-Levy JB, Girardot MN, Derby CD, Daniel PC (1988) Differential associative conditioning and olfactory discrimination in the spiny lobster *Panulirus argus*. Behav Neural Biol 49:315–331

Forward RB Jr, Tankersley RA, Rittschof D (2001) Cues for metamorphosis of brachyuran crabs: An overview. Am Zool 41:1108–1122

Forward RB Jr, Tankersley RA, Smith KA, Welch JM (2003) Effects of chemical cues on orientation of blue crab, *Callinectes sapidus*, megalopae in flow: implications for location of nursery areas. Mar Biol 142:747–756

Gherardi F, Acquistapace P, Hazlett BA, Whisson G (2002) Behavioural responses to alarm odours in indigenous and non-indigenous crayfish species: a case study from Western Australia. Mar Freshwater Res 53:93–98

Harvell CD (1990) The ecology and evolution of inducible defenses. Q Rev Biol 65:323–340

Havel JE, Dodson SI (1984) *Chaoborus* predation on typical and spined morphs of *Daphnia pulex*: behavioral observations. Limnol Oceanog 29:487–494

Hazlett BA (1972) Ritualization in marine crustacean. In: Winn HE, Olla BL (eds) Behavior of Marine Animals, Vol. 1. Plenum Press, New York, pp 97–125

Hazlett BA (1985) Disturbance pheromones in the crayfish *Orconectes virilis*. J Chem Ecol 11:1695–1711

Hazlett BA (1990a) Source and nature of disturbance-chemical system in crayfish. J Chem Ecol 16:2263–2275

Hazlett BA (1990b) Disturbance pheromone in the hermit crab *Calcinus laevimanus* (Randall, 1840). Crustaceana 58:314–316

Hazlett BA (1994) Alarm responses in the crayfish *Orconectes virilis* and *Orconectes propinquus*. J Chem Ecol 20:1525–1535

Hazlett BA (1995) Behavioral plasticity in crustacean: why not more? J Exp Mar Biol Ecol 193:57–66

Hazlett BA (1996a) Organisation of hermit crab behaviour: responses to multiple chemical inputs. Behaviour 133:619–642

Hazlett BA (1996b) Reproductive behavior of the hermit crab *Clibanarius vittatus*. Bull Mar Sci 58:668–674

Hazlett BA (1997) The organization of behaviour in hermit crabs: Responses to variation in stimulus strength. Behaviour 134:59–70

Hazlett BA (1999) Responses to multiple chemical cues by the crayfish *Orconectes virilis*. Behaviour 136:161–177

Hazlett BA (2000a) Responses to single and multiple sources of chemical cues by New Zealand crustaceans. Mar Freshw Behav Physiol 34:1–20

Hazlett BA (2000b) Information use by an invading species: do invaders respond more to alarm odors than native species? Biol Invasions 2:289–294

Hazlett BA (2003a) The effects of starvation on crayfish responses to alarm odor. Ethology 109:587–592

Hazlett BA (2003b) Predator recognition and learned irrelevance in the crayfish *Orconectes virilis*. Ethology 109:765–780

Hazlett BA (2004) Alternative tactics and responses to conflicting inputs in the porcellanid crab *Petrolisthes elongates*. Mar Freshw Behav Physiol 37:173–177

Hazlett BA (2007) Conditioned reinforcement in the crayfish *Orconectes rusticus*. Behaviour 144:847–859

Hazlett BA, Acquistapace P, Gherardi F (2002) Difference in memory capabilities in invasive and native crayfish. J Crustacean Biol 22:439–448

Hazlett BA, Burba A, Gherardi F, Acquistapace P (2003) Invasive species of crayfish use a broader range of predation-risk cues than native species. Biol Invasions 5:223–228

Hazlett BA, McLay C (2000) Contingencies in the behaviour of the crab *Heterozius rotundifrons*. Anim Behav 59:965–974

Hazlett BA, McLay C (2005a) Responses to predation risk: alternative strategies in the crab *Heterozius rotundifrons*. Anim Behav 69:967–972

Hazlett BA, McLay C (2005b) Responses of the crab *Heterozius rotundifrons* to heterospecific chemical alarm cues: phylogeny vs. ecological overlap. J Chem Ecol 31:683–689

Hazlett BA, McLay C (2005b) Responses of the crab *Heterozius rotundifrons* to heterospecific chemical alarm cues: phylogeny vs. ecological overlap. J Chem Ecol 31:683–689

Hazlett BA, Lawler S, Edney G (2007) Agonistic behaviour of the crayfish Euastacus armatus and Cherax destructor. Mar Freshw Behav Physiol 40:257–266

Hazlett BA, Winn HE (1962) Sound production and associated behavior of Bermuda crustaceans (*Panulirus, Gonodactylus, Alpheus* and *Synalpheus*). Crustaceana 4:25–38

Hews DK (1988) Alarm response in larval western toads, *Bufo boreas*: release of larval chemicals by a natural predator and its effect on predator capture efficiency. Anim Behav 36:126–133

Hirvonen H, Holopainen S, Lempiäinen N, Selin M, Tulonen J (2007) Sniffing the trade-off: effects of eel odours on nocturnal foraging activity of native and introduced crayfish juveniles. Mar Freshw Behav Physiol 40:213–218

Keller TA, Powell I, Weissburg M (2003) Role of olfactory appendages in chemically mediated orientation of blue crabs. Mar Ecol Prog Ser 261:217–231

Koops MA (2004) Reliability and the value of information. Anim Behav 67:103–111

Katz JN, Rittschof D (1993) Alarm/investigation responses of hermit crabs as related to shell fit and crab size. Mar Freshw Behav Physiol 22:171–182

Lima SL, Dill LM (1990) Behavioral decisions made under the risk of predation: a review and prospectus. Can J Zool 68:619–640

Mathis A, Smith RJF (1993) Chemical alarm signals increase the survival time of fathead minnows (*Pimephales promelas*) during encounters with northern pike (*Esox lucius*). Behav Ecol 4:260–265

Mima A, Wada S, Goshima S (2003) Antipredator defence of the hermit crab *Pagurus filholi* induced by predatory crabs. Oikos 102:104–110

Mirza RS, Chivers DP (2001) Learned recognition of heterospecific alarm signals: the importance of a mixed predator diet. Ethology 107:1007–1018

Mirza RS, Chivers DP (2003) Response of juvenile rainbow trout to varying concentrations of chemical alarm cue: response thresholds and survival during encounters with predators. Can J Zool 81:88–95

Moore P, Crimaldi J (2004) Odor landscapes and animal behavior: tracking odor plumes in different physical worlds. J Marine Syst 49:55–64

Moore PA, Zimmer-Faust RK, Bement SP, Weissburg MJ, Parrish JM, Gerhardt GA (1992) Measurement of microscale patchiness in a turbulent aquatic odor plume using a semiconductor-based microprobe. Biol Bull 183:138–142

Persons MH, Walker SE, Rypstra AL, Marshall SD (2001) Wolf spider predator avoidance tactics and survival in the presence of diet-associated predator cues (Aranead: Lycosidae). Anim Behav 61:43–51

Rittschof D (1992) Chemosensation in the daily life of crabs. Am Zool 32:363–369

Rittschof D (1993) Body odors and neutral-basic peptide mimics: a review of responses by marine organisms. Am Zool 33:487–493

Rittschof D, Cohen JH (2004) Crustacean peptide and peptide-like pheromones and kairomones. Peptides 25:1503–1516

Rittschof D, Hazlett BA (1997) Behavioural responses of hermit crabs to shell cues, predator haemolymph and body odour. J Mar Biol Assoc UK 77:737–751

Rittschof D, Tsai DW, Massey PG, Blanco L, Kueber GL Jr, Haas RJ Jr (1992) Chemical mediation of behavior in hermit crabs: alarm and aggregation cues. J Chem Ecol 18:959–984

Rosen E, Schwarz B, Palmer AR (2009) Smelling the difference: hermit crab responses to predatory and nonpredatory crabs. Anim Behav 78:691–695

Sakamoto R, Ito A, Wada S (2006) Combined effect of risk type and activity rhythm on anti-predator response of the shore crab *Gaetice depressus* (Crustacea: Grapsidae). J Mar Biol Assoc UK 86:1401–1405

Shabani S, Kamio M, Derby CD (2008) Spiny lobsters detect conspecific blood-borne alarm cues exclusively through olfactory sensilla. J Exp Bio 211:2600–2608

Shave CR, Townsend CR, Crowl TA (1994) Anti-predator behaviours of a freshwater crayfish (*Paranephrops zealandicus*) to a native and an introduced predator. New Zeal J Ecol 18:1–10

Sih A (1992) Prey uncertainty and the balancing of antipredator and feeding needs. Am Nat 139:1052–1069

Smee DL, Ferner MC, Weissburg MJ (2008) Alteration of sensory abilities regulates the spatial scale of nonlethal predator effects. Oecologia 156:399–409

Smith RJF (1989) The response of *Asterropteryx semipunctatus* and *Gnatholepis anjerensis* (Pisces, Gobiidae) to chemical stimuli from injured conspecifics, an alarm response in gobies. Ethology 81:279–290

Smith RJF (1992) Alarm signals in fishes. Rev Fish Biol Fisher 2:33–63

Spitze K (1992) Predator-mediated plasticity of prey life history and morphology: *Chaoborus americanus* predation on *Daphnia pulex*. Am Nat 139:229–247

Stabell OB, Ogbebo F, Primicerio R (2003) Inducible defences in *Daphnia* depend on latent alarm signals from conspecific prey activated in predators. Chem Senses 28:141–153

Tomba AM, Keller TA, Moore PA (2001) Foraging in complex odor landscapes: chemical orientation strategies during stimulation by conflicting chemical cues. J N Am Benthol Soc 20:211–222

Van de Meutter F, Stoks R, De Meester L (2004) Behavioral linkage of pelagic prey and littoral predators: microhabitat selection by *Daphnia* induced by damselfly larvae. Oikos 107:265–272

Williams DD, Moore KA (1985) The role of semiochemicals in benthic community relationships of the lotic amphipod *Gammarus pseudolimnaeus*: a laboratory analysis. Oikos 44:280–286

Wisenden BD (2000) Olfactory assessment of predation risk in the aquatic environment. Phil Trans R Soc B 355:1205–1208

Wisenden BD, Cline A, Sparkes TC (1999) Survival benefit to antipredator behavior in the amphipod *Gammarus minus* (Crustacea: Amphipoda) in response to injury-released chemical cues from conspecifics and heterospecifics. Ethology 105:407–414

Wisenden BD, Rugg ML, Korpi NL, Fuselier LC (2009) Estimates of active time of chemical alarm cues in a cyprinid fish and am amphipod crustacean. Behaviour 146:1423–1442

Zulandt Schneider RA, Moore PA (2000) Urine as a source of conspecific disturbance signals in the crayfish *Procambarus clarkii*. J Exp Biol 203:765–771

Part IV
Towards Identification
of Chemical Signals

Chapter 19
Identification of Crustacean Sex Pheromones

Joerg D. Hardege and John A. Terschak

Abstract Odor and fragrances can carry information about an organism and have long been suggested as vital mechanisms that affect or control animal behavior. The use of such chemical signals is widespread in the aquatic environment, and crustaceans such as lobsters, shrimps, crabs, barnacles, and crayfish are known to utilize odor for predator–prey interactions, mating, establishing of dominance or social hierarchies as well as hatching of young and settlement of larvae. Nevertheless, the chemical identity of these behavior-modifying odors remains largely unknown. Here, we briefly review the literature on crustacean chemical signals and describe our approach to identify these using the example of the shore crab (*Carcinus maenas*) sex pheromones. We describe the principles of a bioassay-driven purification that is combined with a metabolomic approach where differences in the odor profiles of sexually active and inactive crabs are examined. Using such an integrated approach, we identified the female produced signal to be a nucleotide with its production being linked with the female molt (ecdysis). The pheromone enables males to detect the optimal time to mate just after the female molt with the timing of the reproductive event enabling crabs to use a simple, not species-specific, chemical as a sex pheromone. Based on our recent findings, we discuss the implications for future studies on crustacean chemical signaling.

19.1 Background and Historical Overview

It has been known for many decades that odors influence animal behavior, including foraging, predator avoidance, alarm response, social dominance, cohort recognition, and courtship. Darwin (1871) initially proposed chemical signals as a key mechanism in mate choice by which sexual selection is promoted. However, it was not until the discovery of the silkworm moth pheromone "bombykol" by Butenandt et al. (1959)

J.D. Hardege (✉)
Department of Biological Sciences, University of Hull, HU6 7RX, Hull, UK
e-mail: j.d.hardege@hull.ac.uk

T. Breithaupt and M. Thiel (eds.), *Chemical Communication in Crustaceans*,
DOI 10.1007/978-0-387-77101-4_19, © Springer Science+Business Media, LLC 2011

that "chemical ecology," the study of chemical signals that modify an organism's behavior, was born. To date, over 1,800 pheromones have been identified, but despite their widespread importance, few have been characterized in marine organisms (Wyatt 2003, 2009).

In 1986, JDH was approached by Professor E. Zeeck (Oldenburg University, Germany) upon the challenging idea of identifying a sex pheromone in a marine invertebrate – something that had not been done successfully at that time. Fascinated by the idea, JDH, along with Helga Bartels-Hardege, screened the literature to find a species that we could use. Already a year later, we managed to identify the first sex pheromone in a marine invertebrate and published the work in 1988 (Zeeck et al. 1988). Why did this work so fast? Essentially, there is no simple answer as luck was surely on our side when we decided to pursue the route of not neglecting volatile lipophilic compounds as potential pheromones, but the choice of species to attempt the identification was key in this particular case. A species was sought that had a well-described reproductive behavior to develop a "bioassay-driven purification and identification" strategy, where significant evidence for the role of sex pheromones existed, and where substantial amounts of a chemical cue could be collected.

John Terschak combined an ardor for analytical chemistry with that for the ocean. The School of Fisheries and Ocean Sciences at the University of Alaska Fairbanks supported his graduate work, and seminar attendance was encouraged, even though it was often unrelated to a chemist's work. During one such talk on standing stocks of commercial crab species, a passing remark was made about a putative female pheromone that had not been isolated or been identified! Down the rabbit-hole he went! The moral: attend seminar, even if it seems unappealing – you never know what white rabbit you may follow or what curious adventure it may initiate. In this case, it led to our combined effort to identify a crustacean sex pheromone.

The key to bioassay-driven purification and identification strategies (see Fig. 19.1) is the recognition of the desired response expressed by the test subject; all too often, animal behavior is subtle or even ambiguous. Dunham, in two comprehensive reviews (Dunham 1978, 1988), attributed the slow progress in crustacean pheromone research to the unreliability of the bioassays employed. For example, the main problem of the frequently used assays based upon an animal's attraction to an odor source is that it fails to discriminate sex pheromones from other signals such as food cues. Conversely, behavioral assays based on a distinctive sexual behavior, known as a "releaser reaction" (e.g., formation of a mating pair), are a more reliable means of investigating the role of sex pheromones (Dunham 1988; Hardege et al. 2002).

In addition to the ability to differentiate between sexual behaviors (e.g., courtship display) and signals (e.g., food attractants), it is important to employ a receiver that is physiologically and hormonally receptive to a sender's pheromone (Hayden et al. 2007). Most animals live in complex chemical environments and an individual simultaneously detects multiple chemical signals from a variety of sources (Hazlett 1999, this volume), such as living and dead conspecifics, predators, competitors,

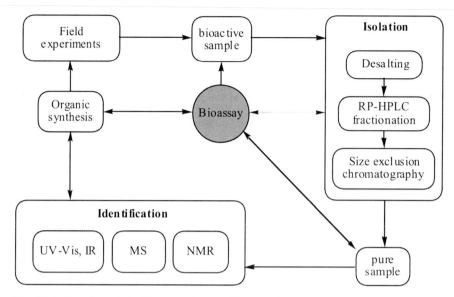

Fig. 19.1 Flowchart of the bioassay-driven purification of chemical signals. Reproduced from Hardege (1999) with kind permission from Springer Science+Business Media

and potential mates. From this cacophony, the receiver filters particular messages from the continuous chemical background noise to initiate behavioral responses; as such an animal's response to chemical stimuli can vary.

Sensitivity to chemical stimuli varies with the internal physiological status of an animal. In crustaceans, this includes molt stage, reproductive status, dominance status, and hunger, all of which are affected by hormonal mechanisms. Physiological differences are often linked to the seasonality of events and controlled via environmental factors such as temperature, light intensity, photoperiod, and circadian cycles, which therefore also influence both the sender and receiver of the chemical signals (Tierney and Atema 1988; Derby 2000; Koehl 2006).

To maximize the effectiveness of signaling, especially for reproductive events, many organisms have developed mechanisms to increase the percentage of individuals in a population that are physiologically in a state to produce and respond to signals as seen, for example, in gregariousness of spiny lobsters (Ratchford and Eggleston 1998) and mass spawning in *Nereis* (Ram et al. 1999). This mechanism of signal enhancement ensures that more individuals produce sufficient signal strength in a turbulent aquatic environment. Such timing of events may be spectacular, such as the mass spawning events in coral reef invertebrates and the Palolo worm spawning in Western Samoa (Caspers 1984). Timing is achieved as all individuals in a population are exposed to the same environmental cues or undergo the same endocrine rhythms, synchronizing development of the same physiological characteristics.

Unfortunately, the complexity of factors that control an individual's sensitivity toward chemical signals provides significant challenges in the development of biological assays producing reliable data that can be compared and reproduced. Ideally, most factors influencing an individual's responses should be understood by the time an attempt is made to identify a pheromone (Fig. 19.2); this is hardly ever possible, but in general, the simpler the mechanisms that control an organism's behavior, the more reliable an assay that can be developed.

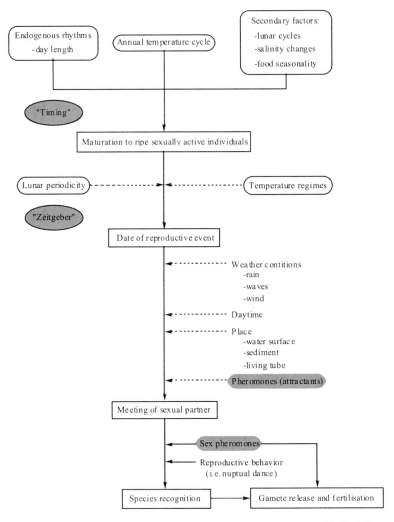

Fig. 19.2 Flow chart of environmental control of spawning in *Nereis*. "Timing" is used to describe factors influencing the maturation of a population; "Zeitgeber" is the terminus used to describe the factors controlling the date and location of the reproductive event; pheromone boxes show involvement of the chemical signals. Reproduced from Hardege et al. (1998) with kind permission from Ecoscience, Kanada

19.2   Nereidid Polychaetes as a Model for Marine Chemical Ecology

We began identifying marine invertebrate sex pheromones using polychaete worms (*Platynereis dumerillii* and *Nereis succinea*). These worms are semelparous, dying after a single reproductive event for which they undergo a metamorphosis into a reproductive stage, the heteronereis. This highly specialized heteronereis leaves its living tube and swims toward the water surface to locate a partner. Upon meeting the opposite sex, gametes are released into the free water column in a "nuptial dance" (Hardege et al. 1990). As a result of having only one opportunity to reproduce, these worms utilize a mass spawning event where a significant proportion of a population participates, thus maximizing the likelihood to meet a partner. The coordination of individual maturation and mass courtship relies on the simultaneous timing of complex environmental and endocrine factors. Because these worms are easy to culture, reproduce in the laboratory under controlled conditions, and their behavior is unambiguous (performance of a nuptial dance with an observable release of gametes), they are ideal organisms to identify sex pheromones through the use of straightforward bioassays (Zeeck et al. 1988; Hardege 1999).

Using a bioassay-driven purification strategy (Fig. 19.1), a number of sex pheromones were identified including such diverse molecules as ketones, uric acid, peptides, and L-ovothiol-A (Zeeck et al. 1988; Röhl et al. 1999; Ram et al. 1999). This has led to recent studies investigating the biological functions of pheromones, such as mate tracking and mate choice (Ram et al. 2008).

19.3   Sex Pheromones in Crustaceans

Crustacean pheromones have been the topic of a large number of studies since Ryan (1966) first demonstrated the existence of a female sex pheromone in *Portunus sanguinolentus*. The list of crustacean species for which there is evidence of a sex pheromone is extensive (see Hardege et al. 2002), but to date the chemical identity of these compounds remains largely unknown. Early studies suggested a pheromone role for the molting hormone 20-hydroxyecdysone (20HE) (Kittredge et al. 1971), as well as the peptide arthropodin (Dunham 1988). Both were dismissed based on experiments using a range of species, including *Carcinus maenas* (Eales 1974; Seifert 1982; Gleeson et al. 1984). The recent tentative identification of a novel ceramide in the hair crab *Erimacrus isenbeckii* (Asai et al. 2000) also proved inconclusive since behavioral assays did not support this theory and ceramides were not detectable at biologically relevant concentrations in urine (Asai et al. 2000). Biogenic amines, such as serotonin and dopamine, have been found to influence behavior in crustaceans (Beltz 1999). Wood and Derby (1996) showed that when injected directly into the hemolymph, dopamine induced the mating posture in the male blue crab (*Callinectes sapidus*), characteristic of its courtship display. Huber et al. (1997), through injection of radiolabelled serotonin, determined that it is not excreted in the urine, but its three major metabolic compounds are; as such, the role of biogenic amines as sex pheromones remains unclear. Some crustaceans, especially those that occur in high-density populations such as shrimps, may not require signals sent over a distance. Instead, they may use contact pheromones coating on the body surface for mate recognition (Zhang and Lin 2006), the nature of which remains unknown (see Bauer, Chap. 14).

In many crustaceans, females carry their young until the larvae hatch in unison with the female contracting its abdomen rapidly, a behavior known as the pumping response. Forward et al. (1987) identified tri-peptides from the eggs of the mud crabs (*Rhithropanopeus harrisii*) that induced this stereotypical hatching behavior and found the peptide to work at extremely low concentrations (Pettis et al. 1993).

Once the larvae of a marine crustacean are reaching the final stages of their planktonic life, they must find a suitable place to settle. Attractions toward and assessment of potential settlement sites possibly involve chemical settlement cues. Barnacles often form dense populations on manmade structures such as dock pilings as well as the hulls of ships causing increased drag and, consequently, significant operating costs in terms of an increased use of energy. Barnacle settlement was studied intensively, and the cuticular glycoprotein arthropodin could be identified as a gregarious settlement cue (Clare and Matsumura 2000).

The behavioral use of chemoreception in crustaceans is best investigated with respect to feeding stimulants with a large number of these cues identified in various species including shore crabs (Hayden et al. 2007; Weissburg, Chap. 4).

The use of novel compounds as chemical signals is actually unlikely, as this would require a de novo synthesis for the purpose of producing a chemical cue. With the multitude of behaviors mediated via chemical cues and the large number of species using these, especially insects, this would presumably require millions of unique compounds each with their own synthesis and receptor system (Wyatt 2003). Most chemical signals are related to the physiological state of the sender, such as *freshly molted* in many decapod crabs (Hardege et al. 2002). Signal specificity in such situations is often achieved through the use of multicomponent pheromone bouquets (Wyatt 2003) that may also include multimodal communication system such as visual or acoustic signals. Equally important could be that a complex behavior that involves multiple steps such as crustacean mating (i.e., attraction from distance, formation of a mating pair or mating stance, attempted copulation) is coordinated through a series of signals that provide a species-specific signaling code. This would prevent mistakes and allow for the use of a combination of simple, nonspecific compounds for every individual step of the process. This "chemical combination lock" type hypothesis is discussed recently in a review by Hay (2009; see also Hay, Chap. 3) and could, for example, involve contact pheromones as shown in shrimps (Caskey et al. 2009). Alternatively, as we have shown in nereidid polychaetes (Ram et al. 2008) external (mainly environmental) cues may bring about species specificity by separating breeding species in space and time, with "timing" negating the need for highly specific sex pheromones.

It seems evident that to achieve pheromone identification, a clear biological assay relying upon unambiguous behavior coupled with a chemical identification strategy, testing each successive purification step, is critical. Additionally, the use of biologically relevant samples such as "conditioned seawater" that contains compounds released into the environment at biologically relevant concentrations should be used.

19.4 *Carcinus maenas*, an Ideal Organism to Attempt Pheromone Identification

As described above, mating in the shore crab (*C. maenas*, synonym: European green crab) is restricted to the time of female molting (ecdysis) when males mate with soft-bodied females. The male guards the female several days before mating, defending her against predators and other competitor males by holding her under his abdomen in a cradling position (Fig. 19.3). It was hypothesized that guarding behavior is induced by sex pheromones emitted by the female (Ryan 1966; Ekerholm and Hallberg 2005). Copulation usually occurs a few hours after the molt (Hartnoll 1969), and it is therefore expected that elaborate methods to signal a potential mate have evolved to take advantage of the limited opportunity (Bamber and Naylor 1996).

Pair formation is a clear, unmistakable behavioral response, which can form the basis of a behavioral assay for pheromones. The coincidence of molt and reproduction further reduces the likelihood of other signals, such as feeding, to interfere with this behavior. The short window of mating opportunity also has two other important advantages. First, collecting pre-copula pairs provides specimens that are known to be both releasing as well as receptive to the signals; as such, the availability of a source and sender (female) for pheromones and a receiver (male) that is receptive to the cues is guaranteed (Hardege et al. 2002). The second major advantage of pair

Fig. 19.3 Pair of *Carcinus maenas* in the field with male grasping female and in the lab in precopula (photos kindly provided by Drs. N. Fletcher and R. Bublitz)

formation at the time of molt is the capacity to investigate potential chemical compounds, which are a direct result of the biochemistry associated with the current physiological state of the sender. For crustaceans such as *C. maenas*, the signal molecule could potentially be a metabolic by-product of the female molting process. This would focus the search to those compounds that change in the female's body during the molt period, thus enabling a "metabolomic" approach (see Kamio and Derby, Chap. 20) as well as a bioassay-driven purification of the cue(s) concurrently.

19.5   Sex Pheromones in *Carcinus maenas* – The Current State of Knowledge

Our efforts to identify the female sex pheromone in the shore crab were not the first attempts to do so. As such, we initially focused on evaluating previously hypothesized cues (candidate approach). The molting hormone 20-hydroxyecdysone (also known as 20HE and crustecdysone) is linked to ovarian development and was proposed to function as a sex pheromone signal in several decapod species (Kittredge et al. 1971).

Although early studies indicated that 20HE functioned as a pheromone for species such as *C. sapidus*, later evidence showed that this is not the case; we also tested the compound and could not elicit pair formation (Hardege et al. 2002). Male crabs do detect 20HE, as is often reported; however, it is not the compound responsible for mate attraction. Instead, 20HE can be linked to a different ecological consequence of its molt cycle-related release into the environment; it functions as a sex-specific feeding deterrent emitted by females preventing cannibalism by males (Fig. 19.4; Tomaschko 1994; Hayden et al. 2007). As such, 20HE plays a role in the pheromone bouquet controlling the reproductive behavior of the shore crab, albeit not as an initiator of pair formation.

To identify the female-produced sex pheromone, we used a bioassay-driven purification similar to that for *Nereis* pheromones. We combined this with a literature-driven metabolomic approach focused primarily on the biochemistry associated with ecdysis. This was based on the hypothesis that compounds linked to the physiological changes that appear during the molt process would be prime candidates to signal the event to other individuals. As outlined earlier, the bioassay to unambiguously test for the sex pheromone activity of any purified compounds presented the first major hurdle. We chose the reproductive-specific cradling behavior (Fig. 19.3) to develop such an assay (Hardege et al. 2002). The premise was to replace the female with an object that bears as little resemblance to her as possible (to eliminate any possible visual cues) and "mask" this object with the "scent" of a female that is about to reproduce (i.e., has just molted) (see Fig. 19.5). Using only pairs collected in the field should then almost guarantee bioactive samples from the females and positive behavioral responses in the males.

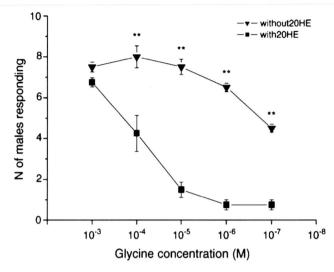

Fig. 19.4 The role of 20HE as feeding deterrent in males. The crabs were exposed to increasing concentrations of the food attractant glycine (from 10^{-6} to 10^{-2} M) against a constant concentration of the potential feeding deterrent 20HE (10^{-4} M). Ten specimens per sex were tested for each data point, and the experiments were repeated four times ($n = 40$ per data point). Data presented show the mean and standard error of mean. 20HE increasingly inhibits feedings response at lower concentrations of glycine (Chi-Square, two-tailed, *asterisk* $P < 0.01$; *double asterisk* $P < 0.001$). Reproduced from Hayden et al. (2007) with kind permission from Elsevier

Fig. 19.5 Artist's drawing – a male crab attempting to mount a golf ball – saying "Oi, I never knew these girls are so difficult to hang onto." Drawing by Jorge A. Varela Ramos

Unlike the nereidid polychaetes that devote all their resources to reproduction, shore crabs maintain other behavioral and physiological functions. They are continuously exposed and respond to a variety of chemical stimuli, such as other individuals, feeding stimulants, and predator odor. Consequently, hunger, social interaction, and seasonality have significant effects upon pheromone communication in crustaceans (Hayden et al. 2007), introducing complex challenges for biological assays. Male responses to female cues are rarely completely efficient, mainly because the physiological states of receiving males vary greatly. For example, social hierarchies as described in lobster and crayfish (e.g., see Breithaupt,

Chap. 13) result in dominant and subordinate males that potentially respond differently to females (Johansson and Jones 2007). A male's age and size significantly affects its success in aggressive encounters and, potentially, in mate choice (Huntingford et al. 1995; Sneddon et al. 1997; Smallegange and van der Meer 2007). Exposure to pheromones, in turn, also affects male–male aggressive encounters (Sneddon et al. 2003; Fletcher and Hardege 2009).

With so many environmental, seasonal, and physiological factors affecting both sender and receiver, there exists an overarching need to standardize as many external parameters as possible when collecting pheromone samples and testing for pheromone activity. As such, we restricted bioassays to the main reproductive season (June to September) using mainly large, dominant males, as they were more likely to respond to female signals (Sneddon et al. 2003). Both negative (filtered sea water) and positive (female conditioned water) controls were used when testing fractions. Bioassays were also undertaken "blind," with the observer of a test not informed as to the exact nature of the sample being tested. This was done to reduce observer bias. We used a variety of different behavioral and physiological assays to record pheromone activity, including electrophysiology, signal visualization, heart rate, and ventilation rate, to verify the different aspects of pheromone responses. Since pair formation and cradling behavior (see Fig. 19.3) represent a late step in the mating process prior to which the mating partners must first meet, we also included a classical olfactometer bioassay (see Fig. 10.2, Thiel, Chap. 10) to test distance functions (i.e., attraction) of the sex pheromone.

19.6   The Identification of Crustacean Sex Pheromones

There exist a number of plausible approaches to the chemical characterization of signal molecules. The definition of these is not entirely strict, and often an approach is chosen that includes elements from all these, but in principle, one can distinguish between:

(a) *Candidate approach*: This method has been used in a number of pheromone studies in aquatic organisms, including fish (Stacey and Sorensen 2002), as well as in examining levels of biogenic amines in urine of crayfish during antagonistic behaviors. For this, synthetically available compounds that potentially function as pheromones are tested by exposing the target species to various cue concentrations. For example, sex hormone levels of fish increase during maturation in both sexes, and it was plausible that these steroids could signal status of maturity to the opposite sex and function as sex pheromones, and they should be released into the environment. Starting with goldfish (*Carassius auratus*), a number of steroid sex pheromones have now been characterized in a variety of fish species (Stacey and Sorensen 2006).

(b) *Metabolomic approach*: This approach is based on significant advances over the past 20 years in the chemical analysis of natural products and the ability to use complex computational methods to compare odor fingerprints (Kamio and Derby, Chap. 20). It involves chemically screening a bioactive sample for compounds that are different (present, missing, or significantly changed in concentration) to a sample of similar origin that is not bioactive. Using crustaceans as an example, this could mean comparing urine of female crabs before the molt, at the time of the molt, and after the molt and examine which chemicals change. Once identified, these compounds would then be the ideal candidates for inclusion in bioassays. Despite its simplicity as an idea, examples of successful implementation are limited, as this approach is still in its infancy (Kamio and Derby, Chap. 20).

(c) *Bioassay-driven purification*: This approach was used for the vast majority of pheromone identifications in terrestrial insects where the transmission medium (air) limits the variety of molecules that can be potential pheromones and where the purification processes, extraction and analysis via gas chromatography (GC), and GC-MS, are long established (see Wyatt 2003). Aquatic signals are less well understood; they are often water-soluble (polar) molecules and are difficult to isolate from the transmission medium (water). This is especially true for seawater where its ionic strength and buffer capacity interfere with standard chromatographic methods. The philosophy of a bioassay-driven purification is that the ultimate detector for a chemical signal is not an electronic sensor, but the animal itself. For purifications testing against the potential to invoke a response in the whole organism, it is the animal's receptors that are validating a chemical's bioactivity.

Although this seems logical, such an approach also bears a number of problems such as:

– A proven bioassay is a prerequisite and with many organisms reproducing seasonally this could reduce the timeframe for analytical work.
– Extraction of organic molecules often involves solvents that could interfere with an animal's response when used in a bioassay.
– Isolating separate compounds can reduce the bioactivity in a sample as a significant percentage of signals are made of a bouquet of compounds having synergistic effects; removing just one can reduce or even eliminate a response, making assays difficult to interpret.

Regardless of which strategy is used to identify pheromones, all share common problems that have created significant disagreements related to published pheromone structures. In chromatography for example, the largest peak is often not the most important in terms of pheromone activity, but likely to be an unrelated compound of an organism's physiology; in bioassays the most active cue in the tests may not be the one found in the environment. For example, an active compound might be isolated from body fluids, but that compound may not be the one actually released into the environment.

To avoid the many pitfalls, we sought to acquire basic chemical knowledge on the stability, solubility, and size of the compounds before attempting pheromone identification. We designed a biological assay that tests for an unambiguous, fast releaser-type reaction, namely, the induction of the mate-guarding behavior. It is important to highlight that any bioassay needs to be fit for its purpose; when long-term changes in hormonal status are the biological effect (as in the induction of maturation via sex hormones in fish), such effects cannot be tested with an immediate releaser-type assay such as an animal's behavior. To avoid this mistake, long-term assays with controlled exposure to cues over time are required, thus making these studies more complex.

Purification and concentration of chemical signals from aquatic sources carry a number of potential problems. A basic decision is to either start with a body fluid (urine, bile, macerates) that is likely to contain the chemical signal, or use "conditioned water" samples. The use of conditioned water has the advantage that, theoretically, only those compounds that are released into the environment can act as a chemical signal. Unfortunately for marine organisms, conditioned water samples contain substantial amounts of ionic compounds, mainly sodium chloride, which interferes with most chemical extraction techniques. Since the concentrations of released chemical signals are quite low, they are often below detection limits for analytical tools such as nuclear magnetic resonance (NMR) spectroscopy and even Mass spectrometry (MS). This requires the concentration of samples and the removal of salts, preferably at the same time. As seen for the *Carcinus* pheromone, the use of desalting media and gel chromatography unfortunately also removes the bioactive compounds (Hardege et al. 2002). Solid phase extraction is a promising technology, but cue stability (Röhl et al. 1999) and the limited solubility of potential pheromone compounds in organic solvents may continue to make sample preparation difficult for some time. Furthermore, many compounds show little or no UV or fluorescence absorbance commonly used in high performance liquid chromatography (HPLC) peak detection, and others, such as carbohydrates, are difficult to ionize in MS.

The extraction of pheromones from body fluids brings different challenges, as these are literally a chemical cocktail containing a massive diversity of compounds. A majority of these compounds are unrelated to the pheromone or may never be released into the environment. Additionally, some compounds may interfere with each other or serve as receptor antagonists or agonists and change receptor responses. Moreover, cues are often only produced during specific communication events such as the induced alarm pheromone production in aphids (Pickett and Griffiths 1980) and sex pheromones in polychaetes (Röhl et al. 1999), thus all that can be found in body fluids are the biological precursors of the actual cues.

For the purification of compounds, methods including molecular filtration, solid phase extraction (SPE, SPME), solvent extraction, and a variety of basic chromatographic techniques (thin layer, low pressure, ion exchange, size exclusion, etc.), HPLC, and GC (with derivatization of nonvolatile compounds) can be used. Additionally, instrumentation to identify compounds is available, such as the different spectrometric applications, including infrared (IR), mass (MS), ultraviolet and visible (UV-Vis), and NMR spectroscopy. In recent years, the so-called "hyphenated techniques" (combined chromatographic and spectral methods such as

GC-MS and LC-MS) have produced powerful new analytical tools providing the separation, detection, identification, and quantification of analytes. Specifically for chemical signals, these applications can even be combined with biological assays employing electrophysiological detectors that make direct use of an organism's olfactory system, as is done with electro-antennograms (EAG), electroantenno-graphic detectors (EAD), and electro-olfactograms (EOG).

For anyone attempting pheromone identification, this diverse range of techniques and potential pitfalls represents a difficult and complex challenge to select the appropriate purification, fractionation, and identification strategies relevant to the compounds under investigation.

19.7   The Identification of Sex Pheromones in the Shore Crab, *Carcinus maenas*

For the purification of the female sex pheromone in *C. maenas*, we began with "conditioned seawater" as well as whole urine samples (Hardege et al. 2002). Aliquots of these samples were then ultrafiltered to eliminate large molecules, such as proteins, that would most likely not serve as a pheromone compound. The filtered samples were then separated via HPLC using a Phenomenex Synergi Fusion RP column (4.6 × 250 mm). This provided separation from the unretained inorganic salt peak when using an isocratic mobile phase of 0.2 M KH_2PO_4 (pH 5.5) with a flow rate of 1 mL/min at 28°C (Hardege, unpulished data). We used an Agilent 1100 HPLC system with quaternary pump, degasser, autosampler (100 μL loop), temperature controlled column compartment, and DAD (diode-array detector) scanning between 200 and 300 nm. Once we had purified and identified the cue using supplemental data from LC-MS and NMR, we verified its identity using a synthetic analog by coinjecting it onto a number of different HPLC columns (Hardege, unpublished data). This coinjection of synthetic compounds enabled a positive identification of the compound as the nucleotide, uridine diphosphate (UDP) (Fig. 19.6).

Synthetic UDP and nucleotide analogs were tested using both olfactometry and guarding stance assays (Hardege et al. 2002) with responses being elicited at biologically appropriate concentrations of 10^{-5} M (Hardege unpublished data). UDP was found to be not only an attractant but also induces the mating behavior, thus representing the first chemically characterized crustacean sex pheromone (Hardege, unpublished data). When examining the seasonality of feeding and sex pheromone responses in shore crabs, we had found earlier that only during the summer mating season do significant percentages of males respond to female cues (Hayden et al. 2009). Similarly, synthetic nucleotides elicit a full response in males only within the summer reproductive season (Hardege, unpulished data). Through the use of electrophysiological assays, we found that males are still able to detect UDP in autumn and spring, but no behavioral responses were elicited upon exposure to UDP outside the normal summer mating period.

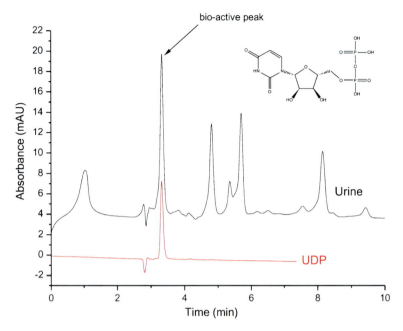

Fig. 19.6 HPLC analysis of the female-produced sex pheromone isolated from female conditioned seawater (Hardege, unpublished data). The chromatogram shows the HPLC analysis of female *C. maenas* urine at 2-day post-moult (*black line*), and synthetic pheromone, UDP (*red line*). HPLC conditions used were: Phenomenex RP Fusion column (4.6 × 250 mm); mobile phase: 0.2 M KH_2PO_4 buffer, pH 5.5; 1 mL min^{-1}. Synthetic pheromone UDP (for structure see insert) was at a concentration of 10^{-5} M. The unresolved shoulder peak in the female sample at 3.3 min represents both tautomeric forms of UDP

Willig (1974) measured the molting hormone (20HE) levels in the hemolymph and urine of female shore crabs over the course of the molt cycle. Unlike the pheromone release pattern that increases slightly after molt (Bamber and Naylor 1996), 20HE levels drop dramatically in the days just prior to ecdysis. As male shore crabs detect both compounds, the receiver (male) could gain information on the molt stage of the female and use it for mate choice. The concept of the receiver of a chemical message being able to utilize the information to time a response is known in male fish (Sorensen and Stacey 1999) and is considered a form of "chemical spying." Chemical spying and male mate choice are quite plausible in *Carcinus* since it has been shown that males differ in quality (mainly size) and position in social hierarchies (Sneddon et al. 1997, 2003).

Since female pheromone production levels are directly linked to the molt stage, we also investigated whether the female pheromone UDP could be a metabolic by-product of the female molting process – specifically chitin biosynthesis. Figure 19.7 shows that the final step of chitin biosynthesis from UDP-N-acetylglucosamine releases UDP. As such, increasing levels of UDP in female urine should exist postmolt and signal to males that a female is approaching the end of the period

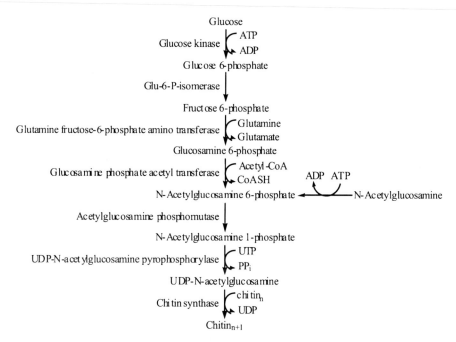

Fig. 19.7 Biosynthesis of Chitin – a potential source of the nucleotide pheromone UDP. Modified and corrected from Mansecal (1999)

during which they can mate. As shown by Bamber and Naylor (1996), this is the period when females are most attractive to males.

As social hierarchies are well known in crustaceans, we also tested the responses of males toward synthetic pheromone after they were exposed to male–male encounters. We found that winners and loser of fights respond differently to female sex pheromone (Fletcher and Hardege 2009). While winners of fights took slightly longer after a fight before responding to female cues than if not subjected to a bout, losers often did not respond to female pheromone at all and, even if responding, took significantly longer before approaching (Fig. 19.8). Since female crabs also show multiple paternity of their egg clutches (Thonet, unpublished data), investment in mate guarding may not be the same for males of different social status. Van der Meeren (1994) showed that dominant males gather in mating territories (hotspots) and smaller males roam in the vicinity attempting to intercept females that move toward these mating grounds. After forming a pair, the couple usually moves away from hotspots, and males are known to bury their female partners in the sediment, presumably as a mechanism to prevent further transmission of pheromone that, in turn, would attract male competitors; this movement is particularly important for small males in possession of a female, because they could be easily displaced by large males (Sneddon et al. 2003). It remains to be studied whether the first male to mate fathers the majority of the offspring. Additionally, it is unknown if large, dominant males roam and mate first, last, or just before the

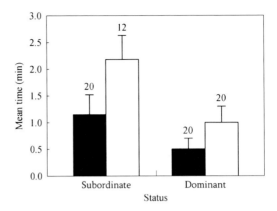

Fig. 19.8 Responses of subordinate and dominant males to synthetic nucleotide pheromone UDP at 10^{-5} M. Data show time interval between pheromone presentation and response for prefight (*filled square*) and postfight males (*open square*). Reproduced from Fletcher and Hardege (2009) with kind permission from Elsevier

Table 19.1 Heterospecific activity of the female produced sex *Carcinus* pheromone on other crustacean species, using behavioral assays with mature males

Species tested	Response to *Carcinus* female produced pheromone
Swimming crab (*Necora puber*)	−
Mud crab (*Panopeus herbstii*)	−
Spider crab (*Maja brachydactyla*)	−
Blue crab (*Callinectus sapidus*)	−
Peppermint shrimp (*Lysmata wurdemanni*)	−
Peppermint shrimp (*Lysmata bogessi*)	−
Mud crab (*Rhithropanopeus harrisii*)	−
Mediterranean shore crab (*Carcinus estuarii*)	+
Snow crab (*Chionoecetes opilio*)	+
Yellowline arrowcrab (*Stenorhynchus seticornis*)	+
Spider crab (*Libinia emarginata*)	+

Adapted from Bublitz et al. (2008)

female hardens. The most effective mating strategy may vary based on a number of factors yet unknown (Fletcher and Hardege 2009). Responses of receivers of signals vary, and the complexities in sender–receiver relationships will be both challenging and fascinating to study using identified synthetic cues (Johansson and Jones 2007).

UDP is a relatively widespread molecule as it occurs in almost all organisms. If produced as a secondary metabolite during the molt process, it should have relatively little species specificity. Table 19.1 shows that a number of species, but not all, responded to UDP (Bublitz et al. 2008). As such, we propose that the *Carcinus* pheromone is reaction-specific, alerting a male to the presence of a molting female. Species-specificity is most likely based on environmental timing ensuring premating species isolation. Since its production is linked to the message (molt), is energetically costly (UDP is an energy carrier) and might even attract potential predators, it should qualify as an honest signal (Zahavi and Zahavi 1997) indicating the physiological state and quality of the sender (Hardege, unpublished data).

19.8   Future Implications and Follow-Up Research Potential

Having identified UDP as the first crustacean sex pheromone, we can now use this knowledge to design experiments that further our understanding of the behavior and evolution of chemical communication in crustaceans. While the identification of nucleotide pheromones represents a major breakthrough in marine chemical ecology, it is evident that more structural information on crustacean pheromones is needed before we can fully address interesting questions such as the evolution of chemical signals. The use of glutathione derivatives as sex pheromones in nereidid polychaetes (Ram et al. 1999) and of nucleotides in a crustacean (Hardege, unpublished data), both of which are known feeding stimulants (Hayden et al. 2007), shows that one potential route for chemical signals to evolve is from feeding cues (Macías-Garcia and Ramirez 2005).

An emerging field with respect to marine organisms is the potential influence of environmental and ecological factors such as pollution on chemical signaling. Interference with chemical signaling systems, or "pheromone disruption," may occur within any of the stages of signaling, either during biosynthesis, release, transmission through the environment, reception, or signal transduction. Endocrine-disrupting chemicals are already known to leach into the environment through waste treatment facilities and, other means (both point- and non-point-sources) and interfere with the hormone systems of a variety of animals (Rodríguez et al. 2007). These well-studied pollutants could potentially also have dramatic conse-quences for chemical communication in crustaceans and other marine species (Beckmann et al. 1995).

Knowledge of the chemical structures of pheromones and their exact biological functions potentially allows their use in population management, especially in aquaculture and pest control (Barki et al., Chap. 25). Well established for insects in terrestrial systems (Wyatt 2003; Baker, Chap. 27), integrated pest management approaches in aquatic systems have not yet been widely pursued, with the current exception of the control of sea lampreys in the Great Lakes (Li et al. 2002; Wagner et al. 2006; Chung-Davidson et al., Chap. 24). This first, promising step for aquatic systems could potentially be followed by a similar endeavor for crustaceans, especially *Carcinus*, which is a major globally invasive marine species threatening endemic species, including those in the pristine environments of Alaska and Australia (Grosholz and Ruiz 1995). Integrated pest management utilizing chemical cues such as attractants and deterrents could be one mechanism to slow their spread.

For decades, pheromone research in aquatic organisms, especially in marine systems, has lagged behind the immense progress made in terrestrial insects and vertebrates. It is encouraging to see that significant progress is now being made in systems such as fish pheromones, brown algae, polychaetes, nudibranchs (Li et al. 2002; Hardege 1999; Painter et al. 1998) and, most recently, in crustaceans. With increasing knowledge of purification and extraction strategies, it can be expected that more breakthroughs will follow, enabling new evolutionary, physiological, neurobiological, and molecular studies that will only add depth to this exciting and fast-growing field of research.

References

Asai N, Fusetani N, Matsunaga S, Sasaki J (2000) Sex pheromones of the hair crab *Erimacrus isenbeckii*. Part 1: isolation and structures of novel ceramides. Tetrahedron 56:9895–9899

Bamber SD, Naylor E (1996) Chemical communication and behavioral interaction between sexually mature male and female shore crabs (*Carcinus maenas*). J Mar Biol Assoc UK 76:691–699

Beckmann M, Hardege JD, Zeeck E (1995) Effects of the volatile fraction of crude oil on the reproductive behaviour of nereids (Annelida, Polychaeta). Mar Environ Res 40:267–276

Beltz BS (1999) The distribution and functional anatomy of amine containing neurons in decapod crustaceans. Microsc Res Tech 44:105–120

Bublitz R, Sainte-Marie B, Newcomb-Hodgetts C, Fletcher N, Smith M, Hardege JD (2008) Interspecific activity of sex pheromone of the European shore crab (*Carcinus maenas*). Behaviour 145:1465–1478

Butenandt A, Beckmann R, Stamm D, Hecker E (1959) Über den Sexuallockstoff des Seidenspinners *Bombyx mori*. Reindarstellung und Konstitution. Z Naturforsch 14:283–284

Caskey JD, Hasenstein KH, Bauer RT (2009) Studies on contact sex pheromones of the caridean shrimp *Palaemonetes pugio*: I. Cuticular hydrocarbons associated with mate recognition. Inv Reprod Dev 53:93–103

Caspers H (1984) Spawning periodicity and habitat of the Palolo worm *Eunice viridis* (Polychaeta: Eunicidae) in the Samoan Islands. Mar Biol 79:229–236

Clare AS, Matsumura K (2000) Nature and perception of barnacle settlement pheromones. Biofouling 15:57–71

Darwin C (1871) The descent of man, and sexual selection in relation to sex. John Murray, London

Derby CD (2000) Learning from spiny lobsters about chemosensory coding of mixtures. Physiol Behav 69:203–209

Dunham PJ (1978) Sex pheromones in Crustacea. Biol Rev 53:555–583

Dunham PJ (1988) Pheromones and behaviour in Crustacea. In: Laufer H, Downer R (eds) Endocrinology of selected invertebrate types. Alan R. Liss, New York, pp 375–392

Eales AJ (1974) Sex pheromone in the shore crab *Carcinus maenas*, and the site of its release from females. Mar Behav Physiol 2:345–355

Ekerholm M, Hallberg E (2005) Primer and short-range releaser pheromone properties of premolt female urine from the shore crab *Carcinus maenas*. J Chem Ecol 31:1845–1864

Fletcher N, Hardege JD (2009) The cost of conflict: agonistic encounters influence responses to chemical signals in the shore crab, *Carcinus maenas*. Anim Behav 77:357–361

Forward RB Jr, Ritttschof D, De Vries MC (1987) Peptide pheromones synchronize crustacean egg hatching and larval release. Chem Senses 12:491–498

Gleeson RA, Adams MA, Smith AB (1984) Characterisation of a sex pheromone in the blue crab *Callinectes sapidus*. J Chem Ecol 10:913–921

Grosholz ED, Ruiz GM (1995) Spread and potential impact of the recently introduced European green crab, *Carcinus maenas*, in central California. Mar Biol 122:239–247

Hardege JD (1999) Nereid polychaetes as model organism for marine chemical ecology: a review. Hydrobiologia 402:145–161

Hardege JD, Bartels-Hardege HD, Zeeck E, Grimm FT (1990) Induction of swarming in *Nereis succinea*. Mar Biol 104:291–295

Hardege JD, Mueller CT, Beckmann M, Bartels-Hardege HD, Bentley MG (1998) Timing of reproduction in marine polychaetes: the role of sex pheromones. Ecoscience 5:395–404

Hardege JD, Jennings A, Hayden D, Müller CT, Pascoe D, Bentley MG, Clare AS (2002) Novel behavioural assay and partial purification of a female-derived sex pheromone in *Carcinus maenas*. Mar Ecol Prog Ser 244:179–189

Hartnoll RG (1969) Mating in the Brachyura. Crustaceana 16:161–181

Hay ME (2009) Marine chemical ecology: chemical signals and cues structure marine populations, communities, and ecosystems. Annu Rev Mar Sci 1:193–212

Hayden D, Jennings A, Mueller C, Pascoe D, Bublitz R, Webb H, Breithaupt T, Watkins L, Hardege JD (2007) Sex specific mediation of foraging in the shore crab, *Carcinus maenas*. Horm Behav 52:162–168

Hazlett BA (1999) Responses to multiple chemical cues by the crayfish *Orconectes virilis*. Behaviour 136:161–177

Huber R, Smith K, Delago A, Isaksson K, Kravitz EA (1997) Serotonin and aggressive motivation in crustaceans: altering the decision to retreat. Proc Natl Acad Sci U S A 94:5939–5942

Huntingford FA, Taylor AC, Smith IP, Thorpe KE (1995) Behavioural and physiological studies of aggression in swimming crabs. J Exp Mar Biol Ecol 193:21–39

Johansson BG, Jones TM (2007) The role of chemical communication in mate choice. Biol Rev 82:265–289

Kittredge JS, Terry M, Takahashi FT (1971) Sex pheromone activity of the moulting hormone crustecdysone on male crabs (*Pachygrapsus crassipes*, *Cancer antennarius* and *Cancer anthonyi*). Fish Bull 69:337–343

Koehl MAR (2006) The fluid mechanics of arthropod sniffing in turbulent odor plumes. Chem Senses 31:93–105

Li W, Scott AP, Siefkes MJ, Yan H, Liu Q, Yun SS, Gage DA (2002) Bile acid secreted by male sea lamprey that acts as a sex pheromone. Science 296:138–141

Macías-Garcia C, Ramirez E (2005) Evidence that sensory traps can evolve into honest signals. Nature 434:501–505

Mansecal R (1999) Chitin. In: Polymer data handbook. Mark JE (ed) Oxford University Press, New York, pp 67–69

Painter S, Clough B, Garden RW, Sweedler JV, Nagle GT (1998) Characterization of *Aplysia* attraction, the first water-borne peptide pheromone in invertebrates. Biol Bull 194:120–131

Pettis RJ, Erickson BW, Forward RB, Rittschof D (1993) Superpotent synthetic tripeptide mimics of the mud-crab pumping pheromone. Int J Pept Protein Res 42:312–319

Pickett JA, Griffiths DC (1980) Composition of aphid alarm pheromones. J Chem Ecol 6:349–360

Ram JL, Mueller CT, Beckmann M, Hardege JD (1999) The spawning pheromone cysteine-glutathione disulfide ("Nereithione") arouses a multicomponent nuptial behaviour and electro-physiological activity in *Nereis succinea* males. FASEB J 13:945–952

Ram JL, Fei X, Danaher SM, Lu S, Breithaupt T, Hardege JD (2008) Finding females: pheromone-guided reproductive tracking behavior by male *Nereis succinea* in the marine environment. J Exp Biol 211:757–765

Ratchford SG, Eggleston DB (1998) Size- and scale-dependent chemical attraction contribute to an ontogenetic shift in sociality. Anim Behav 56:1027–1034

Rodríguez EM, Medesani DA, Fingerman M (2007) Endocrine disruption in crustaceans due to pollutants: a review. Comp Biochem Physiol A 146:661–671

Röhl I, Schneider B, Schmidt B, Zeeck E (1999) L-Ovothiol A: the egg release pheromone of the marine polychaete *Platynereis dumerilii*: Annelida: Polychaeta. Z Naturforsch 54:1145–1147

Ryan EP (1966) Pheromone: evidence in decapod Crustacea. Science 151:340–341

Seifert P (1982) Studies on the sex pheromone of the shore crab, *Carcinus maenas*, with special regard to ecdysone excretion. Ophelia 21:147–158

Smallegange IM, Van der Meer J (2007) Interference from a game theoretical perspective: shore crabs suffer most from equal competitors. Behav Ecol 18:215–221

Sneddon LU, Huntingford FA, Taylor AC (1997) Weapon size versus body size as a predictor of winning in fight between shore crabs, *Carcinus maenas*. Behav Ecol Sociobiol 41:237–242

Sneddon LU, Huntingford FA, Taylor AC, Clare AS (2003) Female sex pheromone-mediated effects on behaviour and consequences of male competition in the shore crab (*Carcinus maenas*). J Chem Ecol 29:55–70

Sorensen PW, Stacey NE (1999) Evolution and specialization in fish hormonal pheromones. In: Johnston RE, Müller-Schwarze D, Sorensen PW (eds) Advances in chemical signals in vertebrates. Kluwer Publishers, Amsterdam, pp 15–47

Stacey N, Sorensen P (2002) Hormonal pheromones in fish. In: Pfaff DW, Arnold AP, Etgen AM, Fahrbach SE, Rubin RT (eds) Non-mammalian hormone-behavior system. Harcourt Publishers, London, pp 375–434

Stacey NE, Sorensen P (2006) Reproductive pheromones. In: Sloman KA, Balshine S, Wilson RI (eds) Behaviour and physiology of fish. Elsevier, Amsterdam, pp 359–412

Tierney AJ, Atema J (1988) Amino-acid chemoreception – effects of pH on receptors and stimuli. J Chem Ecol 14:135–141

Tomaschko KH (1994) Ecdysteroids from *Pycnogonum litorale* (Arthropoda, Pantopoda) act as a chemical defense against *Carcinus maenas* (Crustacea, Decapoda). J Chem Ecol 20:1445–1455

Van der Meeren GI (1994) Sex- and size-dependent mating tactics in a natural population of shore crabs *Carcinus maenas*. J Anim Ecol 63:307–314

Wagner CM, Jones ML, Twohey MB, Sorensen PW (2006) A field test verifies that pheromones can be useful for sea lamprey (*Petromyzon marinus*) control in the Great Lakes. Can J Fish Aquat Sci 63:475–479

Willig A (1974) Die Rolle der Ecdyosne im Häutungszyklus der Crustaceen. Fortschr Zool 22:55–74

Wood DE, Derby CD (1996) Distribution of dopamine-like immunoreactivity suggests a role for dopamine in the courtship display behavior of the blue crab, *Callinectes sapidus*. Cell Tissue Res 285:321–330

Wyatt TD (2003) Pheromones and animal behavior. Cambridge University Press, Cambridge

Wyatt TD (2009) Fifty years of pheromones. Nature 457:262–263

Zahavi A, Zahavi A (1997) The handicap principle: a missing piece in Darwin's puzzle. Oxford University Press, Oxford

Zeeck E, Hardege JD, Bartels-Hardege HD, Wesselmann G (1988) Sex pheromone in a marine polychaete: determination of the chemical structure. J Exp Zool 246:285–292

Zhang D, Lin J (2006) Mate recognition in a simultaneous hermaphroditic shrimp, *Lysmata wurdemanni* (Caridea: Hippolytidae). Anim Behav 71:1191–1196

Chapter 20
Approaches to a Molecular Identification of Sex Pheromones in Blue Crabs

Michiya Kamio and Charles D. Derby

Abstract Molecular identification of sex pheromones in marine crustaceans has proven to be very difficult, and so far no unequivocal identification for any decapod crustacean has been published. Some of these difficulties are common to other animals – pheromones are often blends of molecules at low concentrations. Some difficulties are more specific to marine crustaceans – pheromones are often small and polar molecules that are difficult to separate from salts in their source (often urine) or carrier medium (sea water). These difficulties led us to take on new approaches as we searched for sex pheromones in the blue crab *Callinectes sapidus*. Premolt pubertal female blue crabs that are ready to mate release a pheromone in their urine. This pheromone is detected by male crabs using specific chemical sensors – aesthetasc sensilla on the antennules. Male blue crabs respond to the pheromone with courtship stationary paddling, a distinctive behavior that is useful in bioassays for pheromone identification. We used bioassay-guided fractionation to demonstrate that the pheromone of female blue crabs is of low molecular mass (<1,000 Da) and possibly a mixture. We used liquid chromatography-mass spectrometry (LC-MS), nuclear magnetic resonance, biomarker targeting, and metabolomics approaches to isolate molecules specific to premolt pubertal females and that are thus candidate pheromones. Our working hypothesis is that female blue crabs release a species-specific sex pheromone in their urine that is composed of two functional classes of molecules, both of which are small and polar. One class distinguishes females from males and thus is a sex-specific signal, and a second class distinguishes blue crabs from other species and thus constitutes a species-specific signal.

M. Kamio (✉)
Department of Ocean Science, Tokyo University of Marine Science
and Technology 4-5-7 Konan, Minato-ku, Tokyo 108-8477, Japan
e-mail: mkamio@kaiyodai.ac.jp

T. Breithaupt and M. Thiel (eds.), *Chemical Communication in Crustaceans*,   393
DOI 10.1007/978-0-387-77101-4_20, © Springer Science+Business Media, LLC 2011

20.1 Introduction

Blue crabs are known to all who frequent shallow waters along the east coast of North America. Their abundance and tasty meat have made them a major commercial fishery for hundreds of years. Lifestyles of people, even cultures, have been defined by them, as chronicled in William Warner's book, "Beautiful Swimmers: Watermen, Crabs and the Chesapeake Bay." In fact, the blue crab's scientific name, *Callinectes sapidus*, given in 1896 by Dr. Mary Jane Rathbun, at the time one of the foremost authorities on crustacean systematics, means "good-tasting beautiful swimmer." That they are, but they are much more, as we hope we convey in this chapter.

The life history of blue crabs is well known because of their abundance and economic value. A recent scholarly book, "The Blue Crab, *C. sapidus*," edited by Victor Kennedy and Eugene Cronin, makes the interested reader up to date on what scientists know about blue crabs. Our research focus, and that of this chapter, is on sexual behavior of blue crabs, including pair bonding and mating, and the role of chemical communication in these behaviors. Because the pheromones of reproductive females are only released during a few days of their entire life, as described below, and in very small amounts, we use the advantages of the commercial fishery to obtain these chemicals in amounts sufficient to perform the requisite chemical analyses and behavioral bioassays.

Identifying pheromones of crustaceans, especially sex pheromones, is a challenging prospect, as is obvious to anyone who reads this book. In fact, there are very few successes in molecular identification of pheromones of crustaceans. Examples include larval settlement factors of barnacles, which are large α_2-macroglobulin glycoproteins called "settlement inducing protein complex" (Dreanno et al. 2006a, b, 2007). The pheromone used by male copepods, *Tigriopus japonicus*, to recognize females has been partially characterized – and interestingly, it too has similarity to α_2-macroglobulin (Ting and Snell 2003). Despite many attempts to identify sex pheromones of decapod crustaceans such as crabs, crayfish, clawed lobsters, and spiny lobsters (for example, see Gleeson 1991; Asai et al. 2000, 2001; Hardege et al. 2002; Kamio et al. 2002, 2003; Atema and Steinbach 2007), there are no published papers that present a convincing demonstration of the molecular identity of their pheromones (but see the chapter in this book by Hardege and Terschak on the shore crab *Carcinus maenas*).

Our own individual experiences that have led us to attack this problem are quite different. One of us – Michiya Kamio – has been well trained in chemistry, in addition to behavioral biology, and has applied these skills to the search for crustacean sex pheromones starting with graduate school. The other – Charles Derby – has always been interested in the chemistry of crustacean pheromones, but lacking the training, willing collaborators, and especially in the earlier years, adequate technology, focused on other aspects of crustacean chemical senses. Several years ago, we decided that it was a good time to bring our skills together and start a collaboration focused on blue crab sex pheromones. Julia Kubanek was our first collaborator, and we have since brought in others who have specialized analytical chemistry skills to help in our work. We also were fortunate to have the advice, assistance, and inspiration of Richard Gleeson, who did groundbreaking work on chemical communication in blue crab reproductive behavior (Gleeson 1980, 1982, 1991; Gleeson et al. 1984, 1987). In this chapter, we outline the experimental strategies and techniques that we are using to try to identify the sex pheromones of blue crabs. We also consider chemical signals in the context of blue crab behavioral ecology, chemical ecology, and interspecies ecological interactions in natural habitats. We are working towards a more general understanding of the mating and reproductive behavior of blue crabs.

20.2 Features of Mating Behavior in Blue Crabs and the Role of Chemical Communication

20.2.1 Molting and Mating

Molting and mating are closely linked in blue crabs, as it is in many (but not all) decapod crustaceans (Jivoff et al. 2007). Blue crabs, like other arthropods, grow by molting. Male blue crabs grow continuously throughout their lives and become sexually active after reaching a mature size. Female blue crabs, on the other hand, grow until a final molt, called the pubertal or terminal molt. Following this molt, females are sexually mature and reproductive and are often called "pubertal females." They have an enlarged abdomen for incubating the fertilized eggs.

Female blue crabs mate only at this time of their life (Fig. 20.1). Timing is critical. They must mate just after the terminal molt, before their new hard shell forms. Before mating, females must find a suitable mate. They do this at a severely vulnerable stage of their lives. Both males and females take on a significant risk and energetic investment, along with the promise of reproductive gain, in mating. Molting itself is a risky endeavor. Animals in the stage immediately before molting, i.e., premolt, are not very active, mobile, or strong, and molting animals lose all mobility. The state immediately after molting, i.e., postmolt, is also risky. At this time, they are soft because their new shell has not yet hardened. It is these immediate postmolt crabs, both females and males, that are the gastronomic delicacy known as "soft shell crabs." For females, the risks of being soft-shelled are further compounded by the aggressive behavior of male crabs during mating. How do female crabs survive the risks of mating in this vulnerable state?

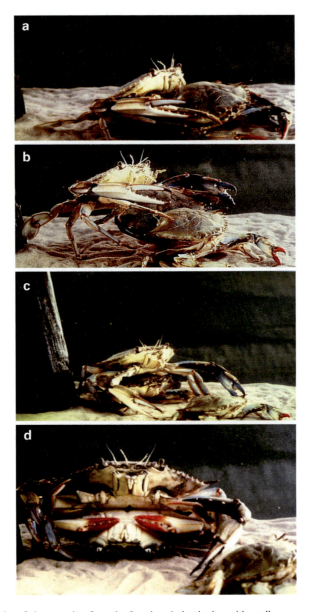

Fig. 20.1 Series of photographs of a male–female pair, beginning with cradle carry and culminating in copulation. (**a**) Male crab (*left*), with his blue-tipped claws, cradle carries the smaller, red-tipped female (*right*) as she is beginning to molt. (**b**) Female continues to back out of her shell, under the protection of the male. (**c**) Female completes molting, with her old shell, or exuvium, littering the substrate in front of her and the male returning to his cradle carrying position. (**d**) Male and female copulate. This previously unpublished series of images is courtesy of Dr. Richard Gleeson and used with his permission

20.2.2 Bidirectional Chemical Communication Between Males and Females During Courtship Behavior in Natural Environment

Most animals mate in a narrow time window of their lives and therefore must signal to each other their readiness to reproduce. Collectively, they do so through a diversity of sensory channels within environmental and evolutionary constraints (Bradbury and Vehrencamp 1998). This is also true for many crabs (e.g. Hardege and Terschak, Chap. 19), including blue crabs, as described here. The female blue crab communicates that she is approaching her terminal molt, and the male communicates his interest in taking her as a mate. They do this through multiple sensory channels. One is visual signaling, including male preferences for the red color on female claws (Teytaud 1971; Baldwin and Johnsen 2009). Another means of communicating sexual readiness is chemical signaling, including reciprocal release and detection of sexual pheromones. As the female approaches her terminal molt, she produces a pheromone that changes the behavior of males. Several scientists, most notably Richard Gleeson, have helped us understand this chemical communication in blue crabs (Gleeson 1980, 1982, 1991; Gleeson et al. 1984, 1987; Bushmann 1999; Jivoff et al. 2007).

What is the effect of the female pheromone on males? Male blue crabs respond to the female pheromone with a series of behaviors. First, they search for the source of the pheromone by walking upstream while sensing features of the chemical plume containing the pheromone (Gleeson 1991; Kamio et al. 2008; Dickman et al. 2009). If the searching male finds a female and she is accessible to him, he immediately grabs her and initiates "cradle carry" behavior (Fig. 20.1c). In this stereotypical behavior, the larger male holds the female under his abdomen and both animals are positioned with their dorsal side upwards. However, if the male detects a female but cannot reach her because she is inaccessible to him, then he performs a highly stereotyped behavior called "courtship stationary paddling" (Fig. 20.2). The term "courtship display" has a broader meaning, since it includes standing high on legs, spreading chelae, holding paddles up over the carapace, and performing courtship stationary paddling. Inaccessibility in the natural environment occurs when the female is hiding, which females do around the time of molting. Inaccessibility can be simulated in the laboratory by caging the female. When a male performs courtship stationary paddling, he elevates his body by standing high on his legs, opens his chelae, and paddles his last pair of legs, which in this species have paddles at the end and thus are called swimming legs.

What is the function of courtship display and courtship stationary paddling? It is probably a multimodal signal to females. It probably contains a visual signal, since sexually receptive females respond to a model of a male crab in the stationary posture of courtship display (Teytaud 1971) and the red color on the female claws can induce courtship by males (Baldwin and Johnsen 2009). The dynamic component of the display, paddling, probably also provides a characteristic and conspicuous visual stimulus, though this has not been experimentally tested. The water current generated

Fig. 20.2 Courtship stationary paddling of a male blue crab (*right*) towards a female (*left*) that is inaccessible to the male due to a barrier between them. This behavioral display has both postural and dynamic components. The male stands erect on extended legs and laterally spreads his chelae. He also rhythmically waves his last pair of legs, which have paddles and also are used for swimming. This behavior includes chemical and visual components. See text for more explanation. Photograph by Peter Essick, used with permission

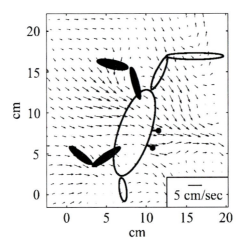

Fig. 20.3 Water currents generated by the courtship stationary paddling of male blue crab. View from above, the male performing courtship stationary paddling. Particle imaging velocimetry was used to visualize water currents generated by this behavior. *Arrows* indicate the direction of the currents, with the length of the *arrows* being proportional to the velocity. Results show that the water current was directed away from the male at a mean velocity of 3.1 cm/s. From Kamio et al. (2008), reproduced with permission of The Journal of Experimental Biology

by paddling may act as a mechanical signal to females. We know for certain that courtship stationary paddling creates a water current directed away from the male and toward the female (Kamio et al. 2008; Fig. 20.3). We also know that males release a pheromone to which pubertal females respond. These females move upstream over a few meters in a flow scented with water from reproductive males (Gleeson 1991), and they will perform other behaviors that facilitate pair formation (Rittschof 2005). We conclude that courtship stationary paddling is a form of chemical signaling that induces females to move towards males, leading to pair formation, cradle carry, and eventually mating.

Our notion is that courtship stationary paddling is an adaptation to the environmental conditions in which mating occurs (Kamio et al. 2008; Fig. 20.4). Females approaching their terminal molt move to areas such as salt marshes, which have plentiful submerged vegetation and other refuges for females to hide from predators. We hypothesize that courtship stationary paddling is a context-dependent behavior that males produce only when females are inaccessible and that functions

Fig. 20.4 This drawing portrays how we envision blue crabs communicate their sexual status to each other, in their natural environment, leading to pair bonding and mating. A pair of blue crabs residing in a salt marsh is shown, with the female on the *left* and the male on the *right*. When premolt females such as this one approach their terminal molt at which time they become sexually mature and mate, they hide in grass beds to avoid predators such as sea turtles, sharks, and egrets, and release a pheromone that attracts males. If a male can easily reach the female, he picks her up and initiates cradle carrying behavior as shown in Fig. 20.1c. But if he cannot easily reach the female, he performs a courtship stationary paddling, as shown in this illustration and in Fig. 20.2. This behavior generates a water current that delivers his pheromone to the hidden female, as depicted in Fig. 20.3. Once the female detects and locates the source of the pheromone, she will move out of hiding so that the male can initiate cradle carrying behavior. Drawing by Jorge A. Varela Ramos

to deliver his sex pheromone to the female, in an effort to entice her to emerge from her refuge and engage in cradle carry.

After pair formation, the male crab will hold and protect the female for a period of time necessary for her to molt, then mate, and then for her shell to harden to the point that she can defend herself and protect their reproductive investment. This period of cradle carrying may last several days. It is possible that contact pheromones mediate interactions of male and female crabs after they find each other, including copulation, as is the case in other decapod crustaceans (Kamio et al. 2002; Díaz and Thiel 2004; Caskey and Bauer 2005), but this has not been studied. The process of molting and mating is intricate, gentle, and seemingly choreographed. A series of photographs of this process is shown in Fig. 20.1. The female backs out of her shell with the male's assistance, still guarded by the male. The male flips her over, and they copulate, a process that can last for many hours. The male then turns her over again, and resumes cradle carrying for 4–5 days until her shell hardens and the male releases her (Jivoff et al. 2007).

20.2.3   Site of Release of Pheromones

The female's pheromone that induces courtship stationary paddling by males is in her urine (Gleeson 1980). Since crabs and other crustaceans can control the release of their urine, they can control the release of pheromone. Female blue crabs may squirt the pheromone towards males at select times to maximize the use of this limited resource, although this has not been experimentally demonstrated. Bushmann's work (1999) suggests that the female's pheromone may be released not only in urine, but also from other, unidentified sources. The site of release of the male pheromone is not known, but we hypothesize that it is also in his urine, as it acts from a distance (Gleeson 1991).

20.2.4   Receptor Organs for Detecting Pheromones

What are the mechanisms responsible for sensing pheromones? The male detects the female's pheromone using only one of his many types of sensors – the aesthetasc sensilla. These are located on the distal end of their first antennae, or antennules. Blue crab aesthetascs are densely packaged on the antennule, and each aesthetasc contains approximately 150 olfactory receptor neurons (Gleeson et al. 1996; Cate et al. 1999). Aesthetascs are a type of sensillum found in many crustaceans, even the more ancient and primitive forms (Hallberg and Skog, Chap. 6; Schmidt and Mellon, Chap. 7). They are the only sensilla known to be exclusively chemosensory. They are considered olfactory because of homologies with olfactory sensors of terrestrial arthropods and analogies with olfactory systems of the vertebrates (Schmidt and Ache 1996). Aesthetascs are innervated by olfactory neurons sensitive to a range of odors, including food-related chemicals. The emerging view is that they are the only antennular sensors containing receptors for pheromones and other intraspecific cues mediating sex, social interactions, aggregations, and alarm (Gleeson 1982; Shabani et al. 2009; Schmidt and Mellon, Chap. 7). Males lacking aesthetascs cannot detect the female's sex pheromone and will not mate with her (Gleeson 1982, 1991). The location of the female's sensors of the male pheromone is not known.

20.3   Experimental Approaches to Identifying Sex Pheromones of Female Blue Crabs

20.3.1   Possible Experimental Approaches

The molecules serving as cues, signals, or pheromones can be identified by a variety of experimental approaches. One approach that has been highly successful, leading to the identification of many bioactive cues, is bioassay-guided fractionation (Hay et al. 1998; Koehn and Carter 2005). Since the pheromone of female blue crabs is not a protein and is present in freeze-dried urine (Gleeson et al. 1984; Gleeson 1991), we focused on identifying relatively small, polar molecules present in the urine. This approach uses isolation of molecules from the natural source of the signals, which are termed natural products, together with bioassays. It begins by separating a natural product with bioactivity into fractions based on physicochemical properties such as polarity, molecular size, or molecular charge (Sarker et al. 2006). An example is using gel filtration to separate compounds based on molecular size, then testing the resultant fractions for bioactivity using behavioral assays. High throughput methods such as electrophysiological screening can speed up the process of identifying molecules that are detected by an animal. Examples of this approach include mouse (Lin et al. 2005), cockroach (Nojima et al. 2005), and sea louse (Ingvarsdóttir et al. 2002a; Fields et al. 2007). But ultimately behavioral assays are required to know the behavioral relevance of the molecules. Bioassay-guided fractionation requires a comparison of the bioactivity of the fractions against the original material and the recombined fractions, the latter to determine if the activity was lost during fractionation either by degradation or loss of bioactive molecules or the effect of trace solvents that may be added to the fractions. A negative control allows identification of fractions containing the bioactivity.

The active components can be extracted from tissues, organs, or secretions. These can then be separated and purified based on differences in polarity, size, and ionic charge using solvent–solvent partitioning, liquid chromatography (LC), or high performance liquid chromatography (HPLC) with a variety of separation modes such as size exclusion, ion exchange, normal phase, and reversed phase. Once separated to sufficient purity, the bioactive molecules can be identified through a combination of spectroscopic methods such as mass spectrometry

(MS), nuclear magnetic resonance (NMR) spectroscopy, infrared spectroscopy (IR), ultraviolet spectroscopy (UV), circular dichroism (CD) spectroscopy, X-ray crystallography, or other approaches including organic synthesis of the candidate compounds, as has been already demonstrated for other pheromones and natural products (Haynes and Millars 1998; Koehn and Carter 2005; Sorensen and Hoye 2007). Databases such as SciFinder Scholar (http://www.cas.org/SCIFINDER/ SCHOLAR/index.html) and Marinlit (http://www.chem.canterbury.ac.nz/marinlit/ marinlit.shtml) are useful in identifying the molecular structures, by comparing spectroscopic data, molecular formulae, and proposed structures of the purified bioactive molecules against those in the databases for identified molecules.

One potential limitation in the use of bioassay-guided fractionation is revealed when a pheromone or other natural product under study is a mixture and the full expression of that mixture's bioactivity requires the simultaneous presence of several or all of its compounds. For example, if more than one fraction has components of the mixture but each fraction contributes a small portion of the mixture's activity, then identification of the bioactive molecules can be difficult. Another example is when there are synergistic interactions among bioactive molecules in different fractions, such that once fractionated, no single fraction has activity.

Although bioassay-guided fractionation has proven successful for identifying many bioactive molecules, other experimental approaches can be useful as well. One of these is biomarker targeting. A biomarker is any molecule that is associated with a specific physiological state or condition, being either unique to or in significantly higher or lower concentration in individuals in that state or condition. Consequently, biomarker targeting is the search for the identity of those distinctive molecules. For example, biomarker targeting in the study of a disease involves identifying molecules in urine, blood, or some other body fluid or odor that indicate that disease's specific occurrence (Soga et al. 2006). Another example of biomarker target is the search for pheromones. Ovulated female masu salmon release L-kynurenine in their urine as a sex pheromone (Yambe et al. 2006). Although L-kynurenine was isolated by bioassay-guided fractionation, direct analysis of urine from mature and immature males and females detected L-kynurenine only in ovulated female urine. Since L-kynurenine has a characteristic ultraviolet absorption, comparative analysis using HPLC equipped with a photodiode array and/or MS identified this compound as a biomarker of a female that is releasing the pheromone. A limitation of biomarker targeting is that it has the potential to lead to false positives because the molecules that are in relatively high or low abundance in the target sample are not necessarily contributing the most, or at all, to that sample's bioactivity. And of course, whatever molecules are identified by this technique must be tested in behavioral assays for their biological relevance.

As described below, we have used biomarker targeting to search for blue crab sex pheromones by seeking molecules that distinguish urine of pubertal females from urines of other females, premolt males, juveniles, and any other urine that does not have pheromonal bioactivity.

20.3.2 Application Towards Identifying Sex Pheromones from Female Blue Crabs

20.3.2.1 Bioassay-Guided Fractionation

Richard Gleeson provided a preliminary characterization of the molecules in urine of pubertal females that induce courtship stationary paddling in males. He reported that the pheromone is 300–600 Da and stable at high temperature, 95°C (Gleeson 1991). This was based on gel filtration using a Sephadex column calibrated with peptide markers. Gleeson also used the other separation methods – ultrafiltration using YC-2 filter andYM-05 filter (Amicon) – to retain the pheromone in a fraction with mass of 500–1,000 Da, then using reverse phase chromatography for chemical separation. Fractions were bioassayed using a behavioral test based on a male crab performing cradle carry behavior towards a target animal in the same aquarium. In nature, the target animal is a premolt pubertal female. Adding to the bioassay, a target animal that does not release the pheromone but does provide a physical presence is important as it increases the responsiveness of the male test animal to the pheromone without increasing the response to sea water, a negative control stimulus. Gleeson used as the target animal another male that had been rendered unresponsive to the female pheromone through antennular ablation. Gleeson et al. (1984) further showed that the active component is not crustecdysone, a molecule previously reported to be a female sex pheromone in other crab species (Kittredge et al. 1971), though this result has been questioned (Hayden et al. 2007).

Our bioassay-guided fractionation began as did Gleeson et al.'s (1984) by separating premolt pubertal female urine into three fractions with molecular size of <500, 500–1,000, or >1,000 Da, using ultrafiltration with YM-1 and YC-05 membranes. Our study differed from that of Gleeson in two ways. First, we used a nonmating postpubertal molt female as the target animal. Second, we used courtship stationary paddling as our behavioral indicator of responsiveness to pheromone, because it is distinct, easily quantifiable, and most importantly, is the only behavior produced by males specifically in response to pubertal female urine. Other behaviors such as grasping and lateral spread of chelae were evoked by pubertal female urine, but not significantly more than by urine from nonreproductive females or urine from males. We found that the <500 and 500–1,000 Da fractions of female urine were the only stimuli to evoke significantly more courtship stationary paddling than the negative control, sea water (Fig. 20.5). This result is partially consistent with that of Gleeson (1991), but differs in revealing activity in a fraction <500 Da. Thus, the sex pheromone of blue crabs includes small, polar molecules. This is also the case with some other decapod crustaceans (see Hardege and Terschak, Chap. 19). However, not all crustacean sex pheromones are small and polar (hydrophilic): those of sea lice appear to be small nonpolar (lipophilic) molecules (Ingvarsdóttir et al. 2002b), and those of copepods are large proteins (Ting and Snell 2003).

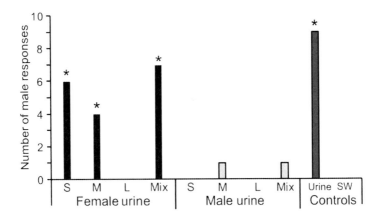

Fig. 20.5 Using bioassay-guided fractionation to identify the molecular size of female sex pheromone. The dependent measure is the number of male blue crabs out of 72 tested that performed courtship stationary paddling. Stimuli were male or female urine fractionated into the indicated molecular sizes (*S* small <500 Da, *M* medium 500–1,000 Da, *L* large >1,000 Da, a mixture of *S+M+L* Mix) and positive control (unprocessed pubertal female urine = Urine) and a negative control (*SW* sea water). Friedman ANOVA shows an overall difference in the responsiveness to these stimuli ($P < 0.0001$, $n = 10$). An asterisk marks stimuli that elicit significantly more males to respond compared to the sea water control (Wilcoxon post hoc tests, $P < 0.05$)

The advantage of this bioassay-guided fractionation method is that it enables discrimination of components with sex pheromone activity from those that do not. Serial repetition of separation guided by bioassay should lead to purification of the sex pheromone molecules. The disadvantage of the application of this experimental approach to blue crabs is that the bioassay is time-consuming. Courtship stationary paddling is strictly specific to courtship, and males produce it only when females are inaccessible (Kamio et al. 2008); consequently, under the conditions of our laboratory assay, only 10% of males showed courtship stationary paddling to the target animal in the presence of premolt pubertal female urine, the positive control. This low occurrence of the courtship stationary paddling is problematic for bioassay-guided fractionation because it slows the process. We are currently modifying our bioassay in an effort to increase the responsiveness of males. For example, we know that visual signals from females (Baldwin and Johnsen 2009) and inaccessibility of females (Kamio et al. 2008) contribute to the male's courtship stationary paddling. Adding these features to our bioassay will, we hope, make it more useful.

20.3.2.2 Biomarker Targeting

Another approach to isolating sex pheromones of blue crabs takes advantage of the economic importance of these animals (Kamio 2009). The significant fishery for hard-shell and soft-shell crabs gives us access to large numbers of crabs of either sex and any molt stage, including pubertal females. Comparison of the chemical composition of urine from these animals allows identification of molecules that are unique to or concentrated in each type, and therefore candidate sex pheromones. In our search for female sex pheromones, we compared the chemical composition of urine from premolt pubertal females against the urine of intermolt females, premolt males, and other stages. The following section describes our use of analytical techniques in biomarker targeting of the female sex pheromone.

Liquid Chromatography-Mass Spectrometry (LC-MS)-Based Biomarker Targeting

LC-MS separates compounds in a mixture using HPLC, identifies the molecular weight of compounds using MS, and determines the relative concentration of each. Applying principal component analysis (PCA) to these data allows us to determine the relative similarity in the overall compositions of mixtures, based on all molecules that are separable, identifiable, and quantifiable. From this, we can identify which components are most different in those mixtures.

In collaboration with Tomoyoshi Soga and Yuji Kakazu of Keio University, Japan, we used LC-MS PCA to compare urine from four premolt males, urine from five premolt females, and three samples of deionized water as a control. The results are shown in Fig. 20.6. The LC-MS detected hundreds of compounds based on molecular weight and retention time in the male and/or female urines and determined their relative concentrations. PCA shows that two principal components (PC1 and PC2) explained 70% of the variance in the data (PC1, 57%; PC2, 13%). Furthermore, the four male urines form a group (represented by a dashed oval in the two-dimensional PC space of Fig. 20.6), the five female urines form a different group (represented by another dashed oval), and both of these are distinct from the group of water controls (Fig. 20.6). This indicates that the overall compositions of the male urines are more similar to each other than to the female urines, and vice versa, and both are very different from water. In other words, despite variations between animals in the chemical compositions of their urine, there is a "male type" urine and a "female type" urine. Further analysis using tandem MS experiments can identify which of the compounds is specific to male urine or female urine, as well as those that are common to both male and female urines.

NMR-Based Biomarker Targeting

An advantage of NMR-based biomarker targeting is that it does not require any separation methods and thus we can use more complex mixtures. But this advantage brings with it a challenge: it can be difficult to identify the compounds in the complex NMR spectrum. A limitation is that NMR, especially 1D-NMR, alone usually yields partial structures even for a pure compound. 2D-NMR gives more information and resolution for structure analysis, but its lower sensitivity is a disadvantage. Complete molecular identification often requires a combination of approaches.

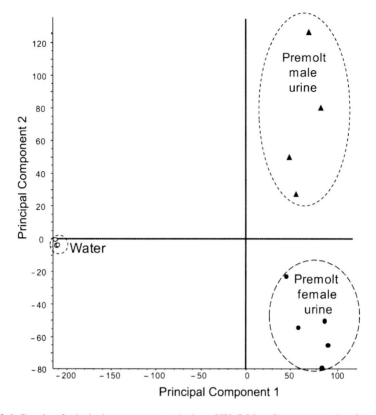

Fig. 20.6 Results of principal components analysis on HPLC-Mass Spectrometry data for premolt male and premolt pubertal female urines. See text for explanation

We used 1D ^{1}H NMR, which analyzes all protons in a mixture. As a first step towards identifying female sex pheromone molecules, we started by detecting features in the spectra, i.e., molecules specific to premolt female urine. ^{1}H NMR spectra of 0.2 mL equivalent of urine from male and female juveniles and premolt, postmolt, and intermolt mature crabs were obtained on 800 MHz NMR equipped with a cold probe. To obtain higher sensitivity in our experiment, we substituted hydrogen atoms with deuterium by using deuterium oxide as a solvent and removing water by freeze-drying. To avoid chemical shift changes of the compounds in urine due to differences in pH, we standardized the pH of the samples by adding phosphate buffer. After spectrum binning, the data for each molting stage were mathematically analyzed by PCA, in collaboration with Koichi Matsumura of Monell Chemical Senses Center. Urines from premolt stage males and females were clearly different in PCA (Fig. 20.7): the first two principal components explained 78% of the variance in the data (PC1, 43%; PC2, 35%), indicating that the NMR spectrum contained enough information to discriminate between the urine of male and female premolt crabs. We are currently pursuing this difference with

Fig. 20.7 Results of principal components analysis of NMR data for premolt male and premolt pubertal female urines. The results are from ¹H NMR data, which observes all protons in the mixture except exchangeable protons. See text for explanation

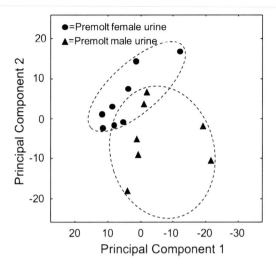

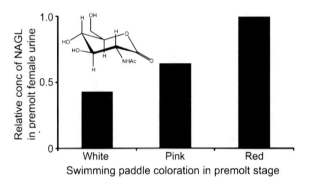

Fig. 20.8 Increase of concentration of N-acetylglucosamino-1,5-lactone, or NAGL (molecular structure shown in *inset*), in urine of premolt females as they approach molting. Coloration of swimming legs of blue crabs indicates their molting stage. *White* indicates early, *pink* means middle, and *red* means late premolt stage (Kennedy and Cronin 2007). Quantification by ¹H NMR showed that the concentration of NAGL increased as crabs progressed in molt stage from *white to pink to red*. Urine sample for each stage is pooled from many females

the aim of identifying the compounds specific to premolt females. We are also performing 2D-NMR on the samples.

This NMR-based biomarker targeting has proven useful for identifying molting biomarkers. Comparison of urine of premolt and intermolt males and females identified a molt-specific compound. We observed a clear difference in the spectra of the 500–1,000 Da fraction of urines from premolt vs. intermolt animals. This signal was isolated using HPLC and identified as N-acetylglucosamino-1,5-lactone (NAGL), whose structure is shown in Fig. 20.8. Its concentration increases as female crabs approach molting (Fig. 20.8). NAGL has never before been reported as a natural product. It is known to be an inhibitor of N-acetylglucosaminidase of limpets, mice, bovines, and *E. coli* (Findlay et al. 1958; Yem and Wu 1976; Legler et al. 1991). It may be part of the metabolic pathway of chitin (Merzendorfer and Zimoch 2003). We bioassayed NAGL using sexually receptive males and found that they detect NAGL, but do not perform courtship stationary paddling to it. Thus, NAGL by itself is not sufficient as the female sex pheromone. It might be a component of a pheromone blend, contributing in an additive or synergistic way with other components of the blend. It is also possible that NAGL is an indicator of molting animals. These possibilities should be experimentally tested. It is also worth noting that P-31 NMR was used to identify the phospholipid 2-aminoethyl phosphonate as being specifically enriched in the gills of male blue crabs (Kleps et al. 2007), but it was not reported to be in the urine and nor is it reported to have any sex pheromone activity.

Our current working hypothesis is that the female sex pheromone is a multicomponent mixture that is constituted by a combination of compounds specific to pubertal females as well as others that are distinctive of blue crabs, i.e., having species specificity but lacking sex specificity. Evidence of species specificity comes from the work by Gleeson (reported in Bublitz et al. 2008) that showed that male blue crabs do not express courtship behavior to the sex pheromone of shore crabs, *C. maenas*. Sex pheromones of many organisms are blends, including insects and mammals (Wyatt 2003). Blends can provide species specificity that is not possible in single molecules. One experimental observation supporting the blend hypothesis for blue crabs is that males responded to two molecular size fractions (see Fig. 20.5), suggesting that more than one molecule is involved. A test of this idea requires formulating an artificial premolt female urine containing female-specific and blue crab species-specific compounds, and bioassaying it against sexually active males.

20.4 Summary and Conclusions

Molecular identification of sex pheromones in marine crustaceans has proven to be very difficult, to the point that no unequivocal identification for any decapod crustacean has been published (but see Hardege and Terschak, Chap. 19). Several factors have made successful identification difficult. Some difficulties are common to searches in other animals: pheromones are likely blends of molecules whose components are at very low concentration. In addition, pheromones of marine crustaceans are often small and polar molecules, making them difficult to separate from salts and other small ions in their background, i.e., sea water or urine. These difficulties have led us to take on new approaches to pheromone identification and to select the blue crab *C. sapidus* as our experimental model.

The blue crab is a good model for pheromone research. Pubertal females approaching the stage where they molt into a mature female release a pheromone in their urine. This pheromone is detected by male crabs using as sensors the aesthetascs on their antennules. Males respond with a very distinctive behavior, called courtship stationary paddling, as well as other behaviors. The distinctiveness of courtship stationary paddling makes it useful as a specific behavioral indicator of the presence of the pheromone; however, this advantage is counterbalanced by the fact that the behavior is not reliably released by the pheromone, and thus, there are no false positives, but many false negatives. Fortunately, several studies of the behavioral ecology of blue crabs have helped to identify why courtship stationary paddling is used by blue crabs in mating. This should lead to modification of bioassays that can decrease the rate of false negatives. The commercial fishery for blue crabs has given us access to animals of all stages of maturity and molting, and thus to sources of the pheromones. We have used a combination of experimental approaches, including commonly used (bioassay-guided fractionation) and new (biomarker targeting), to show that the female pheromone is less than 1,000 Da and possibly a mixture. From the urine, we isolated some molecules that are specific to premolt pubertal females and thus are candidate pheromones. These approaches have yielded promising results, and we hope that more work will uncover the identity of the pheromones.

Generalizations from our results and from other studies of decapod crustaceans must be done with caution, since results for all species are preliminary or fragmentary. Our working hypothesis is that a "species-specific sex pheromone" (though see Wyatt, Chap. 2) is composed of two functional classes of molecules, both of which are small and polar. One class of molecules distinguishes female vs. male, and thus is a sex-specific signal that could be, but is not necessarily, unique to blue crabs. The second set of molecules distinguishes blue crabs from other species, thus constituting a species-specific signal. Thus, we expect the species-specific sex pheromone to be a blend of molecules which contains a variety of messages and information.

Acknowledgments We thank our colleagues who have contributed in many ways to our work, in particular Barry Ache, Peter Anderson, Todd Barsby, Norman Byrd, Sekar Chandrasekaran, Markus Germann, Richard Gleeson, John Glushka, Yuji Kakazu, Julia Kubanek, Koichi Matsumura, Dell Allen Newman, Dan Rittschof, Tomoyoshi Soga, and Siming Wang. Funding was provided by NSF grants IBN 0077474, IBN-0324435, IBN-0614685, Japan Society for the Promotion of Science Postdoctoral Fellowship for Research Abroad, Brains & Behavior Program, and Center for Behavioral Neuroscience through the STC Program of NSF under Agreement No. IBN-9876754.

References

Asai N, Fusetani N, Matsumaga S, Sasaki J (2000) Sex pheromones of the hair crab *Erimacrus isenbeckii*. Part 1: isolation and structures of novel ceramides. Tetrahedron 56:9895–9899

Asai N, Fusetani N, Matsunaga S (2001) Sex pheromones of the hair crab *Erimacrus isenbeckii*. II. Synthesis of ceramides. J Nat Prod 64:1210–1215

Atema J, Steinbach MA (2007) Chemical communication in the social behavior of the lobster, *Homarus americanus*, and other decapod Crustacea. In: Duffy E, Thiel M (eds) Ecology and evolution of social behavior: crustaceans as model systems. Oxford University Press, Oxford, pp 115–144

Baldwin J, Johnsen S (2009) The importance of color in mate choice of the blue crab *Callinectes sapidus*. J Exp Biol 212:3762–3768

Bradbury JW, Vehrencamp SL (1998) Principles of animal communication. Sinauer Associates, Sunderland

Bublitz R, Sainte-Marie B, Newcomb-Hodgetts C, Fletcher N, Smith M, Hardege JD (2008) Interspecific activity of the sex pheromone of the European shore crab (*Carcinus maenas*). Behaviour 145:1465–1478

Bushmann PJ (1999) Concurrent signals and behavioral plasticity in blue crab (*Callinectes sapidus* Rathbun) courtship. Biol Bull 197:63–71

Caskey JL, Bauer RT (2005) Behavioral tests for a possible contact pheromone in the caridean shrimp *Palaemonetes pugio*. J Crust Biol 25:571–576

Cate HS, Gleeson RA, Derby CD (1999) Activity-dependent labeling of the olfactory organ of blue crabs suggests that pheromone-sensitive and food-odor sensitive receptor neurons are packaged together in aesthetasc sensilla. Chem Senses 24:559

Díaz ER, Thiel M (2004) Chemical and visual communication during mate searching in the rock shrimp. Biol Bull 206:134–143

Dickman BD, Webster DR, Page JL, Weissburg MJ (2009) Three-dimensional odorant concentration measurements around actively tracking blue crabs. Limnol Oceanogr Methods 7:96–108

Dreanno C, Matsumura K, Dohmae N, Takio K, Hirota H, Kirby RR, Clare AS (2006a) An α_2-macroglobulin-like protein is the cue to gregarious settlement of the barnacle *Balanus amphitrite*. Proc Natl Acad Sci USA 103:14396–14401

Dreanno C, Kirby RR, Clare AS (2006b) Smelly feet are not always a bad thing: the relationship between cyprid footprint protein and the barnacle settlement pheromone. Biol Lett 2:423–425

Dreanno C, Kirby RR, Clare AS (2007) Involvement of the barnacle settlement-inducing protein complex (SIPC) in species recognition at settlement. J Exp Mar Biol Ecol 351:276–282

Fields DM, Weissburg MJ, Browman HI (2007) Chemoreception in the salmon louse *Lepeophtheirus salmonis*: an electrophysiology approach. Dis Aquat Org 78:161–168

Findlay J, Levvy GA, Marsh CA (1958) Inhibition of glycosidases by aldonolactones of corresponding configuration. 2. Inhibitors of β-*N*-acetylglucosaminidase. Biochem J 69:467–476

Gleeson RA (1980) Pheromone communication in the reproductive behavior of the blue crab, *Callinectes sapidus*. Mar Behav Physiol 7:119–134

Gleeson RA (1982) Morphological and behavioral identification of the sensory structures mediating pheromone reception in the blue crab *Callinectes sapidus*. Biol Bull 163:162–171

Gleeson RA (1991) Intrinsic factors mediating pheromone communication in the blue crab, *Callinectes sapidus*. In: Martin JW, Bauer RT (eds) Crustacean sexual biology. Columbia University Press, New York, pp 17–32

Gleeson RA, Adams MA, Smith AB III (1984) Characterization of a sex pheromone in the blue crab, *Callinectes sapidus*: crustecdysone studies. J Chem Ecol 10:913–921

Gleeson RA, Adams MA, Smith AB III (1987) Hormonal modulation of pheromone-mediated behavior in a crustacean. Biol Bull 172:1–9

Gleeson RA, McDowell LM, Aldrich HC (1996) Structure of the aesthetasc (olfactory) sensilla of the blue crab. *Callinectes sapidus*: transformations as a function of salinity. Cell Tissue Res 284:279–288

Hardege JD, Jennings A, Hayden D, Muller CT, Pascoe D, Bentley MG, Clare AS (2002) Novel behavioral assay and partial purification of a female-derived sex pheromone in *Carcinus maenas*. Mar Ecol Prog Ser 244:179–189

Hay ME, Stachowicz JJ, Cruz-Rivera E, Bullard S, Deal MS, Lindquist N (1998) Bioassays with marine and freshwater macroorganisms. In: Haynes KF, Millars JG (eds) Methods in chemical ecology, vol. 2, bioassay methods. Chapman and Hall, New York, pp 39–141

Hayden D, Jenning A, Müller C, Pascoe D, Bublitz R, Webb H, Breithaupt T, Watkins L, Hardege JD (2007) Sex-specific mediation of foraging in the shore crab, *Carcinus maenas*. Horm Behav 52:162–168

Haynes KF, Millars JG (eds) (1998) Methods in chemical ecology, vol. 1, chemical methods. Chapman and Hall, New York

Ingvarsdóttir A, Birkett MA, Duce I, Genna RL, Mordue W, Pickett JA, Wadhams LJ, Mordue (Luntz) AJ (2002a) Semiochemical strategies for sea louse control: host location cues. Pest Manag Sci 58:537–545

Ingvarsdóttir A, Birkett MA, Duce I, Mordue W, Pickett JA, Wadhams LJ, Mordue (Luntz) AJ (2002b) Role of semiochemicals in mate location by parasitic sea louse, *Lepeophtheirus salmonis*. J Chem Ecol 28:2107–2117

Jivoff P, Hines AH, Quackenbush LS (2007) Reproduction biology and embryonic development. In: Kennedy VS, Cronin LE (eds) The blue crab *Callinectes sapidus*. Maryland Sea Grant, College Park

Kamio M (2009) Toward identifying sex pheromones in blue crabs: using biomarker targeting within the context of evolutionary chemical ecology. Ann NY Acad Sci 1170:456–461

Kamio M, Matsunaga S, Fusetani N (2002) Copulation pheromone in the crab, *Telmessus cheiragonus* (Brachyura: Decapoda). Mar Ecol Prog Ser 234:183–190

Kamio M, Matsunaga S, Fusetani N (2003) Observation on the mating behaviors of the helmet crab *Telmessus cheiragonus* (Brachyura: Cheiragonidae). J Mar Biol Assoc UK 83:1007–1013

Kamio M, Reidenbach M, Derby CD (2008) To paddle or not: determinants and consequences of courtship display by male blue crabs, *Callinectes sapidus*. J Exp Biol 211:1243–1248

Kennedy VS, Cronin LE (2007) The blue crab: *Callinectes sapidus*. University of Maryland Sea Grant Press, College Park, 800 pp

Kittredge JS, Terry M, Takahashi FT (1971) Sex pheromone activity of the molting hormone, crustecdysone, on male crabs. Fish Bull 69:337–343

Kleps RA, Myers TC, Lipcius RN, Henderson TO (2007) A sex-specific metabolite identified in a marine invertebrate utilizing Phosphorus-31 nuclear magnetic resonance. PLoS One 2:e780

Koehn FE, Carter GT (2005) The evolving role of natural products in drug discovery. Nat Rev Drug Discov 4:206–220

Legler G, Lüllau E, Kappes E, Kastenholz F (1991) Bovine N-acetyl-β-D-glucosaminidase: affinity purification and characterization of its active site with nitrogen containing analogs of N-acetylglucosamine. Biochim Biophys Acta 1080:89–95

Lin DY, Zhang S-Z, Block E, Katz LC (2005) Encoding social signals in the mouse main olfactory bulb. Nature 434:470–477

Merzendorfer H, Zimoch L (2003) Chitin metabolism in insects: structure, function and regulation of chitin synthases and chitinases. J Exp Biol 206:4393–4412

Nojima S, Schal C, Webster FX, Santangelo RG, Roelofs WL (2005) Identification of the sex pheromone of the German cockroach, *Blattella germanica*. Science 307:1104–1106

Rittschof D (2005) Male blue crab pheromone originates in semen. Chem Senses 30:A144

Sarker SD, Latif Z, Gray A (2006) Natural products isolation, 2nd edn. Humana Press, Totowa

Schmidt M, Ache BW (1996) Processing of antennular input in the brain of the spiny lobster. *Panulirus argus*. II. The olfactory pathway. J Comp Physiol A 178:605–628

Shabani S, Kamio M, Derby CD (2009) Spiny lobsters use urine-borne signals to communicate social status. J Exp Biol 212:2464–2474

Soga T, Baran R, Suematsu M, Ueno Y, Ikeda S, Sakurakawa T, Kakazu Y, Ishikawa T, Robert M, Nishioka T, Tomita M (2006) Differential metabolomics reveals ophthalmic acid as an oxidative stress biomarker indicating hepatic glutathione consumption. J Biol Chem 281:16768–16776

Sorensen PW, Hoye TR (2007) A critical review of the discovery and application of a migratory pheromone in an invasive fish, the sea lamprey *Petromyzon marinus* L. J Fish Biol 71:100–114

Teytaud AR (1971) The laboratory studies of sex recognition in the blue crab, *Callinectes sapidus* Rathbun. Sea Grant Technical Bulletin, University of Miami Sea Grant Program 15, pp 1–63

Ting JH, Snell TW (2003) Purification and sequencing of a mate-recognition protein from the copepod *Tigriopus japonicus*. Mar Biol 143:1–8

Warner WW (1976) Beautiful swimmers: watermen, crabs and the Chesapeake Bay. Little, Brown, Boston

Wyatt TD (2003) Pheromones and animal behaviour: communication by smell and taste. Cambridge University Press, Cambridge

Yambe H, Kitamura S, Kamio M, Yamada M, Matsunaga S, Fusetani N, Yamazaki F (2006) L-Kynurenine, an amino acid identified as a sex pheromone in the urine of ovulated female masu salmon. Proc Natl Acad Sci USA 103:15370–15374

Yem DW, Wu HC (1976) Purification and properties of β-N-D-acetylglucosaminidase from *Escherichia coli*. J Bacteriol 125:324–331

Chapter 21
The Crustacean Endocrine System and Pleiotropic Chemical Messengers

Ernest S. Chang

Abstract In this chapter, I first present a brief overview of the hormones (intraorganismal chemical signals) and endocrine glands that have been described in crustaceans. This overview focuses on decapods and the physiological processes of molting, metabolism, reproduction, and pigmentation. The hormones include ecdysteroids (molting hormones), molt-inhibiting hormone, methyl farnesoate, crustacean hyperglycemic hormone, androgenic gland hormone (AGH), and chromatophorotropins. I briefly discuss some of the work on the regulation of crustacean behavior by neurotransmitters, especially in respect to aggressive behavior. Evidence is then presented supporting the role of crustacean hormones as pheromones (intraspecific chemical signals). In particular, I describe the experiments demonstrating that ecdysteroids have pleiotropic activities. These experiments include the observations that ecdysteroids (1) are present in urine in varying amounts over the course of the molt cycle; (2) modulate aggressive behavior in a manner that correlates with hemolymph and urinary hormone levels; and (3) can be detected in the environment by olfactory neurons. There is evidence that the AGH may also act as a mating pheromone. I conclude my chapter by describing how the arthropod molting hormone can act as an allelochemical (interspecific chemical signal) in a pycnogonid (sea spider). Pycnogonids can accumulate high concentrations of ecdysteroids that serve as feeding deterrents against crustaceans that are potential predators.

21.1 Introduction

There are many similarities between the hormonal (intraorganismal), pheromonal (intraspecific), and allelochemical (interspecific) communication systems: (1) Discrete organs are responsible for the production of the chemical messengers;

E.S. Chang (✉)
Bodega Marine Laboratory, University of California-Davis,
Bodega Bay, CA 94923, USA
e-mail: eschang@ucdavis.edu

T. Breithaupt and M. Thiel (eds.), *Chemical Communication in Crustaceans*,   413
DOI 10.1007/978-0-387-77101-4_21, © Springer Science+Business Media, LLC 2011

(2) The messengers are released to the circulating medium (hemolymph in the case of crustacean hormones; air or water in the case of crustacean pheromones and allelochemicals); (3) The messengers are detected by target cells that contain specific receptors for the signals; (4) The receiving organism responds to the signals in a distinctive manner. Based upon these similarities between these communication systems, one might suppose that there would be evidence for overlapping signals. There are, however, only a few examples of this overlap throughout the entire animal kingdom (e.g., in fishes; see Chap. 24).

In this chapter, I first briefly review the hormonal system of crustaceans (Sect. 21.2, focusing on the decapods). This review is not comprehensive and serves to present background information on the better-known crustacean hormones. I have generally not provided detailed references for the general section on the crustacean endocrine system and instead refer the reader to the cited reviews in that section. After this background information, I then present some of the evidence supporting the hypothesis that crustacean hormones can have pleiotropic activities by also acting as pheromones or allelochemicals.

My introduction to the field of chemical mediation in crustaceans is yet another example of the importance of individual mentors. My early interest in zoology began with my observations of all sorts of animals in my family's ornamental fish ponds. My first formal exposure to marine biology was in my high school biology class with additional training in the extracurricular marine biology club.

My interests in organismal biology continued as an undergraduate under the mentorship of Profs. Ralph I. Smith and Cadet Hand, Jr. in the Zoology Department at the University of California, Berkeley. I spent several summers taking classes and conducting independent research at (at the time) Berkeley's Bodega Marine Laboratory. I cannot overemphasize the important function that field and research-oriented courses serve in stimulating undergraduate students to pursue scientific careers.

The father of one of my undergraduate friends was Prof. Howard A. Bern, a pioneer in the field of comparative endocrinology. I met him socially and he stimulated me to take his course in chemical mediation. It was through his class and subsequent research opportunity that I became fascinated with the topic and was steered toward arthropod hormones as a long-term academic interest.

21.2   Survey of the Crustacean Endocrine System

21.2.1   Control of Molting

Crustaceans, like all arthropods, must periodically shed their external, confining exoskeletons and take up air or water to expand their new and larger exoskeletons in order to grow in size. The problem of how to increase in body size is even more formidable for crustaceans (compared to insects) due to their mineralized, relatively

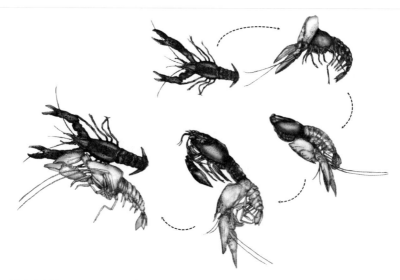

Fig. 21.1 Molting (or ecdysis) of a juvenile lobster (*Homarus americanus*). The entire sequence took about 30 min from start to finish. Molting is the culmination of a cyclic process mediated by the steroid hormone 20-hydroxyecdysone. (*Top, center*) – the lobster has resorbed much of its old exoskeleton's mineralization. The exoskeleton splits at the junction of the thorax and abdomen. (*Top, right*) – the anterior of the animal (*darker shading*) retracts from its old exoskeleton (*lighter shading*) by pulling posteriorly. (*Bottom, right*) – posterior of the lobster pulls anteriorly. (*Bottom, center*) – the lobster is free of its old exoskeleton and begins to take up water to expand its new, flexible, larger exoskeleton. (*Bottom, left*) – the postmolt lobster is above its shed exoskeleton. It will take several days for the epidermis to deposit layers of chitin, protein, and calcium carbonate into the exoskeleton. The molt cycle will continue for several days to months, depending upon the size and age of the animal. It will terminate with another ecdysial event. Drawing by Jorge A. Varela Ramos

rigid exoskeletons. Figure 21.1 shows the molting of the overlying old exoskeleton in the lobster *Homarus americanus*. This ecdysial process lasts about 30 min. However, the entire molt cycle occurred over many weeks and ecdysis was simply the culmination of this process.

21.2.1.1   Ecdysteroids

Horn et al. (1966) isolated and determined the structure of 20-hydroxyecdysone (20E) as the molting hormone from the rock lobster. The 20E was identified as the principal active form of the molting hormone from both insect and crustacean sources. In most species examined, ecdysone (E) is the prohormone secreted by the Y-organ (YO, Fig. 21.2) and it is hydroxylated by target tissues to 20E. Although 20E is the predominant molting hormone in all decapod species examined to date, other ecdysteroids (steroids structurally related to E that have molting hormone activity) have been characterized in hemolymph and tissues of various crustacean species (see Chang 1989, 1993; Chang and Kaufman 2005 for reviews).

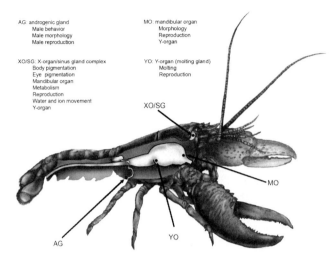

AG: androgenic gland
  Male behavior
  Male morphology
  Male reproduction

MO: mandibular organ
  Morphology
  Reproduction
  Y-organ

XO/SG: X-organ/sinus gland complex
  Body pigmentation
  Eye pigmentation
  Mandibular organ
  Metabolism
  Reproduction
  Water and ion movement
  Y-organ

YO: Y-organ (molting gland)
  Molting
  Reproduction

XO/SG

MO

AG

YO

Fig. 21.2 Approximate locations of the endocrine glands discussed in this chapter. The outline of a schematic decapod crustacean is shown (represented by a lobster). The XO/SG is the X-organ/sinus gland complex located in the eyestalk. The YO is the Y-organ (molting gland) located near the anterior of the branchial cavity. The MO is the mandibular organ located at the base of the mandibular tendon. The AG is the androgenic gland located at the distal end of the sperm duct. Functions or targets for the secretory product of each gland are listed. Drawing of lobster by Jorge A. Varela Ramos

Classical morphological observations and endocrinological experiments indicated that the molting gland in the green crab *Carcinus maenas* was the thoracic YO. Organ culture of the YOs from the crabs *Cancer antennarius* and *Pachygrapsus crassipes* resulted in the characterization of E as the primary secretory product of the YO. Other ecdysteroids are secreted by the YOs in other crab species. It is not clear why different types of ecdysteroids are secreted by the YOs of various species. In almost all cases examined, exogenous ecdysteroids promote progression through the molt cycle.

Like vertebrate steroid hormones, ecdysteroids recognize target tissues by binding with nuclear receptors. The ecdysteroid receptor has been isolated and characterized from the fiddler crab, *Uca pugilator*. It has been sequenced and has homologies with insect ecdysteroid receptors. Transcripts for the receptor were isolated from crab limb buds and developing ovaries.

In addition to their effects upon molting, ecdysteroids also have neuromodulatory activities. Hemolymph was obtained from lobsters at different molt stages and was perfused over the claw opener muscle from lobsters. Significant changes were observed in the excitatory junctional potentials following perfusion with hemolymph obtained from premolt lobsters (Schwanke et al. 1990). Premolt is the stage that has the highest levels of circulating ecdysteroids (Snyder and Chang 1991a). With in vitro neuromuscular preparations, increased amplitudes and frequency of the excitatory potentials were observed in the opener muscle of the lobster claw in the presence of 20E (Cromarty and Kass-Simon 1998). In abdominal muscle, these potentials were significantly smaller in the presence of 20E compared to control

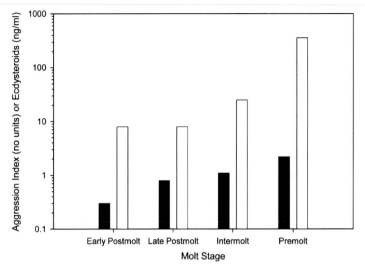

Fig. 21.3 Aggression index (*dark bars*) and hemolymph ecdysteroid concentrations (*light bars*) during the molt cycle of lobsters (*H. americanus*). For determining the aggression index, lobsters were challenged with a threatening stimulus (a piece of plastic pipe) and their responses were videotaped and scored. The lobsters were molt staged according to the method of Aiken (1973). Data are redrawn from Fig. 5a of Cromarty et al. (2000). Reprinted with permission from the Marine Biological Laboratory, Woods Hole, MA. For determining ecdysteroid concentrations, 10 μl of hemolymph was removed for quantification by radioimmunoassay (data are derived from Chang and Bruce 1980). Molt stages were determined by the amount of elapsed time through the molt cycle. The highest ecdysteroid concentration obtained during each particular molt stage is shown. The *x*-axis represents four general molt stages (see Smith and Chang 2007, for a review of molt stages). For the aggression indices and ecdysteroid concentrations, the values in premolt were significantly different from the corresponding values for the other three molt stages. Note the logarithmic scale

abdominal muscle. These observations were consistent with the changes observed in vivo in which premolt lobsters (with high circulating levels of ecdysteroids) had increased aggressiveness relative to other molt stages (Fig. 21.3).

In support of these observations on alterations in neuromuscular activity, experiments were conducted on lobsters using 20E injections. Bolingbroke and Kass-Simon (2001) observed that intermolt females injected with 20E displayed increased aggressiveness relative to saline-injected controls. These experiments on the effects of ecdysteroids on crustacean behavior are especially relevant to the discussion of these hormones as possible pheromones and allelochemicals later in this chapter.

21.2.1.2   Molt-Inhibiting Hormone

Hemolymph ecdysteroid concentration fluctuates dramatically during the molt cycle (e.g., from <10 ng/ml in postmolt to >350 ng/ml in premolt lobster). These changes mediate the various biochemical and physiological processes that occur during the cycle. The rates of synthesis and/or secretion of E by the YO vary during the molt cycle and partially account for these hemolymph fluctuations in ecdysteroid titer. Just prior to the substage of premolt in which the highest concentration of ecdysteroids was observed in the hemolymph, explanted YOs were found to secrete the greatest amount of E. Low hemolymph concentrations were correlated with low secretory rates.

Removal of both stalked eyes of *U. pugilator* resulted in a shortening of the molt interval. This observation lead to the postulation of an endocrine factor present in the eyestalks that normally inhibits molting – a molt-inhibiting hormone (MIH). Detailed microscopical examinations resulted in the description of a neurohemal organ in the eyestalk of several decapod crustaceans. This neurohemal organ is called the sinus gland (SG) and serves as a storage site for neurosecretory products. It consists of the enlarged endings of a group of neurosecretory neurons collectively called the X-organ (XO).

The shortened molt interval observed in eyestalk-ablated decapods is likely due to a rapid elevation in the concentration of circulating ecdysteroids (Fig. 21.4), which is a result of XO/SG (Fig. 21.2) removal. However, recent evidence indicates that the regulation of the YO is more complex than simple inhibition by MIH. Chung and Webster (2005) demonstrated that there were no overt changes in the hemolymph levels of MIH over the molt cycle, except for a large increase during late premolt immediately before ecdysis. This peak of premolt MIH likely mediates the sudden drop in circulating ecdysteroids just prior to ecdysis. MIH from *C. maenas* was among the initial MIHs to be characterized. It is a member of a novel neuropeptide family, representatives of which have so far been found only in arthropods. This neuropeptide family regulates such diverse functions as molting, reproduction, and metabolism (reviewed in Webster 1998 and Böcking et al. 2002).

Fig. 21.4 Effects of eyestalk ablation on lobster growth. The larger animal (*upper*) had both of its eyestalks removed 2 months after hatching. The intact control animal (*lower*) was its full sibling. Both lobsters were kept in the same aquatic system for 1 year and fed excess amounts of food. From Conklin and Chang (1983)

21.2.2 Methyl Farnesoate

The mandibular organ (MO, Fig. 21.2) produces a sesquiterpenoid, methyl farnesoate (MF), related to the insect juvenile hormone (JH). MF was initially isolated from the crab *Libinia emarginata* and from several other decapods. A number of different effects on crustacean development and reproduction have been attributed to MF (for review, see Homola and Chang 1997). One effect is an increase in the molt interval. In contrast to this effect on larvae, the implantation of blue crab MOs into adult white shrimp *Litopenaeus setiferus* resulted in shortened intermolt periods accompanied by more frequent molting. The molt stimulation was not due to ecdysteroid secretion by the MO. In addition, application of MF increased the synthesis and/or secretion of ecdysteroids by the YO. MF may have different functions at different life stages, similar to the actions of JH at different insect life stages.

MF may also be involved in mediating various stress responses. Hemolymph levels of the hormone rise after challenges with heat stress, hypoxia, and hypo- and hypersalinity. It is unknown how MF helps the crustacean cope with these environmental stresses. MF also acts as a sex determinant. In the cladoceran *Daphnia magna*, application of physiological amounts of exogenous MF to egg-maturing females resulted in all-male broods.

21.2.3 Control of Metabolism

Crustacean hyperglycemic hormone (CHH), initially characterized from the crab *C. maenas,* has a high degree of homology with MIH. A primary role of CHH is the regulation of the level of hemolymph glucose (Fig. 21.5). Removal of the SGs from *Orconectes limosus*, leaving the remainder of the eyestalk neural tissue and vision intact, resulted in a permanent decline in hemolymph glucose. Hyperglycemia is commonly observed under a variety of stressful conditions, and there is considerable indirect evidence to suggest that stress-induced hyperglycemia is caused by the release of CHH. For example, under hypoxic conditions, glucose may be released by CHH from carbohydrate stores to serve as a substrate for glycolysis and lactate formation, as is typical for anaerobiosis of decapod crustaceans. After 4 h of emersion, which causes hypoxia, CHH concentration in lobsters rose over 20-fold (Fig. 21.6). In addition to hyperglycemia, other demonstrated actions of CHH in crustaceans are lipid mobilization, secretory activity of the hepatopancreas, and second messenger responses in various tissues.

A dramatic rise in CHH in *C. maenas* was observed in late premolt compared to intermolt values (over 100-fold). This surge in CHH triggers the massive ion and water uptake during ecdysis and mediates the body's expansion necessary for successful ecdysis. The mechanism of this uptake is unknown, but the sources of this premolt surge in CHH are the fore- and hindguts. The structure of this intestinal CHH is different from the eyestalk form of CHH.

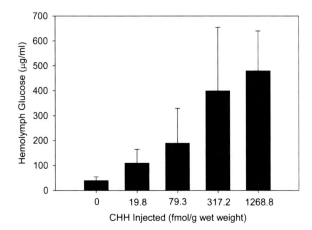

Fig. 21.5 Effects of injections of CHH on hemolymph glucose in juvenile lobsters. The lobsters were injected with the indicated amounts of purified CHH and hemolymph glucose was quantified 1 h later using a colorimetric assay. Mean values are shown ± standard deviations. Unpublished data

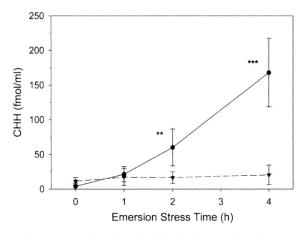

Fig. 21.6 Effects of emersion on hemolymph CHH (*circles, solid line*). Lobsters were removed from the water and placed in an incubator at ambient temperature (13.0°C). Hemolymph was sampled from these animals at various time points. Controls were matched siblings that were left immersed at 13°C and sampled at the same time points. Lobsters ranged in wet weight from 100 to 145 g. Means ± standard deviations are shown. Control data are represented by the triangles and dashed line. Asterisks indicate significant differences from immersed controls at $P < 0.01$ (*double asterisk*) and at $P < 0.001$ (*triple asterisk*). Modified from Chang, Keller, and Chang (1998) with permission from Elsevier

Another member of this MIH/CHH peptide family is MO-inhibiting hormone (MOIH). The MO synthesizes and secretes MF (discussed above). MOIH prevents the synthesis and secretion of MF in vitro. There are indications that additional factors likely regulate MF synthesis by the MO.

21.2.4   Control of Reproduction

21.2.4.1   Male

Charniaux-Cotton (1954) initially identified the crustacean androgenic gland (AG, Fig. 21.2). This gland caused masculinization when it was implanted into a female and feminization or dedifferentiation when it was removed from a male crustacean. More recently, protein blotting experiments with polyclonal antisera raised against androgenic gland hormone (AGH) indicated the presence of a larger, biologically inactive protein. It was deduced that AGH was translated as a prohormone. The gene that codes for the hormone has been sequenced. Secretions from the XO/SG complex may regulate the AG since eyestalk ablation resulted in hypertrophied AGs in crayfish.

An AG-specific gene has been characterized in the red-claw crayfish *Cherax quadricarinatus*. This gene is termed *Cq-IAG* (insulin-like AG factor from *C. quadricarinatus*). In situ hybridization of *Cq-IAG* confirmed the exclusive localization of its expression in the AG. Following cloning and complete sequencing of the gene, the proposed protein sequence coded by the gene is similar to those of members of the insulin/insulin-like growth factor/relaxin family. The peptide and its biological activity remain to be elucidated (Manor et al. 2007).

21.2.4.2   Female

Ecdysteroids likely play a role in female crustacean reproduction. Various ecdysteroids have been identified in follicle cells and oocytes of brine shrimp, amphipods, crabs, crayfish, and shrimp. The function of these ovarian ecdysteroids in crustaceans remains uncertain. Some correlations between vitellogenesis and hemolymph ecdysteroid titers have been reported. For unknown reasons, the effects of ecdysteroid injections or in vitro incubations on vitellogenesis have been either inhibitory or stimulatory.

Varying levels of ecdysteroids have also been found in crustacean embryos (Fig. 21.7). The published studies have shown either a general rise or decline in ecdysteroids from ovulation to hatching, depending on the species. In some studies, a greater sampling frequency has demonstrated multiple peaks of ecdysteroids during embryogenesis.

In addition to its effects on the molt interval previously described, there are observations that MF is a female gonadotropin. Early, indirect studies involved eyestalk ablation and subsequent ovarian maturation. For example, MOs (the

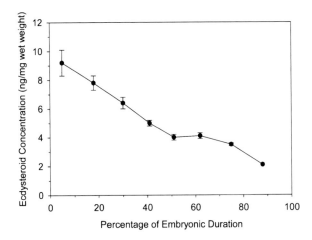

Fig. 21.7 Changes in total ecdysteroid concentrations (means ± standard errors) in developing embryos of the crab *Cancer anthonyi* over the course of one brood cycle. Embryos from individual crabs ($n = 8$) were removed, homogenized, and quantified for ecdysteroids by radioimmunoassay. The data indicate that high levels of ecdysteroids were invested into the embryos by their mother at fertilization and egg extrusion. Presumably, the ecdysteroids were used during embryogenesis for various developmental processes. Modified from Okazaki and Chang (1991) with permission from Elsevier

source of MF) from mature females secreted approximately 6 times as much MF compared to organs from immature female crabs. Direct experiments involved the injection of exogenous MF into eyestalk-ablated crabs (*L. emarginata*). Following injection, these crabs had an increased level of hemolymph yolk proteins. Enhanced egg production was observed in shrimp *Litopenaeus vannamei* that had previously been administered exogenous MF. Similar experiments resulted in increased ovarian maturation in crayfish and the crab *Oziotelphusa senex*.

Another member of the MIH/CHH hormone family is the vitellogenesis-inhibiting hormone (VIH) from *H. americanus*. This peptide was isolated with the use of a heterologous assay involving the inhibition of yolk synthesis in eyestalk-ablated shrimp *Palaemonetes varians*. A VIH was also characterized from an isopod. This peptide should perhaps be renamed "gonad-inhibiting hormone" because it can be found in both sexes of some species, such as the lobster *Nephrops norvegicus*.

More extensive reviews of the hormonal control of crustacean reproduction can be found in Subramoniam (2000), Laufer and Biggers (2001), Sagi and Khalaila (2001), and Chang and Sagi (2008).

21.2.5 Control of Pigments

There are two types of pigmentary effectors in the Crustacea: the chromatophores and the retinal pigment cells. The chromatophores are located primarily in the integument and their activities result in the body pigmentation of the individual animal. The retinal pigments are located in the compound eyes. They regulate the amount of light impinging on the rhabdom, the light-sensitive portion of each ommatidium (the functional unit) of the compound eye (see Rao 2001 for review).

21.2.5.1 Chromatophores

The chromatophores are pigment-containing cells that occur in the integument and internal organs. Their function is to adjust the body coloration to its surroundings, depending on the situation (e.g., protection from predators, mating behavior, antagonistic displays). The pigments within the chromatophores may also aid in thermal regulation and protection from ultraviolet radiation. When the pigment granules are spread out along the extensions of the chromatophore, they are described as being in a dispersed state. In this state, the granules are most visible and give the chromatophore its characteristic color. When the granules are withdrawn from the extensions of the chromatophore, they are in a concentrated state and are not as visible.

In addition to hastening the molt, eyestalk ablation of crustaceans results in a dramatic darkening or lightening of the body color. The fact that injection of eyestalk extracts could reverse the color changes induced by eyestalk ablation indicated that the regulation of body color is under hormonal control. Early observations indicated that the SG was a storage site for neurosecretory material that regulated color change (chromatophorotropins). However, the chromatophorotropins are located not only in the XO/SG complex, but in other neural tissues as well.

There are two general classes of pigmentary effector hormones: the red pigment-concentrating hormone (RPCH) and the pigment-dispersing hormone (PDH) families. Both groups of hormones are neuropeptides. The first crustacean chromatophorotropin to be sequenced was RPCH from the eyestalks of the shrimp, *Pandalus borealis*. In addition to the red pigment chromatophores, however, other types of chromatophores are concentrated with RPCH.

21.2.5.2 Retinal Pigments

There are three retinal pigments in the compound eyes of shrimp and other decapod crustaceans. This distal retinal pigment is found within cells that extend from the distal ends of the retinular cells to the cornea. The proximal retinal pigment is contained within the retinular cells. The reflecting retinal pigment is contained within cells that are variously located between the neighboring ommatidia, depending upon the species.

PDH was first isolated and its structure determined from the eyestalks of *P. borealis*. It was initially referred to as the distal retinal pigment light-adapting hormone because of its action on the distal retinal pigment. Later studies demonstrated that this molecule was able to disperse the pigment granules in several types of chromatophores. This hormone is now called PDH. Several additional members of the PDH family have been isolated, and information has been obtained on the molecular organization of PDH genes.

21.3   Neurotransmitters and Behavior

Neurotransmitters are usually not considered as true hormones. Although they are produced by specific cells, are secreted into the extracellular milieu, and are bound by specific receptors on target cells, they are generally not released into the circulating medium and hence have very localized effects. However, since they are indeed intraorganismal chemical messengers that do affect behavior, I will briefly mention them in this chapter.

After injecting lobsters with the neurotransmitter serotonin (5-hydroxytryptamine; 5-HT), Kravitz et al. (1980) and Livingstone et al. (1980) observed that the animals assumed an apparent dominant posture, which included raised and spread claws. Injection of another neurotransmitter, octopamine, resulted in an opposite effect. Some conflicting observations have been made following 5-HT injection (Peeke et al. 2000; Tierney and Mangiamele 2001; Panksepp and Huber 2002). These postures and behaviors are undoubtedly under highly complex regulation and the dose and method of application of 5-HT likely affect the observed outcomes.

As described in Chaps. 12 (Aggio and Derby) and 13 (Breithaupt), substances in crustacean urine can act as pheromones by signaling dominance (Breithaupt and Atema 2000; Huber et al. 1997; Sneddon et al. 2000; Moore and Bergman 2005). The 5-HT and octopamine could certainly have pleiotropic functions by serving as both neurotransmitters and pheromones.

21.4   Can Crustacean Hormones Act as Pheromones?

The hypothesis that hormones can serve as pheromones has been demonstrated in only a few instances in the animal kingdom. Perhaps, the best known examples are from fish (Stacey et al. 2003; Sorensen et al. 2005; see Chap. 24). Compared to fish, relatively little evidence exists for a pheromonal role of crustacean hormones.

Ecdysteroids are not only mediators of molting, but, as described above, also act as gonadotropins (for review, see Chang and Kaufman 2005). Since molting is usually a prerequisite for insemination in many female decapods, it is not unreasonable to hypothesize that ecdysteroids may act as mating pheromones. There are some early papers in support of this hypothesis. A premating stance is typically displayed by male crabs in the presence of chemical cues released by a premolt female. This stance involves elevation of the cephalothorax with the anterior margin tilted up. Males walk on the tips of their dactyls of the first three pairs of walking legs with the fourth pair extended backwards. The chelipeds are partially extended in a lowered position (Kittredge et al. 1971). In the presence of 20E, male crabs (*P. crassipes*) displayed the premating stance. Kittredge et al. (1971) published a dose-response curve ranging from 10^{-13} to 10^{-5} M. They observed that lower doses of 20E resulted in longer reaction times before the premating stance was displayed. These authors hypothesized that 20E was released by premolt female crabs (presumably in their urine) and initiated mating behavior.

The proposal that pheromones are released through crustacean urine is supported by experiments that are reviewed in several chapters in this volume. There is evidence that urine contains ecdysteroids (Seifert 1982) and that their amounts vary during the molt cycle. Our laboratory (Snyder and Chang 1991b, c) observed that the urinary concentrations of ecdysteroids varied during the molt cycle with the highest levels in late premolt, corresponding to the dramatic decline in hemolymph ecdysteroids that occurs immediately prior to ecdysis.

There have been challenges, however, to the proposal that ecdysteroids are in fact pheromones (reviewed by Dunham 1978). Seifert (1982) was unable to demonstrate any pheromonal activity of various tested ecdysteroids on *C. maenas*. Gagosian and Atema (1973) (Atema and Gagosian 1973) were unable to demonstrate mating behavior when *H. americanus* were bioassayed with 20E or ecdysteroid metabolites. These latter experiments indicated the initiation of alert responses to some of the compounds tested. No mating responses were observed when blue crab *Callinectes sapidus* were tested with 20E (Gleeson et al. 1984; see Chap. 20). Perhaps, the major difficulty in proposing a pheromonal role of ecdysteroids is the lack of species specificity. Since all arthropods examined so far synthesize, secrete, and use ecdysteroids for molting, it is difficult to envision how an ecdysteroid acting as a pheromone can be species specific. One possibility is that only a very limited number of species evolved receptors capable of binding extraorganismal ecdysteroids.

The hypothesis that ecdysteroids have pheromonal activity has recently been revisited (see Hardege and Terschak, Chap. 19). Hayden et al. (2007) observed that male green crabs had decreased feeding responses during the summer reproductive season. During these months, postmolt females are soft-shelled. Males are observed to decrease their foraging activities and are less likely to cannibalize these postmolt females. The application of exogenous 20E deterred male crabs from feeding on bivalve prey items. Female crabs were not deterred. Thus, it appears that 20E may act as a sex-specific feeding deterrent pheromone. Although postmolt female crabs have low circulating ecdysteroid concentrations, presumably there are sufficient quantities excreted via the urine or from exoskeletal pores that can be perceived by males in very close proximity (as in the mating embrace) (see Chap. 19).

A recent study indicates a role for 20E as a chemical mediator of aggressive interactions in premolt lobsters (Coglianese et al. 2008). When physiologically relevant concentrations of 20E were released near the antennules of premolt lobsters, their level of aggression increased relative to controls. The controls consisted of the release of the prohormone E, artificial seawater, or the use of lobsters that could not smell. Presumably, the increased aggression will chase off other lobsters and will ensure that the premolt lobster will deter a subsequent physical confrontation between its future postmolt, defenseless self, and any nearby cannibalistic conspecifics. When 20E is puffed over excised lateral antennual flagellae, a dose-dependent response is observed in the olfactory receptor neurons of female lobsters (Cromarty, personal communication).

As discussed for fish by Chung-Davidson, Huertas, and Li (Chap. 24), androgens can also serve as pheromones. A possible example of this phenomenon in crustaceans has been reported (Barki et al. 2003). When AGs (see Sect. 21.2.4.1) were implanted into young female crayfish, the recipients developed male secondary and tertiary sexual characteristics as expected (Chang and Sagi 2008). Normal females reacted to these AG-implanted females as if the latter animals were males. These normal females displayed copulatory behavior, an observation that implies that the implanted female was producing stimulatory signals characteristic of a male crayfish. One conclusion is that the implanted AG was responsible for the production and release of a male pheromone whose targets are females. Alternatively, the AG hormone itself may have been the responsible factor.

21.5   Ecdysteroids as Allelochemicals

As described above, the primary function of ecdysteroids is to mediate the molting process. Recent work demonstrates that molting hormones have been additionally used by pycnogonids (sea spiders) as defensive compounds against crustacean predators (reviewed in Hoffmann et al. 2006). Pycnogonids are not crustaceans; they are related chelicerate marine arthropods. Eight different ecdysteroids have been isolated from the pycnogonid *Pycnogonum litorale* (Bückmann et al. 1986). Many of these ecdysteroids exist as acetate or glycolate conjugates. These conjugates may affect metabolism of the ecdysteroids and/or their biological activities within the host pycnogonid. These combined ecdysteroids are found in very high concentrations in *P. litorale* (Tomaschko and Bückmann 1993) – as much as two to three orders of magnitude higher than in other arthropods.

By means of feeding choice experiments, Tomaschko (1994b) demonstrated that crabs were deterred from eating either powdered extracts of pycnogonids or food pellets containing high levels of ecdysteroids. Since all arthropods to date have been shown to be sensitive to the molt-promoting effects of ecdysteroids, *P. litorale* must have a mechanism to isolate their allelochemical compounds from their circulating hemolymph. *P. litorale* sequesters its defensive ecdysteroids in epidermal glands. These glands can be selectively stimulated to release their contents (Tomaschko 1994a). Ecdysteroids apparently act as antipredation compounds because potential crustacean predators of pycnogonids are susceptible to the molt-inducing effects of exogenously applied ecdysteroids. The effect of exogenously applied ecdysteroids is frequently an early entry into premolt followed by an unsuccessful ecdysis (Rao et al. 1972).

Just as some terrestrial plants have high concentrations of ecdysteroids that act as antifeeding allelochemicals against herbivorous insects, it will be interesting to determine whether marine algae have high levels of ecdysteroids acting as deterrents against herbivorous crustaceans. The presence of ecdysteroids appears not to be phylogenetically distributed, but rather seems to occur randomly throughout the plant kingdom (Lafont 1997). There is at least one example of ecdysteroid-like molecules in a red alga (Fukuzawa et al. 1986).

Yet, another example of the pleiotropic nature of ecdysteroids has been characterized in a mosquito (*Anopheles gambiae*; Pondeville et al. 2008). Males of this arthropod transfer significant amounts of 20E to females during mating. Since 20E is a female gonadotropin, the contribution of 20E from the male may act as an allohormone by increasing the female's yolk protein production and hence her fertility (an allohormone is a hormone that is transferred between organisms; Koene and ter Maat 2001).

21.6   Future Directions

The study of arthropods is of intense interest for basic research into comparative endocrinology. Active areas of research involve the evolutionary endocrinology of the MIH/CHH family of neuropeptides. Slight alterations in structure result in dramatic changes in function. These peptides evolved from a common ancestor to develop into an extremely multifunctional family of hormones.

Cultured and wild-caught crustaceans (especially shrimp) comprise one of the world's most valuable food commodities. As future protein resources become even scarcer, the intensification of aquaculture and fisheries activities is bound to increase. Research will become even more important in these areas (see Chap. 25). For example, manipulation of the MIH/ecdysteroid hormone axis is an obvious example of increasing growth rates. Attraction pheromones have the possibility of luring desired food species into traps. This mechanism could also be used for the control of invasive species pests (Hardege et al. 2002).

There is growing interest in the field of aquatic endocrine disrupters (deFur et al. 1999). Since much is known about the endocrinology of crustaceans, and since crustaceans are often keystone species in aquatic food webs, there will likely be much more research devoted to the determination of the effects of exogenous chemicals on these various hormonal processes. Crustaceans will be useful indicator species for environmental health.

I presented an overview of the crustacean endocrine system, which historically has focused on decapods. Of course, there are other important crustacean groups. Very little is known about the structures of their endocrines, not to mention their methods of chemical mediation. For both practical and basic interests, future comparative research will need to encompass these other crustacean groups and examine their individual variations of hormonal and pheromonal mechanisms.

21.7   Summary and Conclusion

Although much is known about the crustacean endocrine system, there remains a paucity of information concerning the use of hormones as other chemical messengers, such as pheromones or allelochemicals. Most of the work in this area (pleiotropic effects of crustacean hormones) has focused on the molting hormones (ecdysteroids). Due to the lack of species-specificity, it is unlikely that ecdysteroids are sex recognition pheromones. Strong evidence has been published, however, that ecdysteroids act as feeding deterrents, both within and between species. Ecdysteroids also appear to be modulators of aggressive behavior. Whether other crustacean hormones have such multifunctionality remains to be discovered.

Acknowledgments I thank Drs. Assaf Barki and Stuart Cromarty for helpful comments and Ms. Sharon Chang for editorial and laboratory assistance. I also thank my various mentors, collaborators, and students for their creative stimulants. This is contribution No. 2411 from the Bodega Marine Laboratory, University of California at Davis.

References

Aiken DE (1973) Proecdysis, setal development, and molt prediction in the American lobster. J Fish Res Board Can 30:1334–1337

Atema J, Gagosian R (1973) Behavioral responses of male lobsters to ecdysones. Mar Behav Physiol 2:15–20

Barki A, Karplus I, Khalaila I, Manor R, Sagi A (2003) Male-like behavioral patterns and physiological alterations induced by androgenic gland implantation in female crayfish. J Exp Biol 206:1791–1797

Böcking D, Dircksen H, Keller R (2002) The crustacean neuropeptides of the CHH/MIH/GIH family: structures and biological activities. In: Wiese K (ed) The crustacean nervous system. Springer, Berlin, pp 84–97

Bolingbroke M, Kass-Simon G (2001) 20-hydroxyecdysone causes increased aggressiveness in female American lobsters, *Homarus americanus*. Horm Behav 39:144–156

Breithaupt T, Atema J (2000) The timing of chemical signaling with urine in dominance fights of male lobsters (*Homarus americanus*). Behav Ecol Sociobiol 49:67–78

Bückmann D, Starnecker G, Tomaschko KH, Wilhelm E, Lafont R, Girault JP (1986) Isolation and identification of major ecdysteroids from the *Pycnogonum litorale* Ström (Arthropoda, Pantopoda). J Comp Physiol 156B:759–765

Chang ES (1989) Endocrine regulation of molting in Crustacea. Rev Aquat Sci 1:131–157

Chang ES (1993) Comparative endocrinology of molting and reproduction: insects and crustaceans. Annu Rev Entomol 38:161–180

Chang ES, Bruce MJ (1980) Ecdysteroid titers of juvenile lobsters following molt induction. J Exp Zool 214:157–160

Chang ES, Kaufman WR (2005) Endocrinology of Crustacea and Chelicerata. In: Gilbert LI, Iatrou K, Gill SS (eds) Comprehensive molecular insect science. Elsevier B.V, Oxford, pp 805–842

Chang ES, Keller R, Chang SA (1998) Quantification of crustacean hyperglycemic hormone by ELISA in hemolymph of the lobster, *Homarus americanus*, following various stresses. Gen Comp Endocrinol 111:359–366

Chang ES, Sagi A (2008) Male reproductive hormones. In: Mente E (ed) Reproductive biology of crustaceans. Science Publishers, Enfield, pp 299–317

Charniaux-Cotton H (1954) Découverte chez un Crustacé Amphipode (*Orchestia gammarella*) glande endocrine responsable de la différenciation de caractères sexuels prmaires et secondaires mâles. C R Acad Sci D 239:780–782

Chung JS, Webster SG (2005) Dynamics of in vivo release of molt-inhibiting hormone and crustacean hyperglycemic hormone in the shore crab, *Carcinus maenas*. Endocrinology 146:5545–5551

Coglianese DL, Cromarty SI, Kass-Simon G (2008) Perception of the steroid hormone 20-hydroxyecdysone modulates agonistic interactions in *Homarus americanus*. Anim Behav 75:2023–2034

Conklin DE, Chang ES (1983) Grow-out techniques for the American lobster *Homarus americanus*. In: McVey JP (ed) CRC handbook of mariculture. CRC Press, Boca Raton, pp 277–286

Cromarty SI, Kass-Simon G (1998) Differential effects of a molting hormone, 20-hydroxyecdysone, on the neuromuscular junctions of the claw opener and abdominal flexor muscles of the American lobster. Comp Biochem Physiol A 120:289–300

Cromarty SI, Mello J, Kass-Simon G (2000) Molt-related and size-dependent differences in the escape response and post-threat behavior of the American lobster, *Homarus americanus*. Biol Bull 199:265–277

deFur PL, Crane M, Ingersoll C, Tattersfield L (1999) Endocrine disruption in invertebrates: endocrinology, testing, and assessment. SETAC Press, Pensacola

Dunham P (1978) Sex pheromones in Crustacea. Biol Rev 53:555–583

Fukuzawa A, Miyamoto M, Kumagai Y, Masamune T (1986) Ecdysone-like metabolites, 14[α]-hydroxypinnasterols, from the red alga *Laurencia pinnata*. Phytochemistry 25:1305–1307

Gagosian R, Atema J (1973) Behavioral responses of male lobsters to ecdysone metabolites. Mar Behav Physiol 2:115–120

Gleeson RA, Adams MA, Smith AB (1984) Characterization of a sex pheromone in the blue crab, *Callinectes sapidus*: crustecdysone studies. J Chem Ecol 10:913–921

Hardege JD, Jennings A, Hayden D, Muller CT, Pascoe D, Bentley MG, Clare AS (2002) Novel behavioural assay and partial purification of a female-derived sex pheromone in *Carcinus maenas*. Mar Ecol Progr Ser 244:179–189

Hayden D, Jennings A, Müller C, Pascoe D, Bublitz R, Webb H, Breithaupt T, Watkins L, Hardege JD (2007) Sex-specific mediation of foraging in the shore crab, *Carcinus maenas*. Horm Behav 52:162–168

Hoffmann KH, Dettner K, Tomaschko K-H (2006) Chemical signals in insects and other arthropods: from molecular structure to physiological functions. Physiol Biochem Zool 79:344–356

Homola E, Chang ES (1997) Methyl farnesoate: crustacean juvenile hormone in search of functions. Comp Biochem Physiol B 117:347–356

Horn DHS, Middleton EJ, Wunderlich JA, Hampshire F (1966) Identity of the moulting hormones of insects and crustaceans. Chem Commun 1966:339–340

Huber R, Orzeszyna M, Pokorny N, Kravitz EA (1997) Biogenic amines and aggression: experimental approaches in crustaceans. Brain Behav Evol 50:60–68

Kittredge J, Terry M, Takahashi F (1971) Sex pheromone activity of the molting hormone, crustecdysone, on male crabs. Fish Bull 69:337–343

Koene JM, ter Maat A (2001) "Allohormones": a class of bioactive substances favoured by sexual selection. J Comp Physiol A 187:323–326

Kravitz EA, Glusman S, Harris-Warrick RM, Livingstone MS, Schwarz T, Goy MF (1980) Amines and a peptide as neurohormones in lobsters: actions on neuromuscular preparations and preliminary behavioural studies. J Exp Biol 89:159–175

Lafont R (1997) Ecdysteroids and related molecules in animals and plants. Arch Insect Biochem Physiol 35:3–20

Laufer H, Biggers WJ (2001) Unifying concepts learned from methyl farnesoate for invertebrate reproduction and post-embryonic development. Am Zool 41:442–457

Livingstone M, Harris-Warrick R, Kravitz EA (1980) Serotonin and octopamine produce opposite postures in lobsters. Science 208:76–79

Manor R, Weil S, Oren S, Glazer L, Aflalo ED, Ventura T, Chalifa-Caspi V, Lapidot M, Sagi A (2007) Insulin and gender: an insulin-like gene expressed exclusively in the androgenic gland of the male crayfish. Gen Comp Endocrinol 150:326–336

Moore PA, Bergman DA (2005) The smell of success and failure: the role of intrinsic and extrinsic chemical signals on the social behavior of crayfish. Integr Comp Biol 45:650–657

Okazaki RK, Chang ES (1991) Ecdysteroids in the embryos and sera of the crab. *Cancer magister* and *C. anthonyi*. Gen Comp Endocrinol 81:174–186

Panksepp JB, Huber R (2002) Chronic alterations in serotonin function: dynamic neurochemical properties in agonistic behavior of the crayfish, *Orconectes rusticus*. J Neurobiol 50:276–290

Peeke HVS, Blank GS, Figler MH, Chang ES (2000) Effects of exogenous serotonin on a motor behavior and shelter competition in juvenile lobsters (*Homarus americanus*). J Comp Physiol A 186:575–582

Pondeville I, Maria A, Jacques J-C, Bourgouin C, Dauphin-Villemant C (2008) *Anopheles gambiae* males produce and transfer the vitellogenic steroid hormone 20-hydroxyecdysone to females during mating. Proc Natl Acad Sci USA 105:19631–19636

Rao KR (2001) Crustacean pigmentary-effector hormones: chemistry and functions of RPCH, PDH, and related peptides. Am Zool 41:364–379

Rao KR, Fingerman M, Hays C (1972) Comparison of the abilities of α-ecdysone and 20-hydroxyecdysone to induce precocious proecdysis and ecdysis in the fiddler crab *Uca pugilator*. Z Vergl Physiol 76:270–284

Sagi A, Khalaila I (2001) The crustacean androgen: a hormone in an isopod and androgenic activity in decapods. Am Zool 41:477–484

Schwanke ML, Cobb JS, Kass-Simon G (1990) Synaptic plasticity and humoral modulation of neuromuscular transmission in the lobster claw opener during the molt cycle. Comp Biochem Physiol C 97:143–149

Seifert P (1982) Studies on the sex pheromone of the shore crab, *Carcinus maenas*, with special regard to ecdysone excretion. Ophelia 21:147–158

Smith SG, Chang ES (2007) Molting and growth. In: Kennedy VS, Cronin LE (eds) The blue crab, *Callinectes sapidus*. Maryland Sea Grant, Baltimore, pp 197–254

Sneddon LU, Taylor AC, Huntingford FA, Watson DG (2000) Agonistic behaviour and biogenic amines in shore crabs *Carcinus maenas*. J Exp Biol 203:537–545

Snyder MJ, Chang ES (1991a) Ecdysteroids in relation to the molt cycle of the American lobster. *Homarus americanus*. I. Hemolymph titers and metabolites. Gen Comp Endocrinol 81:133–145

Snyder MJ, Chang ES (1991b) Ecdysteroids in relation to the molt cycle of the American lobster. *Homarus americanus*. II. Excretion of metabolites. Gen Comp Endocrinol 83:118–131

Snyder MJ, Chang ES (1991c) Metabolism and excretion of injected [^{3}H]-ecdysone by female lobsters, *Homarus americanus*. Biol Bull 180:475–484

Sorensen PW, Pinillos M, Scott AP (2005) Sexually mature male goldfish release large quantities of androstenedione into the water where it functions as a pheromone. Gen Comp Endocrinol 140:164–175

Stacey N, Chojnacki A, Narayanan A, Cole T, Murphy C (2003) Hormonally derived sex pheromones in fish: exogenous cues and signals from gonad to brain. Can J Physiol Pharmacol 81:329–341

Subramoniam T (2000) Crustacean ecdysteroids in reproduction and embryogenesis. Comp Biochem Physiol C 125:135–156

Tierney AJ, Mangiamele LA (2001) Effects of serotonin and serotonin analogs on posture and agonistic behavior in crayfish. J Comp Physiol A 187:757–767

Tomaschko K-H (1994a) Defensive secretion of ecdysteroids in *Pycnogonum litorale* (Arthropoda, Pantopoda). Z Naturforsch 49c:367–371

Tomaschko K-H (1994b) Ecdysteroids from *Pycnogonum litorale* (Arthropoda, Pantopoda) act as chemical defense against *Carcinus maenas* (Crustacea, Decapoda). J Chem Ecol 20:1445–1455

Tomaschko K-H, Bückmann D (1993) Excessive abundance and dynamics of unusual ecdysteroids in *Pycnogonum litorale* Strom (Arthropoda, Pantopoda). Gen Comp Endocrinol 90:296–305

Webster SG (1998) Neuropeptides inhibiting growth and reproduction in crustaceans. In: Coast GM, Webster SG (eds) Recent advances in arthropod endocrinology. Cambridge University Press, Cambridge, pp 33–52

Chapter 22
Toward a Characterization of the Chemical Cue to Barnacle Gregariousness

Anthony S. Clare

Abstract Many barnacle species are gregarious. This is an essential behavior for those species that can only reproduce by mating with a neighboring barnacle. Proximity of adult barnacles is achieved by gregarious settlement of the cypris larva. The chemical basis of this behavior was established 60 years ago, but attempts to characterize the cue to settlement met with limited success. This chapter presents evidence obtained in recent years that the cue is an α2-macroglobulin-like cuticular protein, detected by cyprids using a tactile chemical sense as they explore the substratum for a suitable settlement site.

22.1 Introduction

This review will be concerned with an aspect of chemical communication in the marine environment that has long intrigued researchers, namely, how adult benthic invertebrates signal their presence to conspecific larvae during settlement to the substratum (reviewed by Crisp 1984). Most benthic marine invertebrates display a complex lifecycle involving a planktonic larval phase, which may last from minutes to months (during which dispersal occurs), and a benthic juvenile–adult phase (Thorson 1950; Pawlik 1992). The transition from a planktonic to a benthic mode of life is generally accepted as a critical point in the lifecycle (e.g., Raimondi 1988) and fundamental to understanding population and community dynamics (e.g., Connell 1985), and evolution of species (e.g., Dreanno et al. 2006a).

On a relatively large scale ($>$ meters), most larval forms can be viewed as essentially passive particles that are delivered to the substratum by hydrodynamic forces (e.g., Eckman 1990; Abelson and Denny 1997). Even so, we have long known (see Young 1990 for an excellent historical account of marine larval research) that settlement for a number of species is not a chance event. Crucially, larvae may delay

A.S. Clare (✉)
School of Marine Science and Technology, Newcastle University,
Newcastle upon Tyne NE1 7RU, UK
e-mail: a.s.clare@ncl.ac.uk

their settlement (e.g., Wilson 1932; Gebauer et al. 2003), rejecting unfavorable surfaces (Krug 2006). Moreover, as first noted by Thompson (1830) and more recently by numerous researchers (e.g., Abelson and Denny 1997; Jenkins 2005), presettlement behavior can influence larval distribution. Nonetheless, it is a remarkable feat of nature that these comparatively small organisms are able to locate/relocate a substratum that can support their continued existence, even though many species are generalists in their requirements for an adult habitat (Elkin and Marshall 2007). It is even more extraordinary for those species that live in patchy and/or ephemeral habitats, such as hydrothermal vents, or which have stringent, specific requirements for a settlement site, such as species that settle on a future food source, or in gregarious, symbiotic or parasitic associations (reviewed by Pawlik 1992). Nowadays, it is generally accepted that passive transport and active (behavioral) selection of settlement sites are both likely to be involved but that they operate on different scales; the latter occurring at a much smaller scale of centimeters and below (e.g., Pawlik 1992; Koehl 2007).

My own personal inspiration for working on larval settlement was reading, as an undergraduate, the pioneering works of the British researchers Barnes, Crisp, Knight-Jones, and Wilson (see Young 1990 for a review). Much of this research concerned fouling species, in particular barnacles. I was later fortunate to do my graduate studies in the NERC Institute for Marine Invertebrate Biology, while Crisp was the Director. Here several investigators, notably Gabbott, Nott, and Walker (my PhD supervisor), and their research teams, were engaged in research on barnacle larval settlement. While carrying out research on the barnacle egg-hatching pheromone, I witnessed the latter stages of research on the chemical basis of barnacle gregariousness (Gabbott and Larman 1987); a topic that I was eventually to turn to myself.

It is not my intention in this review to give a detailed account of barnacle settlement, as there are several excellent reviews on this topic (Lewis 1978; Crisp 1979, 1984, 1985; Bourget 1988; Walker 1995; Okano and Fusetani 1997), as well as recent contributions from my laboratory (Aldred and Clare 2008; Clare and Aldred 2009). Rather, I intend to focus on recent research on the chemical basis of gregariousness (research up to about 2000 is covered in reviews by Crisp 1984; Gabbott and Larman 1987; Pawlik 1992; Clare 1995; Clare and Matsumura 2000; Hadfield and Paul 2001), particularly on studies that have aimed at identifying an adult glycoprotein that induces settlement of conspecific cyprids (one of only a handful of marine invertebrate settlement cues for which a characterization has been claimed; see Steinberg et al. 2002; Hung et al. 2009) and elucidating its mechanism of action. That said, a brief review of this topic will be presented in the next section to place our own work in context.

22.2   The Chemical Basis of Barnacle Gregariousness

Before discussing the importance of chemical cues to larval settlement, it is first necessary to define the latter term. As noted by Hadfield and Paul (2001), the terms settlement and metamorphosis are usually considered as separate behavioral and developmental processes, respectively. In the context of this review, however, settlement refers to the entire pelago-benthic transition. Hence, settlement for barnacles (Clare and Matsumura 2000) comprises the exploratory phase, involving temporary attachment to the substratum (Aldred and Clare 2008; Clare and Aldred 2009) – a reversible process – during which the settlement-stage cypris larva (Fig. 22.1) senses the physical and chemical characteristics of the substratum; the commitment to permanent attachment, involving the secretion of cyprid cement (Walker 1981); and metamorphosis to the juvenile. This definition conforms to the usage of Pawlik (1992) for marine invertebrate larval settlement in general. It is preferred here because precocious (prior to permanent attachment) metamorphosis can occasionally be observed as the appearance of primordial shell valves within cyprids that have been "aged" by refrigeration at 4–6°C (Crisp 1988) or within plankton netting at 25°C (Clare and Matsumura 2000). There is good evidence to suggest that the cues to attachment and metamorphosis of barnacle cyprids are different (Crisp 1984; Clare and Matsumura 2000), although there may well be interactions between them. Precocious metamorphosis can be induced by phorbol esters (Yamamoto et al. 1995), the neurotransmitter dopamine, albeit with conflicting results (see Zega et al. 2007) and pharmacological concentrations of methyl farnesoate, the crustacean juvenile hormone (Smith et al. 2000). Permanent attachment is induced by settlement pheromones (reviewed by Clare and Matsumura 2000), comprising the adult glycoprotein (the principal subject of this review), adult waterborne pheromone (Rittschof 1985; Clare and Yamazaki 2000) and cyprid temporary adhesive (Walker and Yule 1984; Yule and Walker 1985), and signal

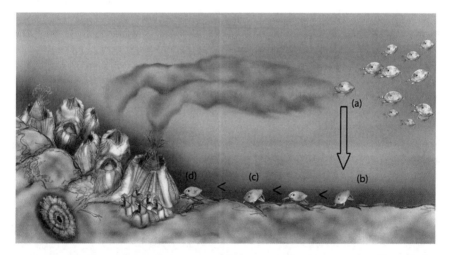

Fig. 22.1 Gregarious settlement of barnacle cypris larvae. (**a**) Cyprids approach the substratum, perhaps after encountering a waterborne cue released by adults; (**b**) contact with substratum and onset of (**c**) searching behavior; (**d**) cyprid contacts an adult conspecific and is stimulated to settle by a cuticular protein – the settlement-inducing protein complex (SIPC). The cyprid may return to the plankton at any stage of the sequence (**b–d**). Drawing by Jorge A. Varela Ramos

transduction molecules, such as cyclic AMP, that are believed to act downstream of settlement pheromones (Clare and Matsumura 2000).

Marine invertebrate larvae are exposed to a multitude of potential cues at settlement including local hydrodynamics, surface roughness, thermal capacity, color, and chemistry (see reviews by Pawlik 1992; Abelson and Denny 1997; Krug 2006; Koehl 2007 and references therein). For many species, biogenic chemical cues, such as those associated with biofilm and other incumbents, including conspecifics, are the primary modulators of larval settlement behavior (see, e.g., reviews by Pawlik 1992; Hadfield and Paul 2001; Steinberg et al. 2002; Krug 2006 and references therein). Barnacles have featured prominently in research that has attempted to understand the relative importance of physical and chemical cues to temporal and spatial variability in larval settlement (e.g., Crisp 1984 and references therein; Larsson and Jonsson 2006). Here, as in marine invertebrate larvae in general, the importance of biogenic chemical cues has gained prominence. In part, this attention to barnacle settlement reflects the economic importance of certain barnacles as fouling species (Christie and Dalley 1987), their abundance, accessibility (Foster 1987), and amenability to both laboratory and field research. Two species in particular, the boreo-arctic *Semibalanus balanoides* (=*Balanus balanoides*) and the subtropical/tropical *Balanus amphitrite* (=*Amphibalanus amphitrite*) have served as important model species (Clare 1996; Jenkins et al. 2000), not least in relation to studies on the chemical basis of barnacle gregariousness.

For those species in which cross-fertilization by the hermaphroditic adults is obligatory (e.g., Walker 1980), there is a requirement for proximity of the mating individuals, at least within a penis length of the nearest neighboring conspecific (e.g., Crisp 1979). While the distance between barnacles will obviously decrease as they grow, gregariousness, in those barnacle species that display this behavior, is achieved at settlement. Gregariousness is so remarkable in barnacles that some early researchers doubted the involvement of a pelagic phase (see Crisp 1979). This behavior has been noted for many species of benthic marine invertebrates (Toonen and Pawlik 1994) and affords many advantages (Pawlik 1992). Apart from enhanced reproductive success, perhaps the most notable advantage afforded by group living is a lower risk from predation through a dilution effect (Mauck and Harkless 2001), but at a cost of increased intraspecific competition, for example, for space and food. For one species, *S. balanoides*, however, there is evidence that living in aggregations may actually enhance foraging success after encounters with predators, as barnacles living in groups responded to signals from active neighbors and emerged from their shells more quickly than did solitary barnacles (Mauck and Harkless 2001).

The phenomenon of gregariousness was first noted for the barnacles *Elminius modestus*, *S. balanoides* and *B. crenatus* by Knight-Jones and Stevenson (1950) and Knight Jones (1953). In the latter now classic paper, the first evidence was presented that a chemical cue, associated with the adult, is recognized by the cyprid at settlement. Moreover, a preliminary characterization of the cue indicated that it is a cuticular protein. Subsequent papers from Crisp's laboratory at Menai Bridge, largely focused on *S. balanoides* (reviewed by Clare and Matsumura 2000), provided further evidence that the cue is a water-soluble glycoprotein associated with the cuticle of adults and only active when isolated after surface absorption. It was assumed that the glycoprotein conformation in its bound state resembled that of the cuticular protein in situ (Crisp and Meadows 1963). Significantly, Knight-Jones (1953) provided evidence that the chemical cue was detected by the cyprid making contact with the adult and not using an olfactory sense. Crisp and Meadows (1963) later coined the term "tactile chemical sense," which is perhaps akin to that discussed by Snell (Chap. 23) for chemoreception of glycoproteins. Crisp and Meadows (1962) coined the term "arthropodin" for the physico-chemical similarity of the cue to insect cuticular protein of the same name (Fraenkel and Rudall 1940). A series of papers by Gabbott and Larman spanning a decade (reviewed by Gabbott and Larman 1987), described a partial characterization of the arthropodin of *S. balanoides*. The cue was found to be a polymorphic system of closely related proteins with a subunit composition of 5–6 and 18 kDa, but whose most stable subunit in boiled extracts had a mass of 23.5–25 kDa. Periodic acid-Schiff staining of native gels and carbohydrate analysis confirmed that arthropodin is a glycoprotein.

Seven years after the last of the papers on *S. balanoides* arthropodin (Larman 1984), researchers in the Fusetani Biofouling Project (1991–1996) in Japan began work on barnacle settlement. One of the species that they chose, *B. amphitrite*, proved particularly tractable for studying cyprid settlement cues. This species, which breeds several times a year, is relatively easy to culture throughout the year, and settlement occurs under static conditions, making it ideal for in vitro settlement assays (see Rittschof et al. 1984). Indeed, a new settlement assay – the nitrocellulose membrane assay (NMA) (Matsumura et al. 1998b) – proved instrumental to isolating the cue to gregarious settlement of *B. amphitrite* and subsequently other species. The NMA assay, utilizing the favorable protein-binding characteristics of the membrane, affords greater control of protein absorption than natural slate (a complex mineral) used by Crisp and colleagues. In a departure from the protocol developed for *S. balanoides* arthropodin, Matsumura et al. (1998a) did not use boiled seawater extracts of crushed barnacles as the starting material for cue isolation; unboiled extracts were shown to be more suitable. Moreover, unlike previous studies, they used a bioassay-directed isolation strategy (Fig. 22.2). It is perhaps not surprising, therefore, that the three major subunits (76-, 88-, and 98-kDa molecular mass) of the *B. amphitrite* settlement-inducing glycoprotein bore no resemblance to the pattern detected for *S. balanoides*. Matsumura et al. (1998a) termed this inductive cue the settlement-inducing protein complex (SIPC)[1].

---

[1]The reader of the literature on the chemical basis of barnacle gregariousness could be forgiven for feeling confused. The nomenclature used for settlement cues alone is enough to confound (Clare and Matsumura 2000). Additionally, different approaches and methodology make comparisons between studies difficult.

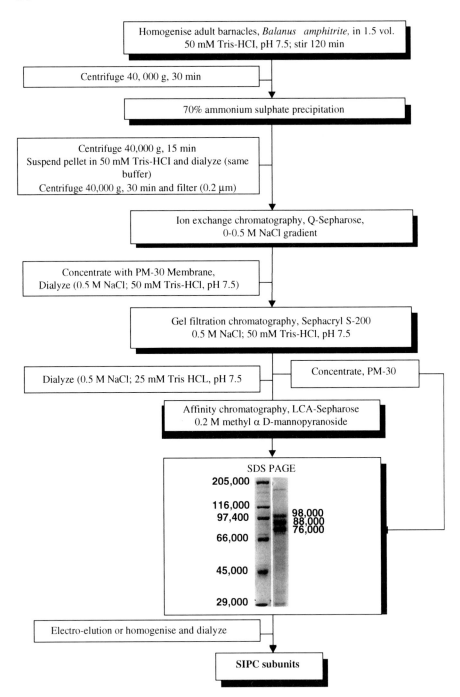

Fig. 22.2 Summary of the extraction procedure for the SIPC. Bands for molecular weight markers on sodium dodecyl sulfate polyacrylamide gel electrophoresis (SDS-PAGE) are to the *left*, and to the *right*, SIPC subunits. After Clare and Matsumura (2000) reproduced by permission of Taylor & Francis Ltd. (www.informaworld.com)

Fig. 22.3 Choice assay used by Crisp to compare cyprid preferences for different substrata. Slate was used as a settlement substratum arranged in glass dishes set on a rotating table. The rotation mitigated against bias from external factors such as light. The air lines were used to impart a flow of seawater in the dishes

22.3   Identity of the SIPC

As already mentioned, the development of the NMA was crucial to progress in identifying the SIPC from adult tissues (Matsumura et al. 1998b). This assay is similar in some respects to the choice assay used by Crisp and Meadows (1962) (Fig. 22.3). Unlike Crisp's assay, however, the NMA used *B. amphitrite* cyprids of known age from laboratory cultures (although Crisp (1990) was later to use his assay with *B. amphitrite* cyprids); the previous focus on *S. balanoides*, which is very difficult to culture to the cyprid, did not allow this degree of control as Crisp later noted (Crisp 1990). Cyprid age has a profound effect on settlement (e.g., Rittschof et al. 1984), which is likely to reflect the physiological condition of the larvae (Miron et al. 2000) rather than their temporal age.

The NMA was used to isolate the three major subunits of the SIPC of *B. amphitrite* for amino acid sequencing by automated Edman degradation (Dreanno et al. 2006a). Degenerate oligonucleotide primers to the 76-kDa subunit were the starting point for successful cloning and sequencing of the full-length SIPC cDNA (Dreanno et al. 2006a). The sequence comprised 5,202 bp. The full-length open-reading frame encoded a protein of 1,547 amino acids. Excluding the 17-residue signal peptide, potential posttranslational modification, and taking account of the estimated 15% carbohydrate content (based on the molecular mass shift following deglycolization by N-glycosidases), the molecular mass of the SIPC is ~200 kDa.

Fig. 22.4 Unrooted phylogenetic tree showing the relationship of the SIPC of *Balanus amphitrite* to the thioester-containing proteins (TEPs) including α2-macroglobulin (A2M). CF indicates the branch leading to the complement factor proteins. The full tree comprising A2M, the SIPC, complement factor, and other TEPs is available as a supplementary figure in Dreanno et al. (2006a). Numbers by major nodes represent bootstrap values for 100 replicates. The scale bar indicates the length of each branch. After Dreanno et al. (2006a)

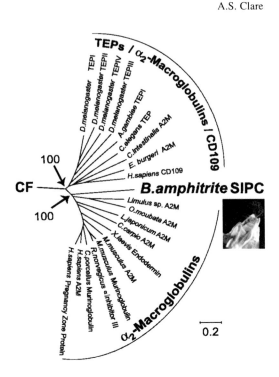

Matsumura et al. (1998a) had previously estimated the molecular mass of the SIPC as ~200–400 kDa by gel filtration, suggesting the possibility that the SIPC is a dimer. Database searches revealed that the SIPC is a previously undescribed protein that shares an overall 25% sequence homology with the thioester-containing family of proteins (TEPs), which includes α$_2$-macroglobulin (A2M). The closest homology was to the A2Ms of the tick, *Ornothodoros moubata* (31%) and the horseshoe crab, *Limulus* sp. (29%) (Fig. 22.4). Available evidence suggests, however, that the *B. amphitrite* SIPC is not an A2M. First, the SIPC does not contain the unique thioester signature sequence – GCGEQ – that defines most TEPs (e.g., Armstrong and Quigley 1999). Second, putative *B. amphitrite* A2M sequences were obtained that *did* contain a thioester region (Dreanno et al. 2006a).

22.4 Origin and Ecological Relevance of the Adult Settlement Pheromone (SIPC)

Further evidence to support the original hypothesis of Knight-Jones (1953) that the SIPC is a cuticular protein was obtained recently (Dreanno et al. 2006b). Polyclonal antibodies raised against peptide sequences from the C-terminus and toward the N-terminus of the SIPC were used to localize the glycoprotein to all tissues lined by cuticle. SIPC RNA expression was colocalized to the cuticle using in situ hybridization. A precise localization could not be achieved at the light microscope level, but it seems reasonable to assume that the epidermis that underlies and secretes the cuticle is the source of the pheromone. If we accept that the SIPC is detected by the cyprid using contact chemoreception, we are naturally led to ask how this is achieved given that the adult barnacle body is enclosed within a calcareous shell. A possible explanation is that the cyprid responds to the SIPC in the epicuticular layer on the outer surface of the shell (Bourget 1977) (see Fig. 22.1).

Using the NMA, an effective concentration of the SIPC required to induce 50% settlement (EC_{50}) of cypris larvae was determined as 100 ng per 0.8 cm^2 (the area of each spot of adsorbed SIPC on the membrane) (Dreanno et al. 2006a; Dreanno et al. 2007); approximately 100-fold lower than estimated previously for crude barnacle extract (Matsumura et al. 1998b). Back-of-the-envelope calculations indicate that the surface concentration of the SIPC, if it is expressed entirely at the surface, is approximately two orders of magnitude higher than the EC_{50} value (Dreanno et al. 2006a). Of course, not all of the SIPC is expressed at the surface, but these rough calculations indicate that the SIPC is active in laboratory assays at ecologically realistic concentrations.

There is considerable evidence from field experiments that extracts of adults that contain the SIPC are able to induce gregarious settlement of cyprids. For example, settlement of *Chthamalus anisopoma* onto Perspex panels could be elevated to 2 m above the normal level on the shore by applying adult extract to the panels (Raimondi 1988). More recently, analysis of cyprid tracks showed that more cyprids explored adult extract-treated surfaces and that they did so for longer than on untreated control surfaces (Matsumura et al. 2000; Prendergast et al. 2008). These results are taken to indicate that adult extract-treated surfaces are more favorable for settlement; a conclusion borne out by the results of other field studies (Prendergast 2007). When reflecting on the ecological relevance of these experiments it is worth noting that there is no consistency between studies in the methods used to prepare adult extracts. The concentration of the SIPC in the extracts is not determined and the adsorbed concentration cannot be controlled on the artificial (e.g., resin tiles) and natural surfaces (e.g., slate) that are commonly used.

As most of the barnacle species that have been studied are obligate cross fertilizers (Barnes and Crisp 1956), settlement of cyprids next to adults of these species is evidently adaptive in terms of future reproductive success (Crisp 1979). It is reasonable to assume that settlement near to recently settled cyprids and juveniles would also be adaptive. A prediction of the latter hypothesis is that expression of the SIPC should not be confined to the adult stage. Indeed, according to Crisp and Meadows (1962), recently settled juvenile barnacles are at least as attractive as adult barnacles to conspecific cyprids. Moreover, Hills et al. (1998) found that extracts of cyprids and adults were equally effective at inducing conspecific cyprid settlement. In contrast, Jeffery (2002) reported that newly settled *Chamaesipho tasmanica* did not induce gregarious settlement of conspecific cyprids, although induction did occur as the recruits aged. Likewise cyprids of both *E. modestus* (=*Austrominius modestus*) (Larman and Gabbott 1975) and *S. balanoides* (Wethey 1984) were reported to avoid conspecific juveniles. Such differences are perhaps to be expected, however, as many factors may act to confound the results of field experiments (see, e.g., Prendergast et al. 2009). For example, as cyprids age in the plankton their discriminatory ability wanes (e.g., Jarrett 1997) and the likelihood of settlement in the absence of adult conspecifics, i.e., pioneer colonizers (these cyprids are arguably the target for antifouling measures) increases. Only when larvae are cultured in the laboratory can any degree of control be exercised over larval age. Whether or not settlement-inducing activity is confined to the adult, the SIPC is expressed throughout the lifecycle (Dreanno et al. 2006b).

Studies of temporary adhesion of cypris larvae have provided good evidence for larva–larva interactions at settlement. Cyprids explore surfaces with their paired antennules (Fig. 22.1) and use a pair of attachment discs to adhere to the substratum. The hairy surface of the attachment discs (Nott 1969) and a secretion discharged via pores on the disc's surface are thought to effect adhesion in a hybrid wet/dry mechanism akin to that used by the fly pulvillus (Clare and Aldred 2009). Depending on the surface that is being explored, the cyprid deposits tracks of footprints that can be visualized by staining for protein with Coomassie Brilliant Blue (Walker and Yule 1984). Surfaces that are attractive to cyprids are explored more than unfavorable surfaces and acquire correspondingly more footprints. In so doing, the cyprids provide a positive stimulus to conspecifics that is detectable as a higher rate of settlement (Yule and Walker 1985; Clare et al. 1994). Walker and Yule (1984) drew attention to a possible relationship between the SIPC and the adhesive secretion, pointing out that they are both cuticular proteins (the antennular secretion is produced by modified integumental glands) that provide a cue to larval settlement. This relationship has since received support in the form of positive immunostaining of footprints with antibodies raised to the 76-kDa subunit and to the C- and N-terminal peptides of the SIPC (Matsumura et al. 1998c; Dreanno et al. 2006c). So even in the absence of adult barnacles, cyprids that are pioneers or "founders" (sensu Toonen and Pawlik 1994) could still be responding to the SIPC as a component of cyprid temporary adhesive.

All barnacle species examined thus far contain protein that is related immunologically to the SIPC of *B. amphitrite* (Kato-Yoshinaga et al. 2000; Dreanno et al. 2007). The likely ubiquity of the SIPC to barnacles could have been predicted from the results of earlier laboratory and field studies that demonstrated that both conspecific and allospecific barnacles induced cyprid settlement (reviewed in Dreanno et al. 2007). Knight-Jones (1953) was the first to demonstrate this effect noting that small differences in the structure of the chemical cue could account for differences in inductive potential and, moreover, that there is a positive relationship between activity and systematic affinity (Crisp and Meadows 1962). We must await sequence information on the SIPCs of other barnacles to examine the first of these claims, but several studies have since obtained evidence in support of the latter hypothesis (Dreanno et al. 2007). As noted by Crisp and Meadows (1962), measures of systematic affinity made on the basis of assaying the settlement-inducing effect of adults would be improved by using known concentrations of the isolated cue. This has been achieved recently by Dreanno et al. (2007), albeit using a limited number of species.

There is clearly strong evidence that the ability of the cyprid to recognize the SIPC of conspecific barnacles is fundamental to gregarious settlement and the advantages afforded by this behavior. Nevertheless, it remains to be established what role this glycoprotein has in solitary species or in gregarious species, such as *Pollicipes pollicipes*, in which the 'SIPC' is not effective in settlement assays. Perhaps in the case of *P. pollicipes*, the absence of other potential cues in the assays, such as the topography of the outer surface of the adult peduncle and a failure to mimic the high wave energy environment of this species in laboratory assays can explain this negative result. Recognition of the SIPC of allospecific barnacles by cyprids will not directly result in future reproduction, but it may be adaptive in that it indicates that the location can support barnacle development. If the local conditions do support postsettlement growth, the pioneer may attract the settlement of conspecific cyprids that are future mates.

Another settlement-inducing protein, unrelated to the SIPC, has recently been isolated from *B. amphitrite* (Endo et al. 2009). This ca. 32-kDa protein rapidly induces searching behavior by cyprids. The authors proposed that the protein acts as a waterborne settlement pheromone but noted that evidence of the release of this protein by adult barnacles needs to be obtained to substantiate this hypothesis. There is, however, evidence of a relatively low molecular mass waterborne cue (Tegtmeyer and Rittschof 1989; Clare and Matsumura 2000; Clare and Yamazaki 2000). Elbourne (2008) reported that the cue is a stable peptide of <1 kDa and that it can be detected by bioassay of seawater sampled adjacent to field populations of *B. amphitrite*. Based on the available evidence, it seems reasonable to suggest that cyprids respond to a waterborne cue(s) by altering their behavior to bring them into contact with the substratum, where they may be induced to settle gregariously by contact with the SIPC (Fig. 22.1).

22.5 Cue Perception

The importance of sensory cue perception to the cyprid is reflected in the nervous system being "most complete" at this stage compared to the nauplius and adult (Harrison and Sandeman 1999). The cyprid engages in elaborate searching behavior (e.g., Lagersson and Høeg 2002) to enable it to locate a suitable place to settle, involving "walking" with the ambulatory antennules, using the attachment discs and a putative temporary adhesive to gain purchase on the substratum (Crisp 1984). The cyprid antennules, and in particular the third (the attachment disc) and fourth articles, are thought to be the location of the SIPC receptor(s) (Crisp 1990), though there is no definitive evidence to support this hypothesis. Darwin (1854) was the first to suggest that parts of the antennule "serve as feelers." From his description, it would seem that he was referring to the two plumose setae of the fourth article of the antennule; terminal setae A and B in the terminology of Nott and Foster (1969) and abbreviated as TS-A and -B by Bielecki et al. (2009). Transmission electron microscopy (TEM) (Nott and Foster 1969; Gibson and Nott 1971; Lagersson et al.

2003) showed that the nine setae of the fourth article are innervated and the presence of an antennular soma cluster of bipolar neurons near the base of the antennules testifies to the important sensory function of the setae (Walley 1969; Walker et al. 1987; Harrison and Sandeman 1999). Indeed, Harrison (1998) recorded electrical activity from the antennular nerve in response to chemical and mechanical stimulation of the antennule. Scanning electron microscopy (SEM) studies of the fourth antennular segment (Clare and Nott 1994; Clare 1995; Lagersson et al. 2003; Chan and Leung 2007) show the setal arrangement to be remarkably similar between species (Fig. 22.5).

In relation to cue detection, Clare and Nott (1994) noted the outward resemblance of some of the setae, notably terminal seta D (TS-D) and the four subterminal setae

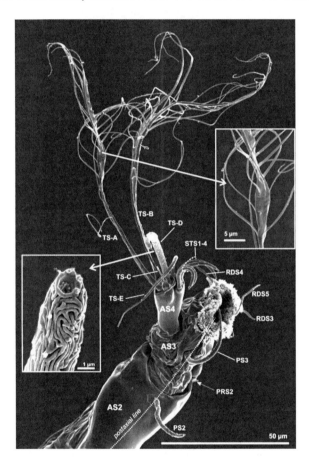

Fig. 22.5 Distal part of an antennule showing segments 2, 3, and 4. Segment 3 includes the prominent attachment disc. Segment 4 bears 9 sensory setae. Lower inset shows the terminal pore in terminal seta D. Upper inset shows the insertions of the thin setules on terminal setae A. *AS2–4* antennular segments 2–4; *PRS2* preaxial seta 2; *PS2* postaxial seta 2; *PS3* postaxial seta 3; *RDS3–5* radial setae 3–5; *STS1–4* subterminal setae 1–4; *TS-A–D* terminal setae A–D. After Bielecki et al. (2009) reproduced by permission of Taylor & Francis Ltd. (www.informaworld.com)

(STS 1–4 in the terminology of Bielecki et al. 2009), to crustacean olfactory aesthetascs (see Chap. 6). Gibson and Nott (1971) had earlier alluded to a possible chemosensory role for TS-D. TEM studies generally supported these interpretations (Harrison and Sandeman 1999; Lagersson et al. 2003; see also Gibson and Nott 1971). Moreover, the fourth antennular segment is flicked through the water column (Gibson and Nott 1971; Clare and Nott 1994; Lagersson et al. 2003), comparable to decapod antennular flicking (see Koehl, Chap. 5) and presumably also facilitating stimulus access to olfactory receptors (Clare and Nott 1994; Clare 1995). It should be noted, however, that TS-D and STS 1–4 all possess a scolopale, which distinguishes them from classical decapod aesthetascs (see Hallberg and Skog, Chap. 6). Indeed, Bielecki et al. (2009) suggested that STS 1–4 are contact chemoreceptors (see below). Although it is difficult to generalize across taxa, Lagersson et al. (2003) stated that "with few exceptions, all the setae are bimodal." Bielecki et al. (2009) later noted that the exceptions were TS-A and -B. These setae possess characteristic features of crustacean plumose setae, including long setules that project from opposite sides of the main shaft and an elaborate articulation (Cate and Derby 2001). Plumose setae have been claimed to be hydrodynamic receptors (Lagersson et al. 2003).

With the exception of TS-E, the terminal setae do not make contact with the substratum during cyprid searching behavior, whereas the four subterminal setae do (Lagersson et al. 2003). The latter have a terminal pore, which may facilitate access of chemical stimuli to the dendrites (see Hallberg and Skog, Chap. 6). This suggests that the subterminal setae may be candidates for detecting the SIPC by the tactile chemical sense proposed by Crisp and Meadows (1963). The centrally placed, open-ended, axial sensory seta of the attachment disc (Nott and Foster 1969) is another possibility. It is tempting to suggest that TS-D has a role in detecting the adult waterborne cue(s). For a fuller description of antennular setae and their putative functions, see Bielecki et al. (2009).

The antennular nerves can be traced to the deutocerebral neuropile of the brain (see Schmidt and Mellon, Chap. 7). While regarded as the putative location for chemosensory processing, glomeruli characteristic of the olfactory lobes of other crustaceans have not been seen (Harrison and Sandeman 1999). Interestingly, the antennular nerves send fine branches to the cement ducts and toward each cement gland and its muscular sac (Harrison and Sandeman 1999), and may control the release of cyprid cement in response to the appropriate sensory cue, to commit the cyprid to permanent attachment. The cement glands are innervated by catecholaminergic neurons (Okano et al. 1998).

22.6 Future Directions

Interest in how barnacles locate a suitable place to settle and the nature of the biogenic cues that are involved is likely to be sustained for the foreseeable future. Barnacles are of major economic importance as fouling species and the means to control their fouling will no doubt benefit from a more comprehensive understanding of the chemical basis of cyprid settlement. Thus far, attention has focused on the proteinaceous character of the SIPC. Research is now underway to understand the nature of the sugar moiety of the SIPC and its role in settlement induction. A complete characterization of the SIPC and its receptor may open the way for the development of antagonists that can be used as part of an antifouling strategy. Conversely, for those species that are an important food resource, such as the stalked barnacle *P. pollicipes* or the giant barnacle *Austromegabalanus psittacus* (Fig. 22.6), research on settlement induction will be important to new aquaculture ventures.

B. amphitrite has become a model species for settlement studies owing to its economic importance, cosmopolitan distribution and relative ease of culture. How representative the findings of these studies are to other barnacle species remains to be fully established. Investigations on the SIPCs of different species are required to address fundamental questions concerning the role of the SIPC in barnacle evolution and the ability of cyprids to distinguish between barnacle species at settlement

Fig. 22.6 The giant barnacle *Austromegabalanus psittacus* from Chile. (**a**) Small individuals are often found on the thick shells of adults. (**b**) Where benthic populations are available this species is sold in local seafood markets. (**c**) Only large and singly grown individuals find their way to the dinner plate. Giant barnacles readily grow on submerged structures but their culture remains as yet largely unexplored, mostly due to unpredictable larval settlement and costly larval culture. Images courtesy of Iván A. Hinojosa

(Dreanno et al. 2007). In this regard, the role of the SIPC in the colonization of invasive barnacle species, particularly those that are gregarious, may emerge as a fascinating area for study. Attempts have been made to examine the relative importance of chemical and physical cues to barnacle settlement in the field (e.g., Prendergast et al. 2008) but not with purified SIPC. As pointed out by Prendergast et al. (2009), however, many intrinsic and extrinsic factors may interact to exert their effects on barnacle settlement and confound the interpretation of field experiments.

Studies on the nature of waterborne cues to barnacle settlement, their relation (if any) to the SIPC and their role in the settlement process are also needed, as are studies on the transduction of cues. The latter will undoubtedly benefit from proteomic (Thiyagarajan and Qian 2008) and genomic (Bacchetti De Gregoris et al. 2009) approaches. What role biofilm associated with barnacle shell has in settlement induction also needs clarification. In the context of waterborne cues, there is evidence for other taxa that biofilm can adsorb chemical cues (e.g., Chan and Walker 1998).

22.7   Summary and Conclusions

For those barnacle species that are gregarious and obligate cross fertilizers, close proximity is required at the time of mating. This proximity is achieved at settlement through recognition of conspecifics. Contact with adults, juveniles, and the footprints of temporary adhesive that cyprids may leave behind as they explore a surface induce cyprid settlement. There is good evidence that the common denominator to this settlement induction is the SIPC, a cuticular glycoprotein that the cyprid recognizes by using a tactile chemical sense. The SIPC has been partially characterized as a previously undescribed glycoprotein related to α_2-macroglobulin. The nature of the sugar moiety of the SIPC, how the SIPC is perceived, and how cyprids distinguish between species are among the questions that remain to be answered. Indeed, 60 years after the seminal study of Knight-Jones and Stevenson (1950) on gregariousness in *E. modestus*, much still remains to be learned about the chemical basis of barnacle settlement.

Acknowledgments ASC gratefully acknowledges Dr Jim Nott for Fig. 22.3. Publications from the author's laboratory on the SIPC would not have been possible without the invaluable contributions of many colleagues, including Drs Kiyotaka Matsumura, Hiroshi Hirota, Richard Kirby, Catherine Dreanno, and Margaret Kirby. Dr Graham Walker, Professor Dan Rittschof and the late Professor Denis Crisp provided much inspiration. This work was funded by the Natural Environment Research Council of the UK (GST/02/1436 and NER/A/S/2001/00532). Research on temporary adhesion and cyprid settlement behavior was funded by the NERC (NER/B/S/2003/00273), the US Office of Naval Research (N00014-02-1-0311; N00014-05-1-0767; N00014-08-1-1240) and EC Framework 6 Integrated Project "AMBIO" (NMP-CT-2005-011827). The views expressed in this publication reflect only those of the author and the Commission is not liable for any use that may be made of the information contained therein.

References

Abelson A, Denny M (1997) Settlement of marine organisms in flow. Annu Rev Ecol Syst 28:317–339

Aldred N, Clare AS (2008) The adhesive strategies of cyprids and development of barnacle-resistant marine coatings. Biofouling 24:351–363

Armstrong PB, Quigley JP (1999) Alpha (2)-macroglobulin: an evolutionarily conserved arm of the innate immune system. Dev Comp Immunol 23:375–390

Bacchetti De Gregoris TB, Borra M, Biffali E, Bekel T, Burgess JG, Kirby RR, Clare AS (2009) Construction of an adult barnacle (*Balanus amphitrite*) cDNA library and selection of reference genes for quantitative RT-PCR studies. BMC Mol Biol 10:62

Barnes H, Crisp DJ (1956) Evidence of self fertilisation in certain species of barnacles. J Mar Biol Ass UK 35:631–639

Bielecki J, Chan BKK, Høeg JT, Sari A (2009) Antenular sensory organs in cyprids of balanomorph cirripedes: standardizing terminology using *Megabalanus rosa*. Biofouling 25:203–214

Bourget E (1977) Shell structure in sessile barnacles. Nat Can 104:281–323

Bourget E (1988) Barnacle larval settlement: the perception of cues at different spatial scales. In: Chelazzi G, Vannini M (eds) Behavioral adaptation to intertidal life. Plenum Press, New York, pp 153–172

Cate HS, Derby CD (2001) Morphology and distribution of setae on the antennules of the Caribbean spiny lobster *Panulirus argus* reveal new types of bimodal chemo-mechanosensilla. Cell Tissue Res 304:439–454

Chan BKK, Leung PTY (2007) Antennular morphology of the cypris larvae of the mangrove barnacle *Fistulobalanus albicostatus* (Cirripedia: Thoracica: Balanomorpha). J Mar Biol Ass UK 87:913–915

Chan ALC, Walker G (1998) The settlement of *Pomatoceros lamarckii* larvae (Polychaeta: Sabellida: Serpulidae): a laboratory study. Biofouling 12:71–80

Christie AO, Dalley R (1987) Barnacle fouling and its prevention. In: Southward AJ (ed) Barnacle biology, crustacean issues 5. A.A. Balkema, Rotterdam, pp 419–433

Clare AS (1995) Chemical signals in barnacles: old problems, new approaches. In: Schram FR, Høeg JT (eds) New frontiers in barnacle evolution. A.A. Balkema, Rotterdam, pp 49–67

Clare AS (1996) Signal transduction in barnacle settlement: calcium re-visited. Biofouling 10:141–150

Clare AS, Aldred N (2009) Surface colonisation by marine organisms and its impact on antifouling research. In: Hellio C, Yebra DM (eds) Advances in marine antifouling coatings and technologies. Woodhead Publishing Limited, Cambridge, pp 46–79

Clare AS, Matsumura K (2000) Nature and perception of barnacle settlement pheromones. Biofouling 15:57–71

Clare AS, Nott JA (1994) Scanning electron microscopy of the fourth antennular segment of *Balanus amphitrite amphitrite*. J Mar Biol Ass UK 74:967–970

Clare AS, Yamazaki M (2000) Inactivity of glycyl-glycyl-arginine and two putative (QSAR) peptide analogues of barnacle waterborne settlement pheromone. J Mar Biol Ass UK 80:945–946

Clare AS, Freet RK, McClary M (1994) On the antennular secretion of the cyprid of *Balanus amphitrite*, and its role as a settlement pheromone. J Mar Biol Ass UK 74:243–250

Connell JH (1985) The consequences of variation in initial settlement vs. post-settlement mortality in rocky intertidal communities. J Exp Mar Biol Ecol 93:11–45

Crisp DJ (1979) Dispersal and re-aggregation in sessile marine invertebrates, particularly barnacles. In: Larwood G, Rosen BR (eds) Biology and systematics of colonial organisms. Systematics association special volume no. 11. Academic, London, pp 319–327

Crisp DJ (1984) Overview of research on marine invertebrate larvae, 1940–1980. In: Costlow JD, Tipper RC (eds) Marine biodeterioration: an interdisciplinary study. Naval Institute Press, Annapolis, pp 103–126

Crisp DJ (1985) Recruitment of barnacle larvae from the plankton. Bull Mar Sci 37:478–486

Crisp DJ (1988) Reduced discrimination of laboratory-reared cyprids of the barnacle *Balanus amphitrite amphitrite* Darwin, Crustacea, Cirripedia, with a description of a common abnormality. In: Thompson M-F, Sarajini R, Nagabhushanam R (eds) Marine biodeterioration. A.A. Balkema, Rotterdam, pp 409–432

Crisp DJ (1990) Gregariousness and systematic affinity in some North Carolinian barnacles. Bull Mar Sci 47:516–525

Crisp DJ, Meadows PS (1962) The chemical basis of gregariousness in cirripedes. Proc R Soc B 156:500–520

Crisp DJ, Meadows PS (1963) Adsorbed layers: the stimulus to settlement in barnacles. Proc R Soc B 158:364–387

Darwin C (1854) A monograph on the sub-class Cirripedia, with figures of all the species. The Balanidae (or sessile cirripedes); the Verrucidae, etc. Ray Society, London

Dreanno C, Matsumura K, Dohmae N, Takio K, Hirota H, Kirby RR, Clare AS (2006a) An α2-macroglobulin-like protein is the cue to gregarious settlement of the barnacle, *Balanus amphitrite*. Proc Nat Acad Sci 103:14396–14401

Dreanno C, Kirby RR, Clare AS (2006b) Locating the barnacle settlement pheromone: spatial and ontogenetic expression of the settlement-inducing protein complex (SIPC) of *Balanus amphitrite*. Proc R Soc B 273:2721–2728

Dreanno C, Kirby RR, Clare AS (2006c) Smelly feet are not always a bad thing: the relationship between cyprid footprint protein and the barnacle settlement pheromone. Biol Lett 2:423–425

Dreanno C, Kirby RR, Clare AS (2007) Involvement of the barnacle settlement-inducing protein complex (SIPC) in species recognition at settlement. J Exp Mar Biol Ecol 351:276–282

Eckman JE (1990) A model of passive settlement by planktonic larvae onto bottoms of differing roughness. Limnol Oceanogr 35:887–901

Elbourne PD (2008) Ecological role of an adult-derived, waterborne cue in cyprid settlement in the barnacle *Balanus amphitrite* Darwin. PhD thesis, Newcastle University, pp 327

Elkin C, Marshall DJ (2007) Desperate larvae: the influence of deferred costs and habitat requirements on habitat selection. Mar Ecol Prog Ser 336:143–153

Endo N, Nogata Y, Yoshimura A, Matsumura K (2009) Purification and partial amino acid sequence analysis of the larval settlement-inducing pheromone from adult extracts of the barnacle, *Balanus amphitrite* (=*Amphibalanus amphitrite*). Biofouling 25:429–434

Foster BA (1987) Barnacle ecology and adaptation. In: Southward AJ (ed) Barnacle biology. A.A. Balkema, Rotterdam, pp 113–133

Fraenkel G, Rudall KM (1940) A study of the physical and chemical properties of the insect cuticle. Proc R Soc B 129:1–35

Gabbott PA, Larman VN (1987) The chemical basis of gregariousness in cirripedes. a review (1953–1984). In: Southward AJ (ed) Barnacle biology. A.A. Balkema, Rotterdam, pp 377–388

Gebauer P, Paschke K, Anger K (2003) Delayed metamorphosis in decapod crustaceans: evidence and consequences. Rev Chil Hist Nat 76:169–175

Gibson P, Nott JA (1971) Concerning the fourth antennular segment of the cypris larva of *Balanus balanoides*. In: Crisp DJ (ed) Fourth European marine biology symposium. Cambridge University Press, Cambridge, pp 227–236

Hadfield MG, Paul VJ (2001) Natural chemical cues for settlement and metamorphosis of marine-invertebrate larvae. In: McClintock JB, Baker BJ (eds) Marine chemical ecology. CRC Press, Boca Raton, pp 431–461

Harrison PJH (1998) The nervous system and settlement of barnacle cypris larvae. PhD thesis, University of New South Wales, pp 150

Harrison PJH, Sandeman DC (1999) Morphology of the nervous system of the barnacle cypris larva (*Balanus amphitrite* Darwin) revealed by light and electron microscopy. Biol Bull 197:144–158

Hills JM, Thomason JC, Milligan JL, Richardson M (1998) Do barnacle larvae respond to multiple settlement cues over a range of spatial scales? Hydrobiologia 375(376):101–111

Hung OS, Lee OO, Thiyagarajan V, He H-P, Xu Y, Chung HC, Qiu J-W, Qian P-Y (2009) Characterization of cues from natural multi-species biofilms that induce larval attachment of the polychaete *Hydroides elegans*. Aquat Biol 4:253–262

Jarrett JN (1997) Temporal variation in substratum specificity of *Semibalanus balanoides* (Linnaeus) cyprids. J Exp Mar Biol Ecol 211:103–114

Jeffery CJ (2002) New settlers and recruits do not enhance settlement of a gregarious intertidal barnacle in New South Wales. J Exp Mar Biol Ecol 275:131–146

Jenkins SR (2005) Larval habitat selection, not larval supply, determines settlement patterns and adult distribution in two chthamalid barnacles. J Anim Ecol 74:893–904

Jenkins SR, Åberg P, Cervin G, Coleman RA, Delany J, Santina PD, Hawkins SJ, LaCroix E, Myers AA, Lindegarth M, Power A-M, Roberts MF, Hartnoll RG (2000) Spatial and temporal variation in settlement and recruitment of the intertidal barnacle *Semibalanus balanoides* (L.) (Crustacea: Cirripedia) over a European scale. J Exp Mar Biol Ecol 243:209–225

Kato-Yoshinaga Y, Nagano M, Mori S, Clare AS, Fusetani N, Matsumura K (2000) Species specificity of barnacle settlement-inducing proteins. Comp Biochem Physiol 125:511–516

Knight-Jones EW (1953) Laboratory experiments on gregariousness during setting in *Balanus balanoides* and other barnacles. J Exp Biol 30:584–598

Knight-Jones EW, Stevenson JP (1950) Gregariousness during settlement in the barnacle *Elminius modestus* Darwin. J Mar Biol Ass UK 29:281–297

Koehl MAR (2007) Mini review: hydrodynamics of larval settlement into fouling communities. Biofouling 23:357–368

Krug PJ (2006) Defense of benthic invertebrates against surface colonization by larvae: a chemical arms race. In: Fusetani N, Clare AS (eds) Antifouling compounds. Springer, Berlin, pp 1–53

Lagersson NC, Høeg JT (2002) Settlement behaviour and antennulary biomechanics in cypris larvae of *Balanus amphitrite* (Crustacea: Thecostraca: Cirripedia). Mar Biol 141:513–526

Lagersson NC, Garm A, Høeg JT (2003) Notes on the ultrastructure of the setae on the fourth antennulary segment of the *Balanus amphitrite* cyprid (Crustacea: Cirripedia: Thoracica). J Mar Biol Ass UK 83:361–363

Larman VN (1984) Protein extracts from some marine animals which promote barnacle settlement: possible relationship between a protein component of arthropod cuticle and actin. Comp Biochem Physiol 77B:73–81

Larman VN, Gabbott PA (1975) Settlement of cyprid larvae of *Balanus balanoides* and *Elminius modestus* induced by extracts of adult barnacle and other marine animals. J Mar Biol Ass UK 55:183–190

Larsson AI, Jonsson PR (2006) Barnacle larvae actively select flow environments supporting post-settlement growth and survival. Ecology 87:1960–1966

Lewis CA (1978) A review of substratum selection in free-living and symbiotic cirripeds. In: Cia F-S, Rice ME (eds) Settlement and metamorphosis of marine invertebrate larvae. Elsevier, New York, pp 207–218

Matsumura K, Nagano M, Fusetani N (1998a) Purification of a larval settlement inducing protein complex (SIPC) of the barnacle, *Balanus amphitrite*. J Exp Zool 281:12–20

Matsumura K, Mori S, Nagano M, Fusetani N (1998b) Lentil lectin inhibits adult extract-induced settlement of the barnacle, *Balanus amphitrite*. J Exp Zool 280:213–219

Matsumura K, Nagano M, Kato-Yoshinaga Y, Yamazaki M, Clare AS, Fusetani N (1998c) Immunological studies on the settlement-inducing protein complex (SIPC) of the barnacle *Balanus amphitrite* and its possible involvement in larva-larva interactions. Proc R Soc B 265:1825–1830

Matsumura K, Hills JM, Thomason PO, Thomason JC, Clare AS (2000) Discrimination at settlement in barnacles: laboratory and field experiments on settlement behaviour in response to settlement-inducing protein complexes. Biofouling 16:181–190

Mauck RA, Harkless KC (2001) The effect of group membership on hiding behavior in the northern rock barnacle, *Semibalanus balanoides*. Anim Behav 62:743–748

Miron G, Walters LJ, Tremblay R, Bourget E (2000) Physiological condition and barnacle behaviour: a preliminary look at the relationship between TAG/DNA ratio and larval substratum exploration in *Balanus amphitrite*. Mar Ecol Prog Ser 198:303–310

Nott JA (1969) Settlement of barnacle larvae: surface structure of the antennular attachment disc by scanning electron microscopy. Mar Biol 2:248–251

Nott JA, Foster BA (1969) On the structure of the antennular attachment organ of the cypris larva of *Balanus balanoides* (L.). Phil Trans Roy Soc B256:115–133

Okano K, Fusetani N (1997) Larval settlement in barnacles. Seikagaku 69:1348–1360 (in Japanese)

Okano K, Shimizu K, Satuito CG, Fusetani N (1998) Enzymatic isolation and culture of cement secreting cells from cypris larvae of the barnacle *Megabalanus rosa*. Biofouling 12:149–159

Pawlik JR (1992) Chemical ecology of the settlement of benthic marine invertebrates. Oceanogr Mar Biol Annu Rev 30:273–335

Prendergast GS (2007) Settlement and succession of benthic marine organisms: interactions between multiple physical and biological factors. PhD thesis, Newcastle University

Prendergast GS, Zurn CM, Bers AV, Head RM, Hansson LJ, Thomason JC (2008) Field-based video observations of wild barnacle cyprid behaviour in response to textural and chemical settlement cues. Biofouling 24:449–459

Prendergast GS, Zurn CM, Bers AV, Head RM, Hansson LJ, Thomason JC (2009) The relative magnitude of the effects of biological and physical settlement cues for cypris larvae of the acorn barnacle, *Semibalanus balanoides* L. Biofouling 25:35–44

Raimondi PT (1988) Settlement cues and determination of the vertical limit of an intertidal barnacle. Ecology 69:400–407

Rittschof D (1985) Oyster drills and the frontiers of chemical ecology: unsettling ideas. Am Malacol Bull Special Ed 1:111–116

Rittschof D, Branscomb ES, Costlow JD (1984) Settlement and behavior in relation to flow and surface in larval barnacles, *Balanus amphitrite* Darwin. J Exp Mar Biol Ecol 82:131–146

Smith PA, Clare AS, Rees HH, Prescott MC, Wainwright G, Thorndyke MC (2000) Identification of methyl farnesoate in the cypris larva of the barnacle, *Balanus amphitrite*, and its role as a juvenile hormone. Insect Biochem Mol Biol 30:885–890

Steinberg PD, de Nys R, Kjelleberg S (2002) Chemical cues for surface colonization. J Chem Ecol 28:1935–1951

Tegtmeyer K, Rittschof D (1989) Synthetic peptide analogs to barnacle settlement pheromone. Peptides 9:1403–1406

Thiyagarajan V, Qian P-Y (2008) Proteomic analysis of larvae during development, attachment, and metamorphosis in the fouling barnacle *Balanus amphitrite*. Proteomics 8:3164–3172

Thompson JV (1830) On the cirripedes or barnacles. Zool Res I (Pt 1) Mem IV:69–88

Thorson G (1950) Reproductive and larval ecology of marine bottom invertebrates. Biol Rev 25:1–45

Toonen RJ, Pawlik JR (1994) Foundations of gregariousness. Nature 370:511–512

Walker G (1980) A study of the oviducal glands and ovisacs of *Balanus balanoides* (L.), together with comparative observations on the ovisacs of *Balanus hameri* (Ascanius) and the reproductive biology of the two species. Phil Trans R Soc B 291:147–162

Walker G (1981) The adhesion of barnacles. J Adhesion 12:51–58

Walker G (1995) Larval settlement: historical and future perspectives. In: Schram FR, Høeg JT (eds) New frontiers in barnacle evolution. A.A. Balkema, Rotterdam, pp 69–85

Walker G, Yule AB (1984) The temporary adhesion of barnacle cyprids: effects of some differing surface characteristics. J Mar Biol Ass UK 64:429–439

Walker G, Yule AB, Nott JA (1987) Structure and function in balanomorph larvae. In: Southward AJ (ed) Barnacle biology. A.A. Balkema, Rotterdam, pp 307–328

Walley LJ (1969) Studies on the larval structure and metamorphosis of *Balanus balanoides* (L.). Phil Trans R Soc B256:237–280

Wethey DS (1984) Spatial pattern in barnacle settlement: day to day changes during the settlement season. J Mar Biol Ass UK 64:687–698

Wilson DP (1932) On the mitraria larva of *Owenia fustiformis* delle Chiaje. Phil Trans R Soc B 221:231–334

Yamamoto H, Tachibana A, Matsumura K, Fusetani N (1995) Protein kinase C (PKC) signal transduction system involved in larval metamorphosis in the barnacle, *Balanus amphitrite*. Zool Sci (Tokyo) 12:391–396

Young CM (1990) Larval ecology of marine invertebrates: a sesquicentennial history. Ophelia 32:1–48

Yule AB, Walker G (1985) Settlement of *Balanus balanoides*: the effect of cyprid antennular secretion. J Mar Biol Ass UK 65:707–712

Zega G, Pennati R, Dahlström M, Berntsson K, Sotgia C, De Bernardi F (2007) Settlement of the barnacle *Balanus improvisus*: the roles of dopamine and serotonin. Ital J Zool 74:351–361

Chapter 23
Contact Chemoreception and Its Role in Zooplankton Mate Recognition

Terry Snell

Abstract Physical constraints cause small aquatic animals like copepods, rotifers, and cladocerans to experience their environment much differently than humans. One solution to sending and receiving signals in a vast, well-mixed environment is to bind the signal to the body surface, requiring receivers to use contact chemoreception for detection. Contact signals permit zooplankters to use large, information-rich molecules like glycoproteins as signals, whereas excreting such chemicals would be energetically costly. Lectin-binding experiments have demonstrated the role of surface glycoproteins in rotifer and copepod mate recognition. Female surface glycoproteins provide males with information about sex, age, species identity, and female anatomy. Mate-guarding in harpacticoid copepods was investigated by binding lectins to surface glycoproteins of females. Male decisions about mate guarding and spermatophore placement are typically determined by a series of stroking behaviors where males probe the body surface of females for chemical cues. Protein receptors on male antennules recognize specific carbohydrate moieties of female surface glycoproteins to identify appropriate partners for guarding and mating. Monoclonal antibody affinity chromatography was used to purify surface glycoproteins from female *Tigriopus japonicus*. A 70-kDa protein was eluted from the column and a partial amino acid sequence was determined using mass spectrometry. This protein has significant similarity to α_2-macroglobulin, a large protease inhibitor found in high concentrations in vertebrate blood and the hemolymph of invertebrates. α_2-Macroglobulin was most highly expressed in late copepodid stage (CV) females, the developmental stage most avidly guarded by males. Based on these observations, a model of the molecular mechanism of *T. japonicus* mate recognition has been developed. Mate recognition in copepods is compared to other malacostracans, cladocerans, insects, and rotifers. Future research should emphasize identification of signal compounds responsible for mate recognition and life cycle regulation. These critical molecules will be pivotal for understanding

T. Snell (✉)
School of Biology, Georgia Institute of Technology, Atlanta 30332-0230, GA, USA
e-mail: terry.snell@biology.gatech.edu

T. Breithaupt and M. Thiel (eds.), *Chemical Communication in Crustaceans*,
DOI 10.1007/978-0-387-77101-4_23, © Springer Science+Business Media, LLC 2011

the evolutionary dynamics of aquatic invertebrates, including the development of reproductive isolation, speciation, and the rich biodiversity of cryptic species.

23.1   Introduction

Because of the physical forces operating at the sub-millimeter scale, small aquatic animals like rotifers, copepods, and cladocerans experience their environment much differently than humans. This sensory world of the very small has been elucidated by Dusenbery (1992, 2009) in two fascinating books. As fate would have it, David Dusenbery was my colleague 1991–2003 and I had the privilege of sharing many lunch conversations about the physical constraints on biological systems and how a biophysicist views the world. Talking about how zooplankton feel, taste, and smell their way through their environment had a big influence on my scientific thinking. It was fascinating to consider how flow, viscosity, Brownian motion, and diffusion influenced the shape and behavior of small animals. This viewpoint seemed useful for understanding many biological phenomena like why size and time are important in signal detection, how to sense and navigate in gradients, and what determines the frequency of encounters with food, mates and predators. Fortunately, the physics of small objects suspended in water is fairly tractable so that reliable and general statements can be made about the selective forces shaping the behavior of zooplankton like the crustacean copepods and cladocerans, and the non-crustacean rotifers. Being a practical experimentalist, I focused on some aspects of rotifer mating behavior that had long puzzled me like why rotifer males were not attracted to females from a distance and instead simply encountered them by random swimming. This question culminated in a paper on the general physical constraints of size on the value of producing attraction pheromones (Dusenbery and Snell 1995). These results apply to all plankton, so it is especially important to develop examples from a variety of taxa like copepods and rotifers.

My thinking on these topics was further influenced by two other developments. Advances in microscale protein chemistry enabled us to extract, isolate, and purify enough protein from the body surface of zooplankters so that we could analyze primary structure (Ting and Snell 2003). Likewise, we have been able to characterize signal molecules excreted into the medium at low concentrations (Snell et al. 2006), that are used as quorum sensing signals to regulate reproduction and diapause in the rotifer life cycle (Kubanek and Snell 2008). The second development was the growing field of ecotoxicology. As the sensory systems of zooplankters become better understood, it became clear that pollutants could interfere with normal channels of chemical communication with dire ecological consequences (Lürling and Scheffer 2007). My lab had been using rotifers and copepods as tools to assess toxicity in a variety of tests from ingestion to mating (Marcial et al. 2003; Snell and Joaquim-Justo 2007), so it was a small extension of ongoing work to include chemical signaling as an endpoint. We were especially interested in the role of contact signals in mating and mate recognition in copepods and rotifers, which will be the focus of this paper. I first describe the contact molecules that copepods use for chemical signaling and compare them to other Arthropods and rotifers.

23.2 Contact Molecules Used for Chemical Communication

Small zooplankters have the difficult task of sending and receiving signals about the location of food and mates, and avoiding predators, while swimming in a vast, relatively well-mixed environment. Chemical signals released from small animals are rapidly diluted, and bioenergetics strongly constrain the amount of signaling molecules that can be excreted, so there is a limited range over which they can be perceived. At the same time, these signals need to contain sufficient information for the identification of sex, age, and fitness of conspecifics, and to recognize species. One solution is to bind the signal to the body surface, thus eliminating the need to release signals into the environment, where most would be lost never contacting a receiver. Surface bound signals require receivers to use contact chemoreception for detection. Contact signals permit zooplankters to use large, information-rich molecules like glycoproteins as signals, whereas releasing such chemicals into the water column would be energetically too costly.

Surface glycoproteins were first discovered to function as contact mating pheromones in single-cell aquatic organisms such as algae (Wiese 1965) and ciliates (Miyake and Beyer 1974). Lectin-binding experiments in rotifers (Snell and Nacionales 1990, see below) demonstrated the role of surface glycoproteins in rotifer mate recognition. Later, it was shown that a 29-kDa glycoprotein serves as a signal for mate recognition and is localized in the corona (head region) and foot of female rotifers of *Brachionus plicatilis* (Snell et al. 1995). These two sites are relatively soft and penetrable, and are most often chosen by the males for insemination (Snell and Hoff 1987). Surface glycoproteins therefore provide rotifers with information about sex, age, species identity, and female anatomy.

Surface glycoproteins also serve as gamete-recognition molecules in some sea urchins and bivalves (Glabe and Clark 1991; Focarelli and Rosati 1995; Capone et al. 1999) and shrimps (Miller and Ax 1990; Wikramanayake and Clark 1994; Glas et al. 1996; Gómez and Dupré 2002). The best characterized gamete recognition proteins in aquatic invertebrates are from broadcast spawners like the abalone, a marine gastropod. A soluble protein called lysin is released by the acrosome reaction when abalone sperm contacts a conspecific egg (Swanson and Vacquier 1995). In a non-enzymatic process, lysin creates a hole in the egg envelope through which sperm swim to activate fertilization. This reaction is species-specific, mediated by lysin binding to a protein receptor on the egg surface. Many papers have reported on the divergence and polymorphism of lysin and a second protein, sp18, also released by the abalone acrosome reaction (Vacquier 1998; Swanson and Vacquier 2002a, b). The lysin receptor also has been isolated and characterized (Swanson and Vacquier 2002a). Perhaps most impressive is that molecular analysis has demonstrated that the evolutionary mechanism driving interspecific divergence of these proteins is positive Darwinian selection. Some of the highest evolutionary rates reported for any proteins have come from this work. Similar results for lysin isolated from the marine snail *Tegula* have been reported by Hellberg et al. (2000). The sea urchin sperm protein bindin, which binds sperm to the surface of eggs, has also been well characterized (Vacquier et al. 1995). Like lysin, it is strongly divergent among populations and this divergence is driven by positive Darwinian selection (Metz and Palumbi 1996).

Gamete and mate recognition proteins often play a key role in diversification and speciation because genes for them often evolve faster than other classes of genes (Civetta and Singh 1998; Swanson and Vacquier 2002b). Cryptic species are common in aquatic invertebrates in which sensory systems are dominated by mechano- and chemosensory modalities (Palumbi 1992; Knowlton 1993, 2000). These species boundaries are invisible to human observers who are biased toward visual recognition.

There is a long history of investigation of chemical signals used by copepods to locate mates (Katona 1973) that continues today (Kiørboe 2007). How the mating biology of copepods has been influenced by sexual selection is reviewed by Titelman et al. (2007). Early observations suggested the existence of species-specific mate-location pheromones that are released into water and attract mates from a distance. These are quite distinctive from contact-recognition pheromones which are not released, but are instead bound to body surface and function in contact mate recognition. Like in many marine invertebrates, mating success in copepods depends on identifying conspecifics of the opposite sex using chemical cues (Lonsdale et al. 1998). Surface glycoproteins may serve as contact mate recognition signals in a variety of copepods, including Calanoida, Cyclopoida, and Harpacticoida (Snell and Carmona 1994). Characteristics of surface glycoproteins have been elucidated through lectin-binding experiments. Lectins are non-immunological proteins or glycoproteins usually isolated from plants that bind to specific sugar residues non-covalently (Sharon and Lis 1989). Probing intact copepods with several lectins of different affinities revealed a rich variety of surface glycoproteins, their location, and the composition of some of their monosaccharides (Snell and Carmona 1994). Prominent lectin-binding sites were found at structures important in mating behavior, including the margin of the genital segment, gonopore, and caudal rami on adult females. There was considerable variation in the type and amount of lectin binding, but binding was observed in all species, implying that surface glycoproteins are common in copepods.

The function of copepod lectin binding sites in mating was demonstrated by binding experiments where blocking of female surface glycoproteins obscured male mate-recognition (Lonsdale et al. 1996). Binding of the lectin *Triticum vulgaris* to surface glycoproteins of females resulted in a significantly reduced male mate guarding in the harpacticoid *Coullana canadensis*. Similarly, treatment of *Coullana* sp. females (a sibling species from Florida) with the lectin *Pisum sativum* resulted in a significant sixfold reduction in male mate guarding. This lectin also binds to male antennules (Fig. 23.1, Lonsdale et al. 1998). Frey et al. (1998) compared lectin binding patterns in four geographical isolates of *C. canadensis* and *Coullana* sp.; females from

Fig. 23.1 Photomicrograph of binding of the lectin *Pisum sativum* on the antennules of *Coullana* sp. males (from Lonsdale et al. 1998)

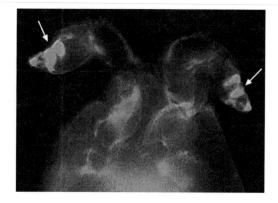

Fig. 23.2 Mate guarding in *Tigriopus japonicus*. Drawing by Jorge A. Varela Ramos

all populations treated with *Triticum vulgaris* lectin elicited less mate guarding from males. Treatment with the lectins *Erythrina corallodendron*, *Ulex europaeus*, and *Glycine max* yielded population-specific patterns of mate guarding inhibition, suggesting population level differentiation in surface glycoprotein composition. This further suggests that surface glycoproteins likely play a key role in the establishment and maintenance of reproductive isolation among copepod populations.

Kelly and Snell (1998) investigated the role of contact chemoreception of surface glycoproteins in mate recognition, mate-guarding, and spermatophore transfer in the marine harpacticoid *T. japonicus* Mori (Fig. 23.2). Male decisions to mate guard are preceded by a series of stroking behaviors where males probe the body surface of females for chemical cues about sex, species, and developmental stage (Fig. 23.3). During stroking behavior, males contact the caudal ramus, abdomen, and lateral prosome of females with their antennules. Throughout mate guarding, male antennules remain in contact with the female lateral prosome, while male swimming legs

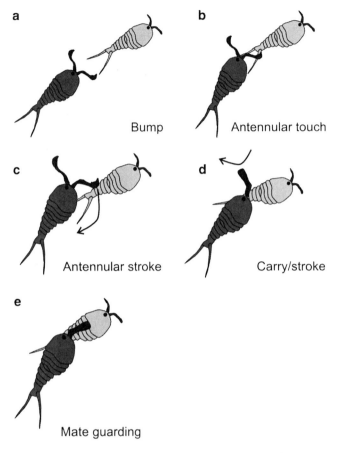

Fig. 23.3 Mating behavior of *T. japonicus* males (*dark gray*) that precedes mate guarding (from Kelly and Snell 1998)

maintain contact with the female's caudal ramus. A survey of the binding sites of twelve lectins to the surface of juvenile and adult females, and adult males demonstrated localized lectin-binding at sites considered important in mating behavior of each sex/age class. *Lens culinaris* lectin was one of three lectins, along with *G. max* and *U. europaeus*, to bind at the genital segment with fluorescence levels 24 times, 3 times, and 3 times higher than the control, respectively. The control in these experiments is a copepod of similar sex and age without lectin fluorescence. Copepodid stage CIII females had significant lectin binding by *Pisum sativium* at the caudal ramus and *Lycopersicon esculentum* at the midlateral prosome (Fig. 23.4). A mate-guarding bioassay demonstrated that lectin binding or exposure to 40 mM concentrations of the monosaccharide glucosamine could competitively inhibit mate-guarding. Also, the treatment of males with proteolytic enzymes or detergents eliminated their mate-guarding without affecting swimming behavior (Kelly et al. 1998). These results

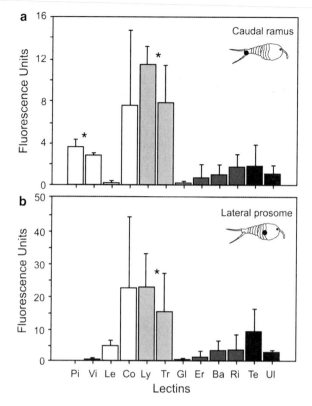

Fig. 23.4 Binding of fluoroisothiocyanate-lectins to the cuticle of CIII *T. japonicus*, showing the four lectin-affinity groups (*open bars* glucose/mannose; *stippled N*-acetyl glucosamine; *hatched* galactose/*N*-acetyl galactosamine; black fucose; lectin). *Pisum satvium* (Pi), *Vicia faba* (Vi), *Lens culinaris* (Le), *Canalvia ensiformis* (Co), *Lycopersicon esculentum* (Ly), *Triticum vulgaris* (Tr), *Glycine max* (Gl), *Erythrina corallodendron* (Er), *Bauhinia purpurea* (Ba), *Ricinus communis* (Ri), *Tetragonolobus purpureas* (Te), and *Ulex europaeus* (Ul). Fluorescence units are arbitrary but self-consistent units of fluorescence intensity. The *asterisk* designates significant ($p < 0.05$) lectin binding above control levels (ANOVA) (from Kelly and Snell 1998)

suggest that protein receptors on male antennules detect glycoprotein signals on the surface of females, recognizing specific carbohydrate moieties of glycoproteins to identify appropriate partners for guarding or mating. The current model of the *T. japonicus* mating system is that there are surface glycoproteins on the prosome and caudal ramus of juvenile females that contain *N*-acetyl glucosamine. These glycoproteins are identified by conspecific males through a complementary receptor concentrated at the distal segment of male antennules. Virgin adult females probably rely on a diffusible pheromone to attract mates, but surface glycoproteins determine species recognition and spermatophore placement.

The chemical characteristics of *T. japonicus* surface glycoproteins were investigated by Ting et al. (2000). The proteolytic enzyme trypsin was used to cleave surface proteins from females, reducing their attractiveness to males in choice experiments. Tryptic fragments were used to make monoclonal antibodies that were used to screen for mate guarding inhibition. One antibody bound to the lateral prosome and terminal urosome of CV females (Fig. 23.5), significantly reducing their attractiveness to males. This monoclonal antibody only bound to *T. japonicus* copepodid stages CV and CVI females, not males, and not *T. californicus* or *T. fulvus* females. In Western blots of tryptic fragments and total protein homogenates from CV females, the antibody bound only to a 48- and 36-kDa protein. Ting and Snell (2003) used this monoclonal antibody in affinity chromatography to purify proteins from *T. japonicus*. A 70-kDa protein eluted from the column and was partially sequenced using mass spectrometry. It was found to have significant similarity to α_2-macroglobulin, a large protease inhibitor found in high concentrations in vertebrate blood and the hemolymph of invertebrates. α_2-Macroglobulin was most highly expressed in CV females, the developmental stage most avidly guarded by males. Treating males with bovine α_2-macroglobulin caused them to make significantly more changes in mate guarding partners than control males. A 36-kDa protein co-purified with α_2-macroglobulin and was shown to be similar to elastase, an enzyme with serine protease activity.

Based on these observations, a model of the molecular mechanism of *T. japonicus* mate recognition has been developed (Fig. 23.6). Mate recognition is initiated by male contact with a female copepidite. Male antennules are thought to contain the receptors for female mate-recognition proteins like the α_2-macroglobulin-like protein designated CSP70. The CSP70 protein is apparently secreted and displayed on the body surface of female copepods, as demonstrated by binding of the monoclonal antibody (Ting et al. 2000). The chitinous exoskeleton of invertebrates contains 30–50% protein (Kumari and Skinner 1993), and CSP70 is found in the soluble fraction (unpublished data). CSP70 is therefore, not tightly bound to the exoskeleton itself. Trypsin treatment of proteins on the surface of *T. japonicus*

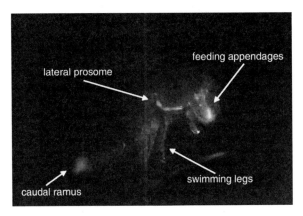

Fig. 23.5 Monoclonal antibody binding to CV *T. japonicus* females. Epifluorescence viewed at $100\times$ from FITC labeled secondary antibody (490 nm excitation, 530 nm emission) (from Ting et al. 2000)

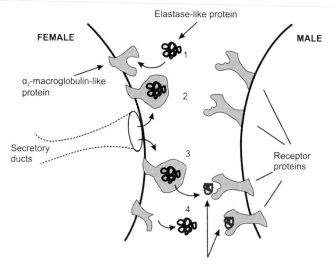

Fig. 23.6 Conceptual model of *T. japonicus* male detection of female mate-recognition proteins. Sequence on the figure from top to bottom. An elastase-like protease from the male may bind to the female, α_2-macroglobulin-like protein (*1*) and become trapped (*2*). Over time, the elastase-like protease digests the α_2-macroglobulin-like protein, releasing a portion (*3*), which then binds to the male receptor (*4*). Binding of the pheromone is likely to be reported by signal transduction pathways. Digestion leaves the female signal protein degraded, unable to bind its cognate protease. This recycles the protease, and requires that the female continually produce mate-recognition proteins throughout mate-guarding (from Ting and Snell 2003)

recovered a protein of the same molecular weight as the antibody-purified CSP70 (Ting et al. 2000). If CSP70 were membrane-bound, antibody purification would have recovered a larger protein than the trypsin fragment. Therefore, it is likely that this protein is secreted and is only loosely associated with the integument. Once CSP70 is secreted, it probably emerges to the surface of the copepod exoskeleton through the many pores in the crustacean integument (for description of pores see Mauchline 1977; Blades and Youngbluth 1979). The nature of the interaction of the CSP70 molecule with its male receptor is not known, but it is possible that a protease trapping mechanism may be involved in activating the male response. This could result in α_2-macroglobulin-like protein CSP70 undergoing a conformational change when attacked by a protease such as elastase, enveloping the elastase, and inhibiting it from attacking other proteins. Male antennules may carry elastase-like proteases that attack, and are enveloped by, the α_2-macroglobulin-like recognition protein CSP70 on females. If the elastase-like 36-kDa protein has a degradative role, the CSP70 signal molecule would have to be refreshed periodically. Degradation by the protease releases a small portion of the recognition protein that interacts with male receptors. Although this system seems complicated, the hydrolysis of signal molecules by proteases is a common theme in marine invertebrate signaling (Rittschof 1990).

The Chap. 22 by Clare describes a settlement-inducing protein complex in the cypris larvae of barnacles that shares an overall 25% sequence homology with the thioester-containing family of proteins, which includes α_2-macroglobulin. Biologically active peptides are commonly generated by serine protease degradation of structural proteins by several marine invertebrates. Crustacean pheromones and kairomones are partial hydrolysis products, generated usually by the action of trypsin-like serine proteases on proteins, and glycosidase enzymes on glycoproteins and proteoglycans (Rittschof and Cohen 2004). Assessment of the generality of protease-based signaling systems awaits the discovery of the molecular details of signaling in other crustaceans.

Contact chemoreception is likely utilized by many zooplankton species for mate and gamete recognition, male assessment of female reproductive condition, and recognition of appropriate food particles prior to ingestion. As the genomes of more zooplankton are sequenced, I expect the discovery of a variety of genes for membrane-bound chemoreceptors. Despite this, there has been more research effort in zooplankton on detecting chemical signals from a distance (e.g., Kiørboe 2007), perhaps because simple, reliable behavioral bioassays are available for many species. However, I expect the use of pheromones to detect objects from a distance to be more restricted than contact chemoreception because of the physical constraints on the former sensory modality. Dusenbery (2009) showed that time available for signal detection and surface area of the detector determine the threshold concentration necessary for chemical signaling. The best signal detection by any receptor is achieved by counting the number of new molecules that diffuse to it. This enables animals to estimate whether the concentration is increasing or decreasing over time or space, and therefore to determine the direction from which the signal is coming. In other words, receptors measure the flux of signal to the receptor and the time available for signal reception and the size of the receptor's surface area determines receptor sensitivity. The sensory systems of most animals have sample times (sniffing/flicking) on the order of 0.1 s and they need to respond in a similar time frame. This places strong constraints on the size of organisms that would be able to detect chemical signals from a distance. The surface area of the detectors of small organisms is generally too small to detect chemicals from a distance in a timely fashion. The short time available for signal detection and small detector surface area pushes the threshold concentration necessary for detection to very low levels (on the order of 10^{-12} moles/L). This concentration is a few orders of magnitude below the capabilities of most known animal sensory system (Dusenbery 2009). Dusenbery makes some further clever calculations to estimate the physical constraints on the size of the target before it can benefit from releasing chemical signals to attract a conspecific searcher. He suggests that organisms smaller than 5–167 µm are not able to improve their encounter rates by releasing pheromones. Dusenbery argues that the ability to detect chemical signals from a distance is rare in organisms smaller than about 1 mm in size because it is difficult to evolve a sensitive enough detector to make the system energetically viable. Thus, many copepods and nearly all rotifers are expected to use contact chemoreception to find mates, but not to release pheromones to detect conspecifics from a distance.

23.3   Comparison with Other Arthropods (Crustaceans and Insects)

Behavioral processes similar to those of copepods have been described in a variety of malacostracans. Some examples include the decapod shrimp *Lysmata wurdemanni* (Caridea, Hippolytidae), where both distance and contact pheromones have been confirmed in mating behavior (Zhang and Lin 2006; Zhang et al. 2009). These euhermaphrodite shrimp mate as females during a short postmolt period, when they secrete both distance and contact sex pheromones. Males use these to track, recognize, and eventually court the receptive females, but males respond aggressively to newly molted male shrimp. Males with their antenna and antennules ablated neither courted nor copulated with female shrimp. Investigation of mating behavior in the invasive Chinese mitten crab, *Eriocheir sinensis*, found no indication that females release a distance pheromone (Herborg et al. 2006). Rather, mate recognition occurs after physical contact between male and female, most likely mediated by a contact pheromone. Female lobsters (*Homarus americanus*) can determine the sex of sheltered conspecifics from a distance, and preferentially enter male-occupied shelters (Bushmann and Atema 2000). Female attraction to male-occupied shelters is based upon chemical signals in the male urine, which contains compounds important for mate location and entering decisions. However, non-urine chemical signals are also utilized during close-up mate evaluation. Mating behavior of the anomopod *Chydorus sphaericus* (Branchiopoda) begins by male attraction to females through a diffusible chemical that is used for reproductive isolation and postcopulatory mate guarding (Van Damme and Dumont 2006). Cuticular hydrocarbons were extracted from the caridean shrimp *Palaemonetes pugio* and at least three were identified that were unique to postmolt parturial females (Caskey et al. 2009a). Male response to these cuticular extracts was inconclusive; however, there currently is no evidence that they play a role in mate recognition. In a second paper, Caskey et al. (2009b) showed that a glycoprotein associated with female shrimp and containing glucosamine or an *N*-acetylglucosamine plays a key role in mate recognition in *P. pugio*. Males in seawater containing 50 mM glucosamine performed significantly fewer copulations, most likely because glucosamine competitively inhibited mate recognition by male lectin-like receptors. These works are typical for crustaceans where behavioral studies are used to confirm the existence of both distance and contact pheromones, but there has been little progress toward identifying the nature of these chemical signals.

Less is known about mating signals in the ecologically important cladoceran *Daphnia*. *Daphnia pulicaria* males determine sex of potential mates on contact (Brewer 1998), but the chemical characteristics of these signals are not known. Mating behavior in *Daphnia pulex* appears to rely on random contact between males and sexual females rather than diffusible pheromones (Winsor and Innes 2002). Males may be able to discriminate sexually receptive females from females in other developmental stages and increase their mating efficiency. Males apparently use chemical signals to avoid mating with female clonemates, thus avoiding intraclonal inbreeding depression. A repertoire of 58 chemoreceptors has been described from bioinformatic analysis of the *D. pulex* genome (Peñalva-Arana et al. 2009). This rich diversity of gustatory receptors equips *D. pulex* for fine discrimination of food, mates, and other chemical features of their environment. Even in *Drosophila*, where mating signals have been known for more than a decade (Markow 1991), the complexity of the system has made progress difficult. Multiple epicuticular hydrocarbons (ECH) serve as mating signals involved in species recognition, but researchers have yet to identify specific ECHs alone or in combination that are responsible for mating success (Etges and Ahrens 2001). Furthermore, the relative importance of ECH variation as compared to mating songs, and the role of sexual selection in driving differentiation is unknown. Ferveur (2005) described the genetic basis of ECH variation between and within *Drosophila* species, the mechanisms of ECH perception, and the role of these compounds in sexual isolation and the evolution of pheromonal communication. It seems reasonable to speculate that cuticular proteins of crustaceans might share some characteristics with ECH found on insect cuticles (Willis 1999), and could similarly be used as signals for mate recognition in some species.

23.4   Comparison with Rotifers

Chemical signals regulate the rotifer life cycle through a chain of events from reception of a mating stimulus, to insemination, to fertilization, to resting egg formation, and to deposition (Snell 1998). This cascade of events and accompanying changes in reproductive physiology is initiated by chemical signals triggering sexual reproduction. In preparation for sex, asexual (amictic) females detect a soluble chemical signal that population density is above the mictic threshold (~70 females/L in *B. plicatilis*, Snell et al. 2006). This causes a change from amictic to mictic reproductive physiology and females begin producing mictic daughters who in turn produce eggs by meiosis (Wallace and Snell 2001). Unfertilized mictic females produce males and fertilized mictic females produce resting eggs. Fertilization necessitates a substantially different type of yolk to be produced by the vitellarium for resting egg formation. This extensive change in reproductive physiology is likely accomplished by expression of a new set of genes for sexual reproduction. The chemical signals triggering sexual reproduction in rotifers (mixis) are analogous to bacterial quorum sensing and its discovery in rotifers is a first for aquatic animals (Kubanek and Snell 2008).

Mictic females have a window for fertilization of only a few hours. Otherwise, they will produce males, even if later inseminated with viable sperm. Once mictic females are produced, they must produce resting eggs, either themselves or through their sons, or their fitness will be zero. Commitment to mictic reproduction is therefore irrevocable. Because of the serious consequences of failure, there is a marked advantage for females to elicit male copulations at every opportunity. Females have little influence on males beyond the glycoprotein signal on their body surface (mate recognition pheromone – MRP). It must be attractive enough to males to elicit mating and stimulate insemination before females are too old to alter their reproductive physiology for resting egg production. Moreover, if inseminated, females must have a high quality diet to form resting eggs, substantially fewer of which are produced than amictic or male eggs (Wallace et al. 2006). Female reproductive performance is therefore largely dependent on the quality of their MRP signal to males (Snell and Stelzer 2005).

Rotifer males have a chemoreceptor in their corona that recognizes on contact the MRP of conspecific females (Snell et al. 1995). Male reception and response to female mating signals is key to understanding variation in the reproductive performance of male rotifers. Males use this sensory information to decide within a few seconds after female contact whether an encounter should result in mating or avoidance. This decision is critical to males since their fitness is dependent on finding and inseminating a mictic female within their short lifetime. A number of factors conspire to make this challenging. Brachionid males live only a few days and do not feed, lacking a mouth and digestive system. All of their energy is derived from their mothers and it is expended frugally to accomplish insemination. Males have relatively few sperm (<40) and therefore can execute only 10–20 copulations before they are expended. Males cannot detect females from a distance and must rely on chance encounters to locate potential mates (Snell and Garman 1986). All of these factors necessitate a finely tuned receptor to discriminate species, sex, and female age. Behaviorally males are choosy, allocating their limited sperm to females with only the highest fitness (Snell et al. 2007).

23.5 Outlook

Going forward, perhaps the first objective should be to identify the molecules responsible for mating communication in a variety of aquatic invertebrates. Although researchers have attempted this, progress has been slow because of instability of the signal molecules, low concentrations, uncertain timing of pheromone secretion, and the presence of blends. However, the reward is substantial because the identification of chemical signals and their receptors will provide the next generation of evolutionary ecologists with the tools to describe the molecular mechanisms of courtship and mating. This will permit investigation of the evolutionary dynamics of genetic variation for mating, the development of reproductive isolation, the process of speciation, and the extent of cryptic species biodiversity. Understanding the molecular basis of signaling also will provide mechanistic models of the evolutionary physiology of life cycle regulation, male and female fertility, and the cause of sexual selection and conflict. Comparing phylogenetic similarities in invertebrate signaling systems could provide insights about the neurobiological origins of animal sensory systems. A molecular description of invertebrate signaling also could help ecotoxicologists to predict when pollutants are likely to cause endocrine disruption, disrupt chemical signaling, and modify critical ecological processes like grazing, predation, secondary production. A mechanistic understanding of copepod signaling systems could also provide insight for controlling ectoparasites that so often plague intensive aquaculture (Bandilla et al. 2008).

Acknowledgment This work was supported by the National Science Foundation grants BE/ GenEn MCB-0412674 to TWS.

References

Bandilla M, Hakalahti-Sirén T, Valtonen ET (2008) Patterns of host switching in the fish ectoparasite *Argulus coregoni*. Behav Ecol Sociobiol 62:975–982

Blades PI, Youngbluth MJ (1979) Mating behavior of *Labidocera aestiva* (Copepoda, Calanoida). Mar Biol 51:339–355

Brewer MC (1998) Mating behaviours of *Daphnia pulicaria*, a cyclic parthenogen: comparisons with copepods. Philos Trans R Soc B 353:805–815

Bushmann PJ, Atema J (2000) Chemically mediated mate location and evaluation in the lobster, *Homarus americanus*. J Chem Ecol 26:883–899

Capone A, Rosati F, Focarelli R (1999) A 140-kDa glycopeptide from the sperm ligand of the vitelline coat of the freshwater bivalve *Unio elongatulus*, only contains o-linked oligosaccharide chains and mediates sperm-egg interaction. Mol Reprod Dev 54:203–207

Caskey JL, Hasenstein KH, Bauer RT (2009a) Studies on contact sex pheromones of the caridean shrimp *Palaemonetes pugio*: I. Cuticular hydrocarbons associated with mate recognition. Invertebr Reprod Dev 53:93–103

Caskey JL, Watson GM, Bauer RT (2009b) Studies on contact sex pheromones of the caridean shrimp *Palaemonetes pugio*: II. The role of glucosamine in mate recognition. Invertebr Reprod Dev 53:105–116

Civetta A, Singh RS (1998) Sex-related genes, directional sexual selection, and speciation. Mol Biol Evol 15:901–909

Dusenbery DB (1992) Sensory ecology. How organisms acquire and respond to information. W.H. Freeman, New York

Dusenbery DB (2009) Living at micro scale. Harvard University Press, Cambridge

Dusenbery DB, Snell TW (1995) A critical body size for use of pheromones in mate location. J Chem Ecol 21:427–438

Etges WJ, Ahrens MA (2001) Premating isolation is determined by larval-rearing substrates in cactophilic *Drosophila mojavensis*. V. Deep geographic variation in epicuticular hydrocarbons among isolated populations. Am Nat 158:585–598

Ferveur J-F (2005) Cuticular hydrocarbons: their evolution and roles in *Drosophila* pheromonal communication. Behav Genet 35:279–295

Focarelli R, Rosati F (1995) The 220-kDa vitelline coat glycoprotein mediates sperm binding in the polarized egg of *Unio elongatulus* through O-linked oligosaccharides. Dev Biol 171:606–614

Frey MA, Lonsdale DJ, Snell TW (1998) The influence of contact chemical signals on mate recognition in a harpacticoid copepod. Philos Trans R Soc B 353:745–752

Glabe CG, Clark D (1991) The sequence of the *Arbacia punctulata* Bindin cDNA and implications for the structural basis of species-specific sperm adhesion and fertilization. Dev Biol 143:282–288

Glas PS, Green JD, Lynn JW (1996) Morphological evidence for a chitin-like glycoprotein in penaeid hatching envelopes. Biol Bull 191:374–384

Gómez D, Dupré E (2002) Monosacáridos terminales presentes en las cubiertas ovocitarias del camarón de roca *Rhynchocinetes typus*. Invest Mar 30:69–74

Hellberg ME, Moy GW, Vacquier VD (2000) Positive selection and propeptide repeats promote rapid interspecific divergence of a gastropod sperm protein. Mol Biol Evol 17:458–466

Herborg L-M, Bentley MG, Clare AS, Last KS (2006) Mating behaviour and chemical communication in the invasive Chinese mitten crab *Eriocheir sinensis*. J Exp Mar Biol Ecol 329:1–10

Katona SA (1973) Evidence for sex pheromones in planktonic copepods. Limnol Oceanogr 18:574–583

Kelly LS, Snell TW (1998) Role of surface glycoproteins in mate-guarding of the marine harpacticoid *Tigriopus japonicus*. Mar Biol 130:605–612

Kelly LS, Snell TW, Lonsdale DJ (1998) Chemical communication during mating of the harpacticoid *Tigriopus japonicus*. Philos Trans R Soc B 353:737–744

Kiørboe T (2007) Mate finding, mating, and population dynamics in a planktonic copepod *Oithona davisae*: there are too few males. Limnol Oceanogr 52:1511–1522

Knowlton N (1993) Sibling species in the sea. Annu Rev Ecol Syst 24:189–216

Knowlton N (2000) Molecular genetic analysis of species boundaries in the sea. Hydrobiologia 420:73–90

Kubanek J, Snell TW (2008) Quorum sensing in rotifers. In: Bassler BL, Winans SC (eds) Chemical communication among microbes. ASM, Washington, pp 453–461

Kumari SS, Skinner DM (1993) Proteins of crustacean exoskeleton II: immunological evidence for their relatedness to cuticular proteins of two insects. J Exp Zool 265:195–210

Lonsdale DJ, Snell TW, Frey MA (1996) Lectin binding to surface glycoproteins on *Coullana* spp. (Copepoda, Harpacticoida) can inhibit mate guarding. Mar Behav Physiol 27:277–286

Lonsdale DJ, Frey MA, Snell TW (1998) The role of chemical signals in copepod reproduction. J Mar Syst 15:1–12

Lürling M, Scheffer M (2007) Info-disruption: pollution and the transfer of chemical information among organisms. Trends Ecol Evol 22:374–379

Marcial HS, Hagiwara A, Snell TW (2003) Estrogenic compounds affect development of harpacticoid copepod *Tigriopus japonicus*. Environ Toxicol Chem 22:3025–3030

Markow TA (1991) Sexual isolation among populations of *Drosophila mojavensis*. Evolution 45:1525–1529

Mauchline J (1977) The integumental sensilla and glands of pelagic Crustacea. J Mar Biol Assoc UK 57:973–994

Metz EC, Palumbi SR (1996) Positive selection and sequence rearrangements generate extensive polymorphism in the gamete recognition protein bindin. Mol Biol Evol 13:397–406

Miller D, Ax RL (1990) Carbohydrates and fertilization in animals. Mol Reprod Dev 26:184–198

Miyake A, Beyer J (1974) Blepharmone: a conjugation-inducing glycoprotein in the ciliate *Blepharisma*. Science 185:621–623

Palumbi SR (1992) Marine speciation on a small planet. Trends Ecol Evol 7:114–118

Peñalva-Arana DC, Lynch M, Robertson HM (2009) The chemoreceptor genes of the waterflea *Daphnia pulex*: many Grs but no Ors. BMC Evol Biol 9:79. doi:10.1186/1471-2148-9-79

Rittschof D (1990) Peptide-mediated behaviors in marine organisms: evidence for a common theme. J Chem Ecol 16:261–272

Rittschof D, Cohen JH (2004) Crustacean peptide and peptide-like pheromones and kairomones. Peptides 25:1503–1516

Sharon N, Lis H (1989) Lectins as cell recognition molecules. Science 246:227–234

Snell TW (1998) Chemical ecology of rotifers. Hydrobiologia 387(388):267–276

Snell TW, Carmona MJ (1994) Surface glycoproteins in copepods: potential signals for mate recognition. Hydrobiologia 292(293):255–264

Snell TW, Garman BL (1986) Encounter probabilities between male and female rotifers. J Exp Mar Biol Ecol 97:221–230

Snell TW, Hoff FH (1987) Fertilization and male fertility in the rotifer *Brachionus plicatilis*. Hydrobiologia 147:329–334

Snell TW, Joaquim-Justo C (2007) Workshop on rotifers in ecotoxicology. Hydrobiologia 593:227–232

Snell TW, Nacionales MA (1990) Sex pheromones and mate recognition in rotifers. Comp Biochem Physiol 97A:211–216

Snell TW, Stelzer CP (2005) Removal of surface glycoproteins and transfer among *Brachionus* species. Hydrobiologia 546:267–274

Snell TW, Rico-Martinez R, Kelly LN, Battle TE (1995) Identification of a sex pheromone from a rotifer. Mar Biol 123:347–353

Snell TW, Kubanek J, Carter W, Payne AB, Kim J, Hicks MK, Stelzer CP (2006) A protein signal triggers sexual reproduction in *Brachionus plicatilis* (Rotifera). Mar Biol 149:763–773

Snell TW, Kim J, Zelaya E, Resop R (2007) Mate choice and sexual conflict in *Brachionus plicatilis* (Rotifera). Hydrobiologia 593:151–157

Swanson WJ, Vacquier VD (1995) Extraordinary divergence and positive Darwinian selection in a fusagenic protein coating the acrosomal process of abalone spermatozoa. Proc Natl Acad Sci USA 92:4957–4961

Swanson WJ, Vacquier VD (2002a) The rapid evolution of reproductive proteins. Nat Genet 3:137–144

Swanson WJ, Vacquier VD (2002b) Reproductive protein evolution. Annu Rev Ecol Syst 33:161–179

Ting JH, Snell TW (2003) Purification and sequencing of a mate recognition protein from the copepod *Tigriopus japonicus*. Mar Biol 143:1–8

Ting JH, Kelly LS, Snell TW (2000) A surface glycoprotein in the marine copepod *Tigriopus japonicus* elicits male mate guarding. Mar Biol 137:31–37

Titelman J, Varpe O, Eliassen S, Fiksen O (2007) Copepod mating: chance or choice? J Plankton Res 29:1023–1030

Vacquier VD (1998) Evolution of gamete recognition proteins. Science 281:1995–1998

Vacquier VD, Swanson WJ, Hellberg ME (1995) What have we learned about sea urchin sperm bindin? Dev Growth Differ 37:1–10

Van Damme K, Dumont HJ (2006) Sex in a cyclical parthenogen: mating behaviour of *Chydorus sphaericus* (Crustacea; Branchiopoda; Anomopoda). Freshwater Biol 51:2334–2346

Wallace RL, Snell TW (2001) Rotifera. In: Thorp JH, Covich AP (eds) Ecology and systematics of North American freshwater invertebrates. Academic, New York, pp 195–254

Wallace RL, Snell TW, Ricci C, Nogrady T (2006) Rotifera: Volume 1: Biology, ecology and systematics, Backhuys, Leiden

Wiese L (1965) On sexual agglutination and mating type substances (gamones) in isogamous heterophallic *Chlamydamondes*. I. Evidence of the identity of the gamone with the surface components responsible for sexual flagellar content. J Phycol 1:46–54

Wikramanayake AH, Clark W Jr (1994) Two extracellular matrices from oocytes of the marine shrimp *Sicyonia ingentis* that independently mediate only primary or secondary sperm binding. Dev Growth Differ 36:89–101

Willis JH (1999) Cuticular proteins in insects and crustaceans. Am Zool 39:600–609

Winsor GL, Innes DJ (2002) Sexual reproduction in *Daphnia pulex* (Crustacea: Cladocera): observations on male mating behavior and avoidance of inbreeding. Freshwater Biol 47:441–450

Zhang D, Lin J (2006) Mate recognition in a simultaneous hermaphroditic shrimp, *Lysmata wurdemanni* (Caridea: Hippolytidae). Anim Behav 71:1191–1196

Zhang D, Lin J, Harley M, Hardege JD (2009) Characterization of a sex pheromone in a simultaneous hermaphroditic shrimp, *Lysmata wurdemanni*. Mar Biol 157:1–6

Chapter 24
A Review of Research in Fish Pheromones

Yu-Wen Chung-Davidson, Mar Huertas, and Weiming Li

Abstract This review provides selected examples of several types of chemical signals and cues important for the social behavior of fish. Alarm substances evoke antipredator behaviors, typified by increased shoaling, refuging, freezing, dashing, area avoidance, and reduced foraging. Migratory pheromones are employed by some fish species that migrate long distances to locate home streams or spawning grounds. Many fishes employ sex pheromones to attract members of the opposite sex or to elicit spawning behavior. Steroids, prostaglandins, bile acids and amino acids have all been shown to serve as sex pheromones in fishes. Pheromones can also be used to recognize kin and establish hierarchies. Nonspecific diet metabolites as well as specific pheromones are important in chemical mediation of social behavior in fish. The use of pheromones in fisheries started long before scientific proof of their existence. Sea lamprey (*Petromyzon marinus*) pheromone is the first vertebrate pheromone that has been tested in the field as a pest control agent. Other potential applications of pheromone usage include conservation of endangered species or aquaculture. Similar approaches could also be useful for other aquatic organisms, including crustaceans.

24.1 Introduction

Chemoreception (chemical senses) is critical for animal survival. In order to find food, avoid predators and synchronize reproduction, animals must coordinate their life functions with environmental cues. For many aquatic animals, vision is sometimes of limited utility since freshwater is frequently turbid. Olfaction, on the other hand, is an advantageous sensory modality since olfactory epithelia are continuously exposed to a mixture of chemicals dissolved in water originating from different sources such as plants, soil and other animals. The ability to detect

Y.-W. Chung-Davidson (✉)
Department of Fisheries and Wildlife, Michigan State University,
East Lansing MI 48824, USA
e-mail: chungyuw@msu.edu

T. Breithaupt and M. Thiel (eds.), *Chemical Communication in Crustaceans*,
DOI 10.1007/978-0-387-77101-4_24, © Springer Science+Business Media, LLC 2011

chemicals and discriminate between mixtures is highly beneficial for aquatic animals (review see Burnard et al. 2008), and it is no coincidence that chemical communication existed long before aquatic animals evolved; even bacteria utilize chemoreception (Winans and Bassler 2008).

Some aquatic animals evolved the ability to release and detect odorants that stimulate specific, innate, and adaptive responses among individuals of their own species (conspecifics). The odorous molecules that mediate communication among conspecifics, known as pheromones, regulate a wide variety of behaviors such as individual identification, group cohesion, parent–offspring recognition, territorial markings, sex attraction, synchronization of reproductive processes, and migration (for a definition of the term "pheromone," see Wyatt, Chap. 2). Pheromone molecules used by aquatic animals are water-soluble and their diffusion rate can be 10,000 times lower than in the air (see also Weissburg, Chap. 4). In the aquatic environment pheromones are further subjected to turbulence and variable flows. The use of water currents, natural or generated by the pheromone donors, amplifies the effectiveness of pheromone cues. However, the action of pheromone molecules can be altered or disrupted by other dissolved compounds such as humic acid, which can bind pheromone molecules or inhibit receptors (Hubbard et al. 2002), or by heavy metals that can disrupt the function of olfactory sensory neurons or cause degeneration of the olfactory system (Brown et al. 1982). See also Olsen, Chap. 26, for a review of the effect of pollutants on chemical communication in fishes and crustaceans.

Crustaceans and fishes both live in aquatic environments, but they are phylogenetically distant and the morphology and mechanisms of their chemosensory systems are quite different. Experimental approaches using members of these two taxa can therefore complement and stimulate each other. This review provides

Fig. 24.1 Some fish species used in pheromone research: European eel (*Anguilla anguilla*), goldfish (*Carassius auratus*), fathead minnow (*Pimephales promelas*), coho salmon (*Oncorhynchus kisutch*), and sea lamprey (*Petromyzon marinus*)

selected examples of fish pheromones (Fig. 24.1) to describe general concepts. Interested readers should refer to original literature for more in-depth discussions of fish pheromones.

Fish pheromones reflect the diversity of evolutionary strategies employed in the aquatic environment. In fact, fishes have the greatest number of reproductive strategies among vertebrate animals (review see Mank and Avise 2006). The complexity of fish pheromone communication requires and facilitates a multidisciplinary approach to conduct meaningful research. Dr. Yu-Wen Chung-Davidson applies neurobiological techniques to study the neuronal and neuroendocrine responses of the fish brain to pheromones. Dr. Mar Huertas utilizes comparative endocrinology to study the subtle interactions between reproductive hormones and pheromones throughout the life history of fish. Dr. Weiming Li, as a fishery biologist, is motivated by the diversity and sensitivity of the responses of fish to pheromones in their natural habitats, and is exploring the possibility of using pheromones to control invasive species.

24.2   Alarm Substances

Whether alarm substances are pheromones is still under debate (Williams 1964, 1992; Magurran et al. 1996; Wisenden and Chivers 2006; Chivers et al. 2007), but early discoveries of fish alarm responses triggered much of the subsequent research in fish pheromones. von Frisch in 1938 reported that European minnows, *Phoxinus phoxinus*, displayed a marked fright reaction to conspecific skin extracts. This was attributed to the presence of an alarm substance or "Schreckstoff." Many fishes in the superorder of Ostariophysi possess specialized epidermal club cells which, when mechanically damaged, release a chemical alarm signal into the water column. When detected by nearby conspecifics, alarm substances elicit antipredator behavior (alarm response), which is typified by increased shoaling, shelter usage (refuging), freezing, dashing, area avoidance, and reduced foraging. Chemical alarm cues can also elicit a variety of covert behavioral responses, including acquired recognition of novel predators, induced morphological and life history changes, and the assessment of local predation risk through predator inspection behavior (Brönmark and Miner 1992; Smith 1992; Stabell and Lwin 1997; Chivers and Smith 1998; for review see Chivers and Mirza 2001).

In the majority of species studied, alarm substances are only released following mechanical damage to the skin as would occur during a predation event; however, there are also examples of nondamage release alarm signaling systems (e.g., Iowa darters, *Etheostoma exile*; Wisenden et al. 1995). The nitrogen oxide (N–O) functional group is a significant component of the ostariophysian chemical alarm signal system. The conservation of N–O groups as the molecular trigger was likely selected because they are easy to produce, allow for reliable cross-species responses among taxonomically related species independent of prior experience, and are stable in natural environment (Brown et al. 2001).

The neural pathways mediating alarm responses were examined in the crucian carp (*Carassius carassius* L.). In these fish, two olfactory tracts convey information from the olfactory bulbs [adjacent to the olfactory organs (i.e., nostrils)] to other parts of the brain. One courses along the midline (the medial olfactory tract) and the other along the side (the lateral olfactory tract). The medial olfactory tract further divides into two bundles (the medial and the lateral bundles of the medial olfactory tract). Severing the medial bundle of the medial olfactory tract eliminated the alarm responses to skin extract, whereas severing the lateral bundle of the medial olfactory tract diminished the feeding behavior (Hamdani et al. 2000).

Williams (1964, 1992), however, argued that there are considerable problems in explaining the evolution of an alarm pheromone. It was assumed that individuals produced alarm substance to warn their school or species of danger, but schools of fish are not composed of closely related individuals (Naish et al. 1993). Magurran et al. (1996) further demonstrated that fright responses in fish were elicited in a context-dependent manner. The alarm responses were likely exaggerated in the laboratory condition where the opportunities for escape were largely reduced. In the natural environment, alarm substances did not produce adaptive behaviors. In crustaceans, behaviors similar to the alarm response in fish can be elicited by the reception of injured conspecifics (Hazlett, Chap. 18).

24.3   Migratory Pheromones

Many aquatic animals guide themselves across open waters, sometimes to specific geographic areas to reproduce or to cope with environmental changes. Geomagnetic, hydrodynamic, and chemical signals (perhaps supplemented in some cases by celestial cues), may provide the foundation for underwater navigation (review see Lohmann et al. 2008). Olfaction is important for salmon to migrate long distance to their home streams for spawning (Ueda and Yamauchi 1995; Ueda and Shoji 2002), and olfactory imprinting may underlie the orientation mechanism of salmon migration (Harden Jones 1968; Hasler and Scholz 1983). Juvenile salmon become imprinted (learning that occurs at a particular life stage) to odorants of abiotic or biotic origin in natal streams, including small inorganic ions such as calcium (Bodznick 1978) and larger organic compounds associated with microbial decay or pheromones (Hasler and Scholz 1983). The imprinting may be a single event or may occur sequentially during downstream migration as smolts (Hansen and Jonsson 1994).

Adult sea lampreys (*P. marinus*) locate spawning streams using a pheromone released by stream-resident larvae (Moore and Schleen 1980; Teeter 1980; Sorensen et al. 2003). Stream selection by migratory adult sea lamprey in the Great Lakes is not species-specific (Morman et al. 1980; Klar et al. 1997). Tagging studies have shown that this selectivity does not reflect a tendency of adults to return to their natal streams (Bergstedt and Seelye 1995). The cue is rather a composite of biologically relevant compounds that migratory adults have developed the ability to recognize (Sorensen et al. 2003), which ensures the reproductive success of lamprey species with poor swimming ability and subject to the movements of their host that might bring them into unsuitable environments for spawning (Polkinghorne et al. 2001).

Petromyzonol sulfate and allocholic acid are commonly produced and released by larval petromyzontid lampreys and likely used as part of a common evolutionarily conserved pheromone. This scenario is reasonable because different lamprey species have similar larval (silt) and spawning (gravel) habitat requirements, and their larvae derive no apparent benefit from producing compounds that serve as an attractant for adults (Fine et al. 2004). However, not all of the activity of the larval pheromone could be explained by these two bile acids released by larvae (Vrieze and Sorensen 2001). A recent study identified two new compounds, petromyzonamine disulfate and petromyzosterol disulfate, released by larval sea lampreys and hypothesized to function as pheromones (Sorensen et al. 2005). Currently, there is no evidence that crustaceans use long distance pheromones.

24.4   Sex Pheromones

Many animals employ sex pheromones during their reproductive seasons. However, most sex pheromones of vertebrates remain unidentified (Marchlewska-Koj et al. 2001). Soon after Døving (1976) hypothesized that fish are predisposed to evolve pheromonal functions for released sex hormones, Colombo et al. (1980) reported that etiocholanolone glucuronide functions as a male pheromone in the black goby (*Gobius niger* or *Gobius jozo*) and van den Hurk and Lambert (1983) proposed that sex steroid glucuronides function as female sex pheromones in zebrafish (*Danio rerio*). Since then, evidence for pheromonal actions of steroids and prostaglandins (Fig. 24.2) has been reported in a diversity of fishes including Salmoniformes (salmons and trouts), Perciformes (perch-like fish) and Cypriniformes (e.g., carps). Other fishes such as Characiformes (e.g., tetras), Siluriformes (e.g., catfish), Elopiformes (e.g., ladyfish), and Osmeriformes (e.g., noodlefishes) also show sensitive and specific olfactory responses to sex steroids and prostaglandins examined by electro-olfactogram (EOG; Fig. 24.3) (for review see Stacey 2003). These pheromones are broadly termed "hormonal pheromones" since they also act as hormones or are metabolites of hormones in these fishes.

Similar to reports from fishes, the crustacean molting hormone 20-hydroxyecdysone (also known as crustecdysone) had also been proposed to function as a sex pheromone in several decapod species (Kittredge et al. 1971). However, later studies have shown that crustecdysone does not play a role as sex pheromone in crab species such as *Callinectes sapidus* and *C. maenas* (Gleeson et al. 1984; Hardege et al. 2002) but rather appears to function as feeding deterrent (Chang, Chap. 21).

In the course of fish pheromone research, different experimental approaches have been employed to identify the pheromone compounds. The first approach is driven by the hormonal pheromone hypothesis and based on the understanding of

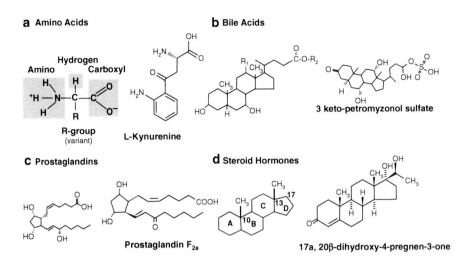

Fig. 24.2 Chemical structure of sex pheromones in fish. For each category of chemicals known to function as fish pheromones, the general structure is followed by an example of a fish pheromone

the reproductive endocrinology of the species of interest. Typically, a hormone molecule would be speculated and examined by EOG and/or behavioral tests to confirm its pheromonal function. A highly successful application of this approach was demonstrated in the goldfish (*Carassius auratus*) pheromone system by Dr. Norm Stacey and his colleagues (Dulka et al. 1987). Female goldfish release preovulatory steroids and postovulatory prostaglandins which affect male physiology, behavior, and mating success. The preovulatory pheromone is a mixture of androstenedione (AD), 17α, 20β-dihydroxy-4-pregnen-3-one (17, 20β-P), and 17, 20β-P-20β-sulfate (17, 20β-P-S), in which steroid ratios change dramatically during the preovulatory luteinizing hormone (LH) surge. At ovulation (~12 h after the onset of LH surge), preovulatory pheromone release drops drastically, and oviductal oocytes stimulate synthesis of prostaglandin $F_{2\alpha}$ ($PGF_{2\alpha}$) that acts hormonally to trigger female sexual behavior. $PGF_{2\alpha}$ and its metabolite, 15-keto-$PGF_{2\alpha}$, are then released in urine as a postovulatory pheromone acting on sensitive olfactory receptors to trigger courtship and increase LH and milt production in males. In addition to complex male responses to female pheromones, other females also respond to 17, 20β-P by increasing ovulation, indicating a mechanism for females to synchronize ovulation (for reviews see Sorensen and Stacey 1999; Stacey and Sorensen 2002, 2006; Stacey 2003).

In species where pheromones are not typical hormones, intensive chemical analyses have been productive in search of pheromone compounds. Such analyses usually involve exhaustive chromatographic and spectrometric methods including high performance liquid chromatography (HPLC), mass spectrometry (MS), and nuclear magnetic resonance (NMR) to concentrate, purify and identify compounds in water conditioned by mature animals, and to confirm the pheromone functions through EOG and/or behavioral tests. Chemical analysis has been used successfully

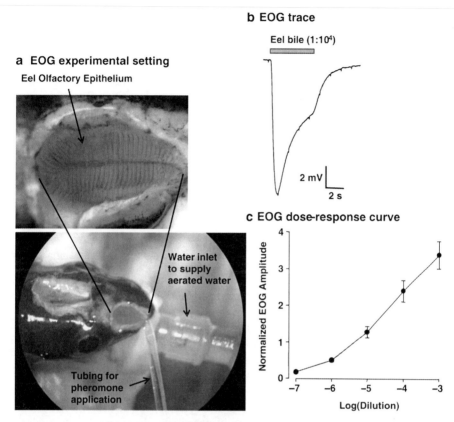

b EOG trace

Eel bile (1:10⁴)

a EOG experimental setting

Eel Olfactory Epithelium

2 mV
2 s

c EOG dose-response curve

Water inlet
to supply
aerated water

Tubing for
pheromone
application

Fig. 24.3 (**a**) An experimental setting for electro-olfactogram (EOG) recording. The fish is anesthetized and its respiration is supported by a constant supply of aerated water flowing through the mouth to the gill. The olfactory epithelium is exposed and an electrode is placed at the surface of the olfactory epithelium. A reference electrode is placed in a different part of the head. Putative pheromones are applied into the olfactory epithelium and the electrical activities are recorded and analyzed. (**b**) A typical EOG trace recorded from an immature eel in response to bile fluid from eels (diluted 1:10,000). The small, regular peaks (~1 Hz) were assumed to be due to the heart-beat, a regular characteristic when recording from eels. A downward deflection of the trace is negative (modified from Huertas et al. 2010). (**c**) Semi-logarithmic plot of pooled normalized EOG amplitudes recorded in immature male eels in response to dilution of bile fluid from eels (modified from Huertas et al. 2010). Data are mean $\pm$ s.e.m. ($n = 6$)

in sea lamprey pheromone research (for review see Li et al. 2003). For example, initial studies showed that only spermiating males released a pheromone which induced search behavior in a two-choice maze and robust upstream movement in a natural spawning stream in ovulatory females (Li et al. 2002; Siefkes et al. 2003, 2005; Johnson et al. 2005, 2006, 2009). Activity-directed fractionation was then used to isolate the active component from the washings of spermiating males (Li et al. 2002). Through integration of HPLC, MS and NMR, the most active compound was identified as 7α, 12α, 24-trihydroxy-5α-cholan-3-one-24-sulfate (3 keto-petromyzonol sulfate, 3kPZS; Fig. 24.2). This compound was subsequently found to be released by mature males from their gills and attract females to spawning nests (Siefkes et al. 2003, 2005; Johnson et al. 2005, 2006, 2009).

Bioassay-guided fractionation also resulted in identification of a male-attracting pheromone from the urine of reproductively mature female masu salmon (*Oncorhynchus masou*). The active compound appears to be L-kynurenine (Fig. 24.2), a major metabolite of L-tryptophan. This pheromone elicits a male-specific behavior at picomolar concentrations (Yambe et al. 2006). For a review of the progress in crustacean sex pheromone detection and identification, interested readers can refer to the chapters of Kamio and Derby (Chap. 20) on identification of blue crab sex pheromones, and of Hardege and Terschak (Chap. 19) on the identification of shore crab sex pheromone.

24.5 Pheromones as Social Cues

Recognition of relatives (kin discrimination) in salmons by olfaction was briefly reviewed by Olsén (1999). Juvenile salmonids such as Arctic char (*Salvelinus alpinus*) and coho salmon (*Oncorhynchus kisutch*) are attracted to the odor of their siblings over the odor of nonsibling conspecifics. When raised from the egg stage with other nonrelated sibling groups, however, the ability of the young salmonids to discriminate their siblings from distant relatives apparently disappears. Young three-spine sticklebacks (*Gasterosteus aculeatus*), however, still discriminate siblings from nonsiblings even when raised along with nonsiblings, but the strength of the discrimination is weakened after exposure to nonsiblings. This indicates that there is a genetic basis for kin recognition in fish, but the ability to discriminate kin can sometimes be masked by social experience (Olsén 1999).

Bullhead catfish (*Ictalurus nebulosus*) detect the body odors of conspecifics indicative of dominant relationships, and increase territorial aggression toward chemical "strangers." Nonspecific diet metabolites as well as specific pheromones are important in chemical mediation of social behavior (Bryant and Atema 1987). Pheromone-mediated social behaviors were also observed in the Nile (*Oreochromis niloticus*) and the Mozambique tilapia (*O. mossambicus*). They use pheromones to establish hierarchies, display elaborate courtship rituals and parental care (mouth-brooding) (Miranda et al. 2005; Barata et al. 2008). In crustaceans, brood pheromone has been associated with maternal behavior in crayfish (Little 1975, 1976). Chemical cues in the urine are also important for crustaceans to recognize individuals and to establish dominance hierarchies (Katoh et al. 2008; Skog et al. 2009). A review on crayfish courtship and dominance pheromones can be found in this volume (Breithaupt, Chap. 13).

24.6 Implications of Pheromones for Population Management

The use of pheromones in fisheries started long before scientific proof of their existence. French fishermen used to bait their sea lamprey traps with mature males to increase their harvest. In the Great Lakes basin, the sea lamprey is an invasive nuisance animal. They dispersed through the Welland Canal, a shipping canal in part of the St. Lawrence Seaway. Upon reaching the Great Lakes, lampreys parasitized the resident trout and caused the collapse of Great Lakes fish populations (Christie and Goddard 2003).

To mitigate such impacts, the U.S. and Canada jointly established the Great Lakes Fishery Commission in 1955 to develop control strategies. Within a few years, lampricides 4-nitro-3-(trifluoromethyl) phenol (TFM) and 2′,5′-dichloro-4-nitrosalicylanilide were developed and used to exterminate lamprey larvae prior to metamorphosis into the parasitic life stage. Though effective, concerns about the costs for lampricide application and possible impacts on non-target species and the environment motivated control agencies to pursue alternative or additional control methods (Christie and Goddard 2003).

The discovery that pheromones regulate specific migratory and reproductive behaviors of the sea lamprey (Fig. 24.4) has prompted interest in testing their potential as nontoxic and cost-effective alternatives to lampricides (Li et al. 2003; Johnson et al. 2005, 2006, 2009). Although pheromones have been employed for over 30 years to control insect pests, research in sea lamprey represents the first attempt to use vertebrate pheromones as part of an integrated pest management strategy (Johnson et al. 2005, 2006, 2009; Wagner et al. 2006).

Pheromones can be used to attract and capture sea lampreys effectively at multiple points during the spawning migration and have the potential to enhance integrated pest management of sea lamprey in the Great Lakes (Li et al. 2003). For example, in streams where trapping has proven effective, catch rates are often low or variable in years following removal of the larval population with lampricide treatment (Moore and Schleen 1980). The addition of a migratory pheromone to an existing barrier-integrated trapping system may increase capture efficiency and stabilize annual variation in capture rates by ensuring that consistent signal strength is maintained in posttreatment years. Although this may ultimately recruit a larger spawning population below the barrier, sex-pheromone-baited traps should be effective at removing ripe females immediately prior to spawning (Fig. 24.5; Johnson et al. 2005, 2006, 2009). Similar attract-and-annihilate strategies have proven effective in managing many insect pests. For attempts to use pheromones for the control of invasive shore crab species in the US, interested readers can refer to the chapter of Hardege and Terschak (Chap. 19).

Although the sea lamprey is a nuisance animal in the Great Lakes, it has great commercial value in European fish markets (Beaulaton et al. 2008). A generic lamprey bile acid pheromone has been proposed to enhance runs of endangered and threatened species of indigenous costal Pacific and European petromyzontid lampreys in their historic spawning habitats (Tuunainen et al. 1980; Close et al. 2002). However, it will be more challenging to apply pheromones in the Columbia basin than in the Great Lakes region since the tributaries in the Columbia basin are usually much bigger than those of the Great Lakes. Pheromones would have to be released in the lower reaches near to the junction points where extant populations could be attracted in.

a Early Spring

b May through July

Fig. 24.4 Schematic diagrams of pheromonal communication in the life history of the sea lamprey. Sea lampreys have a free-living larval stage with a filter-feeding growth phase of variable length which terminates at metamorphosis. In early spring, parasitic sea lampreys in the lakes or oceans cease feeding and begin an upstream spawning migration. During this period, chemoreception of migratory pheromones released by stream-resident larval conspecifics is used to locate a suitable spawning stream. As male sea lampreys arrive at the spawning ground, they build nests and complete final sexual maturation. Reproductively mature males release sex pheromones, which induce preference and searching behavior in ovulating female conspecifics. Drawing by Jorge A. Varela Ramos

The use of pheromones can also enhance the conservation of other endangered species. For example, the European eel, *Anguilla anguilla*, is an ancestral teleost fish with a complex life history which includes a catadromous migration from freshwater to open ocean (Dekker 2003; Starkie 2003). A combination of factors, such as over-exploitation of natural stocks, the changes in oceanographic conditions (e.g., global warming), degradation and reduction of suitable freshwater habitats, pollution and parasitism, has decreased wild glass and adult eel stocks to the point that European eel is expected to be enlisted as a CITES endangered species (http://www.cites.org/). Moreover, the aquaculture industry depends on the capture of wild eel as brood stocks, further increasing the pressure on wild stocks. Several aspects of eel biology, together with their highly developed sense of smell (Silver 1982;

Fig. 24.5 Field investigation of the sea lamprey sex pheromone to develop management application. (**a**) Researchers measure sea lamprey pheromones for a field trial. (**b**) A fluorescent dye (*asterisk*) is applied to measure a pheromone plume in the stream. (**c**) An ovulatory female sea lamprey is attracted to the source of the male pheromone (*double asterisk* tube opening hidden under the rocks). (**d**) Traps baited with sea lamprey pheromones (photos provided by Dr. Nicholas S. Johnson)

Sola 1993; Huertas et al. 2008), suggest that chemical communication could be involved at key stages of their life history, such as the migration across the Atlantic Ocean and complex reproductive processes. Knowledge of eel olfaction could be included in hatchery operation protocols for the aquaculture industry or restoration of European eel. For example, sex pheromones, when identified, could potentially be used to induce reproduction under artificial conditions and produce offspring in a hatchery. Under such conditions the cultured animals could be used in the restocking of eel in European waters or reduce the needs for commercial fishing.

24.7   Future Directions

Crustaceans and fishes share similar habitats but have evolved different chemical communication systems that are adaptive to their life styles. Some fish and crustacean species use alarm substances to avoid predators; some use migratory pheromones to locate home or spawning streams; some use sex pheromones to attract members of the opposite sex for reproductive success; others use pheromones to identify kin or to protect their territories. Contact pheromones (an insoluble coating on the body surface) seem to be more commonly found in crustaceans living in a high-density population as a mate recognition signal (Bauer, Chap. 14), which are detected by antennae and antennules (Reidenbach et al. 2008). In addition to amino acids, bile acids, and hormonal pheromones used in fish species, crustaceans also use peptide pheromones. The release of peptide pheromones from the eggs of mud crab (*Rhithropanopeus harrisii*) during egg hatching cause the female crab to contract her abdomen rapidly (the pumping response). This action breaks open the unhatched eggs and results in the synchronized release of larvae (Forward et al. 1987). Identification of the chemical structure of all these pheromones is equally important in crustaceans and fish.

The initial discovery of Kittredge et al. (1971) that the crustacean molting hormone crustecdysone could act as a pheromone spurred research in fish hormonal pheromones (Sorensen 1992). The advances in the understanding of fish pheromone systems with respect to the chemical identity of pheromones and applications in the management of invasive species and aquaculture can also stimulate further research into crustacean pheromones. Crustaceans have very diverse mating systems and it will be fruitful to examine more species with different mating systems to understand the evolution of the pheromones and their roles in speciation (Paterson 1978; Shine et al. 2002). In addition, many crustacean species are of great economic value. There are also similar problems with invasive crustacean species as have already been confronted with fish species.

Pheromone communication in aquatic animals exemplifies their adaptation to the demands and possibilities of their environment. However, many dimensions of pheromone communication have not been explored. For example, little is known about the effect of pheromone compounds on sympatric species inhabiting the same environment (Olsén 1999; Fine et al. 2004). Therefore, the side effects of pheromone use should be evaluated before its application in the environment. The potential impacts of xenobiotics or environmental pollution on pheromone production and sensitivity should be considered as well (see Olsen, Chap. 26).

References

Barata EN, Fine JM, Hubbard PC, Almeida OG, Frade P, Sorensen PW, Canario AVM (2008) A sterol-like odorant in the urine of Mozambique tilapia males likely signals social dominance to females. J Chem Ecol 34:438–449

Beaulaton L, Taverny C, Castelnaud G (2008) Fishing, abundance and life history traits of the anadromous sea lamprey (*Petromyzon marinus*) in Europe. Fish Res 92:90–101

Bergstedt RA, Seelye JG (1995) Evidence for lack of homing by sea lamprey. Trans Am Fish Soc 124:235–239

Bodznick DD (1978) Calcium ion: an odorant for natural water discrimination and the migratory behavior of sockeye salmon. J Comp Physiol 127:157–166

Brönmark C, Miner JG (1992) Predator-induced phenotypic change in body morphology in crucian carp. Science 258:1348–1350

Brown SB, Evans RE, Thompson BE, Hara TJ (1982) Chemoreception and aquatic pollutants. In: Hara TJ (ed) Chemoreception in fish. Elsevier, Amsterdam, pp 363–393

Brown GE, Adrian JC Jr, Kaufman IH, Erickson JL, Gershaneck D (2001) Responses to nitrogen-oxides by chariciform fishes suggest evolutionary conservation in Ostariophysian alarm pheromones. In: Marchlewska-Koj A, Lepri J, Muller-Schwarze D (eds) Chemical signals in vertebrates. Plenum, New York, pp 305–312

Bryant BP, Atema J (1987) Diet manipulation affects social behavior of catfish: importance of body odor. J Chem Ecol 13:1645–1662

Burnard D, Gozlan RE, Griffiths SW (2008) The role of pheromones in freshwater fishes. J Fish Biol 73:1–16

Chivers DP, Mirza RS (2001) Predator diet cues and the assessment of predation risk by aquatic vertebrates: a review and prospectus. In: Marchlewska-Koj A, Lepri J, Muller-Schwarze D (eds) Chemical signals in vertebrates. Plenum, New York, pp 277–284

Chivers DP, Smith RJF (1998) Chemical alarm signaling in aquatic predator-prey systems: a review and prospectus. Ecoscience 5:338–352

Chivers DP, Zhao XO, Ferrari MCO (2007) Linking morphological and behavioural defenses: Prey fish detect the morphology of conspecifics in the odour signature of their predators. Ethology 113:733–739

Christie GC, Goddard CI (2003) Sea lamprey international symposium (SLIS II): advances in the integrated management of sea lamprey in the Great Lakes. J Great Lakes Res 29(suppl 1):1–14

Close DA, Fitzpatrick MS, Li HW (2002) The ecological and cultural importance of a species at risk of extinction, Pacific lamprey. Fisheries 27:19–25

Colombo L, Marconato A, Belvedere PC, Frisco C (1980) Endocrinology of teleost reproduction: a testicular steroid pheromone in the black goby, Gobius jozo L. Bull Zool 47:355–364

Dekker W (2003) Did lack of spawners cause the collapse of the European eel, Anguilla anguilla? Fish Manag Ecol 10:365–376

Døving K (1976) Evolutionary trends in olfaction. In: Benz G (ed) The structure-activity relationships in olfaction. IRL, London, pp 149–159

Dulka JG, Stacey NE, Sorensen PW, van der Kraak GJ (1987) A steroid sex pheromone synchronizes male-female spawning readiness in goldfish. Nature 325:251–253

Fine JM, Vrieze LA, Sorensen PW (2004) Evidence that petromyzontid lampreys employ a common migratory pheromone that is partially comprised of bile acids. J Chem Ecol 30:2091–2110

Forward RB Jr, Rittschof D, De Vries MC (1987) Peptide pheromones synchronize crustacean egg hatching and larval release. Chem Senses 12:491–498

Gleeson RA, Adams MA, Smith ABIII (1984) Characterization of a sex pheromone in the blue crab Callinectes sapidus crustecdysone studies. J Chem Ecol 10:913–922

Hamdani E-H, Stabell OB, Alexander G, Døving KB (2000) Alarm reaction in the crucian carp is mediated by the medial bundle of the medial olfactory tract. Chem Senses 25:103–109

Hansen LP, Jonsson B (1994) Homing of Atlantic salmon: effects of juvenile learning on transplanted post-spawners. Anim Behav 47:220–222

Hardege JD, Jennings A, Hayden D, Müller CT, Pascoe D, Bentley MG, Clare AS (2002) Novel behavioural assay and partial purification of a female-derived sex pheromone in Carcinus maenas. Mar Ecol Prog Ser 244:179–189

Harden Jones FR (1968) Fish migration. Edward Arnold, London

Hasler AD, Scholz AT (1983) Olfactory imprinting and homing in salmon. Springer, New York

Hubbard PC, Barata EN, Canario AVM (2002) Possible disruption of pheromonal communication by humic acid in the goldfish, Carassius auratus. Aquat Toxicol 60:169–183

Huertas M, Canário AVM, Hubbard PC (2008) Chemical communication in the Genus Anguilla: a mini review. Behaviour 145:1389–1407

Huertas M, Hagey L, Hofmann AF, Cerdá J, Canário AVM, Hubbard PC (2010) Olfactory sensitivity to bile fluid and bile salts in the European eel (*Anguilla Anguilla*), goldfish (*Carassius auratus*) and Mozambique tilapia (*Oreochromis mossambicus*) suggests a 'broad range' sensitivity not confined to those produced by conspecifics alone. J Exp Biol 213:308–317

Johnson NS, Siefkes MJ, Li W (2005) Capture of ovulating female sea lampreys in traps baited with spermiating male sea lampreys. North Am J Fish Manag 25:67–72

Johnson NS, Luehring MA, Siefkes MJ, Li W (2006) Mating pheromone reception and induced behavior in ovulating female sea lampreys. North Am J Fish Manag 26:88–96

Johnson NS, Yun S-S, Thompson HT, Brant CO, Li W (2009) A synthesized pheromone induces upstream movement in female sea lamprey and summons them into traps. Proc Natl Acad Sci USA 106:1021–1026

Katoh E, Johnson M, Breithaupt T (2008) Fighting behavior and the role of urinary signals in dominance assessment of Norway lobsters, *Nephrops norvegicus*. Behaviour 145:1447–1464

Kittredge JS, Terry M, Takahashi FT (1971) Sex pheromone activity of the molting hormone crustecdysone on male crabs (*Pachygrapsus crassipes*, *Cancer antennarius* and *Cancer anthonyi*). Fish Bull 69:337–343

Klar GT, Schlee LP, Young RJ (1997) Integrated management of sea lampreys in the Great Lakes 1996: Annual Report to the Great Lakes Fishery Commission. Ann Arbor, Great Lakes Fishery Commission, p 74

Li W, Scott AP, Siefkes MJ, Yan H, Liu Q, Yun S-S, Gage DA (2002) Bile acid secreted by male sea lamprey that acts as a sex pheromone. Science 296:138–141

Li W, Scott AP, Siefkes MJ, Yun S-S, Zielinski B (2003) A male pheromone in the sea lamprey (*Petromyzon marinus*): an overview. Fish Physiol Biochem 28:259–262

Little EE (1975) Chemical communication in maternal behavior of crayfish. Nature 255:400–401

Little EE (1976) Ontogeny of maternal behavior and brood pheromone in crayfish. J Comp Physiol 112:133–142

Lohmann KJ, Lohmann CMF, Endres CS (2008) The sensory ecology of ocean navigation. J Exp Biol 211:1719–1728

Magurran AE, Irving PW, Henderson PA (1996) Is there a fish alarm pheromone? A wild study and critique. Proc R Soc Lond B 263:1551–1556

Mank JE, Avise JC (2006) The evolution of reproductive and genomic diversity in ray-finned fishes: insights from phylogeny and comparative analysis. J Fish Biol 69:1–27

Marchlewska-Koj A, Lepri JJ, Müller-Schwarze D (2001) Chemical signals in vertebrates, vol 9. Kluwer Academic/Plenum, New York

Miranda A, Almeida OG, Hubbard PC, Barata EN, Canário AVM (2005) Olfactory discrimination of female reproductive status by male tilapia (*Oreochromis mossambicus*). J Exp Biol 208:2037–2043

Moore HH, Schleen LP (1980) Changes in spawning runs of sea lamprey (*Petromyzon marinus*) in selected streams of Lake Superior after chemical control. Can J Fish Aquat Sci 37:1851–1860

Morman RH, Cuddy DW, Rugen PC (1980) Factors influencing the distribution of sea lamprey (*Petromyzon marinus*) in the Great Lakes. Can J Fish Aquat Sci 37:1811–1826

Naish KA, Carvalho GR, Pitcher TJ (1993) The genetic structure and microdistribution of shoals of *Phoxinus phoxinus*, the European minnow. J Fish Biol 43(suppl A):75–89

Olsén H (1999) Present knowledge of kin discrimination in salmonids. Genetica 104:295–299

Paterson HE (1978) More evidence against speciation by reinforcement. South Afr J Sci 74:369–371

Polkinghorne CA, Olson JM, Gallaher DG, Sorensen PW (2001) Larval sea lamprey release two unique bile acids to the water at a rate sufficient to produce detectable riverine pheromonal plumes. Fish Physiol Biochem 24:15–30

Reidenbach MA, George N, Koehl MAR (2008) Antennules morphology and flicking kinematics facilitate odor sampling by the spiny lobster, *Panulirus argus*. J Exp Biol 211:2849–2858

Shine R, Reed RN, Shetty S, Lemaster M, Mason RT (2002) Reproductive isolating mechanisms between two sympatric sibling species of sea snakes. Evolution 56:1655–1662

Siefkes MJ, Scott AP, Zielinski B, Yun S-S, Li W (2003) Male sea lampreys, *Petromyzon marinus* L., excrete a sex pheromone from gill epithelia. Biol Reprod 69:125–132

Siefkes MJ, Winterstein SR, Li W (2005) Evidence that 3-keto petromyzonol sulphate specifically attracts ovulating female sea lamprey, *Petromyzon marinus*. Anim Behav 70:1037–1045

Silver W (1982) Electrophysiological responses from the peripheral olfactory system of the American eel, *Anguilla rostrata*. J Comp Physiol A Sens Neural Behav Physiol 148:379–388

Skog M, Chandrapavan A, Hallberg E, Breithaupt T (2009) Maintenance of dominance is mediated by urinary chemical signals in male European lobsters, *Homarus gammarus*. Mar Fresh Behav Physiol 42:119–133

Smith RJF (1992) Alarm signals in fishes. Rev Fish Biol Fish 2:33–63

Sola C (1993) Bile salts and taurine as chemical stimuli for glass eels, *Anguilla anguilla*: a behavioral study. Environ Biol Fishes 37:197–204

Sorensen PW (1992) Hormones, pheromones and chemoreception. In: Hara TJ (ed) Fish chemoreception. Chapman and Hall, London, pp 199–228

Sorensen PW, Stacey NE (1999) Evolution and specialization of fish hormonal pheromones. In: Johnston RE, Müller-Schwarze D, Sorensen PW (eds) Advances in chemical signals in vertebrates. Kluwer/Plenum, New York, pp 14–47

Sorensen PW, Vrieze LA, Fine JM (2003) A multi-component migratory pheromone in the sea lamprey. Fish Physiol Biochem 28:254–257

Sorensen PW, Fine JM, Dvornikovs V, Jeffrey CS, Shao F, Wang J, Vrieze LA, Anderson KR, Hoye TR (2005) Mixture of new sulfated steroids functions as a migratory pheromone in the sea lamprey. Nat Chem Biol 1:324–328

Stabell OB, Lwin MS (1997) Predator-induced phenotypic changes in crucian carp are caused by chemical signals from conspecifics. Environ Biol Fish 49:145–149

Stacey N (2003) Hormones, pheromones and reproductive behavior. Fish Physiol Biochem 28:229–235

Stacey NE, Sorensen PW (2002) Fish hormonal pheromones. In: Pfaff DW, Arnold AP, Etgen AM, Fahrbach SE, Rubin RT (eds) Hormones, brain, and behavior, vol 2. Academic, New York, pp 375–435

Stacey NE, Sorensen PW (2006) Reproductive pheromones. In: Sloman KA, Wilson RW, Balshine S (eds) Behavior and physiology of fish. Elsevier, Amsterdam, pp 359–412

Starkie A (2003) Management issues relating to the European eel, *Anguilla anguilla*. Fish Manag Ecol 10:361–364

Teeter J (1980) Pheromone communication in sea lampreys (*Petromyzon marinus*) – implications for population management. Can J Fish Aquat Sci 37:2123–2132

Tuunainen O, Ikonen E, Auvinen H (1980) Lampreys and lamprey fisheries in Finland. Can J Fish Aquat Sci 37:1953–1959

Ueda H, Shoji T (2002) Physiological mechanisms of homing migration in salmon. Fish Sci 68 (suppl 1):53–56

Ueda H, Yamauchi K (1995) Biochemistry of fish migration. In: Hochachka PW, Mommsen TP (eds) Biochemistry and molecular biology of fishes. Elsevier Science B.V, Amsterdam, pp 265–279

van den Hurk R, Lambert JGD (1983) Ovarian steroid glucuronides function as sex pheromones for male zebrafish, *Brachydanio rerio*. Can J Zool 61:2381–2387

von Frisch K (1938) Zur Psychologie des Fisch-Schwarmes. Naturwissenschaften 26:601–606

Vrieze LA, Sorensen PW (2001) Laboratory assessment of the role of a larval pheromone and natural stream odor in spawning stream localization by migratory sea lamprey (*Petromyzon marinus*). Can J Fish Aquat Sci 58:2374–2385

Wagner CM, Jones ML, Twohey MB, Sorensen PW (2006) A field test verifies that pheromones can be useful for sea lamprey (*Petromyzon marinus*) control in the Great Lakes. Can J Fish Aquat Sci 63:475–479

Williams GC (1964) Measurement of consociation among fishes and comments on the evolution of schooling. Publ Mus Mich State Univ Biol Ser 2:349–384

Williams GC (1992) Natural selection: domains, levels and challenges. Oxford University Press, New York

Winans SC, Bassler BL (2008) Chemical communication among bacteria. ASM, Herndon

Wisenden BD, Chivers DP (2006) The role of public chemical information in antipredator behaviour. In: Ladich F, Collins SP, Moller P, Kapoor BG (eds) Fish communication. Science, New Hampshire, pp 259–278

Wisenden BD, Chivers DP, Smith RJF (1995) Early warning in the predation sequence: a disturbance pheromone in the Iowa darter (*Etheostoma exile*). J Chem Ecol 21:1469–1480

Yambe H, Kitamura S, Kamio M, Yamada M, Matsunaga S, Fusetani N, Yamazaki F (2006) L-kynurenine, an amino acid identified as a sex pheromone in the urine of ovulated female masu salmon. Proc Natl Acad Sci USA 103:15370–15374

Part V
Applied Aspects

Chapter 25
Chemical Communication and Aquaculture of Decapod Crustaceans: Needs, Problems, and Possible Solutions

Assaf Barki, Clive Jones, and Ilan Karplus

Abstract Chemical communication has received very little attention in the field of crustacean aquaculture research. It has been investigated mainly with regard to the extent that chemical cues might be involved in social control of growth, which results in growth suppression and size variation among individuals under culture conditions, but no consistent conclusion has emerged that stimulated further development and application of solutions based on this knowledge. Implementation of knowledge on chemical communication in crustacean aquaculture has not gone further than some preliminary trials of the incorporation of pheromones as feeding attractants or of their use to facilitate trapping. In this review we attempted to identify those aquaculture procedures in which knowledge of chemical communication might be implemented, to indicate possible uses and to suggest possible solutions related to various aquacultured crustaceans. In most cases solutions based on chemical communication could probably be integrated into current culture techniques and would serve to enhance their efficiency. To achieve this aim, pheromones and potent chemical components that mediate behavioral and physiological processes relevant to aquaculture should be identified and synthetic versions and technical means for their efficient application should be developed.

25.1 Introduction

Decapod crustaceans nowadays comprise a significant portion of the global production value of aquatic animals. The total annual production of crustaceans increased 15-fold during the last 20 years, to exceed four million tons (FAO 2008). Such an impressive increase in production has been achieved through the accumulation, over recent decades, of much knowledge on various basic aspects of the biology of crustaceans. This knowledge has been channeled towards extensive research and development

A. Barki (✉)
Aquaculture Research Unit, Institute of Animal Science, Agricultural
Research Organization, Volcani Center, Dagan 50250, Israel
e-mail: barkia@volcani.agri.gov.il

T. Breithaupt and M. Thiel (eds.), *Chemical Communication in Crustaceans*,
DOI 10.1007/978-0-387-77101-4_25, © Springer Science+Business Media, LLC 2011

activities that address reproduction in captivity, genetic improvement and breeding programs, nutrition, disease control, and culture management. Having studied the growth of freshwater prawns (*Macrobrachium rosenbergii*) under aquaculture conditions, we were stunned by the dramatic effect of the social environment on growth in this commercially important species: the growth of sibling males may vary by an order of magnitude within a few months! Moreover, the growth-suppressed small males would resume growing if the dominant large males were removed from their vicinity. It was clear to us that a prerequisite for resolution of the problem of size variation is the understanding of the basic mechanisms underlying this common phenomenon among aquacultured crustaceans. We were thus motivated to explore the behavioral mechanisms that mediate growth suppression in crustaceans, which involve, among other means of communication, the chemical modality.

A wealth of knowledge regarding chemical communication in crustaceans has also been accumulated. Most of the studies in this area addressed releaser pheromones that elicit specific social or alarm behaviors; much less emphasis has been placed on primer pheromones that control physiological processes and thereby influence important factors related to aquaculture, such as reproduction and growth. Despite the wealth of knowledge on chemical communication, very little of it is being applied to crustacean culture. However, this by no means implies that the existing knowledge cannot contribute to the solution of some currently unresolved problems or to the improvement of current practices in crustacean culture. Any improvement that can be done must be based on a sound knowledge of the cultured species, which includes, among other aspects, social interaction and chemical communication.

The main objective of this chapter is to present ideas on how chemical communication might be used in the various steps of crustacean aquaculture. In order to get a wider perspective, we will start with examples from two groups of economically important animals that in some manner are related to crustaceans and for which the knowledge on chemical communication is more advanced and is, to some degree, implemented in culture. The first is the honeybee, a representative of the related terrestrial arthropods, the insects; the second is the goldfish, a representative of the vertebrates, which inhabits the same aquatic environment in which chemical communication has evolved in crustaceans. Next, we will present a brief overview of crustacean aquaculture and the principle cultured species. We will then proceed chronologically through the steps of the culture procedure, for each of which we will attempt to highlight problems or needs, review previous research, and finally, suggest how chemical communication might be used in finding solutions.

25.2   Odors of Honey (Bee) and Gold (Fish)

The honeybee (*Apis mellifera* L.) is a social insect. The complex communication network underlying the social organization and behavior of honeybee colonies is primarily governed by pheromones released by the queen, the adult workers, and the brood. Various primer and releaser pheromones have been identified in honeybees. One of the major pheromones is the queen retinue pheromone (QRP), which is composed of a mixture of substances that includes the five-component queen mandibular pheromone (QMP) plus at least four additional identified components that do not originate from the mandibular gland. By releasing this pheromone, the queen signals her presence and maintains her control over the colony (for a review see Slessor et al. 2005).

On the basis of the knowledge about pheromonal communication among honeybees, several commercial products have been developed for applications in bee keeping and crop pollination. For example, QMP-based products are used to enhance mating success and queen survival in commercial operations, as substitutes for queen bees when shipping queen-less packs of worker bees, and to attract and hold honey bees to flowering crops in order to achieve more complete pollination and consequently improved fruit quantity, size, and quality (Winston and Slessor 1998). In addition, a product based on the Nasanov pheromone, a seven-component pheromone released by worker bees to attract other workers during swarming or at colony entrances, is used to attract swarms to unoccupied hive equipment or to a swarm-catching box.

In fishes, chemical cues are involved in various key aspects of life; they serve as antipredation alarm cues, as nonreproductive aggregants (in schooling and migration), and as reproductive aggregants and stimulants. A wide variety of substances have been implicated in these functions, but the best understood, with regard to comprehensive knowledge of their biochemical identity, synthesis, release, olfactory detection, and biological/behavioral effects, are hormonal pheromones in the goldfish (*Carassius auratus*). In goldfish and other carps, reproductive females release at least three hormonally derived pheromones that exert priming and releasing effects on males. These female pheromones bring about behavioral changes and mediate increased milt production and enhanced fertility in males. Furthermore, they may also have a priming effect on other females that results in synchrony of ovulations among females (for a review see Sorensen and Stacey 2004, see also Olsen, Chap. 26, and Chung-Davidson, Huertas, and Li, Chap. 24).

It appears that pheromones might be used in fish culture (at least in carps) to stimulate gonadal development and spawning, to synchronize spawning, to manipulate the length of breeding cycles, and to attract responsive males or females (e.g., for trapping or brood-stock purposes). However, despite our extensive knowledge about chemical communication among fish, including knowledge of the identity of specific substances that serve as pheromones, the use of pheromones has been attempted mainly within integrated pest control programs for invasive fish (the sea lamprey in the North American Great Lakes, see Chung-Davidson, Huertas, and Li, Chap. 24), but not for aquaculture purposes (Sorensen and Stacey 2004).

The honey bee example shows that there is promising potential for the use of pheromones in crustacean culture. However, the use of pheromones in the aquatic environment, even for fish culture, is in its infancy, and it should be noted that our knowledge of most aspects of chemical communication in crustaceans lags behind our knowledge of that in fishes.

25.3 Crustacean Aquaculture Overview

Worldwide, aquaculture involves a broad range of decapod crustaceans, and although they represent only 6.2% of total aquacultural animal production, in terms of economic yield they are the second most valuable (20.4%). Production is dominated by the marine penaeid shrimps, primarily *Penaeus vannamei*, the white-legged shrimp, and *P. monodon*, the black tiger shrimp, for which the combined world production (as of 2006) was about 3.2 million tons, worth US$ 12.5 billion, out of a total crustacean production of about 4.5 million tons, worth US$ 17.9 billion (FAO 2008). After the marine penaeid shrimp, the greatest aquacultural production is of freshwater species: crabs of the genus *Eriocheir*, known commonly as river or mitten crab, and freshwater prawns, primarily *Macrobrachium* species. Production of river crabs is restricted to China and South Korea, whereas freshwater prawns are widely cultured in tropical and subtropical regions. Marine crabs (*Scylla* and *Portunus* species) and freshwater crayfish of the families Astacidae and Parastacidae are produced in smaller volumes. Aquaculture of spiny lobsters (Palinuridae) and slipper lobsters (Scyllaridae) is under development and may contribute a significant proportion of production value in the near future (Phillips 2006; Jones 2009). Clawed lobsters have been the subject of considerable aquacultural research and development over many decades, but there has been no commercial outcome to date. In recent years the focus of research has been on stock enhancement programs for both American lobster (*Homarus americanus*) and European lobster (*H. gammarus*).

Decapod crustacean farming uses a variety of approaches, ranging from extensive to superintensive, although the bulk of production comes from semiintensive, pond-based systems (Wickins and Lee 2002) (Table 25.1). Most species have free-living larval stages and require hatchery production of seed, for which hatching and on-growing are managed separately. The general steps in farming of crustaceans are illustrated in Fig. 25.1.

Selective breeding programs are being developed for a range of species, including both of the main marine shrimp species, but there remains a necessity to source at least some of the required brood-stock from the wild for the production of young. With the exception of the freshwater crayfish, whose juveniles emerge directly from the hatched egg, all aquacultured crustaceans require a hatchery for the management of reproduction and larval rearing. Reproduction generally involves the selection of suitable brood animals, and their stocking at an appropriate sex ratio and density, to promote mating and spawning. As most crustaceans have high fecundity, relatively small numbers of brood-stock are usually required. Breeding tanks often have some environmental control that manages temperature, salinity, and water quality, and animals are often fed enriched diets, to stimulate reproduction. Once females have spawned and are carrying eggs, they are generally isolated for incubation, management of hatching, and capture of the larvae.

Rearing of the larvae is then managed in specifically designed tanks, often necessitating provision of specific larval diets that include both live feeds (micro-algae, rotifers, and *Artemia*) and manufactured feeds. For most of the crustaceans,

Table 25.1 Culture systems employed for the aquaculture of crustaceans

Species group	Typical culture system
Marine shrimp[a]	Hatchery production of postlarvae. Semiintensive grow-out in earthen ponds, 1–5 ha in surface area. Free range, high density, manufactured diet
River or mitten crabs[b]	Wild-caught and hatchery-produced seed. Extensive grow-out in rice paddies or cages within lakes, moderate density, supplementary feeding. Semiintensive grow-out in earthen ponds, high density, manufactured diet
Freshwater prawns 	Hatchery production of postlarvae. Extensive grow-out in rice paddies, low density. Semiintensive grow-out in earthen ponds, moderate to high density, manufactured diet
Swimming crabs[a]	Mostly wild-caught seed, developing hatchery sector. Extensive to semiintensive grow-out in earthen ponds, low to moderate densities, manufactured diets
Freshwater crayfish	No hatchery required. Extensive to semiintensive grow-out in earthen ponds, moderate densities, supplemental feeding, and manufactured diets
Lobsters[a]	Currently wild-caught seed only, although development of hatchery technology is well advanced. Semiintensive grow-out in sea cages, using trash fish

[a]Drawings reproduced from Wickins and Lee (2002) with permission from Blackwell publishing
[b]Drawing courtesy Jen Cooper with permission from Project UFO

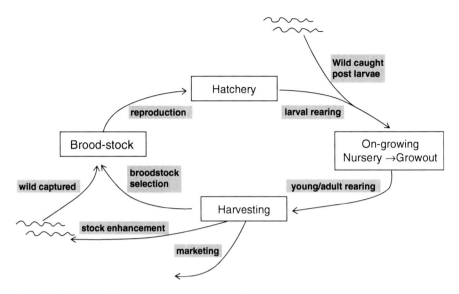

Fig. 25.1 Main possible steps in crustacean aquaculture

the larval duration is on the order of 25–40 days, although in spiny lobsters it may extend to over 400 days. After larvae metamorphose into the postlarval stage, they may be nursed in tanks or in specific nursery ponds, until they are large and robust enough to be stocked for on-growing.

The on-growing of crustaceans is most often performed in earthen ponds, into which postlarvae or juveniles are stocked. The stocks are not otherwise contained and are free to roam throughout their environment and to interact with each other. For the more gregarious species (shrimp, river crabs), relatively high densities can be maintained. However, many aquacultured crustaceans display hierarchical or territorial behavior, and the extent of this influences stocking density and husbandry practices.

Unlike fish aquaculture, there are very few examples of large-scale production of any crustaceans in tank systems. The bulk of crustacean rearing systems involve natural settings, i.e., earthen ponds, in which there is natural productivity and some environmental complexity, in contrast to a highly managed tank system. From a chemical communication perspective, this may present challenges in developing "chemical products" that must work within a chemically complex setting.

25.4 Brood-Stock Acquisition

Needs and problems – In general, the source of sexually mature animals for offspring production depends on how readily they reproduce in captivity. In species that reproduce easily and whose life cycle can be closed in captivity, e.g., freshwater prawns and crayfish, sexually mature animals are usually obtained from grow-out ponds, in order to establish a brood-stock. However, to avoid inbreeding, animals caught in the wild may sometimes be added to the brood-stock. In both these cases, reproductively active, sexually mature animals are sought after; therefore, sex pheromones should be particularly appropriate for selectively attracting such animals.

To shorten the brood-stock holding time, many hatcheries prefer to obtain females that already carry fertilized eggs from the grow-out ponds. This is the common practice in tropical regions, where such females are available year round and the grow-out season is unlimited. In clawed and spiny lobsters, it is sometimes preferred to use wild-caught ovigerous females rather than rearing immature wild or captive-bred females, because better results are usually obtained with the wild ones. In all these cases sex pheromones would not be useful for attracting berried females; in fact, sex pheromones may drive such females away, because berried females tend to avoid other animals during the egg incubation period (e.g., in crayfish Berrill and Arsenault 1984). Thus, sex pheromones might act in either direction, depending on the reproductive state of the female; they might attract ready-to-mate females or males and deter berried females.

In species that do not readily become reproductive in captivity, sexually mature animals, specifically egg-carrying females, are often captured from the wild, as in the case with the penaeid shrimps that form the bulk of farmed crustaceans. Most brood-stocks of shrimps are still obtained from catches by means of trawl fishing (Wickins and Lee 2002), a method that is highly stressful and may result in crushing of animals. Additional stress may be imposed by transporting the selected brood-stock animals and introducing them into a new and confined environment. The whole process of capture, transportation, and introduction could lead to ovary resorption in females. Thus, a low-stress capture technique would certainly be valuable in this case.

Potential benefits – Using sex pheromones not only would attract adult females, but also would enable selection of receptive females whose mating and spawning are fairly synchronized. This would be useful for hatchery operations, because holding synchronized females together would populate the spawning tank with postlarvae or juveniles that were relatively uniform in age and size, at a predictable time, and consequently would reduce growth variation and the development of stunted individuals in the population during the subsequent growth phases in the nursery (see Sect. 25.7).

Possible solutions – Can knowledge of chemical communication help in designing less stressful methods for catching suitable brood-stock animals? The straightforward solution should be based on a "pheromone trap" to which the required animals would be attracted and which they would enter voluntarily. However, the effective range of pheromones is usually unknown.

Several questions regarding chemical communication should be considered in this context. Firstly, are both sexes chemically attracted to each other, and if not, which is the attracting or attracted sex? Furthermore, are animals chemically repelled or attracted by same-sex conspecifics? Cultured crustaceans vary in their mating systems: for example, penaeid shrimp males actively search for receptive females after migrating to spawning grounds. Conversely, in the mating systems of freshwater prawns and some crayfish, clawed lobsters and spiny lobsters, the females primarily search for dominant males. In all these cases chemical cues play a role in mate attraction and mate location. Nevertheless, this characteristic of mating systems does not necessarily imply that only one of the sexes would actively respond to the scent of individuals of the other sex and that the latter would be totally passive. For example, receptive American lobster females are chemically attracted to male shelters and prefer shelters inhabited by dominant males, but males, too, are attracted to shelters that are occupied by pairs of lobsters, and sometimes try to evict the resident male. Indeed, according to Atema and Steinbach (2007), "lobstermen have observed that if a recently molted female finds her way into a trap, the trap filled with males." In summary, depending on the species, sex pheromone traps may be equally useful for capturing males and females, or more useful for capturing one of the sexes, and this would determine whether the appropriate type of sex attractant should be derived from the male or from the female. In most cases the sex ratio in broodstocks is female-biased, so that a female-attractive sex pheromone should be more important. Research in this direction is certainly necessary.

Such research should also address the question of whether sexually mature animals can be attracted by pheromones other than sex pheromones; more specifically, whether there is a pheromone that specifically attracts egg-carrying or gravid (probably nonreceptive) females. For example, spiny lobsters of both sexes are chemically attracted to conspecifics in dens by aggregation cues (see Aggio and Derby, Chap. 12). In the lobster fishery in Florida, traps with live baits of sublegal-size spiny lobsters commonly have been used to capture larger conspecifics; live lobsters were found to be more effective attractants than food (Heatwole et al. 1988). Although chemical aggregation cues are generally efficient attractants for any sexually mature spiny lobster, the response to the attraction may vary with season, sex, and reproductive condition (Childress 2007). Thus, egg-bearing females migrate to the deep seaward edge of the reef or beyond on to deep sand flats and may form almost all-female aggregations to hatch their larvae (MacDiarmid and Kittaka 2000), perhaps guided by chemical cues.

In addition, there are questions concerning the means and techniques by which sex pheromones, and/or any other type of attractant, can be applied in traps. Solutions can range from baiting traps with live ready-to-mate individuals (e.g., newly molted females) to using specific sex attractants. Chemical cues can be detected from some distance, because they are dispersed by currents, including short-range currents generated by the animals. A submersible system capable of dispersing odors of live animals, similar to that developed for research on spiny lobsters in coral reefs (Nevitt et al. 2000), could be used. In this system live animals are kept in "odor source tubes" which are installed with a submersible pumping system that draws water past the bait lobsters and through the trap (Fig. 25.2). One such apparatus can have several outlets that serve several traps, so that, instead of placing live animals in each trap, a few bait animals can be made to serve multiple traps.

As many chemical cues and signals in crustaceans are released in the urine, several studies used urine from catheterized individuals as the source of sex attractants. The urine was applied as a directional flow, e.g., in Y-maze

Fig. 25.2 Experimental
setup of bait containers and
shelter traps (not to scale)
used for capturing spiny
lobsters (*Panulirus argus*)
in the field. Modified from
Nevitt et al. (2000) with
permission from Inter-
Research

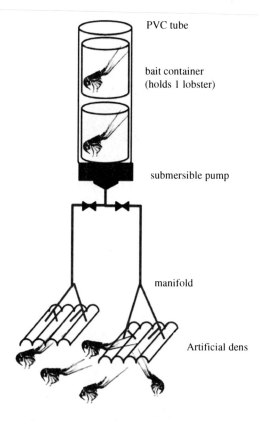

PVC tube

bait container
(holds 1 lobster)

submersible pump

manifold

Artificial dens

experiments, or soaked into a sponge that simulated a concentrated source with
slow release (Kamio et al. 2000) (Fig. 25.3). Stebbing et al. (2004) evaluated
various gels as media for absorbing pheromones and facilitating their slow release
into the environment. Female sex pheromones embedded into Phytagel were
subsequently tested in the field in the UK for pest-control purposes and were
found effective at trapping males of the invasive crayfish *Pacifasticus leniusculus*.
However, practical use of sex pheromones as attractants ultimately requires that
specific active substances be isolated, identified, and produced synthetically (see
Hardege and Terschak, Chap. 19). Sex pheromones of cultured species have yet to
be identified, and research in this direction is still needed.

25.5   Reproduction

Needs and problems – As in any industry, farmers of crustaceans strive to control the
timing and quantities of production, in accordance with demand. Most important in
this regard is the ability to control reproduction, including gonad maturation, mating,
and spawning. To obtain postlarvae or juveniles for staged production of large

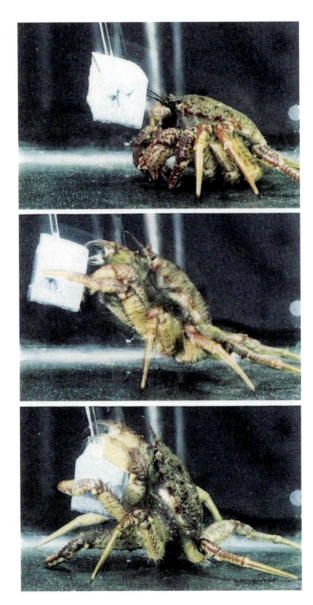

Fig. 25.3 Sequential images (*top to bottom*) presenting the reaction of a male helmet crab, *Telmessus cheiragonus*, to a sponge containing urine from receptive females. The urineborne female sex pheromone released from the sponge attracted the male and elicited a female-guarding-like behavior. Reproduced from Kamio et al. (2000) with permission from the Copyright Council of the Academic Societies, Japan

numbers of animals, spawning within a brood-stock population should be synchronized to provide enough animals at each stocking, whereas spawning of different brood-stock populations should be asynchronized over an extended time frame. The latter might be practicable mainly in controlled indoor systems or in areas where crustaceans have a prolonged breeding season or breed throughout the year.

The timing of natural breeding cycles of crustaceans is controlled by some environmentally induced mechanisms that match spawning to the most favorable conditions to offspring survival. To control the timing of reproduction, breeders can manipulate either the environmental cues that regulate reproduction, notably temperature and photoperiod, or the endocrine factors that mediate the physiological response to the environmental cues. Among cultured species, temperature and/or photoperiod manipulations have been shown to induce spawning in crayfishes, clawed and spiny lobsters, and crabs. In penaeid shrimps, endocrine manipulation, i.e., removal of the X organ/sinus gland complex by means of unilateral eyestalk ablation, is routinely practiced to induce vitellogenesis and spawning in females. However, this endocrine manipulation fails to induce gonad maturation in a large fraction of the females. Furthermore, mating success of penaeids in captivity seems to be poor as large proportions of females with seemingly mature ovaries remain unmated (Parnes et al. 2008).

Possible solutions – Crustacean species employ various reproductive strategies in terms of the temporal relationship between gonad maturation, mating, and spawning, and these reproductive stages are coordinated with the molt cycle in a variety of ways (reviewed by Raviv et al. 2008). Can chemical cues be one of the inputs to the female sensory system that regulate any of these stages of a reproductive cycle?

As in the goldfish example, pheromones might exert their influence on reproduction by priming physiological processes and/or by eliciting mating and spawning behaviors in the receiving animals. Evidence for primer pheromones in crustaceans is lacking. However, in some species, females can delay their premating molt to a limited extent in the absence of males, and it may be that the lack of male chemical cues facilitates this response. Laboratory observation of American lobsters revealed a phenomenon of molt staggering, in which females serially cohabited a shelter with the dominant male, and molted and mated (Cowan and Atema 1990). It was hypothesized that both male and female chemical cues were involved in the mechanism that regulates molt staggering among female lobsters: neighboring females would release a pheromone that mutually suppress each others' premating molt; a female cohabiting with a male would not be exposed to this molt-inhibiting pheromone, but rather to the male's primer pheromone that induced premating molting (Atema and Steinbach 2007). While in American lobsters pheromones might desynchronize molting and mating among females, in other species, e.g., freshwater prawns (Howe 1981), pheromones might induce molt synchronization. Thus, primer pheromones with inhibiting or accelerating effects on female molting might be used to manipulate the timing of spawning in species in which molting and reproduction are temporally linked. To maximize its effectiveness, pheromone application should be considered together with environmental or endocrine manipulations.

Numerous studies have demonstrated, in both males and females, the releasing properties of pheromones on various behavior patterns that are necessary for successful mating, both before copulation, i.e., mate location, courtship, female precopulatory guarding, cradle carrying, and cohabitation, and after copulation, i.e., postcopulatory guarding and cohabitation. It thus appears that the reproductive step most likely to be influenced by chemical communication is the initiation and progression of mating behavior. Since the mating window usually comprises a relatively short phase within the reproductive cycle, and the releasing effect of sex pheromones on mating behaviors is immediate and short-term, such pheromones might be applied to fine-tune the synchronization of mating and spawning. However, the chemical components of crustacean sex pheromones still need to be identified and characterized; therefore, the application and integration of pheromones in reproductive manipulation techniques remains a far goal and requires a vast research effort.

25.6   Hatching and Rearing Larvae

Needs and problems – Although often overlooked during the larval phase, the problem of size-uniformity during the subsequent on-growing phases may already have originated during these early phases. Initial size variation arising, for example, from inherent genetic variability is usually enhanced with passing time, mainly because of the effects of competitive interactions among individuals. In order to slow down the process of size divergence, animals of uniform size should be introduced at the start of the on-growing phases. To this end, size variation initially can be minimized by synchronization of larval hatching and release and, at the end of the larval phase, by synchronization of metamorphosis to the postlarval phase. Chemical communication might be involved in both these processes.

Farming of species in which breeding or larval rearing in captivity is problematic could skip these steps and start the culture process with wild-caught postlarvae or juveniles. Chemical communication might be used to develop methods for efficiently controlled capturing of animals at these early stages of life.

Possible solutions – In subtidal brachyuran crabs, homarid lobsters, and spiny lobsters, the time of larval release is controlled by the developing embryos (see example Ziegler and Forward 2007, and references therein). Pheromones released from hatching eggs, probably small peptides, induce larval release behavior (termed pumping behavior) in the ovigerous female. This behavior involves rapid abdominal extensions and vigorous beats of the pleopods, which cause the embryo cases to break open. The pumping behavior results in synchronized release of the larvae. Thus, application of pumping pheromones at appropriate concentrations might synchronize the timing of larval release among ovigerous females with late-stage eggs. Such manipulation of the timing of larval release would enable prompt transfer of larvae from hatching tanks to rearing facilities.

Settlement and metamorphosis from the planktonic postlarval stage to the benthic juvenile stage of crabs and lobsters are controlled by specific physical and chemical cues associated with preferred habitats (Forward et al. 2001; Jeffs et al. 2005; O'Connor 2007). In addition to habitat cues, such as chemicals released from aquatic vegetation or biofilm (algae and bacteria) covering the substratum, chemical cues that are considered in chemical communication have been shown to influence the time to metamorphosis in brachyuran crabs. For example, odors of adult conspecifics and of related heterospecifics, as well as prey odors, reduced the time to metamorphosis, whereas predator odors increased the time to metamorphosis in crab post-larvae (megalopae). Exposure of megalopae of several crabs to cues for acceleration can shorten the time to metamorphosis by 15–25% (Forward et al. 2001). Thus, chemical cues that accelerate or delay metamorphosis might be used to synchronize the times of metamorphosis among larval rearing tanks. However, it is possible to manipulate the time of metamorphosis only in competent postlarvae, i.e., those that are physiologically capable of metamorphosing. Moreover, the vast majority of species studied thus far metamorphose within a few days – 20 days at most – even in the absence of habitat cues (Forward et al. 2001), which restricts the possibility of using chemical cues to manipulate metamorphosis time to within this time frame.

In crayfish, the larval stages occur within the egg capsule; after hatching, the postembryonic young stages remain on the pleopods of the mother and do not transform into planktonic larvae. While the yolk reserves are depleting, the young occasionally leave the maternal pleopods for excursions of progressively longer duration and distance, to feed independently, but they still return to the mother and cling to her pleopods for refuge. During this period the brooding female is not cannibalistic, in contrast to nonbrooding females and males. Little (1975) demonstrated that pheromones released by the brooding mother are involved in attracting and guiding the returning young.

Maternal pheromones could be used both in indoor hatchery tanks and in outdoor brood-stock ponds to facilitate efficient collection of stage 2 and 3 craylings for transfer to nursery tanks and to decrease loss of juveniles through cannibalism. In farms that maintain the brood-stock in large earthen ponds, e.g., in red-claw crayfish (*Cherax quadricarinatus*) farms in Australia, juveniles are periodically harvested by means of bundles of plastic mesh to which the juveniles cling for refuge. Although sufficient production of juveniles may be achieved, this method is inefficient: the growers are unable to control the age and size of juveniles in the pond, and there is substantial loss of newly released juveniles through cannibalism by larger individuals. Use of maternal pheromones to attract early-stage juveniles to the bundles of mesh might increase the efficiency of this method and its selectivity for juveniles in stages 2 and 3.

25.7 On-growing

Needs and Problems – The main challenge faced by crustacean growers during on-growing is to achieve fast and uniform growth and maximal survival of high-quality animals. Cultured crustaceans, particularly clawed species, are notorious for their heterogeneous growth and relatively low survival rates, which result from social interactions and the cannibalistic nature of crustaceans. These problems are evident under the conditions prevailing in most culture methods and they are common to all stages of on-growing, from nursery to grow-out ponds.

Social control of growth – Clear evidence of the importance of social interactions for growth regulation in crustaceans was yielded by controlled experiments that compared the growth of individuals raised in isolation with those raised in pairs or in groups. The socially induced changes in growth rate differed among species: in clawed crustaceans such as the freshwater prawn *M. rosenbergii* (Karplus 2005 and references therein) and the red-claw crayfish *C. quadricarinatus* (Karplus and Barki 2004), growth of small males was much lower in the presence of larger dominant males than in isolation; conversely, in the gregarious spiny lobster, *Panulirus longipes*, growth rates of isolated individuals were lower than those of grouped individuals by about 30% (Chittleborough 1975). Four social mechanisms by which dominant individuals directly or indirectly suppress growth of subordinates have been suggested: direct competition for food, appetite suppression, altered food-conversion efficiency, and increased motor activity in the subordinate individuals (see review by Karplus 2005).

There is a special interest for crustacean culturists in analysis of the sensory modalities that mediate social suppression of growth: knowledge of the effective cues would enable culture systems to be designed so as to minimize their impact. The involvement of chemical cues in social control of growth in *H. americanus* and *H. gammarus* was demonstrated under circulating water conditions (Nelson and Hedgecock 1983; Nelson et al. 1983). A 40% weight reduction of small lobsters in compartments immediately downstream from larger individuals was found, and this effect disappeared further downstream. Flow rate manipulations revealed that the waterborne growth-inhibiting substances most probably decayed rapidly, with a half-life of about 1 min.

Studies on the role of chemical cues in regulating growth variation in juveniles of the freshwater prawn *M. rosenbergii* yielded differing results (reviewed by Karplus 2005), depending on the source of the chemical cue. Juvenile prawns housed in individual compartments in a flow-through system that allowed chemical communication grew homogenously in water that originated from isolated conspecifics, whereas those that grew in water that had passed over freely interacting prawns showed increased size variation.

Karplus and Barki (2004) investigated the role of chemical, visual, and tactile sensory modalities in mediating growth suppression in juvenile red-claw crayfish, *C. quadricarinatus*. Pairs comprising small and relatively large male juveniles were reared under the following social conditions: in full contact, with either intact or immobilized claws; separated by a partition that allowed transmission of chemical and/or visual cues; and without any contact (Fig. 25.4a). The growth of the small individual was reduced by 50% only when in full contact with the large one (Fig. 25.4b); there was no growth suppression when only chemical or visual cues were involved. Similar results were obtained when the large individual was a several-times larger, sexually mature male. In light of behavioral observations,

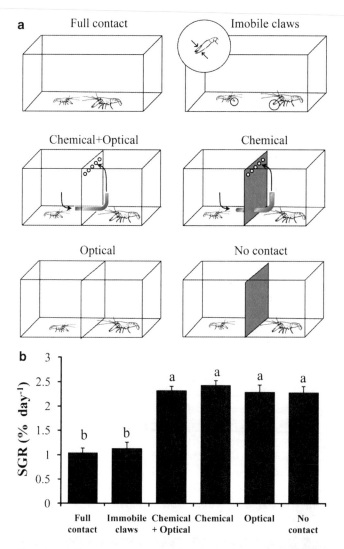

Fig. 25.4 (a) Experimental setup for testing the effects of chemical, visual, and/or tactile cues on growth of juveniles in the red-claw crayfish *Cherax quadricarinatus*. (b) Specific growth rate (SGR) of small juvenile *C. quadricarinatus* exposed to various types of social contact with a relatively large juvenile, or to no contact whatsoever. Treatments marked by different letters yielded significantly different results. Reproduced from Karplus and Barki (2004) with permission from Elsevier

growth suppression was largely attributed to direct competition for food. Although chemical cues from isolated individuals were not found to affect growth in this study, it is still possible that chemical cues released only during direct physical contact are involved in the mediation of growth control in freely interacting crayfish.

Overall, species-specific differences were found among cultured crustaceans with regard to the role played by chemical cues in growth suppression. However, tactile stimulation particularly that involving the claws, the major weapons of many of the cultured crustaceans, seemed to be more important than chemical stimulation in regulating growth.

Cannibalism – Cannibalism is one of the most important causes of mortality among cultured crustaceans. It occurs mainly during early growing stages, because of the high rearing densities and frequent molting. Cannibalism represents a major obstacle in the way of profitable communal culturing of crustaceans, especially in the development of intensive-culture technologies.

Cannibalism among crustaceans is mainly directed towards relatively small or soft-shelled individuals. During and after ecdysis, a variety of chemical cues, specifically amino acids, are released into the water, and they stimulate chemoreceptor activity and enhance feeding behavior in conspecifics. However, chemical cues also seem to be involved in two important mechanisms that might reduce cannibalism in cultured crustaceans: molt synchronization (Fig. 25.5) and mate guarding. Howe (1981) demonstrated partial synchronization of molting among juvenile prawns (*M. rosenbergii*) reared in separate but adjacent chambers that shared the same circulating water. The majority of molts occurred at night in nonrandom peaks and in spatial aggregations that coincided with the measured pattern of water flow among chambers. Some of the prawns that molted within these peaks did so sooner than predicted on the basis of their size-related molt interval. Howe (1981) suggested that waterborne chemical cues released by juvenile prawns were responsible for partial synchrony of molting, as a risk-reducing strategy; since prawns are most vulnerable to cannibalism at ecdysis, the safest time to molt is when other conspecifics are molting.

In many species that mate when the female is soft-shelled, the male and the female form temporary associations prior to and following ecdysis. Chemical cues released by premolt receptive females may play an important role in inducing mate guarding behavior in males. This behavior may take various forms: protection inside a shelter, as with *H. americanus*; caging the female between the chelipeds, as with *M. rosenbergii*; or cradle carrying in the claws, as with various brachyuran crabs. Although conflict may arise regarding the duration of guarding, both the female and

Fig. 25.5 An illustration of chemically mediated molt synchronization in a decapod crustacean. Drawing by artist Jorge A. Varela Ramos

the male benefit from these temporary associations: the female gains protection against cannibalism and predation, and the male monopolizes copulations, prevents sperm competition, and protects his future offspring from cannibalism. Mate guarding by males may also serve as a means for mate choice in both sexes (Barki 2008 and references therein). Understanding the proximate mechanism that inhibits cannibalism in guarding males might help to achieve control of cannibalism among adult crustaceans. In the shore crab *Carcinus maenas,* the molting hormone 20-hydroxyecdysone was recently found to have a sex-specific function of deterring feeding in males but not in females, thus facilitating the safe guarding of soft-shelled females, without cannibalism (Hayden et al. 2007; Chang, Chap. 21).

Possible solutions – Water circulation systems in crustacean nurseries and culture systems should be constructed so as to avoid direct flow from one tank that holds juveniles or adults into another, so that the water cannot transmit chemical cues that might suppress growth. If this cannot be avoided, water transfer at least should be delayed in order to allow rapidly decaying substances to disappear. The discovery of a specific mate-guarding pheromone could lead to control of cannibalism by inducing guarding behavior. The feeding-deterrent properties of the molting hormone 20-hydroxyecdysone might be used to reduce cannibalism among male crabs that are temporarily stored at high densities.

Finally, a potentially important application of pheromones in the on-growing phase of crustaceans is the use of pheromones as food attractants. Since formulated feed incurs the highest proportion of the total costs in aquaculture, efficient feeding is crucial for profitable aquaculture. The incorporation of attractants in the feed enhances growth rates, improves food conversion ratios, and reduces waste of uneaten feed pellets, which may cause significant environmental damage. Mendoza et al. (1997) coated feed pellets with extracts of blue crab urine or of green glands from receptive female prawns and tested their effect on food intake by freshwater prawns (*M. rosenbergii*). Males, but not females, ingested extract-coated feed pellets faster than control pellets under laboratory conditions. This attractant might be applied to increase feeding and trapping in monosex culture of male prawns. Ongoing development of pheromone-based feeding attractants for aquaculture was recently reported in the technological literature (Moore 2007). A preliminary commercial-scale trial with the white-legged shrimp (*Litopenaeus vannamei*) revealed higher growth rates and better food conversion ratios in shrimp ponds that received feed with pheromone attractants than in control ponds; the application of the feeding attractant produced shrimp, which were 30% larger on average than the control shrimp. The identity of the pheromone used in this product was not specified.

25.8 Harvesting

Needs and problems – Cultured crustaceans are benthic animals that slowly forage on the substratum during their active hours, usually at night, and tend to spend most of the rest of the time in shelters between rocks and stones on the pond bank slopes, in burrows, or buried in the soil. These habits make complete harvesting of crustaceans from large grow-out ponds difficult and laborious, and often many animals are left behind, are lost, or die from lack of oxygen. The most common technique for complete harvesting is to drain the pond, often in combination with seining, which forces the animals to walk to the collection basin at the pond exit gate (Wickins and Lee 2002). However, a significant fraction of the animals may remain in pools in bottom depressions, between bank rocks, or buried in the muddy bottom. In the cases of astacid and cambarid crayfishes, most harvesting uses traps baited with fish or artificial baits. A harvesting technique for red-claw crayfish (*C. quadricarinatus*) combines flow-traps with pond draining: fresh water flows from aerated catch boxes down a ramp into the draining pond and thereby stimulates the crayfish to walk upstream into boxes (Fig. 25.6).

Fig. 25.6 (a) Flow trap employed to harvest red-claw crayfish (*Cherax quadricarinatus*) in Australia. The trap is submerged in the full pond. As the pond is slowly drained, a current of water is directed from the trap down an enclosed ramp. The crayfish move against the flow-trap current, eventually up the ramp and into the trap. (b) A male red claw (*C. quadricarinatus*)

Possible solutions – Harvesting has much in common with brood-stock acquisition, in that it deals with concentrating and capturing sexually mature, i.e., marketable animals. In fact, the brood-stock is often sorted out from the harvested animals to start a new culture cycle. Attractive chemical cues, such as sex pheromones involved in chemical communication between adult crustaceans, might be integrated into the aforementioned harvesting techniques, to increase the efficiency and rapidity of animal capture and to reduce the stress and loss of animals associated with this action. For example, such chemical cues might be released from traps or the collection basin.

25.9 Summary and Conclusions

Chemical communication has received very little attention in the field of crustacean aquaculture research. It has been investigated mainly with regard to the extent that chemical cues might be involved in social control of growth, which results in growth suppression and size variation among individuals under culture conditions, but no consistent conclusion has emerged that stimulated further development and application of solutions based on this knowledge. Implementation of knowledge on chemical communication in crustacean aquaculture has not gone further than some preliminary trials of the incorporation of pheromones as feeding attractants or of their use to facilitate trapping.

In this review we attempted to identify those aquaculture procedures in which knowledge of chemical communication might be implemented, to indicate possible uses and to suggest possible solutions related to various aquacultured crustaceans (for a summary see Table 25.2). In most cases solutions based on chemical communication would probably be integrated into current techniques and would serve to enhance their efficiency. To achieve this aim, however, vast research is still needed. To enable widespread use of such solutions, further research is required, to identify and characterize pheromones and potent chemical components that mediate behavioral and physiological processes relevant to aquaculture; synthetic versions should be developed. In addition, relevant properties such as half-life and environmental impact should be elucidated. Furthermore, technical means for pheromone application and dispersion should be developed, so that crustaceans will be able to track the chemical cues.

In conclusion, in light of the prominent role of chemical communication during most life stages of crustaceans, including those that are important for aquaculture, the prospects for beneficial use of pheromones in aquaculture seem promising. We hope that this review will open a new avenue of interdisciplinary research in the fields of aquaculture and chemical communication of crustaceans; an avenue that would eventually lead to implementation of some of the proposed ideas in crustacean aquaculture.

Table 25.2 Summary of aquaculture steps in which knowledge of chemical communication might be implemented, possible uses, and solutions in various aquacultured crustaceans

Aquaculture step	Possible use	Possible solution
Brood-stock	Acquisition of sexually active animals	Sex pheromone traps
	Reproduction:	
	Mating induction/inhibition synchronization/ asynchronization	Application of primer pheromones with (premating) molt-accelerating or inhibiting properties
	Increase male attraction and fertility	Application of urineborne pheromones of premating females
Hatchery	Synchronization of hatching and larvae release	Application of pumping pheromones
	Synchronization of metamorphosis	Application of settlement-accelerating (conspecific, heterospecific, prey) or delaying (predator) cues
	Capturing wild postlarvae	Attraction by settlement cues
	Collection of early-stage juveniles and decrease filial cannibalism	Application of maternal pheromones in crayfish
On-growing	Decrease growth suppression	Aqua-technical solutions to remove waterborne dominance cues
	Reduction of cannibalism	Application of chemical cues for molt synchronization (also useful for management of soft-shell production). Application of the feeding-deterring molt hormone or cannibalism-deterring female pheromone in tanks temporarily storing crowded males
	Feeding stimulation to increase feed intake and decrease feed waste	Incorporation of pheromone-based attractants in the feed
Harvesting	Concentration and capture of animals	Application of sex pheromones at the collection basin or in traps

References

Atema J, Steinbach MA (2007) Chemical communication and social behavior of the lobster *Homarus americanus* and other decapod crustaceans. In: Duffy JE, Thiel M (eds) Evolutionary ecology of social and sexual systems: crustaceans as model organisms. Oxford University Press, Oxford, pp 115–144

Barki A (2008) Mating behaviour. In: Mente E (ed) Reproductive biology of crustaceans: case studies of decapod crustaceans. Science Publishers, Enfield, pp 223–265

Berrill M, Arsenault M (1984) The breeding behaviour of a northern temperate orconectid crayfish, *Orconectes rusticus*. Anim Behav 32:333–339

Childress MJ (2007) Comparative sociobiology of spiny lobsters. In: Duffy JE, Thiel M (eds) Evolutionary ecology of social and sexual systems: crustaceans as model organisms. Oxford University Press, Oxford, pp 271–293

Chittleborough RG (1975) Environmental factors affecting growth and survival of juvenile western rock lobsters *Panulirus longipes* (Milne-Edwards). Aust J Mar Freshwater Res 26:177–196

Cowan DF, Atema J (1990) Moult staggering and serial monogamy in American lobsters, *Homarus americanus*. Anim Behav 39:1199–1206

FAO (2008) Fisheries and aquaculture department. FAO, Rome

Forward RB Jr, Tankersley RA, Rittschof D (2001) Cues for metamorphosis of brachyuran crabs: an overview. Am Zool 41:1108–1122

Hayden D, Jennings A, Müller C, Pascoe D, Bublitz R, Webb H, Breithaupt T, Watkins L, Hardege J (2007) Sex-specific mediation of foraging in the shore crab, *Carcinus maenas*. Horm Behav 52:162–168

Heatwole DW, Hunt JH, Kennedy FS Jr (1988) Catch efficiencies of live lobster decoys and other attractants in the Florida spiny lobster fishery. Fla Mar Res Publ 44:1–15

Howe NR (1981) Partial moulting synchrony in the giant Malaysian prawn *Macrobrachium rosenbergii*: a chemical communication hypothesis. J Chem Ecol 7:487–500

Jeffs AG, Montgomery JC, Tindle CT (2005) How do spiny lobster post-larvae find the coast? N Z J Mar Freshwater Res 39:605–617

Jones CM (2009) Advances in the culture of lobsters. In: Burnell G, Allan GL (eds) New technologies in aquaculture: improving production efficiency, quality and environmental management. Woodhead and CRC Press, Cambridge, pp 822–844

Kamio M, Matsunaga S, Fusetani N (2000) Studies on sex pheromones of the helmet crab, *Telmessus cheiragonus* 1. An assay based on precopulatory mate-guarding. Zool Sci 17:731–733

Karplus I (2005) Social control of growth in *Macrobrachium rosenbergii* (De Man): a review and prospects for future research. Aquacult Res 36:238–254

Karplus I, Barki A (2004) Social control of growth in the redclaw crayfish, *Cherax quadricarinatus*: testing the sensory modalities involved. Aquaculture 242:321–333

Little EE (1975) Chemical communication in maternal behavior of crayfish. Nature 255:400–401

MacDiarmid AB, Kittaka J (2000) Breeding. In: Phillips BF, Kittaka J (eds) Spiny lobsters fisheries and culture. Blackwell Science, Oxford, pp 485–507

Mendoza R, Montemayor J, Verde J (1997) Biogenic amines and pheromones as feed attractants for the freshwater prawn *Macrobrachium rosenbergii*. Aquacult Nutr 3:167–173

Moore A (2007) Pheromone-based feeding attractants for sustainable aquaculture. Feed Technology Update 2: 3–6. http://aquafeed.com/newsletter_pdts/nl_000278.pdf. Accessed 21 October 2008

Nelson K, Hedgecock D (1983) Size-dependence of growth inhibition among juvenile lobsters (*Homarus*). J Exp Mar Biol Ecol 66:125–134

Nelson K, Hedgecock D, Heyer B, Nunn T (1983) On the nature of short-range growth inhibition in juvenile lobsters (*Homarus*). J Exp Mar Biol Ecol 72:83–98

Nevitt G, Pentcheff ND, Lohmann KJ, Zimmer RK (2000) Den selection by the spiny lobster *Panulirus argus*: testing attraction to conspecific odours in the field. Mar Ecol Prog Ser 203:225–231

O'Connor NJ (2007) Stimulation of molting in megalopae of the Asian shore crab *Hemigrapsus sanguineus*: Physical and chemical cues. Mar Ecol Prog Ser 352:1–8

Parnes S, Raviv S, Sagi A (2008) Male and female reproduction in penaeid shrimps. In: Mente E (ed) Reproductive biology of crustaceans: case studies of decapod crustaceans. Science Publishers, Enfield, pp 365–390

Phillips BF (ed) (2006) Lobsters: biology, management, aquaculture and fisheries. Blackwell Publishing, Oxford

Raviv S, Parnes S, Sagi A (2008) Coordination of reproduction and molt in decapods. In: Mente E (ed) Reproductive biology of crustaceans: case studies of decapod crustaceans. Science Publishers, Enfield, pp 365–390

Slessor KN, Winston ML, Le-Conte Y (2005) Pheromone communication in the honeybee (*Apis mellifera* L.). J Chem Ecol 31:2731–2745

Sorensen PW, Stacey NE (2004) Brief review of fish pheromones and discussion of their possible use in the control of non-indigenous teleost fishes. N Z J Mar Freshwater Res 38:399–417

Stebbing PD, Watson GJ, Bentley MG, Fraser D, Jennings R, Rushton SP, Sibley PJ (2004) Evaluation of the capacity of pheromones for control of invasive non-native crayfish. English Nature Research Reports No. 578, English Nature, Peterborough, UK.

Wickins JF, Lee DO'C (2002) Crustacean farming: ranching and culture. Blackwell Science, Oxford

Winston ML, Slessor KN (1998) Honeybee primer pheromones and colony organization: gaps in our knowledge. Apidologie 29:81–95

Ziegler TA, Forward RB Jr (2007) Control of larval release in the Caribbean spiny lobster, *Panulirus argus*: role of chemical cues. Mar Biol 152:589–597

Chapter 26
Effects of Pollutants on Olfactory Mediated Behaviors in Fish and Crustaceans

K. Håkan Olsén

Abstract Streams, lakes and the sea are the final sinks of various pollutants which means that aquatic organisms are exposed to many different chemicals present in the ambient water. Several studies demonstrate that these pollutants may interfere with chemoreception of aquatic animals. Many aquatic vertebrates and invertebrates depend on chemical senses for their survival and reproduction. In fish and crustaceans, olfactory and taste receptors are exposed directly to the pollutants and their function can be disturbed. This can result in behavioral changes or lack of proper behaviors. Anthropogenic compounds can have detrimental effects on reproduction and survival of aquatic organisms. Relatively little is known about the effects of pollutants on crustacean chemoreception and behavior, but studies on fish might help elucidate the problems faced by crustaceans. This chapter gives examples of pollutant effects on pheromone detection in fish in connection with reproductive behavior, shoaling and dominance interactions, fright responses and effects on detection of food odors and foraging. Results from crustacean studies are compared to the examples given from fish studies. Various studies have shown that pesticides of different kinds affect the function of fish olfactory receptor cells. Crustaceans may face similar problems with less sensitive chemo-receptor cells. This will be especially problematic with compounds designed to combat insects and arthropods that are closely related to crustaceans. Many insecticides have direct effects on the nervous system and this will disturb chemoreception and behavior. Heavy metal ions are also problematic as they can be transported by chemosensory neurons into the brain and affect its normal functions. Crustaceans play important roles in aquatic ecosystems as well as in human food industry. Therefore, it is urgent to increase the knowledge about the effects of pollutants on crustacean chemoreception and behavior, especially compounds with direct effects on the nervous system that will affect the normal behavior and signal perception,

K.H. Olsén (✉)
School of Life Sciences, Södertörn University, SE-141 89,
Huddinge, Sweden
e-mail: hakan.olsen@sh.se

T. Breithaupt and M. Thiel (eds.), *Chemical Communication in Crustaceans,*
DOI 10.1007/978-0-387-77101-4_26, © Springer Science+Business Media, LLC 2011

disturbing social communication and reproduction. Such effects are serious for an individual as its growth and ability to compete and reproduce will be suppressed.

26.1   Introduction

Streams, lakes and the sea are the final sinks of various pollutants which means that aquatic organisms are exposed to many different chemicals present in the ambient water. Various human activities create pollutants, i.e. compounds that, because of their chemical composition or quantity, disturb natural processes (Wright and Nebel 2002). These pollutants enter the aquatic environments through point or non-point sources. Point sources are for instance factories and sewage treatment plants. These sources are much easier to monitor than nonpoint sources such as agricultural and road runoff, and atmospheric deposition (Fig. 26.1). Rainfalls can result in sharp concentration peaks of mobile pesticides in streams and lakes. All these activities may affect the downstream water and habitat quality for aquatic organisms such as fish and crustaceans. Pesticides (insecticides, fungicides and herbicides) are especially critical as they are molecules designed and synthesized to

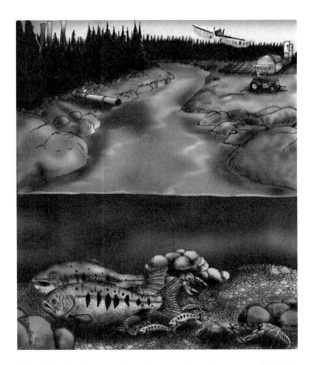

Fig. 26.1 Examples of pollutions entering the aquatic environment. Pollutions may originate from a point source, discharge from a factory, and non-point sources with activities connected to agriculture, i.e. pesticide spraying, fertilizing and cattle. Drawing by Jorge A. Varela Ramos

be biologically active at low concentrations. Furthermore, they appear in streams, lakes, and coastal waters immediately after spraying and sometimes even long time thereafter, often in multi-compound cocktails. Several studies demonstrate that these pollutants may interfere with chemoreception of aquatic animals. Many aquatic vertebrates and invertebrates depend on chemical senses for their survival and reproduction. In fish and crustaceans, olfactory and taste receptors are exposed directly to the pollutants and their function can be disturbed. This can result in behavioral changes or lack of proper behaviors. Relatively little is known about the effects of pollutants on crustacean chemoreception and behavior. Studies on fish might help elucidate the problems faced by crustaceans.

Various synthetic pesticides are present in surface water and streams flowing from or through agricultural areas (e.g. Kreuger 1998). There are diverse molecules included in the group of chemicals designated as herbicides, fungicides and insecticides. They can have significant effects on organisms other than those that they are designed for, e.g. on fish and invertebrates, by interfering with neural or endocrine function or by causing other toxic effects (e.g. Bonde et al. 1998). Some of these molecules may be detrimental for normal function of the primary olfactory receptor cells and lead to disruption of chemical information between organisms (Lürling and Scheffer 2007). This environmental threat has received little attention. The sensitivity and ability of the olfactory sense to detect sex pheromones are under hormonal control (see Stacey and Sorensen 2006). However, there has been little research on endocrine disrupting chemicals (EDCs) affecting the endogenous hormone levels of importance for the detection of sex pheromones during reproduction.

In fish the olfactory sense is important, in a species-specific way, during various situations and behaviors. Chemical cues released from heterospecific and conspecific individuals provide information about the presence of food items, predators, competitors, mates and allow the recognition of kin (e.g. reviews Hara 1994; Olsén 1999). Several species of fish are dependent on the olfactory sense to detect sex pheromones during reproduction (review Stacey and Sorensen 2006). Pollutants can have effects on one or more links in the chemical communication chain between the individuals (see Fig. 26.2). The chemicals can affect the sender of the pheromone or the individual that detects the signal, the receiver. It is also possible that the pheromone is affected by biotic and abiotic factors, such as bacteria, organic materials and pH, when released into the water.

Similar to fish, crustacean species depend on chemical senses for foraging and social interactions (for review see Caprio and Derby 2008). The crustacean chemoreceptors are, as in fish, exposed directly into the environment with risks to be affected by various chemicals in the water. The main chemo-receptors in decapod crustaceans are the aesthetascs on the first antenna (see Hallberg and Skog, Chap. 6), but there are also other chemo-receptors distributed over the surface of the crustaceans body including the major chelae (discussed in Belanger et al. 2008). Some of the chemo-receptors are highly sensitive to free amino acids, taurine and extracts from the prey organisms, and if they are herbivores to sugars (Caprio and Derby 2008). Peptides, proteins and sugar polymers are commonly used as pheromones and kairomones among crustaceans, such as during barnacle settlement

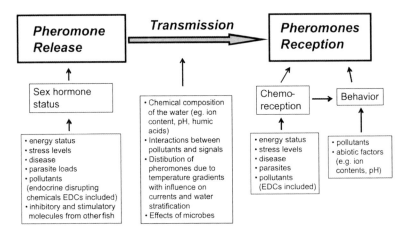

Fig. 26.2 The three parts of sex-pheromone communication and their susceptibility to internal and external factors. Various biotic and abiotic factors influence the three links of chemical communication. In addition to effects on the sender and receiver, the signal molecule/-s can be affected during the transmission step

(protein complex – see Clare, Chap. 22), hermit crab shell acquisition (peptides – see Gherardi and Tricarico, Chap. 15), and sexual interactions (by contact peptide-sugar pheromones in copepods – see Snell, Chap. 23). Anthropogenic compounds intervening with these intricate communication systems can have detrimental effects on reproduction and survival of aquatic organisms, with cascading ecological consequences such as modified community composition and decreased species diversity (cf. Pohnert et al. 2007).

This chapter deals with the effects of pollutants on different chemoreceptive behaviors in fish and crustaceans. Some of the early studies of pollutant effects on fish chemo-reception and behavior have been summarized in reviews of Blaxter and Hallers-Tjabbes (1982) and Brown et al. (1982). Sloman and Wilson (2006) discussed the literature dealing with the effects of chemical pollutants on behaviors mediated by different senses in fish. Since most research on pollutant effects in aquatic systems involves the behavior of fish and crustaceans, in the following paragraphs, I will describe the different vital behaviors that are affected by various types of pollutants.

26.2   Effects on Sex Pheromone Detection and Reproductive Behavior

Most of the information about sex pheromones in fish is based on studies of goldfish and salmonid fishes. The sex pheromones in goldfish are closely linked to internal sex hormones (review Stacey and Sorensen 2006). The goldfish (*Carrassius auratus*) and the closely related crucian carp (*C. carassius*) use both steroids (in free form, conjugated or un-conjugated metabolites) and prostaglandins and are sensitive to the free hormones 17,20β-dihydroxy-4-pregnen-3-one (17,20β-P) and prostaglandin $F_{2\alpha}$ (PGF2α), which are released by the females into the ambient water. Goldfish, crucian carp and common carp (*Cyprinius carpio*) seem to share the same pheromone system. Exposure to 17,20β-P increases male milt volumes in both goldfish and crucian carp (a priming effect, see Wyatt, Chap. 2) (Stacey and Sorensen 2006). In crucian carp the priming effects in males have also been demonstrated during studies in net pens in their natural environment (Olsén et al. 2006). Similarly, in salmonid fishes both endocrine and behavioral effects by female odors have been revealed in males (e.g. Olsén and Liley 1993; Olsén et al. 2002). Pollutants can interfere with pheromone detection thereby producing negative effects on reproduction in crucian carp, goldfish and salmonid fishes (Fig. 26.3). In the example, pheromones from a brown trout female enter the olfactory pits where the olfactory rosette (only one of the two shown) with its olfactory receptor cells is located. After sex pheromones are bound to the receptors the receptor cells depolarize and a signal is transmitted to one of the olfactory bulbs (one on each side of the brain). The signal is reconnected and transmitted in the olfactory tract with effects (primer effect) on the hormone levels through the hypothalamus-pituitary-gonad axes (HPG axes), or it results in a

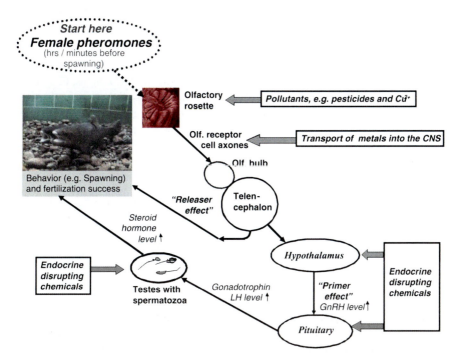

Fig. 26.3 The process from detection of pheromones to the resulting behavior and physiological responses is shown in the figure and the possible negative effects of pollutants on the pheromone communication

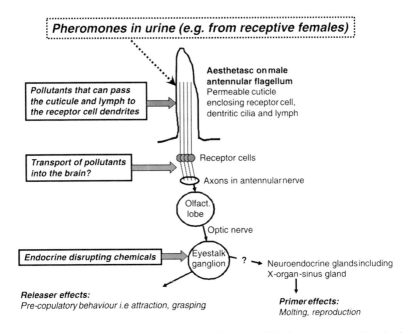

Fig. 26.4 Pheromone detection in crustaceans and the possible factors that can disturb signal transmission and behavior responses. For information on functional organization of olfaction in crustaceans see Schmidt and Mellon, Chap. 7

spontaneous behavior effect (releaser effect), i.e. attraction to a female. Pollutants can have effects at different levels of this process (see arrows and rectangles, Fig. 26.3). On the receptor cell level, pollutants disturb the pheromone's binding to the receptor or the function of the nerve cell. Chemicals can also have negative effects on the sex hormone plasma levels (EDCs) by affecting the function of the hypothalamus or the pituitary and the gonads. This can suppress priming effects, which are important for milt production and the function of the spermatozoa. Increased release of sex steroids from the gonads may also be important to the males' behavior. As in fish there are several levels of pheromone detection that can be affected by pollutants in crustaceans (Fig. 26.4).

26.2.1 Copper Ions

Copper is very toxic to fish and their olfactory sense. Copper compounds are used as pesticides, against fungi in wood preservatives, in anti-fouling paints, and as algaecides (e.g. Gatidou and Thomaidis 2007), and it can get into surface water. Further, plumbing with copper pipes causes significant amounts of copper to enter sewage treatment plants from where it reaches streams and lakes via the effluents or through sludge dumping grounds (Clark 2001).

Copper exposures at 20 μg/L or higher induce degenerating effects on the olfactory receptor cells in fish (Saucier and Astic 1995). Since it is a normal process that receptor cells are regenerating in the olfactory epithelium of fish and other vertebrates as long as basal cells are present, new functional olfactory cells will be continuously produced and the animal can recover its sense of smell (e.g. Zippel 1993). There will, however, be problems if the fish remains in contaminated water and the olfactory epithelium does not acclimate and protect the receptor cells from metal toxicity (e.g. by metallothioneins, mucus production). It has been shown that olfactory receptor neurons can be a transport route of metal ions and organic molecules to the olfactory bulbs and the brain in vertebrates, fish included, with severe disturbing effects on the function of the CNS (e.g. Tjälve and Henriksson 1999; Persson et al. 2002).

Effects of copper sulfate on reproductive behavior and physiology were studied in mature brown trout (*Salmo trutta*) parr (Jaensson and Olsén 2010). The study clearly demonstrated that exposure to copper can disturb the reproductive behavior and endocrinology of the males. Male brown trouts that were exposed for 4 days to copper were then used in a priming experiment with 5-h exposure to the sex pheromone PGF2α. In groups exposed to 10 or 100 μg/L copper the amount of expressible milt was significantly lower, less than half, than in control parr exposed only to water. During the behavioral experiment in the big stream tank (Fig. 26.5),

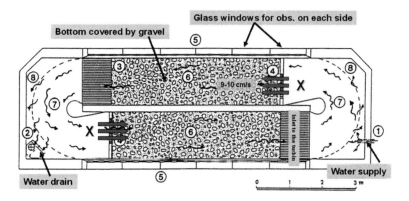

X = Turbines to adjust the speed of the water current

Fig. 26.5 The big stream tank used by Jaensson et al. (2007) and Jaensson and Olsén (2010) to study reproductive and general behaviors in brown trout. Groups of eight small mature males were placed with four adult anadromous males and two nest digging females. Half the numbers of the small males were treated with the insecticide cypermethrin or copper sulphate and the other half only to the solvent ethanol or water before placed in the stream. The behaviors of all fish were individually recorded at regular intervals during 24 h. *Arrowheads* show direction of the water currents. (*1*) Ground water inlet; (*2*) water outlet; (*3*) outlet screen for circulating water. The screen was 100 cm long and 150 cm wide and leant at an acute angle to the spawning area. The distance between the bars was 9 mm; (*4*) inlet screen for circulating water. The screen was at right angles to the bottom and 20 cm high; (*5*) observation windows; (*6*) spawning area with gravel; (*7*) holding area for fish; (*8*) netting wall eliminating waves in the bend. Figure from Olsén et al. (1998); reprinted with kind permission from Springer Science + Business Media

exposed parr spent less time with the female and had a lower number of courting events. Blood plasma levels of the male hormone 11-ketotestostorone (11-KT) were, however, significantly higher in the group exposed to 100 µg/L copper compared to the control group. Further, the exposed group spent significantly less time swimming upstream than the control group.

In a study with shore crabs (*Carcinus maenas*) exposed to copper (CuCl$_2$), Krång and Ekerholm (2006) observed effects on pheromone responses and mating behavior of male shore crabs. The authors found that exposed males required more than twice the time to start searching for a female pheromone source and their movements were less directed than in control males. The exposed males also showed a lower ability to discriminate between dummy females, sponges, with or without female phero- mones. During interactions with receptive females the exposed males showed lower frequency of the courtship behavior striking and higher numbers of aggressive acts (Fig. 26.6). The authors stressed that the mating behaviors, compared to pre- mating responses to pheromones, seem to be more robust and less sensitive to the copper exposure once the males had found a female. The total nominal concentra- tions of copper were rather high, 100 and 500 µg/L (stated as 0.1 and 0.5 mg/L), but as the Cu^{2+} complex binds anions, the effective concentrations are not known.

26.2.2 Synthetic Pesticides

Studies of pesticide effects on the olfactory sense in connection to spawning have mostly concentrated on salmonid fish (Table 26.1). Even low concentrations of pesticides can affect pheromone mediated endocrine function and reproduction in salmon. For instance, exposure of precociously mature male salmon to environ- mental levels of certain common pesticides inhibited the olfactory detection of the female reproductive priming pheromone, which is considered to be involved in the synchronization of spawning between the two sexes (Moore and Waring 1996, 1998; Waring and Moore 1997). A 30 min exposure of the olfactory epithelium to the organophosphate diazinon suppressed the EOG (electro-olfactogram) response to L-serine and PGF2α at a concentration as low as 1 µg/L (Moore and Waring 1996). The authors also observed that physiological responses of males to the urine of ovulated females (i.e., increased milt volumes and blood levels of the sex hormones 17,20β-P and LH) were reduced after 120 h exposure to 0.3 µg/L or higher concentration of diazinon. In a subsequent study, the same authors (Moore and Waring 1998) demonstrated reduced priming effects of female urine in mature Atlantic salmon parr (milt volumes, blood levels of sex steroids) after exposure to the herbicide atrazine at the same low levels that led to reduced EOG responses to PGF$_{2α}$ (0.04 µg/L, or above). The study also showed, in vitro, that the release of sex steroids from the testes of atrazine-exposed parr was affected.

There is a great risk that impairment of sex pheromones reception leads to a strong decrease in the spawning willingness and success of pesticide-exposed fish. Environ- mental levels of the synthetic pyrethroid insecticide *Cypermethrin* (0.1 µg/L)

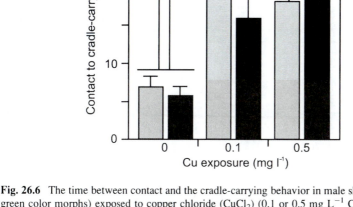

Fig. 26.6 The time between contact and the cradle-carrying behavior in male shore crabs (red or green color morphs) exposed to copper chloride ($CuCl_2$) (0.1 or 0.5 mg L^{-1} Cu) and in control males. The male stands guard over the female and this guarding behavior is induced by pheromones in the female urine. Kruskal-Wallis nonparametric variance analysis followed by Mann Whitney U-test, demonstrated that the latency between contact and cradle-carrying was longer in copper exposed male compare to the unexposed controls (figure from Krång and Ekerholm 2006; photo by K. Reise). Figure reprinted with permission from Elsevier

were shown to disturb salmon reproduction directly by reducing fertilization rates in vitro (Moore and Waring 2001). The authors demonstrated that *Cypermethrin* concentrations $\ll$ 0.1 µg/L eliminated the endocrine effects of the female priming pheromone (Moore and Waring 2001).

In a recent study the effects of *Cypermethrin* on the reproductive behavior of brown trout males were studied in a large stream tank (Fig. 26.5). Males that had been pre-exposed to 1 µg/L *Cypermethrin* during 4 days spent significantly less

Table 26.1 Effects of synthetic pesticides on the endocrine and behavior responses to sex odors in male salmonid fish

Species	Pestic.	Physiology/behavior	Results	References
Salmon (*Salmo salar*) mature male parr	Diazin. (insect.)	Female urine induced enhanced milt volumes and plasma sex hormone levels	No or reduced endocrine responses	Moore and Waring (1996)
Salmon mature male parr	Atrazine (herb.)	Enhanced milt volumes and plasma sex hormone levels after $PGF_{2\alpha}$ exp.	See previous exper.	Moore and Waring (1998)
Salmon mature male parr	Carbof. (insect.)	See previous experiment	See previous exper.	Waring and Moore (1997)
Salmon mature male parr	Cyperm. (insect.)	See previous experiment	See previous exper.	Moore and Waring (2001)
Brown trout (*S. salar*) mature male parr	Cyperm. (insect.)	Priming with female ovarian fluids and urine. See previous experiment	Reduced endocrine response	Jaensson et al. (2007)
Brown trout mature male parr	Cyperm. (insect.)	Courting of females. Blood sex steroid levels and milt volumes	Reduced behavior, milt and 11-KT levels	Jaensson et al. (2007)

insect. insecticide; *herb.* herbicide; *Diazin.* diazinon; *Carbof.* carbofuran; *Cyperm.* cypermethrin; *exp.* exposure; *exper.* experiment

time with the nest digging female and courted her less frequently than control males exposed only to the solvent ethanol (Jaensson et al. 2007). The exposed males also had lower plasma levels of 11-KT and lower volumes of strippable milt. In a priming experiment, males pre-exposed to *Cypermethrin* maintained lower blood plasma levels of 11-KT and 17,20β-P after 5 h exposure to ovarian fluids, indicating that the olfactory sense was affected. *Cypermethrin* is used to treat salmonid fish in aquaculture for sea lice (*Lepeophteirus salmonis*), a commonly occurring copepoid parasite (Ernst et al. 2001). The insecticide is also used in forestry for the treatment of spruce seedlings against pine weevil and there is high risk that streams can be contaminated. An additional concern is the effect on the crustaceans living in adjacent areas (see below).

26.2.3 Various Pollutants

Hubbard et al. (2002) observed that humic acids might block the pheromonal effects of 17,20β-P in goldfish, probably by adsorbing the molecule. Bjerselius et al. (2001) found that male goldfish exposed to the hormone 17β-estradiol (commonly entering the environment via sewage treatment plants) through the food or the ambient water showed very low frequency of various reproductive behaviors that are induced by PGF2α released from ovulated females. Furthermore these males had much lower volumes of expressible milt (spermatozoa and sperm plasma) than control males and very few or no spawning tubercles were present on the operculum (gill cover). EOGs have shown that low pH negatively affects detection of female odors in male Atlantic salmon (Moore 1994), but no behavioral studies were done.

Recent studies with crustaceans have demonstrated behavioral changes to chemical signals after exposure to various pollutants, causing concerns about the possible negative effects (Table 26.2). Krång and coworkers have done a series of studies concerning the effects of pollutants on the detection of odors in crustaceans. Females of the tube-dwelling amphipod species *Corophium volutator* release odors to attract males but if males are exposed to certain pollutants they were less efficient in finding receptive females (Fig. 26.7). One of the chemicals,

Table 26.2 Effects of various pollutants on behaviors dependent on chemical cues in Crustaceans

Species	Pollut.	Behavior	Result(s)	References
Amphipod (*Corophium volutator*)	Medeto. (antifo.)	Male attraction to female odors	Fewer males searching for females	Krång and Dahlström (2006)
C. volutator	Naphth. (fuels to motorb.)	See previous exper.	Could not find females	Krång (2007)
Shore crab (*Carcinus maenas*)	CuCl$_2$	Male attraction to female odors, mating behavior	Problem finding female stroking behavior freq. low	Krång and Ekerholm (2006)
Crayfish (*Cambarus bartonii*)	Cu salt not stated	Foraging	Could not find food	Sherba et al. (2000)
Shore crab (*C. maenas*)	Location with estrogen effects	Male behavior to female odors	Reduced responses to odors	Lye et al. (2005)
Cladoceran species (*Bosmina fatalis*)	Carbaryl (insect.)	Induced defense to predator water	No morpholog. changes	Sakamoto et al. (2006)
Cladoceran species (*Daphnia magna*)	Various insect.	Induced defense	Insect. gave same morpholog. as predator odors	Hanazato (1999)
Crayfish (*Orconectes rusticus*)	Metolac. (herbic.)	Activity to alarm cue and food	Higher activity and could not find food	Wolf and Moore (2002)

Pollut. pollutant; *Medeto.* (*antifo.*) medetomidine (antifouling coating); *Naphth.* (*fuels to motorb.*) naphthalene (fuels to motor boat); *Metho.* metholachlor; *motorb.* motorboat; *exper.* experiment; *morpholog.* morphological

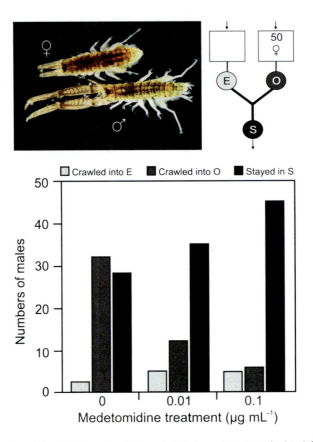

Fig. 26.7 Number of *C. volutator* males that crawled upstream into the arrival tank E or O (males could also stay in the start tank S). The tank O contained 50 females but E was empty. Males were exposed for 24 h (0.01 or 0.1 μg mL^{-1}) to the antifouling compound medetomidine or to control seawater. Statistical tests (G-test) revealed highly significant differences in distribution depending on the treatment. Fewer exposed males left the starting tank than the controls and after exposure to the higher concentration no significant discrimination between scented and unscented water was observed (figure from Krång and Dahlström 2006; photo by D. Lackschewitz). Figure reprinted with permission from Elsevier

medetomidine, is a new promising candidate and a very effective antifouling compound to inhibit settlement of barnacles (*Balanus improvisus*) in Swedish waters. Krång and Dahlström (2006) found that in the two concentrations of medetomidine tested (0.01 and 0.1 mg L^{-1} during 1 day), significantly fewer male *C. volutator* searched for a female compared to unexposed individuals (Fig. 26.7). The males showed the same kind of problems when exposed to low concentrations of the poly-cyclic hydrocarbon naphthalene, present in motorboat fuels (Krång 2007). The clear effects of the chemicals on the amphipod males in these two studies raised concerns about contamination having detrimental effects on reproduction.

In another study with shore crabs (*C. maenas*), individuals were caught at various places around the coast of Great Britain (Lye et al. 2005), where wild caught flounders (*Platichthys flesus*) were previously found to be affected by estrogenic contamination (e.g. Lye et al. 1997). The authors observed that the crabs were affected in a way that was similar to the effects of endocrine disruption. The males showed, among other things, reduced behavioral responses to female pheromones and vitellin-like proteins were detected in their hepatopancreas. These observations indicate that shore crabs and probably other crustaceans are affected by estrogenic effluents in similar ways as fish, and reproduction in certain areas can be disturbed.

26.3 Homing, Shoaling and Dominance Interactions

In salmonid fish the olfactory sense is important for the homing of adult fish to their natal river to spawn (e.g. Hasler and Scholz 1983). If the olfactory sense is affected the homing behavior should be less accurate as the fish will have problems recognizing their home river (cf. Bertmar 1982). The identity and source of the home river odors are still not known, but the odors appear to be learned before the fish leaves the river (e.g. Hasler and Scholz 1983). There are suggestions of odors being released by the juvenile fish of the same population present in the river (e.g. Nordeng 1977). Bile acids have been suggested as migration pheromones guiding salmonids back to their home river to spawn (Døving et al. 1980). In fact, bile acids from sea lamprey (*Petromyzon marinus*) larvae have been shown to act as pheromones attracting adult mature individuals up into streams to spawn (e.g. Bjerselius et al. 2000; Chung-Davidson et al., Chap. 24).

Salmonids are able to discriminate not only between odors from their own and an alien population but also between siblings and non-siblings within the same population (reviewed in Olsén 1999). It has been suggested that sibling-specific odors are used not only during social interactions and shoaling in salmonids, but also during choice of mate. A recent study has shown that during downstream migration to the sea Atlantic salmon juveniles are more synchronized with siblings than with unrelated fish, even when reared separately (Olsén et al. 2004).

26.3.1 Metal Ions

Salmonid fish avoid low concentrations of copper, 5 µg L^{-1} Cu or less (see review by Brown et al. 1982), which can block upstream swimming during spawning migrations (Saunders and Sprague 1967) and can make hatchery water ("home water") less attractive and repellent (Atchison et al. 1987). Rainbow trout could not discriminate between conspecific and heterospecific water after ca. 40 days exposure to 22 μg L^{-1} Cu (Saucier et al. 1991). Some recovery of the olfactory discrimination was observed in fish tested two or 10 weeks after exposure.

26.3.2 Synthetic Pesticides

Eight hours exposure to the imidazole fungicide prochloraz or the carbamate insecticide carbofuran (both 50 and 100 μg L^{-1} gave effects) decreased attraction to a mixture of four amino acids in juvenile goldfish compare to control fish (Saglio et al. 2003). The relative proportions of the L-amino acids (glycine, alanine, valine, taurine) were the same as found in the urine of goldfish. The sulfonylurea herbicide nicosulfuron did not have any significant effect on behavior.

It is known that aggressive interactions between crayfish are mediated by chemical signals (Breithaupt, Chap. 13) and exposure to chemicals disturbing chemoreception may interfere with the behavior responses (Cook and Moore 2008). In fact, individuals exposed to non-lethal levels of the herbicide metolachlor tended to initiate fights later than control individuals and their chance to win encounters decreased significantly (Cook and Moore 2008).

26.3.3 Various Pollutants

During the peak period of metamorphosis for life in the sea, Lower and Moore (2007) exposed juvenile Atlantic salmon for 30 days to low concentrations of a brominated flame retardant (hexabromocyclododecane, HBCD; water concentration ca. 10 ng/L). The olfactory responses of fish to urine from fish of the same ontogenetic stage were tested weekly with EOG. The exposure to HBCD suppressed the responses to the urine by ca. 60% of the original baseline responses. In a study with juvenile Arctic charr, attraction to water scented by fish from the same population was significantly reduced by a surfactant, linear alkyl benzene sulfonate (LAS), used in detergents (Olsén and Höglund 1985). Addition of ammonia and ammonium ions (NH_4Cl) to a water current scented by conspecific attractants made the current repellent to juveniles of Arctic charr (Olsén 1986a, b). In a recent study, Ward et al. (2008) showed that low concentrations of the surfactant 4-nonylphenol impaired the ability of juvenile banded killifish (*Fundulus diaphanus*) to orient to conspecific scented water and the distances between shoaling individuals were longer in exposed fish.

Munday et al. (2009) observed that after exposure to lower pH (pH 7.8 or 7.6) than to control seawater (pH 8.15) larval clownfish (*Amphiprion percula*) lost their ability to detect and respond to olfactory cues from adult habitats that are obligate for their survival. The increasing atmospheric carbon dioxide levels will affect the carbon dioxide – bicarbonate system in the sea water, resulting in lowering of the pH.

26.4   Effects on Detection of Food Odors and Foraging

Foraging in many fishes depends on intact chemical senses (olfactory sense or external taste buds) alone or in combination with each other or with other senses. The ecology of the fish species determines to a great deal if the fish depends on vision, hearing or on a chemical sense or a combination of different senses. Amino acids are known to induce foraging in fish (e.g. Olsén et al. 1986; Valentinčič 2005) and many crustaceans (e.g. Zimmer-Faust 1987), and the olfactory sense or the gustatory sense or both, depending on the species, are involved in the response (e.g. Hara 1994; Valentinčič 2005). If the olfactory receptors or the gustatory sense is affected by chemicals in the environment food intake might be affected negatively, followed by depleting energy reserves, less growth and decreasing fitness.

26.4.1   Metal Ions

Early studies with electrophysiological techniques (EOG, EEG, electro-encephalogram), by Hara and collaborators, have shown that copper and some other heavy metals suppress or completely block the olfactory response to amino acids that are detected as food odors (e.g. Brown et al. 1982). Sandhal et al. (2004) demonstrated with EOG and EEG that a 7 day exposure of juvenile coho salmon to copper-enriched water led to a reduction or total inhibition in the response to the amino acid L-serine or to the bile acid taurocholic acid. The authors found a significant loss of sensory function (at least 50% lower response than in control fish) at 10 or 20 μg Cu/L.

In an earlier study with the freshwater crayfish *Cambarus bartonii*, a 7-day exposure to copper at concentration as low as 20 μg L^{-1} (0.02 mg L^{-1}) disturbed their food search behavior in a Y-maze (Sherba et al. 2000). The exposed individuals could not find their way to the food.

26.4.2   Synthetic Pesticides

Several studies have shown effects of pesticides on the olfactory function in fish. Tierney and collaborators studied the changes of coho salmon parr in responses to amino acids, both with EOG and in behavioral tests. The fungicide iodocarb (IPBC) and the herbicide glyphosate (the active ingredient of the commonly used Roundup[®]) suppressed the olfactory response to L-serine at relatively low concentrations (Tierney et al. 2006b), while the other three tested pesticides (chlorothalonil, endosulfan, 2,4-dichlorophenoxyacetic acid) only had effects at high concentrations. In the following study, the authors concentrated their work on

IPBC and glyphosate plus the herbicide atrazine (Tierney et al. 2007). After 30 min of exposure, all three pesticides eliminated the attraction behavior to L-histidine. The chemicals also reduced the EOG response to the amino acid. Interestingly Roundup® was also detected by the olfactory sense and a behavioral response to the pesticide was observed.

Crayfish (*Orconectes rusticus*) exposed to non-lethal levels of the herbicide metalochlor experienced a decrease in the ability to perceive chemical stimuli; as a consequence they were less efficient in locating food and did not behave appropriately to alarm cues (Wolf and Moore 2002).

26.4.3   Various Pollutants

Kasumyan (2001) reviewed studies concerning the effects of pollutants on foraging behavior in fish. Heavy metals such as copper, low pH levels of water, oil hydrocarbons and pesticides are all pollutants shown to have negative effects on the chemical senses (smell or taste or both) employed during foraging and food intake. Lemly and Smith (1986) found that low pH suppressed the behavioral responses of fathead minnows (*Pimephales promelas*) to chemical feeding stimulants.

After 48 h in acidified water (pH 4.8 and 3.5), crayfish (*Orconectes virilis* and *Procambarus acutus*) showed reduced behavioral responses, i.e. antennule and feeding movements, to food odors (Tierney and Atema 1986). Several earlier studies on crustaceans focused on the effects of drilling fluids and crude oil on the chemoreception and behavior in connection with foraging and food search (e.g. Derby and Atema 1981). Blue crabs (*Callinectes sapidus*) and Dungeness crabs (*Cancer magister*) detected low concentrations of the hydrocarbon naphthalene (Pearson and Olla 1980).

26.5   Effects on Detection of Alarm Cues, on Fright Responses and Inducible Defense

Several fish and crustacean species use chemical alarm cues to avoid predators. The odors are either released by damaged conspecifics or in the feces by the predator preying on the fish or crustacean species (e.g. Ferrari et al. 2007; Hazlett, Chap. 18). Detection of predators is under strong selection as it is important for prey to be able to detect, avoid the predator and assess the risk of being in a certain environment.

26.5.1   Copper

Copper exposure of Colorado pike minnow (*Ptychocheilus lucius*) reduced the fishes' ability to show a fright response to conspecific skin extract (Beyers and Farmer 2001) and the negative effect increased with concentration. Interestingly the authors found that after copper exposure (66 μg/L) for 24 h the fish had a lower response to skin extract than after 96 h exposure. The authors suggested that after 96 h the fishes had developed a physiological adaptation and increased tolerance to copper. In a field study, McPherson et al. (2004) observed that Iowa darters (*Etheostoma exile*) from a contaminated lake (mainly copper, but also zinc and nickel) did not avoid conspecific skin extract containing fright odors, which strongly contrasts with the efficient avoidance reaction of fish from a clean lake.

26.5.2   Synthetic Pesticides

In experiments with chinook salmon (*Oncorhynchus tshawytscha*), the insecticide diazinon significantly inhibited the fright response to skin extract: instead of decreasing their swimming and foraging behavior fish remained highly active after exposure to the pesticide at 1 and 10 μg/L for 2 h and recovery for 1 h (Scholz et al. 2001). The alarm reaction of coho salmon parr to conspecific skin extract was impaired by the carbamate fungicide IPBC (Tierney et al. 2006a). The exposed fish did not show any freezing behavior.

In planktonic crustacean, odors from predators can induce changes in morphology and life history to decrease their vulnerability to predation (e.g. Hanazato and Dodson 1995; Lass and Spaak 2003). In many cladoceran crustaceans inducible defenses (e.g. long spines and helmets) are developed in response to molecules (kairomones) released by predators. The induction of a morphological change in the species *Bosmina fatalis* to kairomones of its predator, the cladoceran species *Leptodora kindtii*, was blocked by the carbamate insecticide carbaryl at concentrations as low as 1–2 μg L^{-1}, but did not affect the predatory behavior of *L. kindtii* (Sakamoto et al. 2006). These results suggest that low concentrations of the pesticide could have a great impact on the survival rate of the *Bosmina* population.

In an earlier study with the cladoceran *Daphnia magna* and its predator (larvae of the midge *Chaoborus* sp.), various carbamate and organophosphorus insecticides, but not the two herbicides and one fungicide tested, had surprisingly the same effect as the kairomone, inducing protuberant structures or accentuating the effect of the kairomone (Hanazato 1999). These two studies highlight that anthropogenic chemicals can disturb chemical interactions in the planktonic communities and influence population structures. The possibility of xenobiotics interfering with chemical signals and mimicking their effects to induce morphological defenses has been suggested as a form of endocrine disruption that has hardly been studied (Barry 1999).

26.5.3   pH

Decrease in pH in the water have been shown to affect the responses to alarm cues in fish either due to suggested changes in the molecules themselves (e.g. Leduc et al. 2004) or possibly due to effects on the olfactory receptors (Thommesen 1983).

26.6 Future Directions

Studies investigating the effects of pesticides on crustaceans should be of high priority. Crustaceans are very important consumers and prey in various aquatic systems and there are delicate relationships between crustacean plankton prey and fish predators in the pelagic zone that can and have been shown to be disturbed. It is known that pesticides are present in surface waters and it is especially urgent to study the effects of insecticides on freshwater species and species that are present in estuaries and coastal waters with high risks of contamination due to vicinity to the sources. In acute toxicity tests crustaceans were much more (often 10-times more) sensitive to insecticides than fish (Maltby et al. 2005), and some of the chemicals probably affect behaviors at very low concentrations. As there are very few studies done on pesticide effects on crustacean chemoreception it is not possible to compare their sensitivity with fish, but it is likely that there are differences. The few crustaceans studied concerning effects of copper indicate that they are less sensitive to the metal compared with fish.

Pesticides designed and developed to act on the nervous system in insects will probably have effects on crustacean olfactory receptor nerve cells. It should be fruitful to combine electrophysiological techniques with behavioral studies. Electrophysiological methods will initially demonstrate what kinds of chemicals disturb detection of odors, and these chemicals should then be tested for their behavioral effects. It would also be interesting to investigate if metals and molecules are transported by olfactory cells into the brain of crustaceans and affect brain functions, similar to those reported for fish (Tallkvist et al. 2002). Accumulation of manganese in the brain of crustaceans has been discussed by Kråvg and Rosenqvist (2006). Chemicals acting on different parts in the central nervous system can affect various behaviors and affect parts of the brain that are important to reproduction, including the endocrine system (Rodríguez et al. 2007). The occurrence and effects of chemicals disrupting the functions of the endocrine systems (EDCs), directly or indirectly, are still much less known in crustaceans than in fish and should be given high priority in research.

In future studies it will be important to investigate the effects of decreasing pH levels in the sea due to higher atmospheric carbon dioxide levels (cf. Munday et al. 2009). As in fish, the chemo-receptors and the detection of signals are dependent on a proper pH. Possibly even more important, are pH-related impacts on both the exo- and endoskeleton, which are extensively calcified (Anderson 2001).

26.7 Conclusions

Both in fish and crustaceans the olfactory receptor cells are exposed directly to the environment which makes them targets for various chemicals in the ambient water. Several studies have demonstrated effects of heavy metal ions, pesticides, pH changes and other xenobiotics on fish olfactory and taste receptor cells with effects on the detection of chemical cues important to reproduction, social interactions, alarm responses and foraging. However, not much is known about the effects of these pollutants on crustacean chemical senses important for survival and reproduction. It is also urgent to study the effects of commonly used insecticides (mostly targeting non-crustacean arthropods) that may affect chemosensory abilities and behavioral reactions of crustaceans at very low concentrations.

References

Anderson DT (2001) Invertebrate zoology. Oxford University Press, Oxford

Atchison G, Henry M, Sandheinrich M (1987) Effects of metals on fish behaviour: a review. Environ Biol Fish 18:11–25

Barry M (1999) Chemical communication in planktonic organisms: environmental contaminants can mimic the effects of natural chemical signals. SETAC Europe News 10:6–8

Belanger R, Ren X, McDowell K, Chang S, Moore P, Zielinski B (2008) Sensory setae on the major chelae of male crayfish, *Orconectes rusticus* (Decapoda: Astacidae) – impact of reproductive state on function and distribution. J Crust Biol 28:27–36

Bertmar G (1982) Structure and function of the olfactory mucosa of migrating Baltic trout under environmental stresses, with special reference to water pollutions, chapter 21. In: Hara TJ (ed) Chemoreception in fishes, developments in aquaculture and fisheries science, vol 8. Elsevier, Amsterdam, pp 395–433

Beyers DW, Farmer MS (2001) Effects of copper on olfaction of Colorado pikeminnow. Environ Toxicol Chem 20:907–912

Bjerselius R, Li W, Teeter JH, Seley JG, Johnsen PB, Maniak PJ, Grant GC, Polkinghorne CN, Sorensen PW (2000) Direct behavioral evidence that unique bile acids released by larval sea lamprey (*Petromyzon marinus*) function as a migratory pheromone. Can J Fish Aquat Sci 57:557–569

Bjerselius R, Lundstedt-Enkel K, Mayer I, Olsén KH (2001) Male goldfish reproductive behaviour and physiology are severely affected by exogenous exposure to 17β-estradiol. Aquat Toxicol 53:139–152

Blaxter JHS, Hallers-Tjabbes CCN (1982) The effect of pollutants on sensory systems and behaviour of aquatic animals. Neth J Aquat Ecol 26:43–58

Bonde JP, Comhaire F, Ombelet W (eds) (1998) Environmental influences on spermatogenesis and sperm quality. Middle East Fertil Soc J 3:1–53

Brown SB, Evans RE, Thompson BE, Hara TJ (1982) Chemoreception and aquatic pollutants, chapter 20. In: Hara TJ (ed) Chemoreception in fishes, developments in aquaculture and fisheries science, vol 8. Elsevier, Amsterdam, pp 363–394

Caprio J, Derby CD (2008) Aquatic animal models in the study of chemoreception. In: Basbaum AI et al (eds) The senses: a comprehensive references: vol. 4, olfaction and taste. Academic Press, San Diego, pp 97–134

Clark R (2001) Marine pollution. Oxford University Press, Oxford, p 248

Cook ME, Moore PA (2008) The effects of the herbicide metolachlor on agonistic behavior in the crayfish, *Orconectes rusticus*. Arch Environ Contam Toxicol 55:94–102

Derby CD, Atema J (1981). Influence of drilling muds on the primary chemosensory neurons in walking legs of the lobster, Homarus americanus. Can J Fish Aquat Sci 38:268–274

Døving KB, Selset R, Thommesen G (1980) Olfactory sensitivity to bile acids in salmonid fishes. Acta Physiol Scand 108:121–131

Ernst W, Jackman P, Doe K, Page F, Julien G, Mackay K, Sutherland T (2001) Dispersion and toxicity to non-target aquatic organisms to pesticides used to treat sea lice on salmon in net pen enclosures. Mar Pollut Bull 42:433–444

Ferrari MCO, Brown MR, Pollock MS, Chivers DP (2007) The paradox of risk assessment: comparing responses of fathead minnows to capture-released and diet-released alarm cues from two different predators. Chemoecology 17:157–161

Gatidou G, Thomaidis NS (2007) Evaluation of single and joint toxic effects of two antifouling biocides, their main metabolites and copper using phytoplankton bioassays. Aquat Toxicol 85:184–191

Hanazato T (1999) Anthropogenic chemicals (insecticides) disturb natural organic chemical communication in the plankton community. Environ Pollut 105:137–142

Hanazato T, Dodson SI (1995) Synergistic effects of low oxygen concentration, predator kairomone, and a pesticide on the cladoceran Daphnia pulex. Limnol Oceanogr 40:700–709

Hara TJ (1994) The diversity of chemical stimulation in fish olfaction and gustation. Rev Fish Biol Fish 4:1–35

Hasler AD, Scholz AT (1983) Olfactory imprinting and homing in salmon. Investigation into the mechanisms of the imprinting process. Springer, Berlin

Hubbard PC, Barata EN, Canario AVM (2002) Possible disruption of pheromonal communication by humic acid in the goldfish, Carassius auratus. Aquat Toxicol 60:169–183

Jaensson A, Olsén KH (2010) Effects of copper on olfactory mediated endocrine responses and reproductive behaviour in mature male brown trout (Salmo trutta Linneaus) parr to conspecific females. J Fish Biol 76:800–817

Jaensson A, Scott A, Moore A, Kylin H, Olsén KH (2007) Effects of a pyretroid pesticide on endocrine responses to female odours and reproductive behaviour in male parr of brown trout (Salmo trutta L.). Aquat Toxicol 81:1–9

Kasumyan AO (2001) Effects of chemical pollutants on foraging behavior and sensitivity of fish to food stimuli. J Ichthyol 41:76–87

Krång A-S (2007) Naphthalene disrupts pheromone induced mate search in the amphipod Corophium volutator (Pallas). Aquat Toxicol 85:9–18

Krång A-S, Dahlström M (2006) Effects of a candidate antifouling compound (medetomidine) on pheromone induced mate search in the amphipod Corophium volutator. Mar Pollut Bull 52:1776–1783

Krång A-S, Ekerholm M (2006) Copper reduced mating behaviour in male shore crabs (Carcinus maenas (L.)). Aquat Toxicol 80:60–69

Krång A-S, Rosenqvist G (2006) Effects of manganese on chemically induced food search behaviour of the Norway lobster, Nephrops norvegicus (L.). Aquat Toxicol 78:284–291

Kreuger J (1998) Pesticides in stream water within an agricultural catchment in southern Sweden, 1990–1996. Sci Tot Environ 216:227–251

Lass S, Spaak P (2003) Chemically induced anti-predator defences in plankton: a review. Hydrobiologia 491:221–239

Leduc AOHC, Kelly JM, Brown GE (2004) Detection of conspecific alarm cues by juvenile salmonids under neutral and weakly acidic conditions: laboratory and field tests. Oecologia 139:318–324

Lemly AD, Smith RJ (1986) A behavioral assay for assessing effects of pollutants on fish chemoreception. Ecotoxicol Environ Saf 11:210–218

Lower N, Moore A (2007) The impact of a brominated flame retardant on smoltification and olfactory function in Atlantic salmon (Salmo salar L) smolts. Mar Freshwater Behav Physiol 40:267–284

Lürling M, Scheffer M (2007) Info-disruption: pollutants and the transfer of chemical information between organisms. Trends Ecol Evol 22:374–379

Lye CM, Frid CLJ, Gill ME, McCormick D (1997) Abnormalities in the reproductive health of flounder Platichthys flesus exposed to effluents from a sewage treatment works. Mar Poll Bull 34:34–41

Lye CM, Bentley MG, Clare AS, Sefton EM (2005) Endocrine disruption in the shore crab *Carcinius maenas* – a biomarker for benthic marine invertebrates. Mar Ecol Prog Ser 288:221–232

Maltby L, Blake N, Brock TCM, Van den Brink PJ (2005) Insecticide species sensitivity distributions: importance of test species selection and relevance to aquatic ecosystems. Environ Toxicol Chem 24:379–388

McPherson TD, Mirza RS, Pyle GG (2004) Responses to wild fishes to alarm chemicals in pristine and metal-contaminated lakes. Can J Zool 82:694–700

Moore A (1994) An electrophysiological study on the effects of pH on olfaction in mature male Atlantic salmon (*Salmo salar*) parr. J Fish Biol 45:493–502

Moore A, Waring CP (1996) Sublethal effects of the pesticide Diazinon on olfactory function in mature male Atlantic salmon (*Salmo salar* L.) parr. J Fish Biol 48:758–775

Moore A, Waring CP (1998) Mechanistic effects of a triazine pesticide on reproductive endocrine function in mature male Atlantic salmon (*Salmo salar* L.) parr. Pest Biochem Physiol 62:41–50

Moore A, Waring CP (2001) The effects of a synthetic pyrethroid pesticide on some aspects of reproduction in Atlantic salmon (*Salmo salar*). Aquat Toxicol 52:1–12

Munday PL, Dixson DL, Donelson JM, Jones GP, Pratchett MS, Devitsina GV, Døving KB (2009) Ocean acidification impairs olfactory discrimination and homing of marine fish. Proc Natl Acad Sci U S A 106:1848–1852

Nordeng H (1977) A pheromone hypothesis for homeward migration in anadromous salmonids. Oikos 28:155–159

Olsén KH (1986a) Emission rate of amino acids and ammonia and their role in olfactory preference behaviour of juvenile Arctic charr, *Salvelinus alpinus* (L.). J Fish Biol 28:255–265

Olsén KH (1986b) Modification of conspecific chemoattraction in Arctic charr (*Salvelinus alpinus* (L.)), by nitrogenous excretory products. Comp Biochem Physiol 85A:77–81

Olsén KH (1999) Review – present knowledge of kin discrimination in salmonids. Genetics 473:1–5

Olsén KH, Höglund LB (1985) Reduction by a surfactant of olfactory mediated attraction between juveniles of Arctic charr (*Salvelinus alpinus* (L.)). Aquat Toxicol 6:57–69

Olsén KH, Liley NR (1993) The significance of olfaction and social cues in milt availability, sexual hormone status and spawning behaviour of male rainbow trout (*Oncorhynchus mykiss*). Gen Comp Endocrinol 89:107–118

Olsén KH, Karlsson L, Helander A (1986) Food search behaviour in Arctic charr (*Salvelinus alpinus* (L.)) induced by food extract and amino acids. J Chem Ecol 12:1987–1998

Olsén KH, Järvi T, Mayer I, Petterson E, Kroon F (1998) Spawning behaviour and sex hormone levels in adult and precocious brown trout (*Salmo trutta* L.) males and the effect of anosmia. Chemoecology 8:9–17

Olsén KH, Johansson A-K, Bjerselius R, Mayer I, Kindahl H (2002) Mature Atlantic *salmon (Salmo salar)* male parr are attracted to ovulated female urine but not to ovarian fluid. J Chem Ecol 28:29–40

Olsén KH, Petersson E, Ragnarsson B, Lundqvist H, Järvi T (2004) Down stream migration in *Salmo salar* smolt sibling groups. Can J Fish Aquat Sci 61:328–331

Olsén KH, Sawisky GR, Stacey NE (2006) Endocrine and milt responses of male crucian carp (*Carassius carassius*) to periovulatory females under field conditions. Gen Comp Endocrinol 149:294–302

Pearson WH, Olla BL (1980) Threshold for detection of naphthalene and other behavioral responses by the blue crab, *Callinectes sapidus*. Estuaries 3:224–229

Persson E, Larsson P, Tjälve H (2002) Cellular activation and neuronal transport of intranasally instilled benzo(a)pyrene in the olfactory system of rats. Toxicol Lett 133:211–219

Pohnert G, Steinke M, Tollrian R (2007) Chemical cues, defence metabolites and shaping of pelagic interspecific interactions. Trends Ecol Evol 22:198–204

Rodríguez EM, Medesani DA, Fingerman M (2007) Endocrine disruption in crustaceans due to pollutants: a review. Comp Biochem Physiol A 146:661–671

Saglio P, Bretaud S, Rivot E, Olsén KH (2003) Chemobehavioral changes induced by short-term exposures to prochloraz, nicosulfuron and carbofuran in goldfish. Arch Environ Contam Toxicol 45:515–524

Sakamoto M, Chang K-H, Hanazato T (2006) Inhibition of development of anti-predator morphology in the small cladoceran *Bosmina* by an insecticide: impact of an anthropogenic chemical on prey-predator interactions. Freshw Biol 51:1974–1983

Sandhal J, Baldwin D, Jenkins J, Scholz N (2004) Oder-evoked field potentials as indicators of sublethal neurotoxicity in juvenile coho salmon (*Oncorhynchus kisutch*) exposed to copper, chlorpyrifos, or esfenvalerate. Can J Fish Aquat Sci 61:404–413

Saucier D, Astic L (1995) Morpho-functional alterations in the olfactory system of rainbow trout (*Oncorhynchus mykiss*) and possible acclimation in response to long-lasting exposure to low copper levels. Comp Biochem Physiol 112A:273–284

Saucier D, Astic L, Rioux P (1991) The effects of early chronic exposure to sublethal copper on the olfactory discrimination of rainbow trout, *Oncorhynchus mykiss*. Environ Biol Fish 30:345–351

Saunders RL, Sprague JB (1967) Effects of copper-zinc mining pollution on a spawning migration of Atlantic salmon. Water Res 1:419–432

Scholz NL, Truelove NK, French BL, Berejikian BA, Quinn TP, Casillas E, Collier TK (2001) Diazinon disrupts antipredator and homing behaviors in chinook salmon (*Oncorhynchus tshawytscha*). Can J Fish Aquat Sci 57:1911–1918

Sherba M, Dunham DW, Harvey HH (2000) Sublethal copper toxicity and food response in the freshwater crayfish *Cambarus bartonii* (Cambaridae, Decapoda, Crustacea). Ecotoxicol Environ Saf 46:329–333

Sloman KA, Wilson RW (2006) Anthropogenic impacts upon behaviour and physiology, chapter 10. In: Sloman A, Wilson RW, Balshine S (eds) Fish physiology, volume 24: behaviour and physiology in fish. Elsevier, Amsterdam, pp 413–468

Stacey N, Sorensen P (2006) Reproductive pheromones, chapter 9. In: Sloman A, Wilson RW, Balshine S (eds) Fish physiology, volume 24: behaviour and physiology in fish, vol 24. Elsevier, Amsterdam, pp 359–412

Tallkvist J, Persson E, Henriksson J, Tjälve H (2002) Cadmium-metallothionein interactions in the olfactory pathways of rats and pikes. Toxicol Sci 67:108–113

Thommesen G (1983) Detection of a blocking effect of low pH in the trout olfactory organ. In: Døving KB (ed) Chemoreception in studies of marine pollution, pp. 44–49. Reports from a workshop at Oslo, July 13 and 14, 1980. The Norwegian Marine Pollution Research and Monitoring Programme

Tierney AJ, Atema J (1986) Effects of acidification on the behavioral response of crayfishes (*Orconectes virilis* and *Procambarus acutus*) to chemical stimuli. Aquat Toxicol 9:1–11

Tierney KB, Taylor AL, Ross PS, Kennedy CJ (2006a) The alarm reaction of coho salmon parr is impaired by the carbamate fungicide IPBC. Aquat Toxicol 79:149–157

Tierney KB, Ross PS, Jarrard HE, Delaney KR, Kennedy CJ (2006b) Changes in juvenile coho salmon electro-olfactogram during and after short-term exposure to current-use pesticides. Environ Toxicol Chem 25:2809–2817

Tierney KB, Singh CR, Ross PS, Kennedy CJ (2007) Relating olfactory neurotoxicity to altered olfactory-mediated behaviors in rainbow trout exposed to three currently-used pesticides. Aquat Toxicol 81:55–64

Tjälve H, Henriksson J (1999) Uptake of metals in the brain via olfactory pathways. Neurotoxicology 20:181–195

Valentinčič T (2005) Olfactory discrimination in fishes, chapter 3. In: Reutter K, Kapoor BG (eds) Fish chemosenses. Science Publishers, Enfield, pp 65–85

Ward AJW, Duff AJ, Horsfall JS, Currie S (2008) Scents and scents-ability: pollution disrupts chemical social recognition and shoaling in fish. Proc R Soc B 274:101–105

Waring CP, Moore A (1997) Sublethal effects of a carbomate pesticide on pheromonal mediated endocrine function in mature Atlantic salmon (*Salmo salar*) parr. Fish Physiol Biochem 17:203–211

Wolf MC, Moore PA (2002) The effects of the herbicide metolachlor on the perception of chemical stimuli by *Orconectes rusticus*. J North Am Benthol Soc 21:457–467

Wright RT, Nebel BJ (2002) Environmental science. Towards a sustainable future. Upper Saddle River, Pearson Education, Prentice-Hall Inc

Zimmer-Faust RK (1987) Crustacean chemical perception: towards a theory on optimal chemore-ception. Biol Bull 172:10–29

Zippel HP (1993) Regeneration in the peripheral and the central olfactory system: a review of morphological, physiological and behavioral aspects. J Hirnforsch 34:207–229

Chapter 27
Insect Pheromones: Useful Lessons for Crustacean Pheromone Programs?

Thomas C. Baker

Abstract Insect pheromones, especially sex pheromones, have successfully contributed to pest management programs around the world since the 1970s. In this chapter I examine some of the ways in which pheromones have been used in insect management programs and introduce some of the real-world issues that have promoted, and hindered, their adoption for use against different species. These include biological differences in the mate-finding behaviors of different species, the chemistries of the pheromones that they use, the successful engineering of controlled-release dispensers for different compounds, as well as the different political, economic, and use-pattern-related situations that exist even just in the United States to more strictly regulate the use of some types of pheromones and techniques than others. The experiences of entomologists who have witnessed insect pheromones finding their place in integrated pest management systems over the past four decades should be instructive in helping crustacean biologists develop crustacean pheromone systems into useful management tools. By far the greatest use of insect pheromones has been for monitoring existing populations and detecting the presence of invasive species. Monitoring with pheromone traps allows for other, curative measures such as insecticide applications or cultural/biological controls to be implemented intelligently. Mating disruption has taken its place in different cropping systems around the world, especially in fruit orchards. Research has now shown that pheromone does not need to completely shut off mating, but merely impede it to the point of delaying first and second matings in females to reduce their fecundity by ca. 50%. Mass trapping by deploying large numbers of pheromone traps regularly spaced in the cropping area has reemerged as a viable and effective technique for using pheromones directly for population suppression. This use of pheromones has been most effective in male-emitted pheromone systems such as in pest species of weevils (snout beetles). In these systems females can be trapped out using the synthetic version of the male-emitted pheromone, and thus egg-laying is

T.C. Baker (✉)
Department of Entomology, 105 Chemical Ecology Laboratory, Penn State University, University Park, PA 16802, USA
e-mail: tcb10@psu.edu

T. Breithaupt and M. Thiel (eds.), *Chemical Communication in Crustaceans*,
DOI 10.1007/978-0-387-77101-4_27, © Springer Science+Business Media, LLC 2011

directly reduced by mass trapping of females. Although the experimental, techno-logical, and legal hurdles of developing pheromones for widespread use in the field can seem daunting, experience has shown that with determination and a real need, these hurdles can be overcome.

27.1 Introduction

As has been articulated by many authors in other chapters of this volume, relatively few crustacean pheromones have been isolated and identified, and their behavioral effects demonstrated. The reasons for this are manifold, involving methodological difficulties inherent in working in aquatic environments and availability of suffi-cient numbers of animals in the proper physiological state to allow the acquisition of sufficient amounts of pheromone and bioassaying their activities.

At this point, the field of crustacean pheromone research seems to be appropri-ately focused on the necessities of developing basic techniques for chemically identifying pheromone components. Ideas for practical application of these pher-omones have only recently begun to be proposed, for instance, in the trapping of invasive species (Hardege and Terschak, Chap. 19), preventing fouling of ships by barnacles (Clare, Chap. 22), and reducing fish parasitization in aquaculture (c.f. Mordue Luntz and Birkett 2009). But if identifications can be successful, there likely will follow many actual efforts and field tests to put pheromones to use. These uses, if crustacean applied pheromone research is to proceed along the successful path that insect pheromone application followed, will depend on a thorough understanding of the behaviors of the animals in their natural environments. Knowledge of how individuals move in response to pheromone signals carried by currents along sandy bottoms or among complex reef structures will allow the most effective strategies to be designed and tested for manipulating crustacean behavior.

Over 35 years ago, I became interested in learning more about arthropod chemical communication while working as a research technician in Wendell Roe-lofs' laboratory at Cornell University. I was struck by the precision of insect sex pheromones and their ability to evoke full-blown, hard-wired sexual responses so reliably that if one had the correct blend of compounds, it could be used to sensitively attract and trap males of the target species and determine whether or not suppressive action needed to be taken against them. I was involved in the electrophysiological and behavioral bioassaying of prospective new pheromone blends for scores of pest species of moths and saw the power of sex pheromones, first in laboratory assays and then under field conditions in upstate New York apple orchards. My sticky traps sometimes overflowed with male moths in a seemingly magical process by which the males could not help but be lured to their demise from hundreds of meters away.

Successful use of pheromones in insect control depends upon successfully controlling insect behavior through the emission of synthetic pheromone blends. The behaviors that have turned out to be most amenable to and important for manipulation are those that involve long-distance attraction. In this chapter I will outline some of the lessons we have learned from decades of research on insect sex pheromones and the intense efforts that have been made to develop successful integrated insect pest management systems with them. Such systems now utilize insect sex pheromones for monitoring, mass trapping, and mating disruption. My goal is to set the stage (an aquatic one) for efforts to develop and use crustacean pheromones in a similarly successful manner. I believe that the concepts developed for insect systems can provide bright illumination for many successful underwater pheromone applications.

27.2   Pheromone Emission and Pheromone Identity

In moths, sex pheromone communication occurs at a particular time of the day or night depending on the species. Females, usually more than 1 day old, take up a "calling" posture (Fig. 27.1a) by raising the abdomen and everting glandular tissue at the end of the abdomen associated with the ovipositor from which pheromone is emitted at a rate of a few tens of picograms per second.

Sex pheromones in insects include a variety of molecules from large, but volatile, aliphatic molecules to small cyclic monoterpenoids. Moth pheromones are the first pheromones identified and are the most widely studied of all the insect pheromones. They include a huge collection of mostly female-emitted chemical blends that are composed of typically fatty acid-derived molecules that function

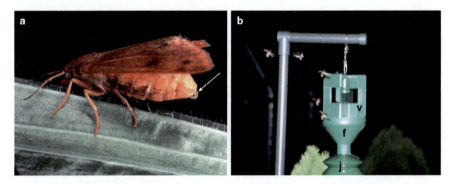

Fig. 27.1 (a) Photograph of a calling female arctiid moth. Notice the everted pheromone gland tissue at the end of the abdomen (*arrow*) associated with the ovipositor. (b) An example of a typical pheromone trap for male scarab beetle pests of turf and ornamentals, including cranberries and blueberries. The trap has intercepting vanes (v) designed to stop flight and cause the males to fall into a funnel (f) and finally into a collecting jug (j) containing a small amount of ethylene glycol-water from which they cannot escape. Males can be seen here being attracted to the trap by the synthetic female-emitted sex pheromone of this species. (a) From Roelofs, W.L. and Cardé, R.T., 1971, Science 171:648–686. Reprinted with permission of AAAS; (b) From Leal, W.S., Sawada, M., Matsuyama, S., Kuwahara, Y., and Hasegawa, M., 1993, J. Chem. Ecol. 19:1381–1391. Reprinted with permission of Springer Science & Business Media

over long distances (10–100 m). Recently, the literature regarding moth pheromone blends as well as other insect pheromones and chemical structures has been compiled into one website, (http://www.pherobase.com), an extremely useful free information resource.

The majority of moth pheromones identified thus far are blends of 10–18-carbon-long straight-chain primary alcohols, acetates, or aldehydes (Fig. 27.2a, b). However, one interesting second major class of moth pheromone components is found in many noctuids, geometrids, arctiids, and lymantriids and is composed of blends of polyene epoxides (Fig. 27.2c) and hydrocarbons (Fig. 27.2d). For the majority of moth species, after synthesis of the fatty acyl chain to either 16 or 18 carbons in length, these pheromone component precursors begin to achieve some species-specific structural differences that are imparted by one or more desaturases that place double bonds at specific locations in the fatty acyl chain. The desaturation step is, depending on the species, either preceded or followed by enzymatic beta-oxidation to reduce the chain length of the molecule.

Merely reversing the two-step sequence in which these two enzyme systems work creates most of the major differences in monounsaturated pheromone component structures in moths. Diunsaturated compounds with double bonds in two

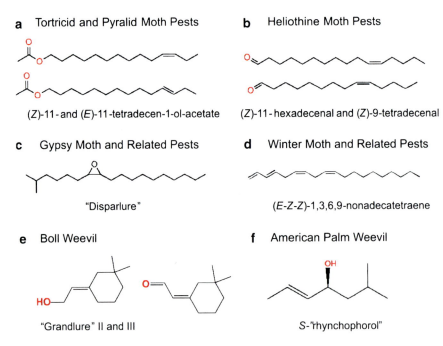

Fig. 27.2 Pheromone structures of some groups of moth and beetle pests. The names of the chemical structures in the blend of boll weevil pheromone components (2e) are (Z)-2-(3,3,-dimethyl)-cyclohexylideneethanol (grandlure II) and (Z)-(3,3-dimethyl)-cyclohexylideneacetalde-hyde (grandlure III). The chemical name of S-rhynchophorol (2f) is 4S-(E)-6-methyl-2-hepten-4-ol. From "The Pherobase" (http://www.pherobase.com)

locations on the chain are created by the desaturase acting twice, both before and after the chain-shortening step. Final species-specific structural differences result from the final biosynthetic step, conversion of the fatty acyl molecule to its active functional group by reductases and oxidases. A subset of two or three compounds from the collection of precursors of different chain lengths and double bond positions may all be converted as a group to comprise the unique blend of that particular species. Precise blend ratios of such two- and three-component sex pheromones are the rule in moth systems, and deviations from these ratios elicit lower levels of upwind flight in males (Baker 1989; Linn et al. 1986a,b; see below).

27.3 Behavioral Responses to Sex Pheromones are Indirect and Depend On Flow-Detection

Males of each species are synchronized in their daily activity to the behavior of females and are optimally responsive to pheromone at the same time of the day as females emit pheromone. They are capable of responding to pheromone by flying upwind in the plume from tens, even hundreds of meters downwind to locate the source (Fig. 27.1b).

Steering with respect to a time-averaged pheromone concentration gradient has often been erroneously invoked as an orientation mechanism used by pheromone-responding insects. This type of *direct* chemical-gradient steering (chemotaxis) is in fact NOT used by moths to locate females. Rather, two *indirect* reactions to sex pheromone, "optomotor anemotaxis" combined with "self-steered counterturning," are the mechanisms that result in the successful location of sex pheromone sources by insects. The anemotactic response (steering with respect to the wind) in response to flow information is performed via optical feedback and not wind mechano-sensors. Detection of wind-induced motion across the eyes results in compensatory motor responses to stabilize this motion; hence, the name optomotor anemotaxis. The result is a pheromone-*modulated* (not pheromone-steered) upwind displacement of the insect toward the pheromone source (Kennedy 1983).

We have learned in more detail since the initial landmark studies by J.S. Kennedy's group (Kennedy and Marsh 1974) how these two programs are performed when a flying insect encounters a series of pheromone strands of the right quality. In optomotor anemotaxis, the moth compensates with changes in its direction of thrust and corrections in its airspeed for any off-course displacement or groundspeed maintenance difficulties caused by changes in wind direction or velocity. During pheromone stimulation, the optomotor anemotactic response keeps the insect heading and progressing more or less directly upwind. The other, nonanemotactic program (Baker and Kuenen 1982), is a "counterturning" oscillatory motor pattern of rapid left–right reversals of direction (Kennedy 1983) that is performed concurrently with anemotaxis. Counterturning frequency increases upon contact with pheromone and coincides with more upwind-oriented optomotor anemotactic steering (Baker and Haynes 1987). Counterturning frequency slows down upon loss of pheromone at the same time as clean air causes the anemotactic program to relax and allow more cross-wind, transverse optical image flow. The result in clean air is a slower-reversing, greater amplitude cross-wind left–right "casting" flight (Kennedy 1983; Baker 1989).

For crustaceans walking or swimming in a plume of sex pheromone, we should anticipate that a similar set of indirect responses (i.e., rheotaxis and looping or side-to-side zigzagging) might be switched on by pheromone that help the animal execute the optimal set of maneuvers to steer up the flow to the sex pheromone source. For instance, swimming male copepods have been shown, like moths, to "cast" cross-current when their pheromone odor is lost (Yen and Lasley, Chap. 9). Benthic crustaceans like lobsters, crabs, and crayfish walk along the substratum to locate an odor source. We can expect that for such animals flow direction can be discerned by using mechanoreceptors on the body or involving the antennules allowing them to orient by odor-gated rheotaxis (Weissburg, Chap. 4).

For macroscopic crustaceans such as shrimp that swim in response to sex pheromones, we might expect that optical feedback would be essential in allowing up-current progress to be made toward the pheromone emitter via optomotor rheotaxis. Obviously sufficient light and light-gathering abilities of crustacean eyes would be needed for sufficient edge- and motion-detection to occur. Over shorter distances, other mechanisms not known in insects might be employed. For instance, in some copepod species (e.g., *Acartia tonsa*) males have been shown to be able to locate females by using their wake structure, which is facilitated by the viscous properties of water (Yen and Lasley, Chap. 9).

Some differences in orientation behavior between crustaceans and flying insects are due to the slower ambulatory speeds of some of the animals that have been researched thus far (lobsters and crabs). In addition, the narrowness of close-range food-odor plumes that crustaceans respond to along the ocean bottom creates opportunities for left–right chemotaxis decisions that do not exist for tiny insects flying in very wide and fast-shifting pheromone plumes far from the source.

Many of the potential differences between the orientation mechanisms used by crustaceans compared to insects will have to do with fluid dynamics that dictate plume dispersion directions as well as microturbulence that determines fine plume structure due to the shedding and shredding of pheromone strands from the source. Whether or not individual strands can be detected and resolved by a crustacean's receptor system will determine whether the animal can only time-average the odor concentration at a particular distance from the source. Some crustaceans can use "temporal sampling" of odor flux in strands with chemosensor responses of up to 5 Hz (Gomez and Atema 1994). However, as pointed out by Weissburg (Chap. 4), time-averaging may be all that is needed for slower-moving, large crustaceans such as lobsters.

27.4   Insect Sex Pheromone Olfaction Systems: Flux Detection and Mixtures

Wright (1958) recognized that an odor plume consists of strands of odor that issue from the surface of the odor source. The strands are sheared from the release surface and drift downwind where they become stretched, twisted, and more tortuous as they are ripped apart into substrands during their journey farther downwind (Murlis 1986). These disjointed substrands interspersed with pockets of clean air comprise what we call the odor plume. However, it was not until intermittency of stimulation from the plume was shown to promote sustained upwind flight (Baker et al. 1985) and subsequently when moths were shown to have the ability to react during upwind flight to the subsecond changes in pheromone concentration (Baker and Haynes 1987) that we began to understand that these individual pheromone strands and pockets of clean air are what are producing the sustained upwind flight behavior that we call attraction (Vickers and Baker 1994; Mafra-Neto and Cardé 1994).

It has become clear that, as emphasized by Kaissling (1998), the insect olfactory system is designed to optimize flux detection and not to measure concentration. This is a major reason why chemotaxis is not used by pheromone-responding insects. Rapid increases in flux from individual plume strands and decreases from the clean-air pockets between strands are essential to producing reiterative upwind surges that result in the attraction of male moths to pheromone sources. The speed of the reaction to pheromone ON and pheromone OFF is related to the need to change behavior according to swings in the wind direction to prevent erroneous upwind movement during zero odor conditions. In crustaceans we might expect behavioral reaction time to pheromone ON and OFF to match the shifts occurring in current flow direction and speed (see Fig. 4.2 in Weissburg, Chap. 4) and the walking or swimming speeds of the respective species.

27.4.1   Subsecond Reaction Times of the Two Programmed Responses to Pheromone

The timescale over which the moths' reactions to pheromone strands and clean air occur is remarkably small. The behavioral responses to both the onset (upwind surge) and loss of filaments (cross-wind casting) can be as fast as 0.15 s in *Grapholita molesta* (Busck) (Baker and Haynes 1987), but usually are between 0.3 and 0.6 s; Vickers and Baker 1997). In studies of host-odor responses by flying female moths, only the latency of the casting flight response to loss of the odor has been measured, and its time course is similar to that of the latency for pheromone loss, 0.7 s (Haynes and Baker 1989).

Recent findings with terrestrial crustaceans seem to indicate that their olfactory systems have adapted to detect and process airborne volatiles (and thus are quite similar to insect olfactory systems), while maintaining the general olfactory pathways of aquatic crustaceans (Schmidt and Mellon, Chap. 7; Hansson et al, Chap. 8). Crustaceans' peripheral receptor neurons appear to be flux detectors, just as in insects, with fast on–off responses. The giant robber crab's aesthetascs on its antennules respond with high-speed depolarizations similar to insects' electroantennograms when challenged with puffs of airborne odor from odor cartridges (Stensmyr et al. 2005). Terrestrial hermit crabs possess pathways in their olfactory systems that appear to parallel those of insects in terms of both high-speed odor processing and odor classification (see Hansson et al., Chap. 8). Marine crustaceans such as the American lobster or spiny lobsters are known to flick their aesthetascs to increase flux and increase detection sensitivity (Koehl, Chap. 5).

27.4.2   Relevance of Fast Pheromone-Strand Reaction Times to Field Attraction

The rapid response to an odor strand is important because the strand has been shed from the odor source, and in higher velocity airflow, continues to travel in a more or less straight line away from the source (Fig. 27.3) (David et al. 1982; Elkinton and Cardé 1987). This trajectory allows the responding insect to move in a straight line toward the source if it flies upwind each time it detects an odor strand: the insect can eventually locate the source by steering into and progressing upwind through reiterative upwind surges in response to strands (Fig. 27.3b). It is, however, just as important for the insect to go into cross-wind casting quickly when it encounters clean air between pheromone strands as it is to surge upwind in response to the strands, because any such clean-air pocket can turn out to be a very large expanse of clean air due to large swings in the wind direction (Baker 2008; David et al. 1983) (Fig. 27.3b). If the insect continues to move upwind into a large pocket of clean air for any length of time, it will, to its detriment, distance itself rapidly from the odor plume that it has just lost contact with while steering increasingly off-line away from the odor source (David et al. 1982).

A subsecond, rapid cessation of upwind progress in response to odor loss and a shift to a more cross-wind track (usually after only one left-to-right reversal of flight direction), and a coupling of these anemotactic reactions to an internal motor program of increasingly lengthy side-to-side oscillations, allows the insect to increase its ability to quickly regain contact with the odor strands that have swung to one side or the other of its body (Figs. 27.3b and 27.4). In some species of moths, the surge reaction in response to pheromone strands and the initiation of casting in response to clean air can be fast enough that the moths can often "scoot" over in response to a shifting wind-line plume, using a sawtoothed-shaped succession of left–right surge-cast tracks (Fig. 27.4). Normally the subsecond alternation between surging and casting in response to strands and clean-air pockets in a plume is seen as "zig-zagging upwind flight" (Figs. 27.3b, 27.4, and 27.5). When the insect encounters a large pocket of clean air due to a wind-swing, the long-duration

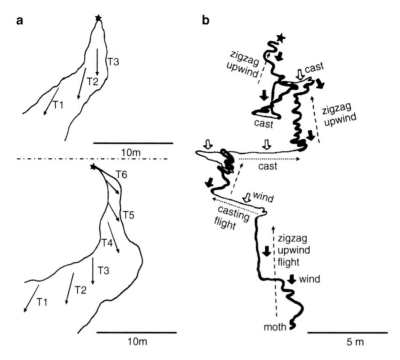

Fig. 27.3 (a) *Top*: The different directions of wind within a smoke plume in the field recorded on videotape by David et al. (1982), illustrating how odor-bearing air parcels have been sheared from the source (*black star*) and are transported away in straight lines (*arrows*) at different time periods (T1–T3), regardless of the instantaneous shape of the time-averaged plume (*black lines* outlining the sinuous snapshot-image of the plume). *Bottom*: Depicted here, ca. 5 s later, the wind has completed a swing so that new odor-bearing air parcels are flying away in straight lines as they were sheared from the source at times T4–T6. The air parcels sheared at times T1–T3 continue on along their previous trajectories. Adapted from David et al. (1982). (b) Fifty-second-long flight track of a male gypsy moth that was video-taped by David et al. (1983) as it approached a source of sex pheromone (*black star*) in a wind-field (0.8–2.0 m/s) that shifted to a similar degree as the wind in (a). Thick black flight tracks indicate periods when the moth was in contact with pheromone, with wind directions at these times shown as *thick black arrows*. Thin-lined flight tracks indicate periods when the male was not in contact with pheromone, with wind directions shown as *hollow arrows*. Progress directly towards the source was due to upwind flight in the pheromone plume (*dashed lines*) and not due to cross-wind casting (*dotted lines*). Adapted from David et al. (1983)

left–right oscillations of casting flight provide a behavioral bridge over clean air that helps position the moth in an unbiased left–right fashion to recontact pheromone strands. Thus, the more typical pheromone-mediated flight track of an insect exhibits an alternation between long-duration casting in response to large clean air pockets and upwind zigzagging (rapid surge-casts) in response to the finely structured plume with its intermittent pheromone strands (Figs. 27.3b and 27.5). The result is a halting succession of upwind flight tracks that gets the insect to the source in a shifting wind-field (Figs. 27.3b and 27.5).

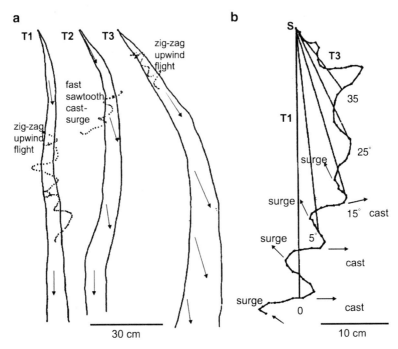

Fig. 27.4 (a) Snapshots at three times ca. 0.5 s apart, T1, T2, and T3, of the flight track (*dotted lines*) of a male oriental fruit moth flying upwind in response to pheromone in a wind-field that is shifting in direction (*arrows*) by 35° in 2 s. The time-averaged outline of the pheromone plume is shown as *solid wavy lines. Dots* indicate the location of the male every 1/60 s. This male "scoots" over with a sawtooth track during T2, alternating between upwind surges and cross-wind casting on a left–right basis every 1/6 s, approximately. (**b**) Flight track (enlarged) of another male during the T2–T3 section that reacted every 1/6 s to pheromone and then clean air on each successive asymmetric track reversal. Dots indicate the male's position every 1/30 s. Long straight lines indicate each 5° shift in the direction of the wind line. "S" denotes the location of the pheromone source (adapted from Baker and Haynes 1987)

27.4.3   Precise Pheromone Blend Composition Determines Intensity of Upwind Flight Behavior

In insects, the optimal blend ratio of synthetic pheromone components for attraction and source contact by males has been shown to be the one that most closely approximates the natural female-emitted ratio. Linn et al. (1986a, b) convincingly demonstrated in field experiments that the optimal blend causes the initiation of upwind flight in more males from greater distances from the source than do blends comprised of suboptimal ratios or those from which some components are missing. The duration and length of males' upwind surging reactions to single strands of pheromone depend on their quality (Vickers and Baker 1997; Quero et al. 2001). Strands of the correct pheromone blend tainted with small amounts of heterospecific antagonist are poorer in intensity than are the surges to strands of the pure

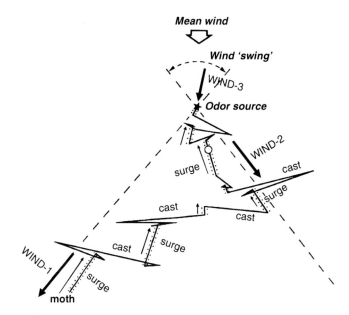

Fig. 27.5 A depiction of the average types of changes in flight track movements of gypsy moths flying upwind to a pheromone source (*black star*) over short grass in a plume that swings nearly 90° in a matter of seconds due to shifts in wind direction (*solid arrows* labeled Wind-1 and Wind-2). The distance from the source to the start of the flight track at lower left is ca. 15 m. Upwind surges by the moth (*dashed arrows*) in response to pheromone-bearing wind (small stippling on either side of the flight track) occur periodically and are interspersed with short periods of left–right casting across the wind line in clean air (no stippling). Wind direction at the time the male locates the source is Wind-3 (adapted from David et al. 1982)

pheromone blend alone (Vickers and Baker 1992). The blend of pheromone molecules must arrive simultaneously on the male's antenna, i.e., they must be perfectly blended in each strand, to have their optimal effect. Plumes from females of two different heliothine moth species (Fig. 27.2b) calling from ca. 5 cm lateral spacing apart can reduce the successful attraction of males by 80% or more, even though only 50% of the plume strands from the females are "mixed", i.e., arriving on the male antenna within 120 ms of each other (Lelito et al. 2008). For crustaceans (e.g., crayfish), in habitats populated by closely related species, males may need to discriminate between odor plumes from conspecific and heterospecific females.

27.5 Pheromone Dispenser and Trap Design: Much Technological Development Needed

Decades of experience with attempts to develop insect sex pheromones as effective pest management tools can be instructive as to the possible ways crustacean pheromones might be used. Strong plumes of the correct blend of components that create above-threshold plume strands far downwind of the pheromone source have been the key to the optimal use of pheromones in detection, mass trapping, and mating disruption. It should be anticipated that crustacean pheromones will find their optimal uses when pheromone plumes in aquatic environments can be optimized to be above behavioral thresholds at maximal distances downstream for months at a time. In large-scale monitoring/detection programs involving sometimes thousands of traps, the visiting of traps by human scouts and replacing their lures every week or two is prohibitively expensive.

For such plumes to be created and maintained for long time periods, much work on controlled release systems in aquatic environments will be needed. In the insect world, dispensers have had to be optimized on a trial-and-error basis, based on behavioral reactions (trap catch) to various dispenser types and pheromone loading rates. Optimization has had to be assessed on a species-by-species trial system because each species has different pheromone components comprised of differing molecular weights and functionalities such as aldehydes, alcohols, and acetates. The field longevity of these same dispensers had to be similarly assessed based on trap capture levels over time.

In aquatic environments the challenges will be great in designing effective, long-lasting pheromone dispensers for emitting possibly highly labile molecules of varying solubilities and migration rates through dispenser walls. The experimentation needed to achieve such optimization will involve, just as it has for insect pheromones, relatively thankless trial-and-error experimentation that will be viewed as "engineering," not science, by many crustacean pheromone researchers' colleagues.

As with the experimentation needed for optimizing dispenser-controlled emissions technology, much experimentation will be needed to create the most effective pheromone traps for each targeted crustacean species. For insects, optimizing the pheromone blends and dosages to create the best lures for different pest species has only been part of the task at hand to make pheromones useful for monitoring and detection. The efficacies of traps cannot be predicted, because even within genera, different species of insects respond differently to different trap shapes, sizes, and colors (c.f., Fig. 27.1b). All of these variables must be tested independently for each new species. In addition, for insects, further testing always needed to be done to optimize deployment schemes for traps to optimize capture rates. These assays involve extensive testing to identify seasonal trap attractiveness, the most efficient trap density, and favorable trap positions. The latter includes habitat differences, e.g., in sun or shade as well as higher vs. lower positions above the substratum or the vegetation canopy. For optimizing uses of pheromones in crustacean systems, similar methodical experimentation will no doubt need to be performed that relates to such real-world issues as how the availability of natural shelters and harborages can affect trap capture levels and also how capture can be affected by other complexities of the underwater terrain (see Aggio and Derby, Chap. 12).

These are time-consuming and difficult-to-publish types of studies, but have proven necessary for optimal decision-making involving curative measures to be taken that depend on the correct interpretation of trap catch levels. Crustacean pheromone researchers should be ready to perform similar types of experimentation in order to use crustacean pheromones most effectively. They should be aware of their colleagues' potential disdain for these types of applied, nonhypothesis-driven studies.

27.6   Pheromones for Monitoring, and for Detection and Eradication of Invasive Species

Monitoring the population densities of endemic pests with pheromone traps has been done successfully for decades against scores of species worldwide. When pheromone traps are used for monitoring, they are widely spaced and are at such a low density that they have no effect on reducing population density by themselves. They only serve as sentinels to trigger, at some threshold capture level, the application of other pest control techniques such as insecticide sprays against larvae at some later date that is often based on accumulated thermal units (Baker 2008).

Insect sex pheromone traps also have played a large role in detecting movements of adult pest species from one region to another and also even from noncrop areas into crop areas. Survey programs involving grids of pheromone traps are used routinely to report and track the yearly arrival of migrating adult populations of insects such as the black cutworm *Agrotis ipsilon* (Hufnagel) in the Midwest (Showers et al. 1989) or expanding populations such as the gypsy moth *Lymantria dispar* L. (Fig. 27.2c) (Elkinton and Cardé 1981). This approach has become an essential part of our arsenal of tools for tracking and mitigating pest-movement-related threats.

As an example, boll weevil pheromone traps were used in a highly successful boll weevil population suppression program in the southeastern United States (Fig. 27.2e). Release of sterile boll weevils and also diapause control procedures (insecticide applications to kill adult weevils before cotton begins fruiting) were triggered if pheromone trap captures exceeded two to five weevils per trap (Ridgway et al. 1990; Baker 2008). Over the years of this program across six southeastern states, boll weevil capture levels declined to near zero and it was economically advantageous to grow cotton again in this region (Baker 2008). The program was extensive and not cheap. In 1988 alone, approximately 590,000 traps were deployed, in which more than 8.25 million dispensers were used (Ridgway et al. 1990). Many more lures than traps were needed because the dispensers were replaced every 2 weeks over the year in order to keep lure attractiveness at optimal levels. The sensitive and reliable detection capabilities provided by pheromones for insect monitoring and detection should also be found in crustacean pheromones. Even if only one sex is attracted, the extra sensitivity of responders to these pheromones compared to food and other general odors should make them a preferred lure to use.

27.7   Pheromone Mass Trapping: Successful Rebirth of an Old Idea

Mass trapping is the deployment of a sufficiently high density of pheromone traps that ensnares and eliminates enough adults from the population and reduces subsequent larval damage. Mass trapping differs from trapping for monitoring or detection in that mass trapping alone directly diminishes the pest population. Although in the early days of sex pheromone research mass trapping for control of pest populations was discounted as impractical on many fronts, a few decades of further experience have shown that this technique can be highly successful against certain species. For instance, research and development of mass trapping systems using male-produced sex pheromones for a variety of highly damaging tropical weevil species has resulted in successful commercial mass trapping systems for suppressing populations of these species (Oehlschlager et al. 2002; Baker 2008). Successive years of mass trapping using pheromone (Fig. 27.2f) dispensed in inexpensive bucket traps at a density of only one trap per 7 ha reduced the damage to trees from the American palm weevil to near zero, as illustrated on thousands of hectares of oil palms on two plantations (Fig. 27.6).

If mass trapping for direct population control is to be used against crustacean pests, the target species likely will need to have some of the same attributes as these weevil species (Oehlschlager et al. 2002). First, the crustacean species should use a sex pheromone system that attracts females, the egg-layers, so that mass trapping will directly reduce egg-laying. Secondly, the adults of the target species should be present in relatively small numbers, live a long time before egg-laying, and lay a small number of eggs whose emerging larvae cause a relatively large amount of damage. Thus, as in the case of the weevils, trapping out females steadily over a period of weeks will remove a large proportion of the egg-layers in any generation and have a large impact on subsequent populations. Trapping males will be less likely to reduce populations because male insects are usually capable of mating once per day and females only once or twice per lifetime. Thus, a single male can mate with multiple females over his lifetime (for crustaceans see, e.g., Gosselin et al. 2005), and this feature allows any remaining males to "replace" the missing inseminations that would have been achieved by the other males had they not been trapped out. Third, if the individuals are sufficiently vagile and the attractive power of the pheromone strong enough that the females can be trapped over long distances, the traps can be widely spaced and thus be deployed at economically favorable numbers in terms of both traps and the pheromone lures that are needed throughout the control program (Oehlschlager et al. 2002).

27.8   Pheromone Mating Disruption Delays Mating

Mating disruption is the technique by which pheromone is dispensed into a pest habitat in sufficient amounts to reduce the ability of males to find females, or vice versa. Here the searching sex is attracted toward extra-high release-rate synthetic

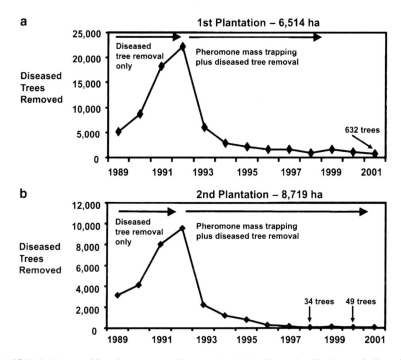

Fig. 27.6 Incidence of *R. palmarum* weevil-vectored red ring disease in oil palms as indicated by the number of diseased palms that had to be removed in two large plantations in Costa Rica before and after nine successive years of mass trapping of the weevils. From 1989 to 1992 with no mass trapping, disease incidence steadily rose and tremendous economic losses were incurred due to the removal of mature fruit-bearing trees. After pheromone mass trapping was begun in 1992, the incidence of tree removal due to red ring disease declined dramatically due to the elimination of adult weevils from the population and near-elimination of egg-laying, larval damage, and disease spread (adapted from Oehlschlager et al. 2002)

pheromone dispensers, and in the process, habituated by a prolonged exposure to the strong flux in these strands, rendering these individuals incapable of smelling (and finding) their mates. In the earliest mating disruption trials (early 1970s), it was thought that it was essential to dispense the pheromone such that a ubiquitous cloud of pheromone pervaded every cubic meter of habitat to disrupt mate-finding communication. The earliest trials concentrated on the use of microcapsules sprayed through conventional pesticide sprayers to achieve uniform coverage. These trials were mostly failures because the tiny microcapsules expended their allotment of pheromone from their small reservoirs in only a few days because the capsule membranes allowed pheromone to escape too fast. Making the walls thicker or a slower-releasing matrix resulted in too slow release rates again causing failure to impede mating.

Then, the first EPA-registered mating disruption formulation, the Conrel hollow fibers, was developed in 1978. It used hollow microfibers ca. 2 cm in length. These fibers were distributed over cotton fields by using airplanes equipped with specialized applicator technology. These dispensers, once they landed, stuck to the cotton leaves, from where they emitted volatilized pheromone at sufficiently high rates in the desert heat for over 2 weeks by means of an open capillary tube-type technique. The fibers were effective at disrupting mating when applied at an approximate density of one fiber per square meter of cotton crop. The amount of active ingredient (pheromone) applied per acre was only 15 g/acre applied every 2 weeks (Cardé and Minks 1995; Baker 2008).

The evolution of dispensers into a yet stronger point-source-emission mode and further away from the original uniform fog mode occurred in the mid-1980s with the commercial appearance of the Shin-Etsu "ropes." These were made up of the pheromone of the target species residing in sealed polyethylene tubes that were hand-applied in a twist-tie manner around plant stems. The reduction in the number of point sources per hectare from tens of thousands (fibers) to now only 1,000 or even 500/ha was achieved with fewer, but higher-strength point sources with no loss of efficacy.

Finally, new types of aerosol dispensers ejecting pheromone at very high rates onto pads were developed so that formulations were developed in which the number of dispensers could be reduced to only 10–15 dispensers per hectare (Mafra-Neto and Baker 1996; Shorey and Gerber 1996; Shorey et al. 1996; Baker et al. 1997; Fadamiro et al. 1999). Efficacy in disrupting sex pheromone communication of a number of different species was clearly demonstrated.

27.8.1   Successful Mating Disruption Involves both Attraction and Habituation

Why could much higher-dose dispensers deployed at a much wider spacing turn out to be more successful at disrupting female–male communication than the uniform fog created by microcapsules? The main reason turns out to be that a fog only can cause habituation (desensitization) up to a certain background level, and not higher. A strong plume, on the other hand, causes males to be attracted and spend lots of time in the plume, while at the same time they become habituated to the very highly concentrated plume strands of pheromone (Cardé and Minks 1995; Cardé et al. 1997; Stelinski et al. 2004, 2005). They get huge ups and downs in flux on a subsecond basis from the strands interspersed with pockets of clean air and this optimally stimulates the central nervous system synapses to weaken them more than would a constant low-level fog.

27.8.2   Delayed (not Elimination of) Mating is Key to Mating Disruption Success

It had routinely been assumed that successful mating disruption can only occur if the majority of females in a population are prevented from mating after the application of a mating disruption formulation. In reality, the females' abilities to obtain their first or second matings merely needs to be impaired and delayed (Knight 1997). Delayed mating was directly confirmed in studies on European corn borer *Ostrinia nubilalis* (Hübner) using very high release rate aerosol mating disruption dispensers (Fadamiro et al. 1999). During each of the two summer flights, 100% of the females eventually became mated despite the application of high release rate, low-point-source density dispensers of the two-component sex pheromone (Fig. 27.2a) (Fadamiro et al. 1999). Damage, however, was reliably reduced by 50% every time the disruptant was used at a dispenser density of only 1 per acre of corn (Baker, unpublished). Population effects of perturbing reproductive timing could be achieved in highly seasonal crustacean species (see Breithaupt, Chap. 13; Hardege and Terschak, Chap. 19; Kamio and Derby, Chap. 20).

Analyses involving spermatophore counts showed that the mating disruptant was impairing the ability of females to attract and mate with males on a constant, daily basis (Fadamiro et al. 1999). Thus, as demonstrated in the oriental fruit moth (Rice and Kirsch 1990) and codling moth studies (Knight 1997), mating disruption success does not require keeping the population of females virgin, but rather just needs to impede females' ability to attract males and retard the dates at which they achieve their first or even second matings. Retarding the dates at which first or second matings are achieved significantly reduces fecundity in the European corn borer and codling moth by over 50% (Fadamiro and Baker 1999; Knight 1997).

27.9   Outlook and Conclusions

There is a good deal to be optimistic about the prospects for successful implementation of crustacean pheromones in population management programs involving trapping for monitoring and detection, and possibly even for mass trapping. Careful studies of behavioral responses, plume dispersion, and dispenser performance/longevity will help inform the decision-making needed for choosing optimal trap deployment locations and trap density. For mating disruption to be considered a useful tool against crustacean pests, some key aspects that are important in insect mating disruption programs should be kept in mind. First, mating disruption works best when applied on an area-wide basis. There is a larger surface-area-to-edge ratio in larger disruption plots, which is advantageous in preventing gravid females from moving in from untreated areas. Second, intense testing of dispenser technology for optimal lifetime and efficiency of the pheromone is critical. Emission rates of the target species' particular pheromone blend need to persist for as long a period as possible while economically using up as much of the active ingredient in the dispenser as possible. Third, knowledge of the behavior of the target species will help in developing strategies for deployment grid or other spacing patterns of the dispensers to ensure that most of the potential responders are contacted by the pheromone plumes from dispensers.

Mating disruption is not a curative procedure used like pesticides that can be applied to immediately reduce population levels. Like many insect pheromone products, crustacean pheromone mating disruption may tend to get outcompeted by pesticides that can be applied on a wait-and-see, curative basis. With no other control measures taken such as insecticide sprays, it often takes three generations of successive use of disruption in order to get populations back to acceptable levels (Baker 2008). Another factor to anticipate for crustacean pheromone usage is that the biggest and most consistent successes involving insect pheromone mating disruption in North America have been in federally supported programs that have subsidized the cost of pheromone for large "area-wide" grower participation programs over periods of several years. Thus, for future crustacean systems, government agencies may have to fund the large-scale initiatives that are needed for area-wide trapping or mating disruption successes. Finally, pheromone usage for insect mating disruption has had to meet environmental regulations that have caused frustration in small companies and been daunting in the scope of requirements such as toxicity testing and label restrictions. Thus, for future uses of crustacean pheromones, registering these compounds and slow-release formulations with government agencies may be a long and difficult process, depending on the country involved, even though these are naturally occurring compounds that are emitted at extremely low rates. Despite this somewhat sobering outlook regarding pheromone regulatory issues, it is exciting to look ahead and envision the many ways in which crustacean pheromones may possibly be used to manipulate behavior and have an impact on populations for the benefit of society.

References

Baker TC (1989) Sex pheromone communication in the Lepidoptera: new research progress. Experientia 45:248–262

Baker TC (2008) Use of pheromones in IPM. In: Radcliffe T, Hutchinson B (eds) Integrated Pest Management. Cambridge University Press, Cambridge, pp 273–285

Baker TC, Haynes KF (1987) Manoeuvres used by flying male oriental fruit moths to relocate a sex pheromone plume in an experimentally shifted wind-field. Physiol Entomol 12:263–279

Baker TC, Kuenen LPS (1982) Pheromone source location by flying moths: a supplementary non-anemotactic mechanism. Science 16:424–427

Baker TC, Willis MA, Haynes KF, Phelan PL (1985) A pulsed cloud of sex pheromone elicits upwind flight in male moths. Physiol Entomol 10(2):57–265

Baker TC, Mafra-Neto A, Dittl T, Rice MA (1997) Novel controlled release device for disrupting sex pheromone communication in moths. IOBC/wprs Bulletin 20:141–149

Cardé RT, Minks AK (1995) Control of moth pests by mating disruption: successes and constraints. Annu Rev Entomol 40:559–585

Cardé RT, Mafra-Neto A, Staten RT, Kuenen LPS (1997) Understanding mating disruption in the pink bollworm moth. IOBC/wprs Bulletin 20:191–201

David CT, Kennedy JS, Ludlow AR, Perry JN, Wall C (1982) A re-appraisal of insect flight towards a point source of wind-borne odor. J Chem Ecol 8:1207–1215

David CT, Kennedy JS, Ludlow AR (1983) Finding of a sex pheromone source by gypsy moths released in the field. Nature 303:804–806

Elkinton JS, Cardé RT (1981) The use of pheromone traps to monitor distribution and population trends of the gypsy moth. In: Mitchell ER (ed) Management of insect pests with semiochemicals. Plenum, New York, pp 41–55

Elkinton JS, Cardé RT (1987) Pheromone puff trajectory and upwind flight of the male gypsy moth in a forest. Physiol Entomol 12:399–406

Fadamiro HY, Baker TC (1999) Reproductive performance and longevity of female European corn borer, *Ostrinia nubilalis*: effects of multiple mating, delay in mating, and adult feeding. J Insect Physiol 45:385–392

Fadamiro HY, Cossé AA, Baker TC (1999) Mating disruption of European corn borer, *Ostrinia nubilalis* by using two types of sex pheromone dispensers deployed in grassy aggregation sites in Iowa cornfields J Asia-Pacific Entomol 2:121–132

Gomez G, Atema J (1994) Frequency filter properties of lobster chemoreceptor cells determined with high-resolution stimulus measurement. J Comp Physiol 174:803–811

Gosselin T, Sainte-Marie B, Bernatchez L (2005) Geographic variation of multiple paternity in the American lobster, *Homarus americanus*. Mol Ecol 14:1517–1525

Haynes KF, Baker TC (1989) An analysis of anemotactic flight in female moths stimulated by host odour and comparison with the males' response to sex pheromone. Physiol Entomol 14:279–289

Kaissling KE (1998) Flux detectors versus concentration detectors: two types of chemoreceptors. Chem Senses 23:99–111

Kennedy JS (1983) Zigzagging and casting as a programmed response to wind-borne odour: a review. Physiol Entomol 8:109–120

Kennedy JS, Marsh D (1974) Pheromone-regulated anemotaxis in flying moths. Science 184:999–1001

Knight AL (1997) Delay of mating of codling moth in pheromone disrupted orchards. IOBC/wprs Bulletin 20:203–206

Leal WS, Sawada M, Matsuyama S, Kuwahara Y, Hasegawa M (1993) Unusual periodically of sex pheromone production in the large black chafer *Holotrichia parallela*. J Chem Ecol 19:1381–1391

Linn CE Jr, Campbell MG, Roelofs WL (1986a) Male moth sensitivity to multicomponent pheromones: critical role of female-released blend in determining the functional role of components and active space of the pheromone. J Chem Ecol 12:659–668

Linn CE Jr, Campbell MG, Roelofs WL (1986b) Pheromone components and active spaces: what do moths smell and where do they smell it? Science 237:650–652

Mafra-Neto A, Baker TC (1996) Timed, metered sprays of pheromone disrupt mating of *Cadra cautella* (Lepidoptera: Pyralidae). J Agric Entomol 13:149–168

Mafra-Neto A, Cardé RT (1994) Fine-scale structure of pheromone plumes modulates upwind orientation of flying moths. Nature 369:142–144

Mordue Luntz AJ, Birkett MA (2009) A review of host finding behaviour in the parasitic sea louse, *Lepeophtheirus salmonis* (Caligidae: Copepoda). J Fish Dis 32:3–13

Murlis J (1986) The structure of odor plumes. In: Payne TL, Kennedy CEJ, Birch MC (eds) Mechanisms in Insect Olfaction. Clarendon Press, Oxford, pp 27–39

Oehlschlager AC, Chinchilla C, Castillo G, Gonzalez L (2002) Control of red ring disease by mass trapping of *Rhynchophorus palmarum* (Coleoptera: Curculionidae). Fla Entomol 85:507–513

Quero C, Fadamiro HY, Baker TC (2001) Responses of male *Helicoverpa zea* to single pulses of sex pheromone and behavioural antagonist. Physiol Entomol 26:106–115

Rice RE, Kirsch P (1990) Mating disruption of oriental fruit moth in the United States. In: Ridgway RL, Silverstein RM, Inscoe MN (eds) Behavior-Modifying Chemicals for Insect Management. Marcel Dekker, New York, pp 193–212

Ridgway RL, Inscoe MN, Dickerson WA (1990) Role of the boll weevil pheromone in pest management. In: Ridgway RL, Silverstein RM, Inscoe MN (eds) Behavior-modifying chemicals for insect management. Marcel Dekker, New York, pp 437–472

Roelofs WL, Cardé RT (1971) Hydrocarbon sex pheromone in tiger moths. Science 171:684–686

Shorey HH, Gerber RG (1996) Use of puffers for disruption of sex pheromone communication of codling moths (Lepidoptera: Tortricidae) in walnut orchards. Environ Entomol 25:1398–1400

Shorey HH, Sisk CB, Gerber RG (1996) Widely separated pheromone release sites for disruption of sex pheromone communication in two species of Lepidoptera. Environ Entomol 25:446–451

Showers WB, Whitford F, Smelser RB, Keaster AJ, Robinson JF, Lopez JD, Taylor SE (1989) Direct evidence for meteorologically driven long-range dispersals of an economically important moth. Ecology 70:987–992

Stelinski LL, Gut LJ, Pierzchala AV, Miller JR (2004) Field observations quantifying attraction of four tortricid moth species to high-doage pheromone rope dispensers in untreated and pheromone-treated apple orchards. Entomol Exp Et Appl 113:187–196

Stelinski LL, Gut LJ, Epstein D, Miller JR (2005) Attraction of four tortricid moth species to high dosage pheromone rope dispensers: observations implicating false plume following as an important factor in mating disruption. IOBC/wprs Bulletin 28:313–317

Stensmyr MC, Erland S, Hallberg E, Wallén R, Greenaway P, Hansson BS (2005) Insect-like olfactory adaptations in the terrestrial giant robber crab. Curr Biol 15:116–121

Vickers NJ, Baker TC (1992) Male Heliothis virescens sustain upwind flight in response to experimentally pulsed filaments of their sex-pheromone. J Insect Behav 5:669–687

Vickers NJ, Baker TC (1994) Reiterative responses to single strands of odor promote sustained upwind flight and odor source location by moths. Proc Natl Acad Sci USA 91:5756–5760

Vickers NJ, Baker TC (1997) Chemical communication in heliothine moths. VII. Correlation between diminished responses to point-source plumes and single filaments similarly tainted with a behavioral antagonist. J Comp Physiol 180:523–536

Wright RH (1958) The olfactory guidance of flying insects. Can Entomol 98:81–89

Index